Soil Biology

Volume 28

Series Editor
Ajit Varma, Amity Institute of Microbial Technology,
Amity University Uttar Pradesh, Noida, UP, India

For further volumes:
http://www.springer.com/series/5138

Ajay Singh • Nagina Parmar • Ramesh C. Kuhad
Editors

Bioaugmentation, Biostimulation and Biocontrol

Springer

Editors
Dr. Ajay Singh
Lystek International Inc.
107-279 Weber Street North
N2J 3H8, Waterloo, Ontario
Canada
asingh@lystek.com

Dr. Nagina Parmar
Ryerson University
Department of Chemistry & Biology
Toronto, Ontario
Canada
naginap@ryerson.ca

Dr. Ramesh C. Kuhad
University of Delhi South Campus
Department of Microbiology
Benito Juarez Road
110021 New Delhi
India
kuhad85@gmail.com

ISSN 1613-3382
ISBN 978-3-642-19768-0 e-ISBN 978-3-642-19769-7
DOI 10.1007/978-3-642-19769-7
Springer Heidelberg Dordrecht London New York

Library of Congress Control Number: 2011930787

Cover design: SPi Publisher Services

Printed on acid-free paper

Springer is part of Springer Science+Business Media (www.springer.com)

Preface

Soils sustain an immense diversity of prokaryotic and eukaryotic organisms. Microbial functions in ecosystems are as diverse as the microbes themselves. Microbes adapt to these microhabitats and live together in consortia, interacting with each other and with other parts of the soil biota. Microorganisms play an essential role in the functioning and sustaining of all natural ecosystems including biogeochemical cycling of nutrients and biodegradation.

Plant–microbe interactions involving plant growth-promoting rhizosphere microorganisms (PGPRs) are of beneficial agricultural importance, e.g., improve plant productivity, suppress disease-causing microbes and nematodes, and accelerate nutrient availability and assimilation. PGPRs compensate for the stress and reduction in plant growth caused by weed infestation, drought, heavy metals, salt, and other unfavorable environmental conditions and are frequently used as biofertilizers. Biochemical and molecular tools are continuously being developed in an attempt to better appreciate microbial abundance and distribution in natural environments to evaluate community structures with ecosystem functions and to develop appropriate biofertilization and remediation approaches.

Bioaugmentation, biostimulation, and biocontrol approaches using microbial inoculants, biofertilizer, bio(chemicals), and organic amendments have been used for a long time to improve soil biology, fertility, crop productivity, and soil remediation. In comparison with chemical-synthesized pesticides and fertilizers, biofertilizers have several advantages including: they are relatively more safe, potentially reduced environmental damage and human health risk, much more targeted activity, effective in small quantities, multiply themselves but are controlled by the plant and indigenous microbial populations, decompose more quickly than conventional chemical pesticides, and can be used in conventional or integrated pest management systems.

This volume, Bioaugmentation, Biostimulation, and Biocontrol of the Soil Biology Series, is a selection of topics related to biological processes with an emphasis on their application in improving soil health, fertility, and plant productivity. Topics include an overview of the role of bioaugmentation, biostimulation, and biocontrol in soil biology; beneficial interactions of PGPRs and their products; application of biofertilizer technology for pulse production; beneficial role of phosphate-solubilizing microorganisms in soil, composting of lignocellulosic wastes and beneficial utilization of agro-industrial waste material for bioaugmentation and soil amendment; various bioaugmentation strategies for bio- and phytoremediation of contaminated soils, role of biosurfactants in soil biology and remediation, and various aspects of biocontrol strategies for suppression of soil-borne diseases for the protection of agricultural and horticultural plants

Experts in the area of soil science and environmental microbiology from diverse institutions worldwide have contributed to this book. This book should prove to be useful to students, teachers, and researchers in the disciplines of soil and environmental sciences, microbiology, biochemistry, and biotechnology.

We gratefully acknowledge the cooperation and support of all the contributing authors and valuable advice and encouragement provided by Prof. Ajit Varma and Dr. Jutta Lindenborn throughout the preparation of this volume.

Waterloo, Canada — Ajay Singh
Toronto, Canada — Nagina Parmar
New Delhi, India — Ramesh C. Kuhad

Contents

Contributors

Arora, Anju Division of Microbiology, Indian Agricultural Research Institute, New Delhi 110 012, India

Bansal, R.K. Department of Genetics and Plant Breeding, Institute of Agricultural Sciences, Banaras Hindu University, Varanasi 221 005, Uttar Pradesh, India

Brahmaprakash, G.P. Agril Research Station, R.A.U, Durgapura, Jaipur 302 018, Rajasthan, India

Chandna, Piyush Department of Microbiology, University of Delhi South Campus, Benito Juarez Road, New Delhi 110 021, India

Chandra, Ramesh RAK College of Agriculture, Sehore 466 001, Madhya Pradesh, India

Dhar, Bansi Department of Soil Science, G.B. Pant University of Agriculture and Technology, Pantnagar 263 145, Uttarakhand, India

Dudeja, S.S. Department of Microbiology, CCS Haryana Agricultural University, Hisar 125 004, Haryana, India

Dufresne, Jaimie Department of Chemistry and Biology, Ryerson University, Toronto, Ontario, Canada

Gaikawad, B.G. Agricultural Research Station, Gulbarga, Karnataka, India

Gundappagol, R.C. Pulse Research Unit, Dr. PDKV, Akola 444 104, Maharashtra, India

Gupta, S.C. Department of Plant Breeding and Genetics, Punjab Agricultural University, Ludhiana 141 004, Punjab, India

Kalia, Anu Department of Microbiology, Punjab Agricultural University, Ludhiana 141 004, Punjab, India

Kuhad, Ramesh C. Department of Microbiology, University of Delhi South Campus, New Delhi 110 021, India

Lata, Division of Microbiology, Indian Agricultural Research Institute, New Delhi 110 012, India

Lebeau, Thierry Dpt Génie Biologique, Equipe Dépollution Biologique des Sols (EDBS), EA 3991 LVBE, University of Haute-Alsace, 29 rue de Herrlisheim, BP 50568, Colmar 68008, France

Malina, Grzegorz Department of Hydrogeology and Engineering Geology, AGH University of Science and Technology, Mickiewicza 30, 30–059 Cracow, Poland

Mandal, Sudhamoy Plant Pathology Laboratory, Central Horticultural Experiment Station, Aiginia, Bhubaneswar 751019, Orissa, India

Mishra, Brijesh K. Department of Molecular Biology and Biotechnology, Rajasthan College of Agriculture, Maharana Pratap University of Agriculture and Technology, Udaipur, Rajasthan, India

Mudgil, Poonam School of Natural Sciences, University of Western Sydney, Locked Bag 1797, Penrith South DC NSW, 1797, Australia; 3/60 Alice Street, Harris Park, NSW 2150, Australia

Mudhar, Rajinder K. Department of Microbiology, Punjab Agricultural University, Ludhiana, Punjab 141004, India

Nagaraj, K.S. Pulses Improvement Project, Mahatma Phule Krishi Vidyapeeth, Rahuri, Ahmednagar 413 722, Maharashtra, India

Pandey, Alok K. Division of Microbiology, Indian Agricultural Research Institute, New Delhi 110 012, India

Panda, Smita H. Agri-Bioresource Research Foundation, Khandagiri, Bhubaneswar 751 019, India

Parmar, Nagina Department of Chemistry and Biology, Ryerson University, Toronto, Ontario M5B 2K3, Canada

Potdukhe, S.R. Department of Agricultural Microbiology, University of Agricultural Sciences, GKVK Campus, Banglore 500 065, Karnataka, India

Ray, Ramesh C. Regional Centre, Central Tuber Crops Research Institute, Dumduma, Bhubaneswar, Orissa 751 019, India

Sharma, Poonam Indian Institute of Pulses Research, Kanpur Uttar Pradesh 208 024, India

Sindhu, Satyavir S. Department of Microbiology, CCS Haryana Agricultural University, Hisar 125 004, Haryana, India

Singh, Ajay Lystek International Inc, 107–279 Weber Strcet North, Waterloo, Ontario N2J 3H8, Canada

Singh, N.P. Department of Microbiology, CCS Haryana Agricultural University, Hisar 125 004, Haryana, India

Singh, Surender Division of Microbiology Division, Indian Agricultural Research Institute, New Delhi 110 012, India

Swain, Manas R. Department of Biotechnology, College of Engineering and Technology, Biju Patnaik University of Technology, Bhubaneswar 751003, Orissa, India

Ward, Owen P. Department of Biology, University of Waterloo, Waterloo, Ontario N2L 3G1, Canada

Zawierucha, Iwona Institute of Chemistry and Environmental Protection, Jan Dlugosz University of Czestochowa, Armii Krajowej 13/15, 42–200 Czestochowa, Poland

Chapter 1
Bioaugmentation, Biostimulation, and Biocontrol in Soil Biology

Ajay Singh, Nagina Parmar, Ramesh C. Kuhad, and Owen P. Ward

1.1 Microbial Diversity and Function in Soil

Biological diversity (or biodiversity) can be defined as the set of animal and vegetable species, their genetic material, and the ecosystems they belong to, that is it encompasses diversity at the ecosystem, species, and gene diversity levels (Fontaine et al. 2003; Lynch et al. 2004). Soil organic matter and the associated bioactivity are major contributors to carbon and nutrient cycling in the biosphere: it is the main nutrient source for plant growth (after microbial decomposition) and impacts upon soil quality (soil structure, resistance to erosion). It also represents the major carbon reservoir of the biosphere–atmosphere system.

It is believed that up to one billion bacterial species actually exist in the earth environment and yet only about 5,000 species have been described (Hunter-Cevera 1998; Curtis and Sloan 2004). Only about 1% of the soil bacterial population can be cultured by standard laboratory practices. Similarly, more than 1.5 million species of fungi are thought to exist of which only about 72,000 species have been isolated or described. Microorganisms exist in every conceivable place on earth and soil may harbor up to 10 billion microorganisms per gram. It is estimated that 1 g of soil may contain about 4,000 different bacterial "genomic units" based on DNA–DNA re-association. The tropics are considered to be richer in microbial diversity than boreal or temperate environments. Some microbiologists believe that there is a similar level of microbial diversity in the deserts. Many anthropogenic activities,

A. Singh (✉)
Lystek International Inc, 107-279 Weber Street North, Waterloo, Ontario, Canada, N2J 3H8,
e-mail: asingh@lystek.com

N. Parmar
Department of Chemistry and Biology, Ryerson University, Toronto, Ontario, Canada, M5B 2K3

R.C. Kuhad
Department of Microbiology, University of Delhi South Campus, Benito Juarez Road, New Delhi 110 021, India

O.P. Ward
Department of Biology, University of Waterloo, Waterloo, Ontario, Canada, N2L 3G1

A. Singh et al. (eds.), *Bioaugmentation, Biostimulation and Biocontrol*, Soil Biology 28,
DOI 10.1007/978-3-642-19769-7_1, © Springer-Verlag Berlin Heidelberg 2011

such as city development, agriculture, dispersal of pesticides, and other chemical pollutants can potentially affect soil microbial diversity (Forney et al. 2004; Blagodatskaya and Kuzyakov 2008).

Numerous factors are known to affect diversity: trophic interactions, spatial and temporal habitat heterogeneity, disturbance, and eutrophication. Ecosystem stability, productivity, and resilience toward stress and disturbances are influenced by the microbial functional diversity (Table 1.1). Differences in microbial community structures reflect the abilities of microorganisms to respond to specific environmental factors and substrates (Kuhad et al. 2004; Little et al. 2008). For example, the fluorescent pseudomonads are attracted to plant roots and show diversity between soil and plant surfaces. Species of *Penicillium* are abundant in temperate and cold climates, whereas *Aspergillus* species predominate in warmer regions. Cyanobacteria are commonly found in neutral to alkaline soils. Depending on the nature of the metabolites present in the soil, nitrogen-fixing, sulfur- and hydrogen-oxidizing, and nitrifying bacteria are often found together with denitrifiers, sulfate-reducers, and methanogens.

Further, microbial functions in ecosystems are as diverse as the microbes themselves. Microbially digested organic materials enhance plant growth and improve soil structure and nutrient status of soil. Denitrifying bacteria utilize nitrous oxides (NO_x) as the terminal electron acceptor. These denitrifiers produce NO_x reductase and can metabolize NO_x in aerobic and anaerobic conditions. Varieties of microhabitats with different physicochemical gradients and discontinuous environmental conditions are found in soil. Microbes adapt to these microhabitats and live together in consortia, interacting with each other and with other parts of the soil biota that control microbial community structure and diversity. Competitive interactions are influenced by soil structure and water regimes. Particle size and other factors, such as pH, together with type and amount of available organic compounds, may affect microbial community structure. Soil microbes are also subjected to considerable seasonal fluctuations in environmental conditions, particularly those conditions known to affect microbial activity, such as temperature, water content, and nutrient availability.

Plant growth-promoting rhizobacteria (PGPR) are free-living bacteria of beneficial agricultural importance, for example, plant health and growth, suppress disease-causing microbes, and accelerate nutrient availability and assimilation. PGPR compensate for the stress and reduction in plant growth caused by weed infestation, drought, heavy metals, salt, and other unfavorable environmental conditions and is frequently used as biofertilizer. These bacteria belong to the genera *Acetobacter*, *Acinetobacter*, *Alcaligenes*, *Arthrobacter*, *Azoarcus*, *Azospirillum*, *Azotobacter*, *Bacillus*, *Beijerinckia*, *Burkholderia*, *Derxia*, *Enterobacter*, *Gluconacetobacter*, *Klebsiella*, *Ochrobactrum*, *Pseudomonas*, *Rhodococcus*, *Serratia*, *Zoogloea*, etc. and have been subject of extensive research for decades. PGPRs may have more than one mechanism for accomplishing plant growth by production of root exudates, repression of soil-borne pathogens (by the production of hydrogen cyanide, antibiotics, and/or competition for nutrients), siderophore production, nitrate reduction, nitrogen fixation, phosphate solubilization, production of organic

Table 1.1 Functional diversity of microbes in natural environment

Microbial process in soil	Examples of microbes
Organic matter decomposition	*Trichoderma, Fusarium, Bacillus, Streptomyces, Clostridium*
Nitrogen fixation	*Rhizobium, Bradyrhizobium, Frankia, Anabaena, Azotobacter, Beijerinckia, Aerobacter, Chlorobium, Nostoc*
Nitrogen cycles	*Bacillus, Pseudomonas, Serratia, Nitrobacter, Nitrosomonas, Achromobacter, Pseudomonas*
Phosphate solubilization	*Azotobacter, Enterobacter, Bacillus, Aspergillus, Penicillium, Rhizoctonia, Trichoderma, Irwinia*
Sulfur transformation	*Desulfovibrio, Thiobacillus*
Iron transformation	*Ferribacterium, Leptothrix*
Siderophore production	*Neurospora, Trichoderma, Agaricus, Fusarium, Penicillium*, ericoid mycorrhizal fungi, *Nocardia, Pseudomonas, Bacillus, Aeromaonas, Erwinia*
Phytohormone production (auxin, gibberellin, cytokinin)	*Azotobacter, Azospirillum, Pseudomonas, Rhizobium, Bacillus, Flavobacterium, Actinomyces, Nocardia, Fusarium, Gibberella, Aletrnaria, Penicillium*
Vitamins production (biotin, thiamin)	*P. fluorescens, P. putida*
Antibiotics (kanosamine, oligomycin A, oomycin A, phenazine-1-carboxylic acid, pyoluteorin, pyrrolnitrin, viscosinamide, xanthobaccin and zwittermycin A) production	*Bacillus* spp.
Enzymes production (chitinase, cellulase, glucanase, protease, lipase, dehydrogenase, phosphatase, nitrogenase)	Plant growth-promoting rhizobacteria and fungi
Lipopeptide biosurfactants (viscosinamide, tensin)	*Pseudomonas fluorescens*
Metabolites production (HCN, diacetylphloroglucinol)	*P. fluorescens*
Volatile compounds (2,3-butanediol, acetoin, pyoluteorin, auxofuran) production	*Pseudomonas* spp.
Biocontrol	*Agrobacterium, Pseudomonas, Bacillus, Strepetomyces, Trichoderma*, mycorrhizal fungi
Bioremediation	*Acinetobacter, Alcaligenes, Ochrabactrum, Pseudomonas, Flavimonas, Rhodococcus, Stenotrophomonas, Comamonas, Arthrobacter, Burkholderia, Ralstonia, Moraxella, Nocardia, Klebsiella, Phanerochaete, Penicillium, Aspergillus, Fusarium, Cladosporium*
Phytoremediation	*Pseudomonas, Agrobacterium, Enterobacter, Rhizobium, Kluyvera, Glomus*, Rhizobacteria, mycorrhizal fungi

acids, and phytohormones (indole acetic acid or IAA), NH_3, release of enzymes (dehydrogenase, phosphatase, nitrogenase, 1-aminocyclopropane-1-carboxylate (ACC) deaminase), and the induction of systemic disease resistance (Figueiredo et al. 2010).

Plant-associated bacteria include endophytic, phyllospheric, and rhizospheric bacteria. Endophytes are bacteria or fungi that colonize healthy plant tissue inter- and/or intracellular without causing any apparent symptoms of disease (Wang and Dai 2010). They are found in almost every host plant studied so far and the relationship between endophytes and host plants involves both mutualism and antagonism, which beneficially impact upon the symbiotic system. The phyllosphere refers to the above ground external regions of plant parts including leaves, stems, flowers, and fruits. Bacteria residing in the phyllosphere are exposed to large and rapid fluctuations in temperature, solar radiation, and water availability.

Root exudates are believed to have a major influence on the diversity of plant growth-promoting rhizosphere microorganisms. Root exudates are chemical compounds such as photosynthates, organic acids, sugars, polyamine putrescine excreted from root tissues. Indirect interactions between plants and microbes occur in the rhizosphere due to root exudates (Yang 2009).

Plant growth-promoting mycorrhizal fungi play a major role in the induced resistance against diseases and uptake of P, Zn, Fe, and N in organically grown crops (Raviv 2010). Simultaneously, mycorrhizae provide additional benefits, not the least of which being their positive effect on gradual improvement in soil aggregate stability, resulting from the direct effect of mycorrhizae mycelia.

Plant–microbe interactions such as biofertilization, rhizoremediation, biocontrol, and phytostimulation may be quorum sensing (QS) dependent. In QS behavior, small diffusible extracellular signaling molecules meditate cell–cell communication. The signaling molecules for gram-negative bacteria are named autoinducers, usually acylated homoserine lactones (AHLs). The gram-positive bacteria use peptide-signaling molecules for QS. The catabolic response profile (CRP), a measure of short-term substrate-induced respiration, has been used to calculate the catabolic diversity in soil (Gil-Sotresa et al. 2005). After a major disturbance (landslides, volcanic eruptions, chemical or petroleum oil spills, etc.), significant changes in catabolic functional diversity have been reported in soil ecosystems. The major sources of carbon input for soil organisms are the plant roots and organic residues contributed during and following plant growth. The proportions of nitrogen, carbon, and other organic matter alter microbial activity and diversity (Bending et al. 2002). Microorganisms play an essential role in the functioning and sustaining of all natural ecosystems including biogeochemical cycling of nutrients and biodegradation. The types of nutritional substrates available are different in soils with varying soil organic matter quality, and they directly affect the microbial community active in the soil. Native soil organic matter content may also significantly affect enzyme diversity, which is greater in high organic-containing soils.

Microbial functional diversity analysis is important when considering the ability of ecosystems to respond to changing environmental conditions, the need to conserve the microbial gene pool, and utilization of the selective gene pools for useful biotechnological applications relevant to bioremediation and phytoremediation (Ward et al. 2003; Ohtsubo et al. 2004; Zhuang et al. 2007). Fortunately, with development of advanced molecular in situ methods and improved cultivation procedures, better estimates of the microbial functional diversity on earth can be predicted and its role in soil ecosystem can be thoroughly evaluated.

1.2 Characterization of Natural Microbial Communities in Soil

Characterization of natural microbial communities is a daunting task due to the interactions, including those involving substrates and metabolites, possible in soil. Methods for studying microbial diversity and community function can be broadly divided into culture-dependent and culture-independent methods (Dahllöf 2002). Culture-dependent methods are generally based on differential morphological, metabolic, and physiological properties, including use of techniques for isolation and cultivation on solid media, determination of most probable number (MPN), and characterization of substrate utilization patterns. Culture-independent methods include various biochemical and molecular approaches of community analysis involving direct examination of metabolically active microbes using differential stains, phospholipids fatty acid analysis (PFLA), polymerase chain reaction (PCR), and application of DNA microarray to study specific microorganisms or groups of microorganisms, specific genes, and to evaluate overall community profiles.

Soil biochemical properties related to the biocycling of elements (C, N, P, and S) are generally useful indicators of soil quality (Gil-Sotresa et al. 2005). These properties include both general biochemical parameters such as microbial biomass C, dehydrogenase activity and N mineralization potential, and specific biochemical parameters like activities of hydrolytic enzymes, such as phosphatase, urease, and β-glucosidase. Biochemical properties can be used both individually, as simple indices, or in combinations using complex equations derived from mathematical combinations or the application of statistical programs.

Due to the diversity of compounds contained within the soil organic matter, a great diversity of enzymes exists in soil. Because of the diversity of the soil community and of the physical soil matrix, multiple soil enzymes are required to efficiently degrade different compounds. With the recent advances in molecular ecology, the genetic potential of microbial communities to produce enzymes can be identified by the genomic studies targeting functional genes coding for extracellular enzymes (Wallenstein and Weintraub 2008).

Metagenomics is a fast growing and diverse field directed at obtaining knowledge on genomes of environmental microbes and entire microbial communities, omitting the cultivation step (Chistoserdova 2010). Other terms such as environmental genomics, ecogenomics, community genomics, and megagenomics are also used to describe this area of biology. Function-based metagenomics relies on cloning environmental DNA into expression vectors and propagating them in appropriate hosts (Craig et al. 2010). Following appropriate activity screens, an active clone is identified and the sequence of the clone is determined. The gene of interest and its respective product are further analyzed, and their biotechnological potential is explored. Transcriptomic studies of mRNA and emerging proteomic tools can now be used to assess the microbial regulation of extracellular enzymes, pool sizes, diversity, and microbial source of soil enzymes. Furthermore, new mass-spectrometry approaches can be used to quantify the enzymatic degradation products and develop improved models of decomposition.

Biochemical and molecular tools are continuously being developed in an attempt to better appreciate microbial abundance and distribution in natural environments, to evaluate community structures with ecosystem functions, determine the community structure and function in soil, long-term effects of pollution (Prosser 2002; Singh and Ward 2005) and to develop appropriate remediation approaches (Siciliano et al. 2003; Van Hamme et al. 2003; Singh et al. 2009). Molecular methods for the analysis of microbial diversity and community analysis will be greatly advanced if genome projects are initiated to sequence environmentally important microorganisms.

1.3 Microbial Inoculants and Biofertilizers

Traditional use and importance of chemical fertilizers in agricultural production cannot be over-emphasized, but with fertilizer costs going up, generally in parallel to increase in energy costs, these need to be supplemented or substituted with cheaper available alternatives such as beneficial microbial inoculants, biofertilizers, and organic amendments to improve soil quality, fertility, biology, and agricultural productivity (Saleem et al. 2007; Ray et al. 2008; Babalola 2010).

Biofertilizers contain different types of microorganisms, which have an ability to convert nutritionally important elements from unavailable to available form through biological processes in soil. Biofertilizers have emerged as a potentially important component of the integrated soil nutrient supply system and hold great promise to improve crop yields. Microbial inoculants and biofertilizers are an important component of organic farming accounting for about 65% of the nitrogen supply to crops worldwide. In comparison with chemical/synthesized pesticides and fertilizers, microbial inoculants or biofertilizers have several advantages (Berg 2009) including:

(a) Greater relative safety
(b) Potentially reduced environmental damage and human health risk
(c) Much more targeted activity
(d) Effectiveness in small quantities
(e) Capacity for self-multiplication while being controlled by the plant as well as by the indigenous microbial populations
(f) Faster decomposition than conventional chemical pesticides
(g) Ability to be used in conventional or integrated pest management systems

Biofertilizers containing N-fixer (*Rhizobium* spp., *Bradyrhizobium* spp., *Azotobacter chroococcum*), P-solubilizer (*Bacillus megaterium*) and K-solubilizer (*Bacillus mucilaginous*), and arbuscular mycorrhizal fungi (*Glomus mosseae* and *Glomus intraradices*) have been developed for commercial applications. However, in the current economic situation there is a need to get maximum output with minimum cost, which is possible only if chemical fertilizers are supplemented with organic- and bio-fertilizers.

Inoculation of legume seed by dusting with peat culture in the presence of adhesives is an efficient and convenient way to introduce effective rhizobia to soil and subsequently to the rhizosphere of legumes (Deaker et al. 2004). Lime-pelleting of inoculated legume seed with superfine limestone ($CaCO_3$) is used to counteract the acidic effects of soil or superphosphate on the survival of the rhizobia. Co-inoculation studies with PGPR and *Rhizobia* have shown increased plant nodulation and N fixation (Figueiredo et al. 2010). Co-inoculation of some *Bacillus* strains with effective *Bradyrhizobium* resulted in enhanced nodulation and plant growth of green gram (*Vigna radiata* L.).

1.4 Fate of Genetically Modified Organisms

Determining the impact and fate of genetically modified organisms (GMOs) or non-modified organisms on the environment are of great concern today. Genetic exchange between microbes, plants and animals may be promoted by transformation. In general, release of DNA from different organisms occurs by cell lysis after death. However, some microorganisms possess active mechanisms for releasing large amounts of chromosomal or plasmid DNA which can reach concentrations that could support horizontal gene transfer by transformation (Singh et al. 2006). Plant DNA enters the soil continuously, predominantly from the sloughing off of root cap cells, as a result of pathogen colonization of below-ground biomass, through pollen dispersal, and during crop residue decomposition.

Bacteria are the only organisms capable of natural transformations and considered for the genuine bacterial gene transfer process. In bacteria, gene transfer can occur by three mechanisms in the natural environment:

(a) Transformation – extracellular DNA is taken up by recipient bacteria
(b) Conjugation – genetic material is transferred from one bacterium to another by cell to cell contact)
(c) Transduction – the transfer of genetic information between bacteria is mediated by bacteriophages

Both bacteria and free DNA may be dispersed by percolation and flow of water, air and dust, and other soil organisms. Upon entering the soil environment, extracellular DNA is subjected to dynamic biological, physical, and chemical factors that determine its fate (Levy-Booth et al. 2007; Pietramellara et al. 2009). Extracellular DNA up to 20 kb in size may persist through cation bridging onto soil minerals and humic substances, and may be enzymatically degraded and restricted by DNases of microbial origin, and/or enter the microbial DNA cycle through natural transformation of competent bacteria. Lateral gene transfer may disseminate DNA through the microbial community. DNA also tends to adsorb to the clay and sand particles.

The potential risks associated with the release of GMOs into the environment has led to the development and construction of active biological containment systems in which bacteria are killed in a controlled suicide process (Ronchel and Ramos 2001). This strategy has been developed to prevent the undesirable spread of genetically modified microorganisms in the environment after they have completed their intended tasks.

Genetically modified plants (GMPs) have great potential for future agricultural, but also require a well-defined risk assessment. Most of the studies that have been conducted in order to determine the effects of GMPs on soil microorganisms and processes have been able to detect some sort of effect (Bruinsma et al. 2003). GMPs have been found to affect bacteria, non-target fungi, target fungi, enzyme activities, substrate utilization, and decomposition. Natural transformation is the most likely mechanism for horizontal transfer of genes from transgenic crops to bacteria. The single-stranded DNA taken up by the bacteria can either integrate into the bacterial genome by homologous recombination or form an autonomous replicating element. From laboratory experiments, >40 bacterial species from different environments are known to be naturally transformable. Transgenic plant DNA can be degraded during plant senescence and during microbial degradation of the plant residue in soil. However, measured amounts of transgenic plant DNA can escape these degradation processes and the long-term persistence, even of a small percentage of released plant DNA, is assumed to enhance the likelihood of bacterial transformation.

However, the effects of transgenic crops on soil microbial populations are expected to be low or at least less important compared to other biosafety issues of transgenic crops such as out-crossing to weedy species, effects on non-target organisms or the appearance of new viruses (Mercier et al. 2006; Icoz and Stotzky 2008).

1.5 Organic Amendments

Typical organic wastes and amendments that are applied to soil are pulp and paper industrial sludge, municipal wastewater sludge, animal manure, abattoir waste, and compost. Direct application of raw organic wastes is inappropriate for land use due to their unknown compositions with respect to pathogens, toxic compounds, weed seeds, heavy metals, and foul odors. These materials, if not appropriately treated or processed to reduce environmental risks and disposal constraints, may pose a serious threat to the environment and human health and cause toxicity to beneficial microflora in soil. The practice of using landfills for organic waste disposal has to diminish due to large quantities of waste generation, and reduced availability of dumping sites and the associated environmental hazards. Similarly, incineration is expensive and causes air pollution. In contrast, land application of treated organic wastes has emerged as an attractive and cost-effective strategy. These materials have been proved to supply plant nutrients and organic matter to the soil for improved crop production. The beneficial impacts of organic amendments to soil and nutrient composition of a range of organic material are shown in Tables 1.2 and 1.3, respectively.

Use of organic soil amendments is a traditional cultural practice to improve soil fertility and structure. It is also known as a control method for soil-borne diseases, including plant–parasitic nematodes. Organic amendments have also been proposed to control diseases caused by soil-borne pathogens such as *Aphanomyces euteiches*, *Gaeumannomyces graminis*, *Macrophomina phaseolina*, *Rhizoctonia solani*,

Table 1.2 Beneficial impact of organic amendments to soil

Soil property	Beneficial effect to soil
Biological	Microorganisms, earthworm, decomposition, humus production, nutrient availability, production of beneficial chemicals (hormones, amino acids, vitamins, organic acids, antibiotics), suppression of plant pathogens, crop productivity
Chemical	Buffering capacity, chelating capacity, cation exchange capacity, pH
Nutritional	Micronutrients (B, Cu, Mn, Mo, Zn) and macronutrients (Ca, Fe, Mg, K, P, C, N, O, H)
Physical	Soil aggregation, texture, porosity, bulk density, crusting, erosion, water holding capacity, water infiltration and percolation

Table 1.3 Nutrient values of organic waste material

Organic matter	% Nitrogen	% Phosphorus	% Potassium	Availability of nutrients
Alfalfa hay	2–3	0.5–1	1–2	Medium
Cottonseed meal	6	3	1	Slow
Compost	1.5	0.5	1	Slow
Bone meal	1	11	0	Slow
Dried blood	12	1.5	0.5	Rapid
Feather meal	12	0	0	Medium
Fish meal	10	4	0	Slow
Grass clippings	1–2	0–0.5	1–2	Medium
Horn meal	12–14	1.5–2	0	Medium
Kelp	1	0.5	9	Rapid
Leaves	1	0–0.5	0–0.5	Slow
Legumes	2–4	0–0.5	2–3	Medium
Cow manure	0.25	0.15	0.25	Medium
Horse manure	0.3	0.15	0.5	Medium
Sheep manure	0.6	0.33	0.75	Medium
Swine manure	0.3	0.3	0.3	Medium
Pine needles	0.5	0	1	Slow
Poultry manure	2	2	1	Rapid
Sewage sludge	2–6	1–4	0–1	Moderate
Wood ashes	0	1–2	3–7	Rapid

Thielaviopsis basicola, *Verticillium dahlia*, etc. (Bonanomi et al. 2010). Application of organic soil amendments is a traditional control method for plant–parasitic nematodes as well (Oka 2010). A variety of organic amendments, such as animal and green manures, compost, nematicidal plants, and proteinous wastes, are used for this purpose. Combinations of different mechanisms appear to produce nematode suppression in amended soils. Possible mechanisms involved in nematode suppression are:

(a) Release of pre-existing nematicidal compounds in soil amendments
(b) Generation of nematicidal compounds, such as ammonia and fatty acids, during degradation
(c) Enhancement and/or introduction of antagonistic microorganisms
(d) Increase in plant tolerance and resistance
(e) Changes in soil physiology those are unsuitable for nematode behavior

1.5.1 Conventional Compost and Vermicompost

Composting is considered one of the most appropriate options for addressing the constraints associated with organic solid waste materials for agricultural use. However, according to an estimate (Ahmad et al. 2007), 827 million tons of compostable materials are produced each year, largely by agriculture, municipalities, and industry. However, only 140 million tons, or 17%, of those are collected for composting. Composting is a biological process which converts heterogeneous organic wastes (manure, sludge, yard wastes, leaves, fruits, vegetables, and food wastes) into humus-like substances by mixed microbial population under controlled optimum conditions of moisture, temperature, and aeration. Composts provide plant nutrients and improve soil biophysical properties, soil organic matter, and crop yields. Decomposers include bacteria, actinomycetes, and fungi that are widespread in nature and are indigenous to soil, dust, fruit and vegetable matter, and wastes of all sorts, so special organisms are not required.

Vermicompost, like conventional compost, provides many benefits to agricultural soil, including increased ability to retain moisture, better nutrient-holding capacity, better soil structure, and higher levels of microbial activity. Vermicompost may sometimes be superior to conventional aerobic compost in the levels of plant-available nutrients, beneficial microorganisms, ability to stimulate plant growth, ability to suppress diseases, and ability to repel pests. This is a relatively new area and not much information is available. There seems to be strong evidence that worm castings may repel hard-bodied pests probably due to the production of the chitinase enzyme by the worms, which breaks down the chitin in the insects' exoskeleton.

Climate change is one of the most serious and pressing environmental problems of our time. Farms are a significant contributor to climate change, largely through the release of carbon from soils and the generation of methane gas from livestock and their manures. Both composting and vermicomposting address these issues through carbon sequestration, a process of locking up carbon in organic matter and organisms within the soil. Because composts are stable, more carbon is retained in the soil than would occur if raw manure or inorganic fertilizer were applied. The consistent application of compost or vermicompost gradually raises the level of carbon in the soil.

Since, the composting process results in the same level of greenhouse gas (GHG) emissions as if the materials were allowed to decay naturally, it is considered to be neutral with respect to GHG generation. The potential advantages of composting described above also apply to vermicomposting. In theory, however, vermicomposting should provide some potentially significant advantages over composting with respect to GHG emissions. The vermicomposting process does not require manual or mechanical turning, as the worms aerate the material as they move through it. This should result in fewer anaerobic areas within the piles, reducing methane emissions from the process. It also reduces the amount of fuel used by farm equipment or compost turners. It has been suggested that the increased effectiveness of vermicompost relative to compost in promoting plant growth and increasing yield can result in the displacement of 5–7 times as much fertilizer per unit of vermicompost, thereby decreasing the GHG emissions proportionately. Finally, analysis of vermicompost samples has shown generally higher levels of nitrogen than analysis of compost samples made from similar feedstock. This implies that the process is more efficient at retaining nitrogen, probably because of the greater numbers of microorganisms present in the process. This, in turn, implies that less nitrous oxide is generated and/or released during the process, such that less free ammonia is generated. Since N_2O is 310 times as potent a GHG as CO_2, this could be a significant benefit.

1.5.2 *Wastewater Biosolids*

Biosolids are the residual solids remaining after wastewater or sludge has been treated. The need for solids reduction is becoming more evident as the volume of generated wastewater biosolids is growing and municipal plants are choosing to dispose of the nutrient-rich solids through recycling to agriculture fields and thus helping to save space in landfills. Biosolids contain significant amounts of nutrients required by plants, including nitrogen, phosphorus, potassium, and micronutrients, making them an excellent fertilizer for use in agriculture and forestry. Addition of biosolids to soil improves bulk density, increases porosity, soil aggregation, moisture and nutrient retention, and organic carbon. Biosolids have been used in conjunction of phytoremediation technology for landfill remediation as landfill or phytocapping to stabilize soil and simultaneously remediate landfill leachate (Kim and Owens 2010). However, long-term application of biosolids to cultivated land may raise concerns for food safety from contaminants, such as the pathogens, heavy metals, and endocrine disruptive compounds (EDCs) present in the biosolids that may find their way in water streams or accumulate in plant tissues.

A large number of human pathogens, which primarily originate from human feces, can find their way into biosolids. The diversity and number of pathogens in biosolids depend upon the general health of the contributing population, presence of hospitals, farm animals, and abattoir industries in the area (Girones 2006; Sidhu and Toze 2009). Enteric virus, protozoa, and parasites are obligatory parasites and hence unable to multiply in biosolids, whereas bacteria may multiply under favorable conditions. Generally, pathogenic viruses and bacteria die within 1–3 months, whereas protozoan oocysts and helminth ova can survive for up to a year in wastewater and possibly much longer in untreated biosolids. The inactivation of pathogens in the biosolids depends upon a number of factors such as temperature, moisture content and competition from indigenous microflora. Other factors such as predation, pH adjustment, sunlight, oxygen and oxidants, mechanical shearing or abrasion, soil type, and texture also influence pathogen inactivation. A number of different types of pathogens can be present in the biosolids but only a small portion of them are a cause of concern including bacteria: *Escherichia coli* O157:H7, *Listeria*, and *Helicobacter pylori*; the viruses: coxsackievirus, echovirus, hepatitis A, rotavirus, and norovirus; and the parasites: *Cryptosporidium*, *Cyclospora*, *Toxoplasma*, *Microsporidia* and *Giardia*. Currently, the lack of well-developed methods for the detection and enumeration of viral, protozoan, and helminth pathogens is the main cause of non-availability of the data on pathogen behavior in biosolids.

Various stabilization processes for the treatment of wastewater biosolids to remove pathogens have been used commercially; including composting, heat drying, pelletizing, incineration, mechanical and thermal destruction, enhanced thermophilic digestion and chemical (alkali, ammonia, sulfamic acid, fly ash, etc.) treatment (Table 1.4). Treatment technologies involving high temperature or pressure systems are generally more energy extensive and expensive to operate and maintain with high capital/operating costs as compared to technologies involving only chemical and pasteurization processes. Chemical stabilization and heat-drying processes have been used that produce pathogen-free nutrient-rich high solids liquid or dry soil-like organic product for soil enrichment, topsoil blend, and as organic fertilizer amendments.

Municipal wastewater may contain a complex mixture of EDCs, originating from personal care products, pharmaceuticals, excreted hormones, household and industrial chemicals, etc. Different environmental agencies have classified the following compounds as EDC or potential EDC: steroids (17β-estradiol, ethynil estradiol, estrone, diethylstilbestrol), some alkylphenols (nonylphenol, nonylphenol ethoxylate, octylphenol, octylphenol ethoxylate), polychlorinated biphenyls (PCBs), brominated flame retardants, polyaromatic hydrocarbons (PAHs), di-(2-ethyl hexyl) phthalate, bisphenol A (BPA), hexachlorobenzene, pentachlorophenol, polychlorinated dibenzodioxins/furans (PCDD/F), tributylin, and many pesticides including, atrazine, lindane, and dieldrin. Some pre-treatment methods (enzymatic, thermal, oxidation) may facilitate EDC biodegradation in subsequent biological

Table 1.4 Commercially used wastewater biosolids stabilization processes

Stabilization process	Description	Supplier/developer
Alkali/pasteurization	Using a combination of heat, alkali and high shear to cost effectively achieves cell lysis; high solid pathogen-free liquid fertilizer product as soil amendment; recycling to anaerobic digester to improve digestion/biogas production/ solids reduction	Lystek, Canada
Alkali/heat drying treatment	Use of chemicals such as lime, sulfamic acid, cement kiln dust, lime-kiln dust, and/or fly ash, heating; soil-like dry fertilizer product as soil amendment	N-Viro, USA Bioset, USA
Anhydrous ammonia/heat	Mixing by pug mill, concentrated phosphoric, sulfuric acid, ammonia, drying; dry fertilizer product as soil amendment	Vitag, USA
Pelletizing/heat drying	Heat drying sludge and manure to a dry product as soil amendment	Terratec, Canada Veolia, USA Unity Envirotech, USA
Thermal hydrolysis	Pre-treatment of sludges for anaerobic digestion of municipal and industrial sludge and biowaste; dry fertilizer product as soil amendment	Veolia, USA Cambi, Norway
High-pressure homogenizer/ alkali	Chemical pre-treatment and high-pressure homogenizer to weaken and burst cells; application in anaerobic digestion improvement	Paradigm, Canada
Auto-thermophilic aerobic digestion (ATAD) digestion/chemical	Chemical pre-treatment and thermophilic aerobic digestion of organic waste; granular or liquid product for land application	PMC Biotec, USA
Mechanical disintegration	Homogenizing, pressurizing and passing through disintegration nozzle to rupture cell structure; application in anaerobic digestion improvement	Siemens, USA

treatment (mesophilic anaerobic digestion, sequential biological reactors). However, none of the research studies carried out till date report partial breakdown products and/or intermediates of EDCs after pre-treatment, which could be potentially more toxic than the parent compounds. Therefore, systematic studies are required to explore the degradation of toxic intermediates after sludge pre-treatment and subsequent stabilization processes. However, as is discussed further in the next section, studies in environmental technology, including passive bioremediation through natural attenuation and more engineered bioremediation processes, have proven that soil ecosystems respond to the presence of synthetic chemical contaminants by adjusting their microbial populations to facilitate contaminant biodegradation. Indeed, even in cases where the contaminants were perceived to be recalcitrant, organisms have been shown to develop the requisite new biodegradative capacities. Furthermore, the principal nutrients in biosolids act to promote this biodegradation through bioaugmentation.

One of the limiting steps in conventional treatment of wastewater by aerobic or anaerobic processes is the availability of biodegradable organics called as volatile suspended solids (VSS) in the wastewater sludges. Consequently, many research efforts have been made in the recent past to increase the VSS destruction by adopting various types of pre-treatments (chemical, thermal, mechanical, and biological pre-treatments and combinations thereof) and are recently reviewed by Carrère et al. (2010). Sludge pre-treatments rupture suspended solids (microbial cells), liberate the nutrients, partially solubilized the suspended solids, increase the soluble chemical oxygen demand, decrease viscosity, and improve the overall sludge digestibility. Ultrasonication, high pressure homogenizer, shearing, alkaline hydrolysis, thermal hydrolysis, and partial oxidation through chemical reaction or ozonation are efficient pre-treatment methods that have been reported to increase sludge biodegradability.

Bioconversion wastewater sludge into value-added products (VAPs) such as biopesticides or other biocontrol agents (BCAs), microbial inoculants, industrial enzymes, bacterial bioplastics, and other biopolymers have been achieved with successful and encouraging results (Barnabé et al. 2009). However, wastewater sludges are known to contain toxic metals, organic micro-pollutants and pathogens and may pose risks associated with the commercial application of VAPs. Environmental risks related to various contaminants may be reduced by selecting the biosolids with low contaminant levels and by adopting necessary steps to decontaminate or eliminate the contaminants of concern followed by subjecting the product to rigorous ecotoxicological (teratogenicity, carcinogenicity, and mutagenicity) tests during product registration with the government regulatory authorities.

1.6 Bioaugmentation and Biostimulation in Bioremediation

1.6.1 Bioremediation

Soil and aquatic environments are potential target media for thousands of contaminants that vary in composition and in concentration. These contaminants enter the system as a result of a wide range of actions such as intentional applications, inadequate residue disposal, accidental wastes, and inappropriate use. The pollution by inorganic compounds such as nitrates, phosphates, and perchlorates is due to an inadequate disposal of manufacture residues of fireworks and matches; explosives such as hexahydro-1,3,5-trinitro-1,3,5-triazine (RDX) and octahydro-1,3,5,7-tetranitro-1,3,5,7-tetrazocine (HMX) from their manufacture and tests; monoaromatic hydrocarbons such as benzene, toluene, ethylbenzene, and xylene from oil spills and of storage tanks leaking; polycyclic aromatic hydrocarbons from accidental spills; a range of herbicides such as diuron, linuron, and chlorotoluron used in weed control and by heavy metals (Fig. 1.1) (Sheoran et al. 2008; Ward et al. 2009).

Microorganisms can degrade numerous organic pollutants due to their metabolic machinery and to their capacity to adapt to inhospitable environments. The capacity of a microbial population to degrade pollutants within an environmental matrix (e.g., soil, sediment, sludge or wastewater) can be enhanced either by stimulation of the indigenous microorganisms, by addition of nutrients or electron acceptors (biostimulation) or by the introduction of specific microorganisms to the local population (bioaugmentation).

The concept of introduction of non-indigenous or cultured microorganisms into natural or engineered environments is not new, and has been practiced in agriculture, in some wastewater treatment processes, and bioremediation of contaminated sites. Traditional bioaugmentation practice has achieved its greatest results through repeated application of highly competent pollutant degrading bacteria using appropriate microbial strain selection better suited for a particular tasks and environments (Singh et al. 2009). The potential for successful bioaugmentation may be increased by using soil containing indigenous degrader populations exposed previously to contaminants. Introduction of naturally developed microbial consortia may be more effective as compared to single strains isolated and applied as pure cultures.

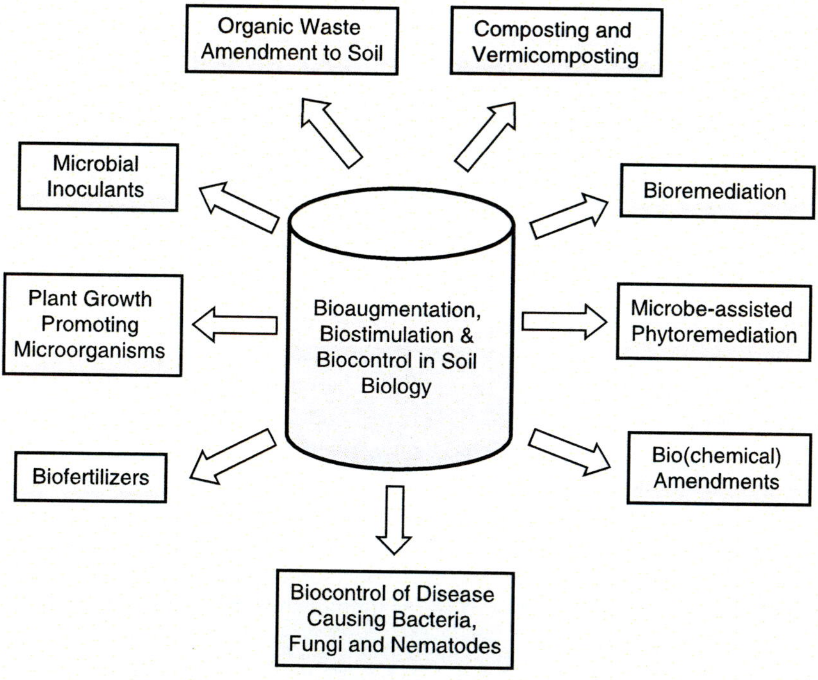

Fig. 1.1 Role of bioaugmentation, biostimulation and biocontrol in soil biology and fertility

Most bioaugmentation studies have been carried out using gram-negative bacteria belonging to species of *Pseudomonas*, *Flavobacterium*, *Sphingomonas*, *Alcaligenes*, and *Achromobacter* (Singh et al. 2004; El Fantroussi and Agathos 2005). Other potential bioaugmentation strains may be gram-positive bacteria species, such as *Rhodococcus*, *Mycobacterium*, and *Bacillus*. Useful fungal species belongs to genus *Absidia*, *Achremonium*, *Aspergillus*, *Verticillium*, *Penicillium*, and *Mucor* (Mrozik and Piotrowska-Seget 2010). No single microbial species or group appears to be universally applicable to bioaugmentation and success in removal of recalcitrant toxic substances from soils using microbial consortia has been demonstrated.

One of the most important strategies for bioremediation efficacy relates to maintaining appropriate viable counts of the introduced strain. There are different approaches in delivering bacterial strain to desired contaminated areas. Immobilization of microorganisms has been suggested as a strategy to improve the effectiveness of bioaugmentation as it provides a protective barrier around microorganisms. Several carriers were tested for soil bioremediation, for example, polyvinyl alcohol, chitin and chitosan, vermiculite, sugarcane bagasse, corncob and corncob powder, gellan gum, wheat straw, organo-mineral carriers consisting of zeolite-clinoptilolite and humic acids. Survival of inoculated microorganisms as well as their bioremediation performances can be most likely enhanced by the supply of carbon substrates co-localized with the inoculum in the immobilization matrix, which could at the same time selectively promote immobilized cell growth and not indigenous microbial populations in soil.

The combination of bioaugmentation and biostimulation might be a promising strategy to speed up the bioremediation process. Both indigenous and exogenous microorganisms could benefit from biostimulation by the addition of energy sources or electron acceptors. The combination of bioaugmentation with surfactant addition might increase its chances of success.

1.6.2 *Microbe-Assisted Phytoremediation*

Phytoremediation for clean-up of contaminated soil has long been recognized as a remediation option. Since the interface between microbes and rhizosphere is considered to greatly influence the growth and survival of plants, alternative phytoremediation methods that exploit rhizosphere bacteria to reduce metal and organic pollutant toxicity to plants have been investigated extensively (Zhuang et al. 2007; Hsieh et al. 2009; Karami and Shamsuddin 2010). Bacterial diversity and the number of culturable heterotrophic bacteria decrease significantly in polyaromatic hydrocarbon- and phenyl urea herbicides-contaminated soils. The number of prokaryotic genomes per gram of wet weight of soil declined eightfold following many years of heavy metal treatment (Kent and Triplett 2002). With the exception of the Proteobacteria, all phylogenetic taxa examined declined as a percentage of the total number of prokaryotes in the soil. The percentage of Proteobacteria doubled with the heavy metal treatment. A significant reduction in the enzyme activities such as dehydrogenase, acid phosphatase, and β-glucosidase is observed following high levels of heavy metal contamination (Nowack 2008).

In phytoremediation plants facilitate favorable conditions for microbial degradation through their root systems and provide O_2 necessary for biodegradative pathways, which represents a simple and inexpensive means of accessing contaminants existing in subsurface soils and water and encouraging degradation, volatilization or immobilization of pollutants (Abhilash et al. 2009; Weyens et al. 2009). The plant and microbial enzymes involved in the primary and secondary oxidations of organics are reasonably well characterized (Kavamura and Esposito 2010). Microbes provide a link between roots and the soil, changing metal availability and toxicity. Some bacteria, such as biosurfactant-producing *Pseudomonas aeruginosa*, are able to decrease metal solubility and its mobility, minimizing its phytotoxicity (Lynch and Moffat 2005; Jing et al. 2007; Juwarkar et al. 2007). Microbe-assisted phytoremediation techniques have been used in the treatment of trichloroethylene, PCBs, and polycyclic aromatic hydrocarbons (Zhuang et al. 2007).

PGPR have recently been used in environmental remediation, particularly to overcome plant stress under flooded, high temperature and acidic conditions. Indirect impact of PGPR is usually achieved by production of plant hormones (gibberellins, cytokinins, and auxins), by increasing the plant tolerance to diseases and releasing heavy metal binding components, such as methallothioneins and phytochelatines, which are useful in the bioremediation of contaminants (Wu et al. 2006).

Recently, the benefits of combining endophytic bacteria with plants for increased toxic metal contaminated soil remediation have been successfully demonstrated (Rajkumar et al. 2009). Endophytic bacteria reside within plant hosts without causing disease symptoms. The metal resistant endophytes are reported to promote plant growth by various mechanisms such as nitrogen fixation, solubilization of minerals, production of phytohormones, siderophores, transformation of nutrient elements, and utilization of 1-aminocyclopropane-1-carboxylic acid as a sole N source.

Fungi have also been used in phytoremediation technology, like arbuscular mycorrhizae fungi or AMF (Khan 2005). In order to enhance phytoremediation efficiency, the effects of dual inoculation with ectomycorrhizal fungi and the ectomycorrhiza-associated bacteria *Micrococcus luteus* and *Sphingomonas* sp. on the growth and metal accumulation of willows in contaminated soil has been investigated with success (Zimmer et al. 2010). The principal role of AMF is to improve the uptake of phosphorus and mineral nutrients for plants and enhance the number and length of root branch; however, the alleviation mechanism of AMF on the phytoremediation of metal is not clear.

Bioavailability of heavy metals can also be increased by decreasing pH by adding sulfuric acid or organic fertilizers or using chelating reagents such as ethylene diamine tetraacetic acid (EDTA) and (*S*,*S*)-*N*,*N*′-ethylenediamine disuccinic acid (EDDS) (Hong et al. 2010). Plant roots produce chelating agents that can complex with lead, gold, and uranium in plants, increasing the metal solubility and translocation from root to shoot. Soil microbes produce small, high-affinity iron-chelating compounds siderophores that are one of the strongest soluble Fe^{3+} binding agents known, which enable them to live in diverse environments (Guenther 2007).

1.6.3 Chemical Additives

Biostimulation of microbiological processes for pollutant degradation usually involves the modification of the contaminated medium by adjusting pH, addition of limiting nutrients to achieve an ideal nutrient C:N:P ratio and improving the soil moisture. Another biostimulation approach is to add microbial products, such as biosurfactants or enzymes, directly as an amendment either alone or in combination with microbial inoculants. Biosurfactants have been used for bioremediation of metal and organic-contaminated material, and they may also have a use in bioaugmentation applications either to protect microbial inoculants from metal toxicity or to increase the amount of organic substrates available for degradation.

Chemical additives such as (bio)surfactants, organic/inorganic acids and alkali, water soluble solvents such as methanol, complexing agents such as EDTA, and oxidizing and reducing agents have been used for soil washing, flushing, and dissolving organic pollutants for soil remediation in conjunction with natural microbial activity. Cationic, anionic, and non-ionic surfactants contain both hydrophobic and hydrophilic compounds making them ideal for solubilization of hydrophobic compounds. Numerous studies at the laboratory, pilot or field scale have indicated enhancement in the process in the presence of these chemicals. Removal efficiencies are affected by pH, soil type, particle size, permeability, and type of contaminants. High clay and organic matter are detrimental for remediation processes. Biosurfactant applications in the environmental industries are promising due to their biodegradability, low toxicity, and effectiveness in enhancing biodegradation and solubilization of low solubility compounds. Studies on application of biosurfactant such as surfactin (*Bacillus subtilis*), rhamnolipid (*Pseudomonas aeruginosa*), and sophorolipid (*Torulopsis bombicola*) in enhancing removal of metals (Cu, Cd, Zn) from contaminated soil have shown some promise in the past (Mulligan 2005).

Commonly employed remediation technologies for treating PAH-contaminated soils include soil washing or solvent extraction, bioremediation, and chemical oxidation. In all PAH remediation techniques, a major influencing factor is the tendency of PAHs to adsorb tightly to organic matter in soil due to their hydrophobic nature. An interesting advancement in the field of soil remediation for hydrophobic organic compounds is the inclusion of vegetable oil in various technologies. Vegetable oil offers multiple applications in soil remediation, ranging from its utilization as a solvent to physically extracting PAHs to its usage as soil amendment to enhance biological and non-biological treatments (Pannu et al. 2004; Yap et al. 2010).

1.7 Biocontrol

Traditional methods used to protect crops from diseases have been largely based on the use of chemical pesticides. Applications of fungicides, fumigants, herbicides, and insecticides can have drastic effects on the environment and consumers. Chemical methods may not be economical in the long run because they pollute the atmosphere, damage the environment, leave harmful residues, and can lead to the development of resistant strains among the target organisms with repeated use. Therefore, a reduction or elimination of synthetic pesticide applications in agriculture is highly desirable. One of the most promising means to achieve this goal is by the use of BCAs for disease control alone, or to integrate with reduced doses of chemicals in the control of plant pathogens resulting in minimal impact of the chemicals on the environment. Biocontrol of pests in agriculture is a method of controlling pests including insects, mites, weeds, and plant and soil-borne diseases. A number of BCAs have been registered and are available as commercial products, including strains belonging to bacterial genera such as *Agrobacterium*, *Pseudomonas*, *Streptomyces*, and *Bacillus*, and fungal genera such as *Gliocladium*, *Trichoderma*, *Ampelomyces*, *Candida*, and *Coniothyrium*.

PGPR produce numerous compounds that are toxic to pathogens, such as HCN, phenazines, pyrrolnitrin, pyoluteorin, and enzymes, antibiotics, metabolites, and phytohormones. The potential of PGPR for biological control can result from one or more mechanisms, including antagonism, competition, and the production of antibiotics or siderophores or by inducing disease resistance and/or direct growth by improving nutrient uptake for plants by the alteration of plant hormone levels. Combining multiple PGPR types can suppress disease development in many crop plants and protect them against a broad range of soil-borne plant pathogens.

During the infection process, phytopathogenic bacteria enter plant tissues either by wounds or natural openings and occupy the apoplast of plant tissues or the xylem where they multiply and spread, a process that often involves the participation of hydrolytic enzymes and toxins (Soto et al. 2009). Pathogenic bacteria and mutualistic rhizobia are able to invade and establish chronic infections within their host plants. The success of these plant–bacteria interactions requires evasion of the plant innate immunity by either avoiding recognition or by suppressing host defenses. The primary plant innate immunity is triggered upon recognition of common microbe-associated molecular patterns. Different studies reveal striking similarities between the molecular basis of rhizobial nodulation factors and microbe-associated molecular patterns from plant pathogens.

The global market for biocontrol has been valued over $580 million (Bolckmans 2008). This was divided into 43.5% of sales in North America, 20.7% in Europe, 12.2% in Asia, 11.2% in Oceania/Australia, 8.3% in Latin America and 3.9% in Africa. While *Bacillus thuringiensis* is used to control most of the economically important insect pests, including American bollworm sp., *Earias* spp., *Spodoptera* sp., and *Plutella* sp., strains of *Bacillus subtilis*, and *P. fluorescens* are used to control bacterial as well as fungal pathogens.

Trichoderma, an asexual fungus found in the soils of all climatic zones, is a secondary opportunistic invader, a fast growing and strong spore producer, an excellent source of cell wall degrading enzymes (cellulases, ligninases, chitinases, glucanases, etc.), and an important antibiotic producer. Numerous *Trichoderma* strains are "rhizosphere competent" and are able to degrade hydrocarbons, chlorophenolic compounds, polysaccharides, and the xenobiotic pesticides used in agriculture. During the interaction of *Trichoderma* with the plant, different classes of metabolites may act as elicitors or resistance inducers.

Mycorrhitic plants are better protected against soil-borne diseases (Pozo and Azcón-Aguilar 2007; Lioussanne et al. 2009), whereby the protection by the AM fungal colonization of the plant root conferred to plants against many soil-borne pathogens such as species of *Aphanomyces*, *Cylindrocladium*, *Fusarium*, *Macrophomina*, *Phytophthora*, *Pythium*, *Rhizoctonia*, *Sclerotinium*, *Verticillium*, and *Thielaviopsis* and various nematodes.

1.8 Conclusions

Bioaugmentation and biostimulation approaches using microbial inoculants, biofertilizers, bio(chemicals), and organic amendments have been used for a long time to improve soil biology, fertility, crop productivity, biocontrol method for soil-borne diseases and plant–parasitic nematodes and soil remediation. Knowledge of microbial diversity and function in soils is limited because of the taxonomic and methodological limitations associated with studying these organisms. With the recent rapid advances in the development of molecular methods we are beginning to understand the astonishing diversity of microbial populations and communities in the environment (Singh and Ward 2009; Christoserdova 2010). Chemical fertilizers increase yield in agriculture but are expensive and harm the environment. They deplete non-renewable energy via side effects, such as leaching out, and polluting water basins, destroying microorganisms, and friendly insects, making the crop more susceptible to the attack of diseases, reducing soil fertility, thereby causing irreparable damage to the overall system. The use of plant growth-promoting organisms or biofertilizers could be a better alternative to chemical fertilizers. They are economical, not harmful to the environment, and could easily be found. Microbial inoculants, which can fulfill diverse functions in plants, lead to promising solutions for a sustainable, environmentally friendly agriculture.

References

Abhilash PC, Jamil S, Singh N (2009) Transgenic plants for enhanced biodegradation and phytoremediation of organic xenobiotics. Biotechnol Adv 27:474–488

Ahmad R, Jilani G, Arshad M, Zahir ZA, Khalid A (2007) Bio-conversion of organic wastes for their recycling in agriculture: an overview of perspectives and prospects. Ann Microbiol 57:471–479

Babalola OO (2010) Beneficial bacteria of agricultural importance. Biotechnol Lett 32:1559–1570

Barnabé S, Brar SK, Tyagi RD, Beauchesne I, Surampalli RY (2009) Pre-treatment and bioconversion of wastewater sludge to value-added products: fate of endocrine disrupting compounds. Sci Total Environ 407:1471–1488

Bending GD, Turner MK, Jones JE (2002) Interactions between crop residue and soil organic matter quality and the functional diversity of soil microbial communities. Soil Biol Biochem 34:1073–1082

Berg G (2009) Plant–microbe interactions promoting plant growth and health: perspectives for controlled use of microorganisms in agriculture. Appl Microbiol Biotechnol 84:11–18

Blagodatskaya E, Kuzyakov Y (2008) Mechanisms of real and apparent priming effects and their dependence on soil microbial biomass and community structure: critical review. Biol Fertil Soils 45:115–131

Bolckmans K (2008) Biocontrol files. Can Bull Ecol Pest Manag 13:1–10

Bonanomi G, Antignani V, Capodilupo M, Scala F (2010) Identifying the characteristics of organic soil amendments that suppress soilborne plant diseases. Soil Biol Biochem 42:136–144

Bruinsma M, Kowalchuk GA, van Veen JA (2003) Effects of genetically modified plants on microbial communities and processes in soil. Biol Fertil Soils 37:329–337

Carrère H, Dumas C, Battimelli A, Batstone DJ, Delgenès JP, Steyer JP, Ferrer I (2010) Pretreatment methods to improve sludge anaerobic degradability: a review. J Hazard Mater 183:1–15

Christoserdova L (2010) Recent progress and new challenges in metagenomics for biotechnology. Biotechnol Lett 32:1351–1359

Craig JW, Chang FY, Kim JH (2010) Expanding small molecule functional metagenomics through parallel screening of broad-host-range cosmid environmental DNA libraries in diverse proteobacteria. Appl Environ Microbiol 76:1633–1641

Curtis TP, Sloan WT (2004) Prokaryotic diversity and its limits: microbial community structure in nature and implications for microbial ecology. Curr Opin Microbiol 7:221–226

Dahllöf I (2002) Molecular community analysis of microbial diversity. Curr Opin Biotechnol 13:213–217

Deaker R, Roughley RJ, Kennedy IR (2004) Legume seed inoculation technology: a review. Soil Biol Biochem 36:1275–1288

El Fantroussi S, Agathos SN (2005) Is bioaugmentation a feasible strategy for pollutant removal and site remediation? Curr Opin Microbiol 8:268–275

Figueiredo MVB, Seldin L, Araujo FF, Mariano RLR (2010) Plant growth promoting rhizobacteria: fundamentals and applications. In: Maheshwari DK (ed) Plant growth and health promoting bacteria, microbiology monographs 18. Springer, Berlin

Fontaine S, Mariottib A, Abbadie L (2003) The priming effect of organic matter: a question of microbial competition? Soil Biol Biochem 35:837–843

Forney LJ, Zhou X, Brown CJ (2004) Molecular microbial ecology: land of the one-eyed king. Curr Opin Microbiol 7:210–220

Gil-Sotresa F, Trasar-Cepedab C, Leirosa MC, Seoane S (2005) Different approaches to evaluating soil quality using biochemical properties. Soil Biol Biochem 37:877–887

Girones R (2006) Tracking viruses that contaminate environment. J Environ Eng Sci 1:19–25

Guenther W (2007) Ecology of siderophores with special reference to the fungi. Biometals 20:379–392

Hong SH, Ryu HW, Kim J, Cho K-S (2010) Rhizoremediation of diesel-contaminated soil using the plant growth-promoting rhizobacterium *Gordonia* sp. S2RP-17. Biodegradation. doi:10.1007/s10532-010-9432-2

Hsieh JL, Chen CY, Chiu MH, Chein MF, Chang JS, Endo G, Huang CC (2009) Expressing a bacterial mercuric ion binding protein in plant for phytoremediation of heavy metals. J Hazard Mater 161:920–925

Hunter-Cevera JC (1998) The value of microbial diversity. Curr Opin Microbiol 1:278–285

Icoz I, Stotzky G (2008) Fate and effects of insect-resistant Bt crops in soil ecosystems. Soil Biol Biochem 40:559–586

Jing Y, He Z, Yang X (2007) Role of soil rhizobacteria in phytoremediation of heavy metal contaminated soils. J Zhejiang Univ Sci B 8:192–207

Juwarkar AA, Nair A, Dubey KV, Singh SK, Devotta S (2007) Biosurfactant technology for remediation of cadmium and lead contaminated soils. Chemosphere 68:1996–2002

Karami A, Shamsuddin ZH (2010) Phytoremediation of heavy metals with several efficiency enhancer methods. Afr J Biotechnol 9:3689–3698

Kavamura VN, Esposito E (2010) Biotechnological strategies applied to the decontamination of soils polluted with heavy metals. Biotechnol Adv 28:61–69

Kent AD, Triplett EW (2002) Microbial communities and their interactions in soil and rhizosphere ecosystems. Annu Rev Microbiol 56:211–236

Khan AG (2005) Role of soil microbes in the rhizospheres of plants growing on trace metal contaminated soils in phytoremediation. J Trace Elem Med Biol 18:355–364

Kim K-R, Owens G (2010) Potential for enhanced phytoremediation of landfills using biosolids: a review. J Environ Manage 91:791–797

Kuhad RC, Kothamasi DM, Tripathi KK, Singh A (2004) Diversity and functions of soil microflora in development of plants. In: Varma AK, Abbott L, Werner D, Hampp R (eds) Plant surface microbiology. Springer, Berlin, pp 71–98

Levy-Booth DJ, Campbell RG, Guldenb RH, Hart MM, Powell JR, Klironomos JN, Pauls KP, SwantonCJ TJT, Dunfield KE (2007) Cycling of extracellular DNA in the soil environment. Soil Biol Biochem 39:2977–2991

Lioussanne L, Jolicoeur M, St-Arnaud M (2009) Role of the modification in root exudation induced by arbuscular mycorrhizal colonization on the intraradical growth of *Phytophthora nicotianae* in tomato. Mycorrhiza 19:443–448

Little AEF, Robinson CJ, Peterson SB, Raffa KF, Handelsman J (2008) Rules of engagement: interspecies interactions that regulate microbial communities. Annu Rev Microbiol 62:375–401

Lynch JM, Moffat AJ (2005) Bioremediation: prospects for the future application of innovative applied biological research. Ann Appl Biol 146:217–221

Lynch JM, Benedetti A, Insam H, Nuti MP, Smalla K, Torsvik V, Nannipieri P (2004) Microbial diversity in soil: ecological theories, the contribution of molecular techniques and the impact of transgenic plants and transgenic microorganisms. Biol Fertil Soils 40:363–385

Mercier A, Kay E, Simonet P (2006) Horizonal gene transfer by natural transformation in soil environment. In: Nannipieri P, Smalla K (eds) Nucleic acids and proteins in soil. Springer, Berlin, pp 35–373

Mrozik A, Piotrowska-Seget Z (2010) Bioaugmentation as a strategy for cleaning up of soils contaminated with aromatic compounds. Microbiol Res 165:363–375

Mulligan C (2005) Environmental applications for biosurfactants. Environ Pollut 133:183–198

Nowack B (2008) Chelating agents and the environment. Environ Pollut 153:1–2

Ohtsubo Y, Kudo T, Tsuda M, Nagata Y (2004) Strategies for bioremediation of polychlorinated biphenyls. Appl Microbiol Biotechnol 65:250–258

Oka Y (2010) Mechanisms of nematode suppression by organic soil amendments: a review. Appl Soil Ecol 44:101–115

Pannu JK, Singh A, Ward OP (2004) Vegetable oil as a contaminated soil remediation amendment: application of peanut oil for extraction of polycyclic aromatic hydrocarbons from soil. Process Biochem 39:1211–1216

Pietramellara G, Ascher J, Borgogni F, Ceccherini MT, Guerri G, Nannipieri P (2009) Extracellular DNA in soil and sediment: fate and ecological relevance. Biol Fertil Soils 45:219–235

Pozo MJ, Azcón-Aguilar C (2007) Unraveling mycorrhiza-induced resistance. Curr Opin Plant Biol 10:393–398

Prosser JI (2002) Molecular and functional diversity in soil microorganisms. Plant Soil 244:9–17

Rajkumar M, Ae N, Freitas H (2009) Endophytic bacteria and their potential to enhance heavy metal phytoextraction. Chemosphere 77:153–160

Raviv M (2010) The use of mycorrhiza in organically-grown crops under semi arid conditions: a review of benefits, constraints and future challenges. Symbiosis. doi:10.1007/s13199-010-0089-8 (Online 15 September 2010)

Ray RC, Ward OP, Singh A, Isobe S (2008) Commercialization of microbial biotechnology in horticulture: summary, outlook of achievements, constraints and prospects. In: Ray RC, Ward OP (eds) Microbial biotechnology in horticulture, vol 3. Science, New Hampshire, pp 341–365

Ronchel MC, Ramos JL (2001) Dual system to reinforce biological containment of recombinant bacteria designed for rhizoremediation. Appl Environ Microbiol 67:2649–2656

Saleem M, Arshad M, Hussain S, Bhatti AS (2007) Perspective of plant growth promoting rhizobacteria (PGPR) containing ACC deaminase in stress agriculture. J Ind Microbiol Biotechnol 34:635–648

Sheoran V, Sheoran AS, Poonam P (2008) Remediation techniques for contaminated soils. Environ Eng Manag J 7:379–387

Siciliano SD, Germida JJ, Banks K, Greer CW (2003) Changes in microbial community composition and function during a polyaromatic hydrocarbon phytoremediation field trial. Appl Environ Microbiol 69:483–489

Sidhu JPS, Toze SG (2009) Human pathogens and their indicators in biosolids: a literature review. Environ Int 35:187–201

Singh A, Ward OP (2005) Relevance of microbial diversity to environmental biotechnology. In: Satyanarayana T, Johri BN (eds) Microbial diversity: current perspectives and potential applications. I.K. International, New Delhi, pp 521–536

Singh A, Ward OP (2009) Molecular methods for monitoring environmental contaminants and bioremediation. In: Varma AK, Verma N (eds) Text book on molecular biotechnology. I.K. International, New Delhi, pp 1265–1274

Singh A, Kuhad RC, Shareefdeen Z, Ward OP (2004) Methods for monitoring and assessment of bioremediation processes. In: Singh A, Ward OP (eds) Biodegradation and bioremediation, vol 2, Soil biology series. Springer, Germany, pp 279–304

Singh A, Billingsley K, Ward OP (2006) Composting: a potentially safe process for disposal of genetically modified organisms. Crit Rev Biotechnol 26:1–16

Singh A, Kuhad RC, Ward OP (2009) Biological remediation of soils: an overview of global market and available technologies. In: Singh A, Kuhad RC, Ward OP (eds) Advances in applied bioremediation, vol 17, Soil biology series. Springer, Berlin, pp 1–19

Soto MJ, Domínguez-Ferreras A, Pérez-Mendoza D, Sanjuán J, Olivares J (2009) Mutualism versus pathogenesis: the give-and-take in plant-bacteria interactions. Cell Microbiol 11:381–388

Van Hamme JD, Singh A, Ward OP (2003) Recent advances in petroleum microbiology. Microbiol Mol Biol Rev 67:503–549

Wallenstein MD, Weintraub MN (2008) Emerging tools for measuring and modeling the in situ activity of soil extracellular enzymes. Soil Biol Biochem 40:2098–2106

Wang Y, Dai C-C (2010) Endophytes: a potential resource for biosynthesis, biotransformation, and biodegradation. Ann Microbiol. doi:10.1007/s13213-010-0120-6 (Online 26 August 2010)

Ward OP, Singh A, Van Hamme JD (2003) Accelerated biodegradation of petroleum waste. J Ind Microbiol Biotechnol 30:260–270

Ward OP, Singh A, Van Hamme JD, Voordouw G (2009) Petroleum microbiology. In: Schaechter M (ed) Encyclopedia of Microbiology. 3rd Edn, Elsevier, Amsterdam, pp 443–456

Weyens N, van der Lelie D, Taghavi S, Newman L, Vangronsveld J (2009) Exploiting plant–microbe partnerships to improve biomass production and remediation. Trends Biotechnol 27:591–598

Wu CH, Wood TK, Mulchandani A, Chen W (2006) Engineering plant microbe symbiosis for rhizoremediation of heavy metals. Appl Environ Microbiol 72:1129–1134

Yang J (2009) Rhizosphere bacteria help plants tolerate abiotic stress. Trends Plant Sci 14:1–4

Yap CL, Gana S, Ng HK (2010) Application of vegetable oils in the treatment of polycyclic aromatic hydrocarbons-contaminated soils. J Hazard Mater 177:28–41

Zhuang X, Chen J, Shim H, Bai Z (2007) New advances in plant growth-promoting rhizobacteria for bioremediation. Environ Int 33:406–413

Zimmer D, Baum C, Leinweber P, Hrynkiewicz K, Meissner R (2010) Associated bacteria increase the phytoextraction of cadmium and zinc from a metal-contaminated soil by mycorrhizal willows. Int J Phytorem 211:200–213

Part I
Bioaugmentation and Biostimulation

Chapter 2
Beneficial Interactions of Plant Growth Promoting Rhizosphere Microorganisms

Nagina Parmar and Jaimie Dufresne

2.1 Introduction

The plant rhizosphere is the major soil ecological environment for plant–microbe interactions involving colonization of different microorganisms in and around growing roots which may either result in associative, symbiotic, neutralistic, or parasitic interactions depending upon plant nutrient status in soil, soil environment, plant defence mechanism, and the type of microorganism proliferating in the rhizosphere zone. Finding the microorganisms very close to epidermis, plants secrete signal molecules for protection against invasion of the heterogeneous microbes in the root zone, and at this stage the differentiation takes place between pathogenic, associative, symbiotic, or neutralistic adaptation of microbes with the plant (Hayat et al. 2010). The plant signal molecules produced in response to microbial adhesion are the flavonoids and flavones which are secreted in the rhizosphere bacteria and some remain attached to plant cell walls to act as antimicrobial agents (phytoalexins).

In legume–*Rhizobium* symbiosis, the rod-shaped soil bacterium, *Rhizobium*, induces nitrogen-fixing nodules on the roots of leguminous plants. In this process, dinitrogen which is chemically inert and makes up approximately 80% of the volume present in the earth's atmosphere is reduced to ammonia by the bacterial enzyme nitrogenase. The plants provide a micro-aerobic environment for the effective functioning of the oxygen-sensitive nitrogenase and carbohydrates for bacterial endosymbionts to support their metabolism. In return, the bacteria fix atmospheric nitrogen used by the plant for the synthesis of organic nitrogenous compounds to meet its biological needs. Due to its agricultural importance, this symbiotic association has been the subject of extensive scientific research, and different laboratories world over are trying to increase the effectiveness of symbiosis through genetic manipulation of the host and the bacterium and to extend the

N. Parmar (✉) and J. Dufresne
Department of Chemistry and Biology, Ryerson University, Toronto, Ontario M5B 2K3, Canada
e-mail: naginap@ryerson.ca

A. Singh et al. (eds.), *Bioaugmentation, Biostimulation and Biocontrol*, Soil Biology 28,
DOI 10.1007/978-3-642-19769-7_2,

Rhizobium–host infectivity to other non-leguminous crops (Stacey et al. 1980; Fisher et al. 1985; Fisher and Long 1992).

In the rhizosphere system, plant growth promoting bacteria (PGPR) and nodule promoting rhizobacteria (NPR) capable of producing growth promoting substances have been identified. These organisms induce phytoalexins production by the plant, creating antibiosis in the rhizosphere for pathogenic forms, siderophores production to chelate insoluble cations and associative action with the plant (Lifshitz et al. 1986; Halverson and Handelsman 1991). The rhizosphere bacteria involved in such type of interactions are species of *Pseudomonas* and *Bacillus* (Capper and Higgins 1993; Guaiquil and Luigi 1992; Parmar and Dadarwal 1997). This chapter will focus on the effect of such bacteria and will provide an insight into plant–microbe interactions.

2.2 Interactions Among Diazotrophs

Rhizobium, a gram-negative bacteria, is able to establish symbiosis with leguminous plants such as *Cicer* as well as many other rhizobacterial strains, and develops positive interactions with legumes by inhabiting root nodules. Within these nodules, nitrogen-fixing bacteria reduce atmospheric nitrogen to ammonia. This provides many plants with a sufficient useable nitrogen source (Sessitsch et al. 2002). Studies on legume rhizosphere bacteria have shown that besides indigenous rhizobia interacting and competing for nodulation with an inoculant strain by antagonistic or synergistic interactions, other diazotrophs such as *Azotobacter* and *Azospirillum* as well as rhizosphere fungi and bacteria especially species of *Pseudomonas* and *Bacillus* do interact with *Rhizobium* affecting nodulation and nitrogen fixation (Bolton et al. 1986; El-Mokadem 1989; Ahmad et al. 2006; Gaind et al. 2007; Rodriguez and Frioni 2003). These diazotrophs manage important biological functions by symbiotically interacting with *Rhizobium* populations within the rhizosphere and help create a beneficiary region where interacting microorganisms benefit from additional nutrient resources (Halbleib and Ludden 2000; Gaind et al. 2007).

2.2.1 *Interaction of Rhizobium with Azotobacter/Azospirillum*

Interactions of *Azotobacter/Azospirillum* with the *Rhizobium* as co-inoculants have been observed to be synergistic in a majority of studies conducted under laboratory, greenhouse or field conditions. Combined inoculation of *Azotobacter* and *Rhizobium* sp. produces a positive response. *Azotobacter* sp. influence *Rhizobium* by significantly increasing nodulation. Increasing N_2 content within roots and shoots of respiring/metabolizing plant cells improves conditions within the rhizosphere and enhances synergistic interactions between host and *Azotobacter* sp. In an open field conditions, *Azotobacter* and *Azospirillum* have both been shown to improve growth yields in various soil mineral compositions. This suggests that a mutualistic relationship exists between *Azotobacter* and *Azospirillum* where both interact with the *Rhizobium* to improve *Cicer arietinum* (chick pea) yields (Parmar 1995; Parmar and Dadarwal 1997).

The beneficial effects of *Azotobacter* and *Azospirillum* on plants are mainly attributed to improvements in root development, an increase in the rate of water and mineral uptake by roots, the displacement of fungi and plant pathogenic bacteria, and to a lesser extent, biological nitrogen fixation (Okon and Itzigsohn 1995). Associative effect of *Azospirillum lipoferum* and *Azotobacter chroococcum* with *Rhizobium* sp. improved the growth of chick pea grown on both loamy sand and sandy soils (El-Mokadem et al. 1989). Associative effect of *A. chroococcum* on *Bradyrhizobium* strains (BM 42 and BM 43) specific to moong bean (*Vigna radiata*) was also observed (Yadav and Vashishat 1991). The effect was more pronounced when *A. chroococcum* was co-inoculated with both the strains of *Bradyrhizobium.*

Certain species of *Azospirillum* have been used to study the relationship between free-living nitrogen-fixing rhizobacteria and legumes. Abundant in the rhizosphere, *Azospirilla* possesses a versatile metabolic system where carbon and nitrogen are metabolized readily. In an unfavourable arid or nutrient-deficient conditions, *Azospirilla* can morphologically transform into what appears to be enlarged cysts and the development of an outer polysaccharide coat by accumulating poly-L-hydroxybutyrate granules which serve as carbon and energy sources. A phenotypic advantage, such as a flagellum, allows the highly motile *Azospirillum* genus to swim toward nutrients via chemotactic attraction thus enhancing growth and increased yields (Steenhoudt and Vanderleyden 2000). The inoculation of legumes with *Azospirillum* prompts enlarged lateral roots and root hairs. This results in improved water uptake and retention with higher nutrient uptake (Steenhoudt and Vanderleyden 2000).

Some of the studies have shown that a relationship exists between chemotactic behaviour and *Azotobacter's* influence on plant growth such as cotton (*Gossypium hirsutum* L.) and wheat (*Triticum aestivum* L.) (Kumar et al. 2007). In the areas of soil where plant root exudates such as sugars, glucose, amino acids and organic acids have been deposited, bacteria mobilize towards these exudates through chemotactic attraction. Increased yields and enhanced growth using *A. chroococcum* indicate a positive response attributed to nitrogen fixation, phosphorus mobilization, bacterial production and the release of phytohormones (Kumar et al. 2007).

2.2.2 *Interaction of Rhizobium* sp. with Rhizobacteria

Rhizosphere bacteria, especially species of *Pseudomonas* and *Bacillus,* have been identified in the rhizosphere of various leguminous and non-leguminous crops that help in plant colonization and suppression of plant pathogens (Table 2.1). Such characteristics have defined rhizobacteria more recently as PGPR or NPR. Interactions of these rhizobacteria with *Rhizobium* may be antagonistic or synergistic and the beneficial effects of these bacteria have been extensively exploited for

Table 2.1 Important plant growth-promoting rhizobacteria (PGPR)

PGPR	Agricultural crop	References
Pseudomonas fluorescens PCL1606	Avocado	Cazorla et al. (2006)
P. fluorescens CHA0	*Arabidopsis* sp.	Iavicoli et al. (2003)
Bacillus subtilis FB17	*Arabidopsis thaliana*	Rudrappa et al. (2008)
Collimonas fungivorans	Tomato	Kamilova et al. (2008)
Bacillus subtilis, *B. amyloliquefaciens* IN 937, *Enterobacter cloaca*	*Arabidopsis* sp.	Ryu et al. (2003)
P. putida KD	Tomato and cucumber	Rezzonoco et al. (2005)
P. fluorescens WCS 365	Tomato	Kamilova et al. (2006)
Comamonas acidovorans	Kiwi	Erturk et al. (2010)
Bacillus cereus UW 85	Grain legumes	Vessey and Buss (2002)
Bradyrhizobium and PGPR	Mungbean	Shaharoona et al. (2006)
Pseudomonas BA-8, *Bacillus* OSU-142, *Bacillus* M-3	Strawberry	Pirlak and Kose (2009)
Agrobacterium amazonense	Rice	Rodrigues et al. (2008)
Bacillus cepacia strain OSU-7	Stored potatoes	Recep et al. (2009)
Pseudomonas brassicacaerum, *P. Marginali*, *P. oryzihabitans*, *P. putida*, *Alcaligenes xylosoxidans*	Indian mustard and rape	Belimov et al. (2007)

economic gains in the recent years (Bolton et al. 1986; Halverson and Handelsman 1991; Parmar 1995).

Parmar and Dadarwal (1999) studied co-inoculation of the rhizobacteria with effective *Rhizobium* strains of chickpea and observed a significant increase in nodule weight, root and shoot biomass and total plant nitrogen when grown either in sterilized chillum jars or under pot culture conditions. The *Rhizobium* stimulatory *Pseudomonas* sp. "CRP55b" showed maximum increase in all the symbiotic parameters. On co-inoculation with *Rhizobium* strains "Ca181" and "Ca313", *Pseudomonas* sp. "CRP55b" and "CRS68" resulted in significant increases in nodule weight, root and shoot biomass and total plant nitrogen (Fig. 2.1). The nodule-stimulating rhizobacteria enhanced levels of flavonoid-like compounds in roots on seed bacterization. In another study, a greater number of nodules per plant were also produced where *Bradyrhizobia* was used with strains of *Pseudomonas aeruginosa* compared with *Bradyrhizobia* used alone (Izhar et al. 1995).

The influence of PGPR on dry matter accumulation and chick pea (*C. arietinum* L.) yield under field conditions has been thoroughly studied (Rokhzadi et al. 2008). Studies have shown that a combined inoculation of *Azospirillum* spp., *A. chroococcum* 5, *Mesorhizobium ciceri* SWR17 and *Pseudomonas fluorescens* P21 improved nodulation, increased dry matter accumulation in roots and shoots, grain yields, biomass and protein yield of chick-pea by a significant margin. This can be attributed to the cumulative effects of an enhanced supply of nutrients, mainly nitrogen and phosphorus and the production of growth promoting substances. In addition, *P. fluorescens* has been found to synergistically interact with additional rhizobacteria to form interactions within the rhizosphere, attributing to

Fig. 2.1 Co-inoculation of *Rhizobium* strain "Ca 313" with *Pseudomonas* sp. "CRS68" showed an increase in nodulation in *Cicer*. (1) Control (no bacteria), (2) Ca 313 alone, (3) Co-inoculated Ca313 + CRS68

phytohormone production, the stimulation of nutrient uptake and the bio-control of deleterious soil bacteria and phyto-pathogenic fungi.

Synergistic effects of plant growth-promoting rhizobacteria and *Rhizobium* on nodulation and nitrogen fixation by pigeonpea (*Cajanus cajan*) were also observed (Tilak and Ranganayaki 2006). Co-inoculation of a variety of PGPR such as *A. chroococcum* and *Pseudomonas putida* with *Rhizobium* sp. (AR-2-2 k) showed increased plant growth, nodulation and improved nitrogenase activity. The association of *Rhizobium* sp. with *P. putida*, *P. fluorescens* and *Bacillus cereus* seems to produce the best agronomical results.

Inoculation of *Rhizobium phaseoli* and PGPR such as *P. fluorescens* P-93 and *A. lipoferum* S-21on bean yield and plant growth parameters yielded promising results (Yadegari et al. 2008). In the dually inoculated plants, there were significant increases in quantity, weight, total dry matter, seed yield, and protein content. All treatment combinations resulted in higher yield; however, *Rhizobium* strain Rb-133 inoculated with *P. fluorescens* P-93 gave the highest number of seeds and pods per plant, seed protein yield, and overall seed quantity.

2.2.3 *Interaction of Rhizobium with Actinomycetes*

The agonistic and antagonistic effects of soil microbes through various interactions of bacteria, fungi, and *Actinomycetes* on *Rhizobium* have profoundly influenced sustainable annual harvests. *Actinomycetes*, a common antagonistic bacterium is often studied for its inhibitory effects on bacteria within the host rhizosphere. There are various studies in literature suggesting the antagonistic effect of Actinomycetes under in vitro and in vivo conditions. Out of 60 isolates of Actinomycetes, bacteria and fungi from pasture soil samples, where no nodulation was observed in clover and 25–70% isolates of *Actinomycetes* were antagonistic toward 12 strains of *Rhizobium trifolii* tested (Patel 1974). Nine lysogenic *Streptomyces* sp. NSA4 were isolated from the nodule surface of black gram which was found to inhibit fast- and slow-growing strains of cow pea and soybean rhizobia. The fast-growing strain of *Rhizobium* (both cow pea miscellany and soybean) was more sensitive to antibiosis as compared to slow-growing stains (Jayaraman et al. 1985; Pugashetti et al. 1992). Another study observed that 90% of the *Actinomycetes* sp. isolated from soil obtained from field plots was antagonistic to *Rhizobium japonicum* (Pugashetti et al. 1992). In addition, 70% of other *Actinomycetes* sp. isolated from soybean rhizosphere were antagonistic to its homologous rhizobia. However, few isolates stimulated growth of *Bradyrhizobium japonicum*.

The isolates of *Streptomyces lydicus* WYEC108 from pea plants (*Pisum sativum*) were originally studied for its properties as an antifungal biocontrol agent. This strain is capable of mycoparasitic colonization of fungal root pathogens and the excretion of antifungal metabolites within plant rhizospheres. WYEC108 is a unique *Streptomyces* strain that has the ability to act as a PGPR. It was also hypothesized that root and nodule colonization is one of the several mechanisms by which *Streptomyces* acts as a naturally occurring plant growth-promoting bacterium in pea and possibly other leguminous plants. *Streptomyces* WYEC108 enhanced nodule growth, bacteroid differentiation and act as an aid in bacteroid assimilation of iron and other inorganic nutrients from soils, resulting in enhanced overall growth (Tokala et al. 2002).

There are some specific interactions in plant rhizosphere among different genera of *Actinomycetes*. *Actinomycete* mycelium makes up to 20% of the total bacterial biomass in the rhizosphere. There is significant lytic activity within the rhizosphere. *Actinomycete* mycelium content within the rhizosphere is significantly higher in root systems of healthy plants compared to those of plants suffering from root rot disease. Inoculating winter rye (*Secale cereale* L.) with *Actinomycetes* has beneficiary growth advantages; however, co-inoculation of *Actinomycetes* with the cow clover plants (*Trifolium pretense* L.) had no effect on growth (Merzaeva and Shirokikh 2006).

2.2.4 Interaction of Rhizobium with Mycorrhiza

The role of mycorrhizal fungi, especially the vesicular-arbuscular mycorrhizae (VAM) belonging to the *Zygomycetes* class in phosphorous mobilization in soils having a relatively low level of available phosphorous, is well established for cereals as well as legumes. AM fungi are obligate symbionts, but differ from VAM as they are not host specific.

Associative action of mycorrhizal fungi in legumes has a great impact on root and shoot development and phosphorous uptake which results in the enhancement of nodulation and nitrogen fixation. There are several studies reporting the interactions between AMF and *Rhizobium* sp. (Adholeya and Johri 1986; Albrecht et al. 1999; Poi et al. 1989; Sivaprasad 1991). Variation in the response of nodulating pigeon pea (*C. cajan*) to different isolates of myccorhizal fungi was also observed (Ianson and Linderman 1993). Inoculation with an effective *Rhizobium* combination with seven VAM fungi (*Glomus* sp.) had a variable effect on plant growth enhancement, nodulation, and N_2 fixation.

There are various studies in the literature describing many significant findings in the synergistic interaction between AMF and asymbiotic N_2-fixing bacteria such as *A. chroococcum*, *Azospirillum* spp. and *Acetobacter diazotrohicus* (Barea et al. 1998; Barea et al. 2002, 2005; Barea and Azcon-Aguilar 1982). The role of AM fungi as P suppliers to legume root nodules is of great relevance when a specific AM fungus, *Rhizobacterium* sp. known for effective nodulation and N_2 fixation was found in a mycotrophic legume *Anthyllis cytisoides* in a Mediterranean semi-arid ecosystem in Spain (Requena et al. 1996, 2001). The strain *Glomus intraaridices* was found to be more effective with *Rhizobium* sp. NR 4, whereas *Glomus coronatum* was effective when co-inoculated with *Rhizobium* sp. NR9 strain. Research has provided evidence that the genetic pathway of AM symbiosis is shared in part by other root–microbe symbioses such as N_2-fixing rhizobia (Peterson and Guinel 2000).

Such specific interactions between AM fungi, *Rhizobium,* and PGPR have provided an insight into specific functional compatibility relationships between AMF and PGPR and their management when used as biofertilizers or biocontrol agents.

2.3 Rhizobacterial Factors in Growth Promotion

Research on the use of rhizobacteria to promote plant growth (legumes as well as non-legumes) has increased dramatically over the last few years due to potential benefits observed in the use of PGPR or NPR, both under cultural conditions as well as under field conditions. A diverse array of bacteria, including species of *Pseudomonas*, *Azospirillum*, *Azotobacter*, *Bacillus*, *Klebsiella*, *Enterobacter*, and *Serratia* has been shown to enhance plant growth. The mechanisms by which these rhizobacteria enhance plant growth are multitudinous which include production of plant growth-regulating substances (PGRs), phytohormones, suppression of plant pathogens through antibiosis, bacteriocinogenic action, siderophore production, nitrogen fixation, mineralization of organic phosphorus, production of phytoalexins/flavonoids-like compounds, enhancement of mineral uptake, etc. (Parmar 1995; Parmar and Dadarwal 1997; Mukerji et al. 2006).

The cumulative effect of these complex interactions among plant roots and various microbial populations can result in plant growth promotion and or pathogenesis and decay. This section will focus on plant growth promotion by rhizobacteria either directly or indirectly.

Rhizobacteria can stimulate growth by producing plant growth regulators known as phytostimulators in the absence of pathogens. Many phytostimulators are plant hormone analogues, meaning PGPR produce an identical or nearly identical compound that mimics the action of a plant hormone. These substances are usually light weight volatile organic compounds. Thus far, known phytostimulators include indole-acetic acid (IAA), gibberelic acid, cytokinins, and in some cases ethylene (Ahmad et al. 2008). The plant responses to PGPR have been excellently reviewed elsewhere (Bakker et al. 2007; Van Loon 2007).

2.3.1 Plant Growth Regulators

Many rhizosphere bacteria produce IAA in culture media especially in the presence of tryptophan (Ek et al. 1983; Strzelceyck and Pokjska-Burdzej 1984) in the rhizosphere and rhizoplane of forage grasses and many economically important cereals such as wheat, barley, and pearl millet (Tien 1979), and tomato and bean plants (Barea and Brown 1974). The accumulation of IAA in the cultural filtrate of rhizobacterial isolates from the rhizosphere of sugar beet has been reported previously (Loper and Schroth 1986).

In particular, the production of IAA seems to be one of the most prevalent plant growth promoting traits among PGPR. Higher auxin levels impair plant defence mechanisms making colonization easier. The biosynthesis of IAA in rhizobacteria is affected by several environmental factors. In particular, IAA production increases in conditions of higher pH, limited carbon and higher quantities of tryptophan (Spaepen et al. 2009). Thus far, six pathways for the biosynthesis of IAA have been identified in rhizobacteria, five of which are tryptophan dependant and one which is tryptophan independent. Instead of tryptophan, this pathway depends on the presence of indole-3-glycerolphosphate. Some rhizobacteria have several IAA biosynthesis pathways. In plants, most IAA are found in a conjugated form that allows for storage and prevents degradation (Spaepen et al. 2007).

It is believed that approximately 80% of rhizobacteria produce IAA (Khalid et al. 2004). Arshad and Frankenberger (1988) showed while studying the production of ethylene by the soil fungi *Acremonium falciforms* that microbially produced ethylene can affect plant growth of etiolated pea seedlings. Etiolated pea seedlings presented a classical triple response, which included reduction in elongation, swellings of the hypocotyls and a change in the direction of the growth (horizontal), when *A. falciforms* was used as an inoculant. Further studies showed the production of PGPRs by many soil microorganisms in the presence of suitable precursors (Arshad and Frankenberger 1990).

It has also been well documented that the biosynthesis of auxins with their excretion into soil makes a major contribution to the bacterial plant growth-promoting effect (Lambrecht et al. 2000; Steenhoudt and Vanderleyden 2000). Not as well understood as IAA, cytokinins and gibberellins have also been shown to stimulate root and shoot development in several ways (van Loon 2007). Cytokinins, for example, have been implicated in cell division and nitrogen fixing nodule development (Murray et al. 2007; Tirichine et al. 2007). They also promote rapid growth of the primary root and enhance branching (Ortiz-Castro et al. 2009). Another class of recently discovered phytostimulating compounds called *N*-acyl-L-homoserine lactones have been shown to modulate gene expression in plants. These molecules are also used by bacteria for cell-to-cell communication (Ortiz-Castro et al. 2009). The cofactor PQQ (pyrroloquinoline quinine) was also identified as a plant growth promoting factor while promoting the growth of tomato and cucumber plants (Choi et al. 2008). The results showed the property of PQQ as an antioxidant; however, the effect is mostly indirect.

These studies suggest that microbially released PGRs in the rhizosphere may affect plant growth and may be subjected to direct uptake by plant roots because of the intimate contact between microbial and plant cells.

2.3.2 Phytoalexins

Phytoalexins are low molecular weight, antimicrobial compounds that are both synthesized by and accumulated in plants after exposure to microorganisms (Dakora 1985; Dakora et al. 1993; Van Peer et al. 1991, 1990). The concept of phytoalexin is expanded because many isoflavonoids (the most widely studied class of phytoalexins) were shown to serve as signal molecules during infection of plant roots by symbiotic microbes (Landa et al. 2002). Phytoalexin synthesis can be used as an indicator of enhanced defence mechanism in bacteria-treated plants. An increase in the production of three phytoalexins, for example, rishitin, lubimin, and solvetivon were observed in potato slices and an inhibition in mycelia growth of *Phytophthora infestans* by culture filtrate of *Streptomyces* (Bochow and Fritzsche 1990). It has also been observed that like other plants, Graminae contain secondary plant metabolites that have been found to be toxic to plant pathogenic fungi and bacteria and are proposed to be responsible for resistance to microbial pathogens (Gross 1991).

Similarly, in another study induction and accumulation of phytoalexins in cow pea roots were observed when infected with myccorhizal fungus and also their resistance to *Fusarium* wilt disease. From the studies in our laboratory (Parmar 1995; Parmar and Dadarwal 1999), it was apparent that the rhizosphere bacteria such as fluorescent Pseudomonads and *Bacillus* sp. produced certain signal molecules which probably enhanced the flavonoid production by plant roots. The enhanced flavonoid production could be an additional factor in nodule promotion. In addition, production of phytoalexins was demonstrated to increase after prior inoculation of chick pea (*C. arietinum* L.) seedlings with non-pathogenic isolates of *Fusarium oxysporum* (inducers) and this was correlated with a delay on the onset of symptom and reduction of *Fusarium* wilt development (Landa et al. 2002).

Indirectly, PGPR can act as biofertilizers via asymbiotic nitrogen fixation and the solubilization of mineral phosphates and other nutrients. Rhizobacteria can also act as biocontrol agents by producing siderophores that compete with pathogenic organisms for iron, by producing antibiotics and bacteriocins that suppress bacterial pathogens and by producing anti-fungal metabolites (Ahmad et al. 2008).

The optimization of these interactions may lead to improvement in the yields of various leguminous and non-leguminous crops. In particular, the extension of symbiotic biological nitrogen fixation to non-legume crops would be of enormous economical and environmental benefit.

2.3.3 Biocontrol Agents

Biocontrol agents include molecules that induce an immune response within the plant or molecules that in some way suppress plant pathogens either indirectly by competing for essential nutrients or directly inhibiting growth of phytopathogens.

2.3.3.1 Stimulation of Host Defence

When a plant comes into contact with a pathogenic microorganism, it responds with a systemic acquired response (SAR) where the plant's immune system is primed to defend itself against disease. Many phytopathogenic fungi, for example, are known to induce systemic acquired responses in plants. The most common parasitic fungi belong to genera *Pythium* sp., *Rhizoctonia* sp. and *Fusarium* sp. (Mukerji et al. 2006). Some *Fusarium* sp. cause root rots and wilts and some feed on dead plant tissues (Mukerji et al. 2006). The exact mechanisms for how the plant immune system primes itself are still unknown; however, certain molecules in the pathway such as salicylic acid, for example, appear to play a critical role as a plant messenger once the plant is exposed to a pathogen (Wildermuth et al. 2001). Some PGPR can stimulate a plant's defence system without the presence of a pathogen by emitting molecules similar to those in the plant's SAR. This response is called induced systemic resistance (ISR). Some of these molecules include methyl salicylate (MeSA), methyl jasmonate (MeJA) and ethylene. Thus far, evidence of PGPR eliciting ISR has been observed in carnations (Van Peer et al. 1991), the common bean, cucumber (Wei et al. 1991) and grapevine (Verhagen et al. 2004, 2010). In other experiments, the colonization of root systems with PGPR, such as *P. fluorescens*, *P. putida*, *Bacillus pumilus* and *Serratia marcescens* was protected against foliar diseases (Pieterse et al. 2002).

2.3.3.2 Siderophores

Siderophores are small molecules excreted by rhizobacteria when deficient in iron. By complexing with available iron in the rhizosphere it becomes less available to competing phytopathogens which also require iron thus inhibiting competitor growth. Siderophore production by *P. fluorescens* F113 has been shown to play a role in biocontrol of potato soft rot under iron limiting conditions (Whipps 2001). In addition, the antifungal activity of test isolates was greatly enhanced when both HCN and siderphores were produced indicating that together these two plant growth-promoting activities work synergistically to inhibit pathogenic fungi and protect plant health (Ahmad et al. 2008).

2.3.3.3 Antibiotics

Many rhizobacteria have been shown to produce antibiotics that inhibit the growth of an antagonistic bacterium. *P. fluorescens* (Trevisan) Migula F113, for example, has been shown to control the soft rot potato pathogen *Erwinia carotovora* subspecies *atroseptica* by producing the antibiotic 2,4-diacetylphloroglucinol (DAPG) (Whipps 2001). Three glucanase-producing actinomycetes, when used separately or more effectively in combination, could significantly promote plant growth and therefore inhibit the growth of *Pythium aphanidermatum* (El Tarabily et al. 2009). Other major antibiotics produced by *B. cereus* are phenazine-e-carboxylic acid and phenazine-1-carboxamide; 2,4-diacetyl phloroglucinol (phl) (Dunne et al. 1998), pyoluteorin (Nowak-Thompson et al. 1999), zwittermicin A (Emmert et al. 2004), gluconic acid (Kaur et al. 2006), 2-hexyl-5-propyl resorcinol (Cazorla et al. 2006) and kanosamine (Milner et al. 1996).

Bacteriocins are proteins that normally kill or inhibit the growth of closely related bacterial strains. *Bacteriocin thuricin* 17 was isolated from the PGPR *Bacillus thuringenisis* NEB17 (Gray et al. 2006). Oddly enough, this novel bacteriocin was able to inhibit the growth of not only related gram positive bacterial strains, but also of a gram negative strains of *Escherichia coli* MM294 (pBS42).

2.3.3.4 Antifungal Metabolites

Many antifungal metabolites have been produced and shown to be effective in vitro. These antifungal metabolites are also suspected to have antifungal activity in vivo. These metabolites include ammonia, butyrolactones, 2-4-diacetylphloroglucinol, HCN, kanosamine, Oligomycin A, Oomycin A, phenazine-1-carboxylic acid (PCA), pyoluterin (Plt), pyrrolnitrin (pln), viscosinamide, xanthobaccin and zwittermycin A (Milner et al. 1996; Whipps 2001). In addition, certain fungi have been shown to be sensitive to particular combinations of metabolites.

2.4 Conclusions

The beneficial effects of the rhizobacteria in enhancing root development associated with increase in nodule biomass by native as well as co-inoculated *Rhizobium* strains are well documented. Some degree of specificity was observed with regard to plant species as the rhizobacteria from *Cicer* rhizosphere were observed as better co-inoculants compared to rhizobacteria isolated from the other crops such as green gram and pigeon pea. Various studies have also provided an evidence of different mechanisms by which there is an increase in crop productivity and the disease suppressive ability of these rhizobacteria. There is still not enough data to suggest the establishment of the newly isolated rhizobacterial strains in the rhizosphere, but further studies using genetically marked strains should make it possible to determine their exact role in rhizosphere establishment.

Acknowledgement We thank Hammad Khan, Department of Chemistry and Biology, Ryerson University, Toronto, for providing support with the literature search during the preparation of this chapter.

References

Adholeya A, Johri RN (1986) Effect of *Glomus coledonium* and *Rhizobium* inoculation on green gram (*Vigna radiata* L. Wilczek). In: Singh R, Nainawatee HS, Sawney SK (eds) Proceedings of national symposium on current status of biological nitrogen fixation research, Haryana Agricultural University, Hisar, India, p 173

Ahmad F, Ahmad I, Aqil F, Ahmed WA, Sousche YS (2006) Plant growth promoting potential of free-living diazotrophs and other rhizobacteria isolated from Northern Indian soil. Biotechnol J 1:1112–1123

Ahmad F, Ahmad I, Khan MS (2008) Screening of free-living rhizospheric bacteria for their multiple plant growth promoting activities. Microbiol Res 163:173–181

Albrecht C, Geurts R, Bisseling T (1999) Legume nodulation and mycorrhizae formation; two extremes in host specificity meet. EMBO J 18:281–288

Arshad M, Frankenberger WTJ (1988) Influence of ethylene produced by soil microorganismson etiolated pea seedlings. Appl Environ Microbiol 54:2728–2732

Arshad M, Frankenberger WT Jr (1990) Response of *Zea mays* and *Lycopersicon esculentum* to the ethylene precursors, L-methionine and L-ethionine, applied to soil. Plant Soil 122:129–135

Bakker PA, Pieterse CM, van Loon LC (2007) Induced systemic resistance by fluorescent Pseudomonas spp. Phytopathology 97:239–243

Barea JM, Azcon-Aguilar C (1982) Production of plant growth-regulating substances by the vesicular-arbuscular mycorrhizal fungus *Glomus mosseae*. Appl Environ Microbiol 43: 810–813

Barea JM, Brown ME (1974) Effects on plant growth produced by *Azotobacter paspali* related to synthesis of plant growth regulating substances. J Appl Bacteriol 37:583–593

Barea JM, Andrade G, Bianciotto VV, Dowling D, Lohrke S, Bonfante P, O'Gara F, Azcon-Aguilar C (1998) Impact on arbuscular mycorrhiza formation of pseudomonas strains used as inoculants for biocontrol of soil-borne fungal plant pathogens. Appl Environ Microbiol 64:2304–2307

Barea JM, Azcon R, Azcon-Aguilar C (2002) Mycorrhizosphere interactions to improve plant fitness and soil quality. Antonie Van Leeuwenhoek 81:343–351

Barea JM, Pozo MJ, Azcon R, Azcon-Aguilar C (2005) Microbial co-operation in the rhizosphere. J Exp Bot 56:1761–1778

Belimov AA, Dodd IC, Safronova VI, Hontzeas N, Davies WJ (2007) Pseudomonas brassicacearum strain Am3 containing 1-aminocyclopropane-1-carboxylate deaminase can show both pathogenic and growth-promoting properties in its interaction with tomato. J Exp Bot 58:1485–1495

Bochow H, Fritzsche S (1990) Induction of phytoalexins biosynthesis by culture filtrate of bacterial antagonists. In: Proceedings of international workshop on PGPR, Switzerland, pp 158–160

Bolton GW, Nester EW, Gordon MP (1986) Plant phenolic compounds induce expression of the *Agrobacterium tumefaciens* loci needed for virulence. Science 232:983–985

Capper AL, Higgin KP (1993) Application of *Pseudomonas fluorescens* isolates to wheat as potential biological control agents against take-all. Plant Pathol 42:560–567

Cazorla FM, Duckett SB, Bergstrom FT, Noreen S, Odik R et al (2006) Biocontrol of avocado Dematophora root rot by the antagonistic *Pseudomonas fluorescens* PCL 1606 correlates with the production 2-hexyl-5-propyl resorcinol. Mol Plant Microbe Interact 19:418–428

Choi O, Kim JG, Joeng Y, Moon J, Park C, Hwang J (2008) Pyrrolquinoline quinine is a plant growth promotion factor by *Pseudomonas fluorescens* B16. Plant Physiol 146:657–668

Dakora FD (1985) Use of intrinsic antibiotic resistance for characterisation and identification of rhizobia from nodules of *Vigna unguiculata* (L) Walp. and *Phaseolus vulgaris* (L). Acta Microbiol Pol 34:187–194

Dakora FD, Joseph CM, Phillips DA (1993) Alfalfa (*Medicago sativa* L.) root exudates contain isoflavonoids in the presence of *Rhizobium meliloti*. Plant Physiol 101:819–824

Dunne C, Moenne-Loccoz Y, McCarthy J, Higgins P, Powell J, Dowling DN, O'Gara F (1998) Combining proteolytic and phloroglucino-producing bacteria for improved control of *Pythium*-mediated damping off of sugar beet. Plant Pathol 47:299–307

Ek M, Ljungquist PO, Stenstrom E (1983) Indole-3-acetic acid production by mycorrhizal fungi determined by gas chromatography. New Phytol 94:401–407

El Tarabily KA, Nassar AH, Hardy GE, Sivasithamparam K (2009) Plant growth promotion and biological control of Pythium aphanidermatum, a pathogen of cucumber, by endophytic actinomycetes. J Appl Microbiol 106:13–26

El-Mokadem MT, Helemish FA, Abou-Bakr ZYM, Sheteaws A (1989) Associative effect of *Azotobacter lipoferum* and *Azotobacter chroococcum* with *Rhizobium* sp. on mineral composition and growth of chick pea (*Cicer arietinum*) on sandy soils. Zentralblatt für Mikrobiologie 144:255–265

Emmert EA, Klimowicz AK, Thomas MG, Handelsman J (2004) Genetics of zwittermicin A production by *Bacillus cereus*. Appl Environ Microbiol 70:104–115

Erturk Y, Ercisli S, Haznedar A, Cakmakci R (2010) Effects of plant growth promoting rhizobacteria (PGPR) on rooting and root growth of kiwifruit (*Actinidia deliciosa*) stem cuttings. Biol Res 43:91–98

Fisher RF, Long SR (1992) Rhizobium-plant signal exchange. Nature 357:655–660

Fisher RF, Tu JK, Long SR (1985) Conserved nodulation genes in *Rhizobium meliloti* and *Rhizobium trifolii*. Appl Environ Microbiol 49:1432–1435

Gaind S, Rathi MS, Kaushik BD, Nain L, Verma OP (2007) Survival of bio-inoculants on fungicides-treated seeds of wheat, pea and chickpea and subsequent effect on chickpea yield. J Environ Sci Health B 42:663–668

Gray E, Lee K, Di Falco M, Souleimanov A, Zhou X, Smith DL (2006) A novel bacteriocin, thuricin 17, produced by PGPR strain *Bacillus thuringiensis* NEB17: isolation and classification. J Appl Microbiol 100:545–554

Gross D (1991) Antimicrobial defense compounds in the Gramineae. Z Pflunzenkr Pflanzenschutz 96:535–553

Guaiquil VH, Luigi C (1992) Plant growth promoting rhizobacteria and their effect on rapeseed (*Brassica napus* L.) and potato seedlings. Microbiol Rev 23:264–273

Halbleib CM, Ludden PW (2000) Regulation of biological nitrogen fixation. J Nutr 130:1081–1084

Halverson LJ, Handelsman J (1991) Enhancement of soybean nodulation by *Bacillus cereus* UW85 in the field and in a growth chamber. Appl Environ Microbiol 57:2767–2770

Hayat R, Safdar Ali S, Amara U, Khalid R, Ahmed I (2010) Soil beneficial bacteria and their role in plant growth promotion: a review. Ann Microbiol 60:579–598

Ianson DC, Linderman RG (1993) Variation in the response of nodulating pigeon pea (*Cajanus cajan*) to different isolates of mycorrhizal fungi. Symbiosis 15:105–119

Iavicoli A, Boutet E, Buchala A, Metraux JP (2003) Induced systemic resistance in Arabidopsis thaliana in response to root inoculation with *Pseudomonas fluorescens* CHA0. Mol Plant Microbe Interact 16:851–858

Izhar I, Ehteshamul-Haque S, Javeed M, Ghaffar A (1995) Efficacy of *Pseudomonas aeruginosa* and *Bradyrhizobium* sp. in the control of root rot diseases in chickpea. Pak J Bot 27:451–455

Jayaraman S, Rangarajan M, Jayachandran S (1985) Metabolism of antibiotic resistant mutants of Rhizobium sp. as influenced by different temperature regimes. Hindustan Antibiot Bull 27:38–41

Kamilova F, Kravchenko LV, Shaposhnikov AI, Makarova N, Lugtenberg B (2006) Effects of the tomato pathogen *Fusarium oxysporum* f. sp. radicis-lycopersici and of the biocontrol bacterium *Pseudomonas fluorescens* WCS365 on the composition of organic acids and sugars in tomato root exudate. Mol Plant Microbe Interact 19:1121–1126

Kamilova F, Lamers G, Lugtenberg B (2008) Biocontrol strain *Pseudomonas fluorescens* WCS365 inhibits germination of *Fusarium oxysporum* spores in tomato root exudate as well as subsequent formation of new spores. Environ Microbiol 10:2455–2461

Kaur R, Macleod J, Foley W, Nayudu M (2006) Gluconic acid, an antifungal agent produced by *Pseudomonas* species in the biological control of take-all. Phytochemistry 67:595–604

Khalid A, Arshad M, Zahir ZA (2004) Screening plant growth-promoting rhizobacteria for improving growth and yield of wheat. J Appl Microbiol 96:473–480

Kumar R, Bhatia R, Kukreja K, Behl RK, Dudeja SS, Narula N (2007) Establishment of Azotobacter on plant roots: chemotactic response, development and analysis of root exudates of cotton (*Gossypium hirsutum* L.) and wheat (*Triticum aestivum* L.). J Basic Microbiol 47:436–439

Lambrecht M, Okon Y, Vande BA, Vanderleyden J (2000) Indole-3-acetic acid: a reciprocal signalling molecule in bacteria-plant interactions. Trends Microbiol 8:298–300

Landa BB, Cachinero-Diaz JM, Lemanceau P, Jimenez-Diaz RM, Alabouvette C (2002) Effect of fusaric acid and phytoanticipins on growth of rhizobacteria and *Fusarium oxysporum*. Can J Microbiol 48:971–985

Lifshitz R, Kloepper JW, Scher FM, Tipping EM, Laliberte M (1986) Nitrogen-fixing pseudomonads isolated from roots of plants grown in the Canadian High Arctic. Appl Environ Microbiol 51:251–255

Loper JE, Schroth MN (1986) Influence of bacterial sources of indole-3-acetic aid on root elongation of sugar beet. Phytopathology 76:386–389

Merzaeva OV, Shirokikh IG (2006) Colonization of plant rhizosphere by actinomycetes of different genera. Mikrobiologiia 75:271–276

Milner J, Silo-Suh L, Lee JC, He H, Calrdy H, Handelsman J (1996) Production of kanosamine by *Bacillus cereus* UW 85. Appl Environ Microbiol 62:3061–3065

Mukerji KG, Manoharachary C, Singh J (2006) Microbial activity in the rhizopshere. Springer, New York

Murray JD, Karas BJ, Sato S, Tabata S, Amyot L, Szczyglowski K (2007) A cytokinin perception mutant colonized by Rhizobium in the absence of nodule organogenesis. Science 315:101–104

Nowak-Thompson B, Chaney N, Wings J, Gould S, Loper JE (1999) Characterization of the pyoluteorin biosynthetic gene cluster of *Pseudomonas fluorescens* Pf-5. J Bacteriol 181:2166–2174

Okon Y, Itzigsohn R (1995) The development of Azospirillum as a commercial inoculant for improving crop yields. Biotechnol Adv 13:415–424

Ortiz-Castro R, Contreras-Cornejo HA, Macias-Rodriguez L, Lopez-Bucio J (2009) The role of microbial signals in plant growth and development. Plant Signal Behav 4:701–712

Parmar N (1995) Interactions of rhizosphere bacteria with Cicer-*Rhizobium* symbiosis. CCS Haryana Agricultural University, Hisar, India

Parmar N, Dadarwal KR (1997) Rhizobacteria from rhizosphere and rhizoplane of chick pea (Cicer arietinum L.). Indian J Microbiol 37:205–210

Parmar N, Dadarwal KR (1999) Stimulation of nitrogen fixation and induction of flavonoid like compounds by rhizobacteria. J Appl Microbiol 86:3–44

Patel JJ (1974) Antagonism if actinomycetes against rhizobia. Plant Soil 41:395–402

Peterson RL, Guinel FC (2000) The use of plant mutants to study regulation of colonization by AM fungi. In: Kapulnik KDY, Douds DD (eds) Arbuscular mycorrhizas: physiology and function. Kluwer, Dordrecht, pp 147–171

Pieterse CM, Van Der Ent S, Van Pelt JA, Van Loon LC (2002) The role of ethylene in rhizobacteria-induced systemic resistance (ISR). In: Ramina A, Chang C, Giovannoni J, Klee H, Perata P, Woltering E (eds) Advances in ethylene research. Springer, Netherlands, pp 325–331

Pirlak M, Kose M (2009) Effects of plant growth promoting rhizobacteria on yield and some fruit properties of strawberry. J Plant Nutr 32:1173–1184

Poi SC, Ghosh G, Kabi MC (1989) Response of chick pea to combined inoculation with Rhizobium phosphobacteria and mycorrhizal organisms. Zentralbl Mikrobiolog 144:249–253

Pugashetti BK, Angle JS, Wagner GH (1992) Soil microorganism antagonistic towards *Rhizobium japonicum*. Soil Biol Biochem 14:45–47

Recep K, Fikrettin S, Erkol D, Cafer E (2009) Biological control of the potato dry rot caused by *Fusarium* species using PGPR strains. Biol Control 50:194–198

Requena N, Jeffries P, Barea JM (1996) Assessment of natural mycorrhizal potential in a desertified semiarid ecosystem. Appl Environ Microbiol 62:842–847

Requena N, Perez-Solis E, Azcon-Aguilar C, Jeffries P, Barea JM (2001) Management of indigenous plant-microbe symbioses aids restoration of desertified ecosystems. Appl Environ Microbiol 67:495–498

Rezzonoco F, Binder C, Defago G, Moenne-Loccoz Y (2005) The type III secretion system of biocontrol *Pseudomonas fluorescencs KD* targets the phytopathogenic chromista *Pythium ultimum* and promotes cucmuber protection. Mol Plant Microbe Interact 9:991–1001

Rodrigues EP, Rodrigues CS, de Oliveira ALM, Baldani VL, Teixeira da Silva JA (2008) *Azospirillum amazonense* inoculation: effects on growth, yield and N2 fixation of rice (*Oryza sativa* L.). Plant Soil 302:249–261

Rodriguez A, Frioni L (2003) Characterization of rhizobia causing nodules on leguminous trees native to Uruguay using the rep-PCR technique. Rev Argent Microbiol 35:193–197

Rokhzadi A, Asgazadeh A, Darvish F, Nour-Mohammed G, Majidi E (2008) Influence of plant growth promoting rhizobacteria on dry matter accumulation and yield of chick pea (*Cicer arietinium* L.) under filed conditions. American-Eurasian J Agric Environ Sci 3:253–257

Rudrappa T, Czymmek KJ, Pare PW, Bais HP (2008) Root-secreted malic acid recruits beneficial soil bacteria. Plant Physiol 148:1547–1556

Ryu CM, Farag MA, Hu CH, Reddy MS, Wei HX, Pare PW, Kloepper JW (2003) Bacterial volatiles promote growth in Arabidopsis. Proc Natl Acad Sci U S A 100:4927–4932

Sessitsch A, Howieson JG, Perret X, Antoun H, Martínez-Romero E (2002) Advances in *Rhizobium* Research. Crit Rev Plant Sci 21:323–378

Shaharoona B, Arshad M, Zahir ZA (2006) Effect of plant growth promoting rhizobacteria containing ACC-deaminase on maize (*Zea mays* L.) growth under axenic conditions and on nodulation in mung bean (*Vigna radiata* L.). Lett Appl Microbiol 42:155–159

Sivaprasad R (1991) Synergistic association between *Glomus fasciculatum* and *Rhizobium* sp. and its effect on pigeon pea. Indian J Agric Sci 61:97–101

Spaepen S, Vanderleyden J, Remans R (2007) Indole-3-acetic acid in microbial and microorganism-plant signaling. FEMS Microbiol Rev 31:425–448

Spaepen S, Das F, Luyten E, Michiels J, Vanderleyden J (2009) Indole-3-acetic acid-regulated genes in Rhizobium etli CNPAF512. FEMS Microbiol Lett 291:195–200

Stacey G, Paau AS, Brill WJ (1980) Host recognition in the rhizobium-soybean symbiosis. Plant Physiol 66:609–614

Steenhoudt O, Vanderleyden J (2000) *Azospirillum*, a free living nitrogen fixing bacterium closely associated with grasses. FEMS Microbiol Lett 24:506

Strzelceyck E, Pokjska-Burdzej A (1984) Production of auxins and gibberllins-like substances by mycorrhizal fungi, bacteria and actinomycetes isolated from soil and the mycorrhizosphere of pine (*Pinus silvestris* L.). Plant Soil 81:185–194

Tien TM (1979) Plant growth substances produced by *Azospirillum brasilense* and their effect on the growth of pearl miller. Appl Environ Microbiol 37:1016–1024

Tilak KV, Ranganayaki N (2006) Synergistic effects of plant-growth promoting rhizobacteria and Rhizobium on nodulation and nitrogen fixation by pigeon pea (*Cajanus cajan*). Eur J Soil Sci 57:67–71

Tirichine L, Sandal N, Madsen LH, Radutoiu S, Albrektsen AS, Sato S, Asamizu E, Tabata S, Stougaard J (2007) A gain-of-function mutation in a cytokinin receptor triggers spontaneous root nodule organogenesis. Science 315:104–107

Tokala RK, Strap JL, Jung CM, Crawford DL, Salove MH, Deobald LA, Bailey JF, Morra MJ (2002) Novel plant-microbe rhizosphere interaction involving *Streptomyces lydicus* WYEC108 and the pea plant (*Pisum sativum*). Appl Environ Microbiol 68:2161–2171

van Loon LC (2007) Plant responses to plant-growth promoting bacteria. Eur J Plant Pathol 119:243–254

van Peer R, Punte HL, de Weger LA, Schippers B (1990) Characterization of root surface and endorhizosphere pseudomonads in relation to their colonization of roots. Appl Environ Microbiol 56:2462–2470

Van Peer R, Nieman GJ, Schippers B (1991) Induced resistance and phytoalexin accumulation in biological control of Fusarium wilt of carnation by *Pseudomonas* sp. strain WCS417r. Phytopathology 81:728–734

Verhagen BW, Glazebrook J, Zhu T, Chang HS, van Loon LC, Pieterse CM (2004) The transcriptome of rhizobacteria-induced systemic resistance in arabidopsis. Mol Plant Microbe Interact 17:895–908

Verhagen BW, Trotel-Aziz P, Couderchet M, Hofte M, Aziz A (2010) Pseudomonas spp.-induced systemic resistance to Botrytis cinerea is associated with induction and priming of defence responses in grapevine. J Exp Bot 61:249–260

Vessey JK, Buss TJ (2002) *Bacillus cereus* UW85 inoculation effects on growth, nodulation, and N accumulation in grain legumes. Controlled environmental studies. Can J Plant Sci 82:282–290

Wei G, Kloepper JW, Tuzun S (1991) Induction of systemic resistance of cucumber on *Colletotrichum orbiculare* by select strains of plant growth-promoting rhizobacteria. Phytopathology 81:1508–1512

Whipps JM (2001) Microbial Interactions and biocontrol in the rhizophere. J Exp Bot 52:487–511

Wildermuth MC, Dewdney J, Wu G, Ausubel FM (2001) Isochorismate synthase is required to synthesize salicylic acid for plant defence. Nature 414:562–565

Yadav AS, Vashishat RK (1991) Associative effect of *Bradyrhizobium* and *Azotobacter* inoculation on nodulation, nitrogen fixation and yield of Moong bean (*Vigna radiata* L. Wilczek). Indian J Microbiol 31:297–300

Yadegari M, Rahmani HA, Noormohammadi G, Ayneband A (2008) Evaluation of bean (*Phaseolus vulgaris*) seeds inoculation with Rhizobium phaseoli and plant growth promoting rhizobacteria on yield and yield components. Pak J Biol Sci 11:1935–1939

Chapter 3
Biofertilizer Technology and Pulse Production

S.S. Dudeja, N.P. Singh, Poonam Sharma, S.C. Gupta, Ramesh Chandra, Bansi Dhar, R.K. Bansal, G.P. Brahmaprakash, S.R. Potdukhe, R.C. Gundappagol, B.G. Gaikawad, and K.S. Nagaraj

S.S. Dudeja (✉)
Department of Microbiology, CCS Haryana Agricultural University, Hisar 125 004, Haryana, India
e-mail: ssdudeja@gmail.com

N.P. Singh
Indian Institute of Pulses Research, Kanpur 208 024, Uttar Pradesh, India

P. Sharma
Department of Plant Breeding and Genetics, Punjab Agricultural University, Ludhiana 141 004, Punjab, India

S.C. Gupta
RAK College of Agriculture, Sehore 466 001, Madhya Pradesh, India

R. Chandra
Department of Soil Science, G.B. Pant University of Agriculture and Technology, Pantnagar 263145, Uttarakhand, India

B. Dhar
Department of Genetics and Plant Breeding, Institute of Agricultural Sciences, Banaras Hindu University, Varanasi 221 005, Uttar Pradesh, India

R.K. Bansal
Agricultural Research Station, Rajendra Agriculture University, Durgapura, Jaipur 302018, Rajasthan, India

G.P. Brahmaprakash and K.S. Nagaraj
Department of Agricultural Microbiology, University of Agricultural Sciences, GKVK Campus, Banglore 500065, Karnataka, India

S.R. Potdukhe
Pulse Research Unit, Dr. PDKV, Akola 444 104, Maharashtra, India

R.C. Gundappagol
Agricultural Research Station, Gulbarga, Karnataka, India

B.G. Gaikawad
Pulses Improvement Project, Mahatma Phule Krishi Vidyapeeth, Rahuri, Ahmednagar 413 722, Maharashtra, India

A. Singh et al. (eds.), *Bioaugmentation, Biostimulation and Biocontrol*, Soil Biology 28,
DOI 10.1007/978-3-642-19769-7_3,

3.1 Introduction

Microorganisms in the soil constitute less than 0.5% (w/w) of the soil mass; yet they have a major impact on soil properties and processes. Soil microflora constantly interact with each other and such interactions are very dynamic in nature. Microbe–microbe, plant–microbe, soil–plant, and soil–plant–microbe interactions are very important. The important microbe–microbe interactions affecting soil microflora are neutralism, commensalism, protocooperation, symbiosis, competition, amensalism, parasitism, and predation. While important plant–microbe interactions are mainly at three interphases, i.e., in the soil with plant roots (rhizosphere), on the leaves (phyllosphere), and on the germinating seeds (spermoshere). Broadly, the interactions may be symbiotic, associative, endophytic, or pathogenic.

In plant–microbe interactions, plant rhizosphere is the major soil environment that affects the different types of microorganisms in and around the growing roots. Influence of plant root is around the root surface on all sides and extends into the soil. It is populated with a high density of microbes supported by organic compounds, such as sugars, amino acids, organic acids, and polysaccharides. Presence of these metabolites allows 5–100 times more organisms per unit volume to be supported in the rhizosphere than in nearby bulk soil. Further, this determines the structure and metabolic activities of the rhizosphere-associated community.

The most commonly studied beneficial interactions are the arbuscular mycorrhizal (AM) symbiosis between the majority of land plant families and fungi in the phylum *Glomeromycota*, and the nodule symbiosis restricted to legumes and bacteria belonging to α- and β-proteobacteria. Several other genera of soil bacteria, including *Pseudomonas* and *Bacillus* species, can stimulate root proliferation or have antagonistic effects on pathogens in the rhizosphere. Although nitrogen fixation is widely distributed among bacteria, rhizobia are distinguished from the rest because they elicit the development of a specialized organ, the nodule, and engage in a symbiotic relationship with their hosts. These processes can be monitored by using microscopy; signals and other components can be assessed by HPLC, GS-MS, or using molecular techniques. Apart from rhizobia, the importance of VA mycorrhizae, PGPR organisms such as *Bacillus*, *Pseudomonas*, *Pantoa*, and other organisms, and P solubilizers such as *Bacillus*, *Pseudomonas*, and other endophytic bacteria can also be not ignored for achieving optimum crop productivity.

Numerous rhizobial strains have been identified that show nitrogen-fixing ability with their target host legume (Dudeja and Narula 2008; Dudeja et al 2009; Dudeja and Singh 2008; Weir 2010). To date, 78 symbiotic nodulating bacterial species have been identified in 13 genera: *Azorhizobium*, *Bradyrhizobium*, *Burkholderia*, *Cupriavidus*, *Ensifer*, *Herbaspirillum*, *Devosia*, *Mesorhizobium*, *Methylobacterium*, *Ochrobactrum*, *Phyllobacterium*, *Rhizobium*, and *Shinella*. Most of the species are in the genera *Rhizobium*, *Bradyrhizobium*, *Mesorhizobium*, *and Ensifer* and are related to one another in the order Rhizobiales. Commonly, all these nitrogen-fixing bacterial biofertilizers that form root nodules on legume plants are called rhizobia. Most of these bacterial species are in the proteobacteria with α- and β-class.

Chickpea (*Cicer arietinum* L.) is the most important legume crop of the arid zones of India, and different cultivars of chickpea including black and white have been grown under field conditions since centuries and most of the soils harbor diverse group of rhizobial populations. These rhizobia are capable of interacting and nodulating all chickpea cultivars to varying extent depending upon the soil, site, and management practices. There is need to develop an efficient symbiosis of host specific rhizobial isolates and also to develop isolates with superior nodulation competitiveness that can overcome the limitations of low nitrogen fixation, poor crop yield, and lower effectiveness under field conditions.

The genus *Mesorhizobium* has been described as in between *Rhizobium* and *Bradyrhizobium* and identified in Africa, Asia, Australia, Europe, South and North America, and the Arctic (Jarvis et al. 1997; Poinsot et al. 2001). Different *Mesorhizobium* species including *M. albiziae, M. alhagi, M. amorphae, M. australicum, M. caraganae, M. chacoense, M. ciceri, M. huakuii, M. loti, M. gobiense, M. mediterraneum, M. metallidurans, M. opportunistum, M. plurifarium, M. shangrilense, M. septentrionale, M. tarimense, M. temperatum*, and *M. tianshanense* nodulate various types of legumes in the Mimosoideae and Papilionoideae subfamilies of the Fabaceae. *Mesorhizobium* has been reported to infect and form nodules in *Acacia* spp, *Albizia kalkora, Alhagi sparsifolia, Amorpha fruticosa, Anthyllis vulneraria, Astragalus, Biserrula pelecinus* L, *Caragana spp, Carmichaelia, Cicer arietinum, Cytisus scoparius* (broom), *Glycyrrhiza, Leucaena, Lotus, Montigena, Oxytropis glabra, Prosopis alba, Robinia pseudoacacia, Sophora, and Ulex europaeus. Mesorhizobium* and *Phyllobacterium* belong to Phyllobacteriaceae family of α-proteobacteria. Different species of mesorhizobia such as *M. ciceri, M. mediterraneum, M. temperadae*, and *M. tianshanense* have been reported to form nodules in chickpea (Dudeja and Narula 2008).

As per the mandate of All India Coordinated Research Programme (AICRP) on pulses, applied aspects of biofertilizer technology were the main component of research and to enhance pulses production. With the inception of the AICRP scheme, different centres located in different parts of India conducted various experiments. The achievements of biofertilizer technology in enhancing chickpea productivity in India are discussed in this chapter.

3.2 Nodulation Status of Chickpea in India

In order to predict the response of rhizobial inoculation in chickpea, it is imperative to assess the native rhizobial population in Indian soils. No suitable media for plate count method of chickpea rhizobia directly from the soil is available yet. However, MPN (most probable number) method specifically developed for small seed legumes (Vincent 1970) has also been used for medium or even large seeded legumes including chickpea. In this, different soil dilutions are used as inoculum for germinating seeds of chickpea in agar tubes with specific media or in small cups using sterilized sand with five replicates. After 30–40 days of nodulation, positive tubes or cups are scored and the number of rhizobia per gram of soil is calculated. Since this technique is very laborious and time-consuming, only limited information could be generated in the project. The numbers of nodule formed by chickpea at farmers' field in response to the presence of native rhizobia are less confirmative, but still provide good information regarding the presence of the number of chickpea rhizobia and that too under those particular ecological and environmental conditions. Further, it has been proposed that rhizobial numbers are the primary determinant of the number of nodules formed.

At different Indian centres and farmer's field, a number of nodules formed by the native/indigenous rhizobia were observed (Khurana and Dudeja 1997). Depending upon the overall nodule formation in chickpea, a rating index was prepared indicating poor (1–10 nodules/plant), moderate (11–20 nodules/plant), good (21–30 nodules/plant), and very good nodulation (>30 nodules/plant). More than 3,200 locations in different parts of India were surveyed for chickpea nodulation than 55.7% of the locations showed poor nodulation, 29.8% of the locations moderate, 12% showing good, and only 2.5% of the locations showed very good nodulation. This indicated that native rhizobia infecting chickpea in the country have generally low population and there is a need to inoculate the seeds with rhizobial inoculants. Soil moisture plays an important role in rhizobial population dynamics in the soil. The locations near to a canal showed better nodulation. On the other hand, in the sand dunes of Loharu/Badra region of Haryana state, where no or sparse nodulation was observed in chickpea, enhanced nodulation was observed with the provision of sprinkler irrigation.

3.2.1 *Nodulation Variability in Different Host Genotypes of Chickpea*

To assess the variability in nodulation within different host genotypes of chickpea with native rhizobial population at any given location is also very important to select high and low nodulating chickpea cultivars or even to select non-nodulating cultivar, which could be used as a source for crossing cultivars for higher nodulation and nitrogen fixation and hence better productivity. Differences in productivity can be related to the extent of nodulation. Nitrogen difference method is used to determine the quantity of N fixed by a legume. Non-nodulating lines in chickpea were also attempted for determination of N_2 fixation by chickpea. At six different centres in India – Hisar, Badnapur, Durgapura, Sehore, Pantnagar, and Gulbarga – more than 1,000 lines of chickpea were assessed for nodulation for the past 25 years and few good nodulating or low nodulating cultivars were identified.

More than 200 lines of chickpea were screened for nodulation up to maturity at 15 days interval for nodulation and N_2 fixation in collaboration with International

Crop Research Institute for Semi-Arid Tropics (ICRISAT) in Hisar. Cultivars showing high or low nodulation and active nodules till maturity of crop were identified. Screening at different centres showed that some cultivars that were good nodulating at all the locations. Other cultivars that showed good nodulation over the years include HK 88-232, PG5, BG 256, RSG 143-1, and JG 74. Interestingly, broad-leafed chickpea cultivars were usually better nodulating as compared to others. This also paved the way for identification of nodulation variants. Within a cultivar of chickpea, nodulation variability was also explored and high and low nodulating selections of cultivars were selected at different centres, but stable variants could not be identified in this project (Dudeja et al. 1982).

Screening of chickpea germplasm and isolation and screening of chickpea rhizobia was a continuous process resulting in the isolation of new rhizobial isolates from chickpea nodules. Effectiveness was evaluated under controlled conditions in chillum jars, pots, or under field conditions to select the most efficient chickpea rhizobial strain. During the last 25 or more years, about 1,000 chickpea mesorhizobia were isolated at different centres and were screened for efficacy in jars, pots, or field conditions, and efficient strains selected at different locations are shown in Table 3.1. Efficient isolates in terms of nodulation/nitrogen fixation/dry matter production/grain yield were identified and 95 efficient rhizobia from 18 centres were selected.

Table 3.1 Effective chickpea mesorhizobia selection after initial screening of about 1,000 fresh isolates at different centres for the past 25 years under sterilized/pot or field conditions

Centres involved in initial screening	Effective mesorhizobia selected
Akola, Badnapur, and Rahauri (Maharashtra)	BCR72
Coimbatore	CoBe 13, CoBe 18
Delhi	F75, F6, B1, G-10-80, G-5-81, SG 3-87, SG-8-87, GTB-5-88, GNA-3-88, SG 3-87, SG-8-87, GTB-5-88, SNG 95-1, GNA-3-88, G33-97, G20-98
Dholi (Bihar)	RG2, RG3, DG48
Durgapura (Rajasthan)	GD, DG34, DG34-11, DG34, DG90, DGB-34C, DG34-2
Gulbarga and Bangalore (Karnataka)	GR2, GR8, GR9, DGW4, UASB 732, UASB 701, UASB 702, UASB 835, UASB 855
Hisar (Haryana)	Ca181, CBH 32, CH777, CV4A, S3, H109, CH1, SP4, CH8115, Ca181Sm, CH8406, CH8410, CH8545, Ca28, Ca1002, Ca1003, P23, CH9116, CH1233, CH458, CP2121, CP2381, CP7006, CP1311, CP1428, CP741
Kanpur	KG31, KG46, KG61, KG46, KG61
Jabalpur, Khargone and Sehore (Madhya Pradesh)	H45, H58, H60, H65, H68, H72, JGRS108, JGRS88, JGRS105, JGRS92
Ludhiana (Punjab)	LGR1370, LRG151, LGR305
Pantnagar (Uttaranchal)	PR15
S.K. Nagar (Gujarat)	GRS4, GRS6,
Varanasi (Uttar Pradesh)	G534, G567, GHUR15, GHUR16, GHUR22, G567SM, G567EMR, G567SMR, GHUR25

3.2.2 *Multilocation Testing of Mesorhizobia with Local Best Cultivar*

The purpose of this testing was to select an efficient chickpea rhizobial inoculant that performs efficiently under different agroecological conditions where chickpea is being grown. A pool of the most efficient strains selected at different centres is shown in Table 3.1. Around 10–18 strains were tested every year with local best cultivar of chickpea. In the initial years, 30 kg N/ha and afterward 20 and 40 kg N/ha in the form of urea were used along with uninoculated control and reference strains. Experiment was repeated for 3–5 years and most efficient strains were selected (Dudeja and Khurana 1999) performing best under all the locations representing different agroecological conditions on overall basis and were released for the production of inoculants. These strains were then further tested under interaction trials. The important outcome of the trials was as follows:

1. The quantum of yield increase by different rhizobial strains varied at any given location and the extent and trend of increase may be same or different at other locations. Further testing for the next 3–5 years showed that during different years there could be variation in the extent and trend of increase in grain yield. Therefore, overall performance of a strain over the locations and years was pooled, and on the basis of overall mean the most efficient stains were selected.
2. Range of increase in grain yield over the locations and years varied from 0 to 40% over the uninoculated control.
3. Usually the quantum of response to rhizobial inoculation was higher than that of 20–40 kg N/ha, indicating that application of rhizobial inoculants could save more than 40 kg of N as urea.
4. Selections of efficient strains based on this study are shown in Table 3.2, which could be used for rhizobial inoculant production on large scale by the inoculant producers.

Table 3.2 Selection of efficient chickpea mesorhizobia strains under different agroclimatic zones under multilocation testing trials at different centres

Centres conducting the multilocation testing	Total number of isolates screened	Effective mesorhizobia
Akola, Badnapur, and Rahauri (Maharashtra); Coimbatore (Tamil Nadu); Delhi; Dholi (Bihar); Durgapura (Rajasthan); Gulbarga and Bangalore (Karnataka); Hisar (Haryana); Jabalpur, Khargone, and Sehore (Madhya Pradesh); Ludhiana (Punjab); Pantnagar (Uttaranchal); S.K. Nagar (Gujarat); and Varanasi (Uttar Pradesh)	Every year 10–18 efficient rhizobia selected at each centre as mentioned in the above table were tested for 3–5 years with local best cultivars of that centre	Ca181, CBH32, CH777, F75, H45, IC76, F-6, KG31, H109, Ca28, IC49, G-10-80, TAL1148, G567, SG-3-87, H65, GRS4, GHUR22, GD, GRS6, G567 SMR, GHUR15, GTB5-88, H58, H60, GR8, GHUR25, CH9116, BCR72, GHUR15, CH1233, CP2121, JGRS105

3.2.3 *Interaction of Efficient Mesorhizobia and Response in Farmer's Field*

All the 34 efficient chickpea rhizobia selected above as detailed in Table 3.2 under different agroclimatic zones belonging to different centres were further tested at different locations with different chickpea varieties recommended for a particular zone. Such widely adapted strains should show enhanced chickpea productivity under all sets of conditions and irrespective of the cultivar being used by farmer. Another objective was to select rhizobial strains showing specificity to a particular cultivar and boosting its yield to a higher extent. Most efficient rhizobia were finally selected after 3 years and were released for use by biofertilizer producing industry. The most efficient rhizobial strains selected and released by the AICRP Microbiology Group for Chickpea are presented in Table 3.2. Inoculants of these strains released from time to time in the past were used to demonstrate their performance under farmers' field conditions (Khurana and Dudeja 1982; Khurana and Dudeja 1997; Chandra and Pareek 1985; Pareek and Chandra 2003).

The above-selected most efficient mesorhizobial strains responsive to all the varieties under different agroecological conditions were then used for assessing their response under farmers' field conditions. More than 200 demonstration trials on this were carried out by seven centres – Badnapur, Durgapura, Hisar, Ludhiana, and Sehore in Maharashtra, Rajasthan, Haryana, Punjab, and Madhya Pradesh states, respectively, and few trials were carried out by Bharari and Pantnagar in Uttar Pradesh, and Uttaranchal states (Table 3.3). These demonstration trials were usually carried out in one acre area, with half an acre being uninoculated and the remaining half being inoculated by the mesorhizobial inoculants. Compilation of the results from all the farmers fields located in different parts of India indicated that the response to mesorhizobial inoculation varied from farmer to farmer, location to location, and state to state, and the range of this benefit was grain yield of 65–401 kg/ha, corresponding to 9–33% increase in grain yield over the uninoculated control.

Table 3.3 Response of chickpea to mesorhizobial inoculation at the farmers' fields during the past 25 years

Location	Number of trials conducted	Increase over uninoculated control			
		Range		Overall average	
		kg ha^{-1}	Percent	kg ha^{-1}	Percent
Badnapur	60	40–190	3.5–19	120	15
Bharari	3	120–400	24–26	250	25
Durgapura	31	20–597	3.7–30	250	22
Gulbarga	12	90–660	13.1–67	250	23
Hisar	24	0–340	0–47	200	15
Ludhiana	45	30–600	8–39	240	17
Pantnagar	4	100–150	10–12	130	11
Sehore	23	120–300	10–20.7	220	18
Overall	202	65–401	9–33	210	18

Overall, mean showed that on an average 120–250 kg increase in grain yield could be obtained under different conditions by applying an efficient inoculant strain costing less than US$0.40.

Further outcome of the project was reflected in the increase in the number of biofertilizer producers. In 1995–1996, there were only 42 producers, whereas at present about 200 units are producing biofertilizers, and recently liquid biofertilizer technology was licensed to Bac India by Department of Microbiology, CCS Haryana Agricultural University, Hisar, India, for production of biofertilizers (Khurana et al. 1997; Khurana and Dudeja 2003; Khurana and Dudeja 1997; Pareek and Chandra 2003).

3.3 Enhancing Productivity Using Multiple Microbial Inoculants

Mesorhizobia can be used in combination with other plant growth promoting organisms such as VA mycorrhizae, *Bacillus*, *Pseudomonas*, *Pantoa*, and phosphate solubilizers for achieving enhanced crop productivity. Free nitrogen fixers such as *Azotobacter* and *Azospirillum* are also important biofertilizers. Apart from these beneficial microbes, biocontrol agents such as *Trichoderma viridi* and *Pseudomonas maltophila* were also tested for their compatibility with the above inoculants.

3.3.1 Phosphorus-Solubilizing Bacteria

Higher P application in the legumes enhances nodulation and nitrogen fixation. Apart from mesorhizobial inoculants, need for using phosphate-solubilizing bacteria (PSB) was realized, since only 16% of P is available to the plant when single super phosphate (SSP) is used as P source and remaining gets fixed in the soil. Some microbial species have the capability to solubilize this insoluble fixed P in the soil to a soluble form, which is then available to the plants. Different fungal phosphate solubilizers such as *Aspergillus awamorii* and bacterium *Pseudomonas striata* and a mycorrhizal fungus were tested with positive results of enhanced nutrient uptake and crop yield of chickpea.

Organized trials were conducted with 20 and 40 kg P_2O_5/ha as SSP with and without mesorhizobial and PSB inoculation in 1994–1996 in different centres. The results showed that the use of SSP with biofertilizer inoculation enhanced the yield by about 20% (Khurana et al. 1999). Beneficial effects of mesorhizobial and PSB inoculation at different levels of SSP encouraged the microbiology group to conduct experiment in collaboration with Agronomists of the centre for 4 years (1996–2000) at Hisar and Sehore centres with two levels of two P sources – rock phosphate (low cost) and diammonium phosphate (DAP, higher cost) with two PSB inoculants. Analysis of the overall data provided interesting results:

- Availability and uptake of P and other nutrients were enhanced by the use of phosphate solubilizers (*Bacillus megaterium* or *B. polymyxa*) inoculation.
- When no P source was used, inoculation with PSB inoculants enhanced the chickpea grain yield by 9.4%.
- Application of rock phosphate with PSB inoculants enhanced the chickpea grain yield up to 34.8%, compared to the yield of only 24% with P source alone.
- Application of DAP with PSB inoculants enhanced the chickpea grain yield by 46.1%, compared to a yield of 30.7% with a P source alone.
- Use of PSB inoculants resulted in saving of about 20 kg P_2O_5/ha.
- Although DAP produced better results than rock phosphate, it is still a better alternate source of P nutrient for pulses.
- Between two PSB inoculants, one performed better at Hisar centre, while other performed better at Sehore.

3.3.2 *Plant Growth Promoting Rhizobacteria*

Rhizosphere microorganisms closely associated with roots with beneficial properties are called plant growth promoting rhizobacteria (PGPR). PGPR includes a diverse group of soil bacteria that can improve host plant growth by interacting with other soil organisms, thereby either by promoting the growth of beneficial microbes such as rhizobia or phosphate solubilizers or plants directly or by inhibiting the growth of pathogenic bacteria. PGPR strains of *Pseudomonas* were tested at different centres – Bangalore, Coimbatore, Dholi, Gulbarga, Durgapura, Hisar, Ludhiana, and Sehore. Rhizobial inoculation increased the yield by 12.4%, but the combined use of PSB and PGPR resulted in an increase by 22.1% in the grain yield of chickpea.

3.3.3 *Biological and Chemical Control Agents*

Treatment of chickpea seeds with mesorhizobial, PSB, and PGPR inoculants was recommended for achieving the higher chickpea productivity. However, at the same time, pathologist and entomologist also recommended the treatment of seeds with different fungicides and insecticides to achieve higher productivity. It became important to assess the compatibility of biocides with various microbial inoculants, since the use of a biocide might be detrimental to other beneficial inoculant. Therefore, the compatibility of *Mesorhizobium* inoculation with different fungicides such as Agrosan, Benlate, Captan, Thiram, and Calixin-M in the seeds of chickpea was tested at recommended dose of application and was found compatible with the rhizobial inoculants. Similarly the effect of repeated use of BHC and aldrin was also determined and was found to be compatible at recommended doses of application.

Compatibility of microbial inoculants (*Mesorhizobium*, PSB, and PGPR) with *Trichoderma viridi* and Vitavax in chickpea was also assessed at four centres for 3 years including a biocontrol strain of *Pseudomonas maltophila* from Hisar. Both *Trichoderma viridi* and *Pseudomonas* were found to be compatible with microbial inoculants and enhanced the chickpea productivity. On the basis of trials, the use of *Trichoderma viridi* and Vitavax along with all the three types of microbial inoculants (*Mesorhizobium*, PSB, and PGPR) were recommended.

3.3.4 VA Mycorrhizal Fungi and Mesorhizobial Inoculation

VA mycorrhizae have multiple functions of solubilizing, making P and other nutrients available, and also protecting against moisture stress, pathogens, and heavy metal toxicity. Major problem with VA mycorrhiza is that it cannot be cultured and is difficult to multiply. Generally, high volume of the inoculant is required for conducting field trials making it impractical. However, field trials at limited number of locations were conducted at Hisar, Ludhiana, and Bangalore for 3 years. Dual inoculation of mesorhizobia and VAM resulted in enhanced seed production by 16.6% based on the overall mean of the three centres.

3.3.5 Evaluation of Liquid and Carrier-Based Inoculants

Shelf life of rhizobial inoculants is an important aspect for successful marketing of the biofertilizers. As per the recommendations of Bureau of Indian Standards (BIS) for biofertilizers, prepared inoculants must show good viable plate counts up to 3 months to cover a period of inoculants' preparation, marketing, and sowing by the farmers. If marketing of fertilizers and seeds is to be done through the dealer, inoculant must be able to survive for the next cropping season. Keeping this in mind, liquid inoculants were prepared by adding different cell protectants to increase shelf life. Since the seed exudates are known to inhibit rhizobial growth at the initial stages, addition of cell protectants is required to avoid inhibition and to enhance shelf life.

The Microbiology group examined the performance of liquid and carrier-based chickpea rhizobial inoculants under field conditions for further recommendation of biofertilizer marketing. Field evaluation of liquid and carrier-based inoculant of the same strains of chickpea rhizobia was carried out at Bangalore, Ludhiana, Hisar, Durgapura, and Sehore centres for 3 years. Cumulative data of ten experiments showed that performance of chickpea liquid inoculation was slightly better than the carrier-based inoculation.

3.4 Problems and Prospects of Biofertilizer Technology in India

Addressing the basic problems in enhancing chickpea productivity, further selected experiments were carried out. Major objectives of these experiments were to convince the farming and scientific communities as well as administrators and planners regarding the importance of N_2-fixing organisms and nodulation in legumes, particularly in chickpea. The following queries were posed by the various participating groups:

(a) How much N_2 is being fixed by *Mesorhizobium*–legume symbiosis
(b) What is the survival and persistence of rhizobia, and whether there is a need to inoculate every year
(c) What is the success of inoculated rhizobia
(d) What types of rhizobia are present in the soil

To address the above queries, other experiments at different centres were conducted under this coordinated programme.

3.4.1 *Quantity of N_2 Fixed by Mesorhizobium*: Chickpea Symbiosis

To answer this query in collaboration with agronomists, the benefit of chickpea crop on succeeding crops of wheat or bajra was determined. Mungbean, urdbean, cowpea, and chickpea were found to leave about 30–40 kg N/ha for the next crop and pigeon pea were found to be an exhaustive crop.

Experiments in collaboration with ICRISAT using nodulation variants and non-nodulating cultivars were conducted at five locations – Akola, Badnapur, Hisar, New Delhi, and Sehore for 2 years (Table 3.4). High nodulating (HN) and low nodulating (LN) variants of two cultivars ICC 4948 and ICC 5003 being referred as parent were used along with two non-nodulating cultivars ICC 4993 and ICC 4918 (Fig. 3.1). All these nodulation variants were evaluated at 0 and 100 kg N/ha. Compilation of data of all the centres showed that all the nodulation variants of both cultivars showed stability in the nodulation character at all the five locations and nodulation was consistent across all the locations. HN selections always formed more number of nodules than the low nodulating selections and parent bulk formed nodules in between of both nodulation variants.

The HN selection ICC 5003 HN produced higher grain yield than ICC 5003 LN in all the seven experiments at five locations, and the increase ranged from 4 to 41% (Table 3.4). In six of the seven experiments, it also yielded higher than its parent ICC 5003. The beneficial effect of higher nodulation of ICC 5003HN compared to its parent cultivar which produced lower number of nodules was also evident in grain yield. Similarly, ICC 4948HN yielded higher than ICC 4948 LN at all the

Table 3.4 Grain yield of chickpea nodulation variants at five locations in India

Nodulation variants	Akola		Badnapur		New Delhi		Hisar				Sehore		Mean	
							1st year		2nd year					
	N0	N100	N0	N100	N0	N100	N0	N100	N0	N100	N0	N100	N0	N100
ICC 4948 HN	1,590	1,850	960	1,000	1,320	1,050	1,440	1,350	1,950	1,020	1,300	1,580	1,427	1,308
ICC 4948 LN	1,060	1,090	750	710	1,270	1,370	2,500	3,130	2,500	1,690	1,040	1,340	1,520	1,555
ICC 4948 P	1,240	1,530	830	820	1,340	1,230	2,570	2,700	1,230	1,510	1,230	1,510	1,406	1,550
ICC 5003 HN	1,340	1,780	1,080	1,130	1,640	1,590	2,120	1,740	1,590	1,170	1,500	1,800	1,545	1,800
ICC 5003 LN	940	1,560	950	950	1,490	1,540	1,680	1,490	1,430	1,170	1,440	1,730	1,322	1,730
ICC 5003 P	1,160	1,590	1,050	1,060	1,270	1,450	1,740	1,700	1,580	1,170	1,460	1,770	1,376	1,770
ICC 4993 NN	600	980	780	780	1,220	960	2,230	2,540	660	920	690	1,060	1,030	1,060
ICC 4918 NN	490	920	690	840	280	380	250	140	520	1,115	870	1,290	517	1,290

Fig. 3.1 Nodulation variants of chickpea

Table 3.5 Grain yield of Nod^+ and Nod^- cultivars of chickpea in a 2-year experiment

N application	Grain yield (kg/ha)					
	Non nodulating (1CC435 M)			Nodulating (1CC435)		
	1st year	2nd year	Mean	1st year	2nd year	Mean
0 kg N/ha	1,240	2,450	1,850	4,100	3,470	3,790
20 kg N/ha	1,580	2,730	2,160	3,930	3,450	3,690
40 kg N/ha	1,580	2,780	2,180			
80 kg N/ha	1,610	2,850	2,230			
Mesorhizobium				4,320	3,700	4,010

locations except at Hisar. The increase in yield ranged from 4 to 106% at the different N levels and locations. At Hisar Centre due to *Fusarium* wilt, a significant number of plants died, which apparently affected its yielding capacity. The selection ICC 4948HN yielded 5–42% higher than its parent in four of the five experiments. ICC 5003 yielded 2–29% higher than its parent ICC 5003 in all the seven experiments. Overall, the extent of nodulation was correlated with chickpea grain yield (Dudeja et al 1997; Khurana et al. 1998).

To estimate the quantity of nitrogen fixation by these nodulation variants, N difference method was used in which the quantity of N fixed by a nodulating line was subtracted by the quantity of N taken up by a non-nodulating line. The grain N concentration was determined at Akola, Badnapur, and Sehore and the quantity of N from stover was determined only at Akola. At all the three locations, ICC 4948 HN grains had more fixed N (8.4–43.3 kg N/ha) than ICC 4948 LN grains and its parent. Similarly, grain of ICC 5003 HN had more fixed N (2.3–12.6 kg N/ ha in all the experiments than ICC 5003 LN. In most of the cases, the superiority of the two HN selections was statistically non-significant. However, it was felt that there was a need to compare the N difference method with a more reliable method such as radioactivity method of ^{15}N enrichment and natural abundance method.

Another experiment was conducted at Hisar centre for 2 years in fixed plots of size 8 × 6 m to quantify the amount of N fixed by chickpea using non-nodulating

Table 3.6 Nodule occupancy in chickpea cultivars in loamy sand field F1 with irrigation and sandy loam field F2 with conserved moisture

Cultivars	Percent nodule formed by CM-1	
	F1	F2
Low nodulating (LN)		
L550	10.7	2.5
H208	10.9	5.8
Moderate nodulating (MN)		
BG209	20.7	18.6
Pant G114	28.0	15.6
C235	15.7	10.7
High nodulating (HN)		
K850	6.0	8.7
H75-35	8.7	7.5

line along with its nodulating parents. The availability of N made by *Mesorhizobium* and particularly by biological N_2 fixation by nodulating cultivar ICC 435 and its non-nodulating mutant 435M was determined (Dudeja and Khurana 2001). The non-nodulating mutant was supplemented with 0, 20, 40, and 80 kg N/ha, while nodulating cultivar was either inoculated with *Mesorhizobium* or supplemented with 20 kg N/ha. Averaged grain yield results of 2 years showed that chickpea *Mesorhizobium* fixed more than 80 kg of N/ha (Table 3.6).

Legumes are known to meet the demand of their N requirement by fixing biologically nitrogen through the nodules (10–95%) and remaining requirement is met by the uptake of N from soil through roots. In *kharif* legmes, particularly ureide producing legumes, the amount of ureide and amide N determination in stem sap can provide good indication that how much nitrogen is being fixed biologically and how much is taken up from the soil. Since chickpea is not a ureide producing legume, biologically fixed nitrogen was calculated by using non-nodulating and no nitrogen-fixing line in a trial conducted at Hisar centre. N contents in soil, roots, shoots, and seed were measured. Up to 76% of the N demand of chickpea was met through biological fixation and through uptake from the soil, indicating that there was still lot of scope for the improvement of N_2 fixation up to 95%. Simultaneously chlorophyll *a* fluorescence measurement of chickpea nodulation variants using a Handy Plant Efficiency Analyser was measured (Fig. 3.2) to assess the N fixation ability in chickpea (Dudeja and Chaudhary 2005).

3.4.2 *Survival and Competitiveness in Chickpea Rhizobia*

Survival, persistence, and competitiveness of chickpea mesorhizobia were determined to evaluate whether there is a need to inoculate the mesorhizobial biofertilizer inoculants every year or if once every other year. In collaboration with ICRISAT, experiments were conducted for 3 years at Hisar centre under dry land

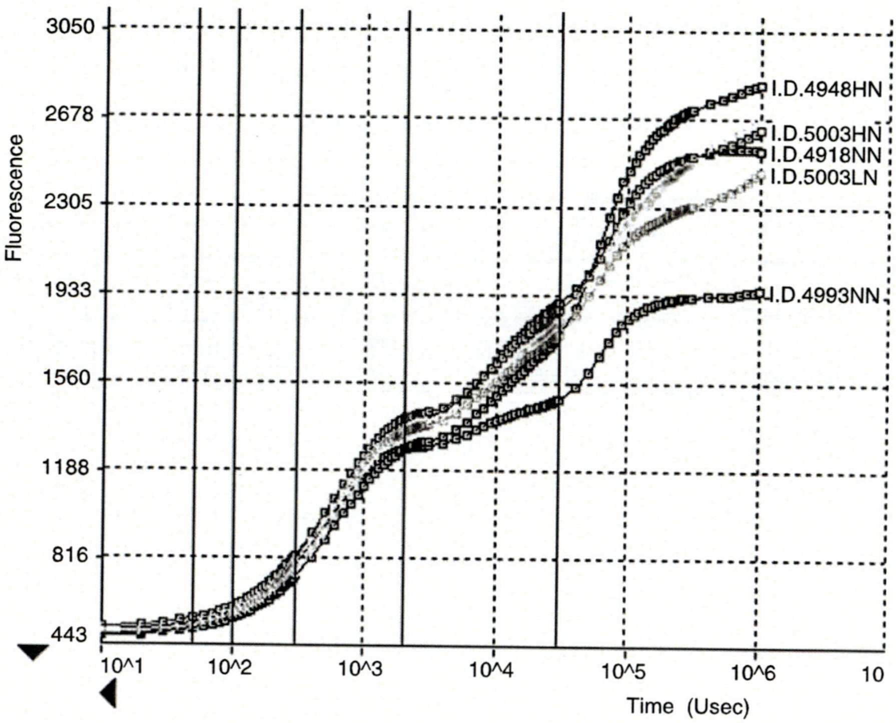

Fig. 3.2 Chlorophyll *a* fluorescence measurement of chickpea nodulation variants using a Handy Plant Efficiency Analyser for assessment of the nitrogen fixation ability in chickpea

conditions. The results indicated that there was poor nodule occupancy (<18%) and no buildup in the population of mesorhizobia in the field. Survival and persistence of chickpea rhizobia under field conditions was very low and rhizobia could not survive till the next cropping season.

To understand the basis of competiveness and to enhance the nodule occupancy, seven varieties of chickpea varying in extent of nodulation in the second set of experiments were tested. To identify the inoculated mesorhizobial strain from soil, an antibiotic resistance chickpea mesorhizobial strain CM-1 was used. Experiments were conducted at Hisar centre under two types of field conditions: irrigated (F1) and dry land conditions (F2) (Khurana et al. 1991). Under irrigated conditions, the nodule occupancy ranged from 6 to 28% with higher grain yield, while under dry land condition it varied from 2.5 to 18.6% (Tables 3.6 and 3.7).

After the development of nodulation variants of chickpea, nodulation variants (HN, LN and parent) of two chickpea cultivars ICC 4948 and ICC 5003 inoculated with two most efficient mesorhizobial strains with antibiotic markers were experimented at Hisar centre to determine nodule occupancy. Two regimes of N levels 0 and 100 kg N/ha were also tested. The results showed that in ICC 4948, both strains showed very poor nodule occupancy of 7.4–7.6%. Parent showed better nodule occupancy as compared to HN and LN selection, and nodule occupancy was

Table 3.7 Overall grain yield of chickpea cultivars varying in extent of nodulation

Fields	Percent increase in grain yield			
	LN	MN	HN	Mean
F1	7.3	10.9	1.0	6.4
F2	4.2	17.1	15.0	12.1
Mean	5.8	14.0	8.0	

Table 3.8 Molecular diversity of native chickpea rhizobia isolated from high and low nodulation variants of two cultivars based on the profiles of DNA fragments generated by PCR with enterobacterial repetitive intergeneric consensus (ERIC) sequences

Chickpea cultivars	Number of rhizobial genotypes trapped by nodulation variants		
	LN	HN	LN and HN
1 ICC 5003	6	4	6
2 ICC 4948	7	5	7
3 ICC 5003 and ICC 4948	7	5	8

low at 100 kg N/ha as compared to the no N application. Similar results were observed in the nodulation variants of another cultivar ICC 5003. Potential for enhancing nodule occupancy by selecting LN cultivars and using their compatible mesorhizobial strains exists (Sheoran et al. 1997).

3.4.3 Mesorhizobial Diversity in Soil and Productivity of Chickpea

Mesorhizobial diversity is considered to be of particular interest due to their strong symbiotic ability with the members of leguminosae. Although host plant is an important factor in shaping the genetic structure of a natural rhizobial population, the question is whether the extent of nodulation can be correlated with the diversity or not? Further, it is quite possible that LN selection remains low nodulating because of the nonavailability of specific rhizobia. Therefore, if a compatible *Mesorhizobium* strain is selected, which could nodulate the LN selection to the level of high nodulating selections, then it could be a better option for enhancing competitiveness through host, *Mesorhizobium,* and their interactions.

The biodiversity of native rhizobia infecting HN and LN nodulation variants of chickpea and the dynamics of native rhizobial population have been studied (Chaudhary et al 2001,2002; Dudeja and Singh 2008; Dudeja et al 2009; Nandwani and Dudeja 2009). On the basis of morphological characteristics, antibiotic resistance pattern, molecular weight, and Rm values of total cellular protein, the rhizobial isolates infecting HN selections belonged to diverse groups, while LN selections were nodulated by restricted groups of rhizobia. The host and its extent of nodulation were responsible for shaping the level of genetic diversity as both LN variants showed the presence of more number of rhizobial genotypes, i.e. seven as compared to high nodulating selections, which showed the presence of four to five genotypes of rhizobia (Table 3.8). Overall, eight chickpea rhizobial genotypes were present in the Hisar centre farm soil.

Potential to develop strategies for enhancing competitiveness is likely to depend on the structure and diversity of the rhizobial population existing in the soil. It has not been established that nodule dominant types possess superior competitiveness traits compared with the minor occupants, but there are reports that the dominant genotypes of *Rhizobium* could be competitive and even 50% nodule occupancy by a dominant strain has been reported. On the basis of nodulation and N uptake by plants, efficient *Mesorhizobium* strains out of all the rhizobia belonging to the two predominant groups I and II were selected, tested under pot culture conditions, and strains showing highest nodule occupancy were selected for conducting field experiment. In this way, *Mesorhizobium mediterraneum strains* LN 707b, LN 1115b for host cultivar ICC 4948 LN, and LN 7007 for cultivar ICC 5003 were selected.

Field experiments were carried out at three different locations – Hisar, Sehore, and Durgapura for 3 years using two low and high nodulating chickpea cultivars ICC 4948 LN, ICC 4948 HN, ICC 5003 LN, and ICC 5003 HN along with one non-nodulating cultivar ICC 4993 NN (Dudeja et al 2007). Nodulation at Hisar location was comparatively better followed by Durgapura and Sehore locations (Fig. 3.3). Both low nodulating cultivars when inoculated with the selected *Mesorhizobium* strains LN 707b, LN 1115b, and LN 7007 formed more number of nodules and produced more nodule and plant biomass. Inoculant strains also resulted in the increased uptake of nitrogen by the plants. Differences in nodulation with rhizobial inoculation treatments were statistically significant. The inoculation with strain LN 707b showed more increase in nodule biomass and plant dry weight at Sehore and Durgapura, whereas LN 1115b showed more increase in the nodule biomass at Hisar. Strain LN 707b showed more N uptake at Sehore and Hisar, whereas strain LN 1115b showed more N uptake at Durgapura. ICC 5003 comparatively formed more nodules and produced more nodule and plant biomass and showed comparatively more N uptake. Non-nodulating line ICC 4993 NN used in this study did not

Fig. 3.3 Competitiveness of chickpea rhizobia inoculated in compatible LN selections of chickpea experiments conducted at Durgapura (**a**, **b**), Sehore (**c**), and Hisar (**d**) centres

form any nodule at any of the location and showed less plant growth and lower N uptake in the plants.

To study nodule occupancy, isolation of rhizobia was done on plates with antibiotic streptomycin, and nodule occupancy by the inoculant strains was further confirmed by the amplification of ERIC sequences by PCR. The nodule occupancy of predominant strain LN 707b, based on antibiotic resistance in its specific host ICC 4948 LN at Hisar, Sehore, and Durgapura centres, was 23, 19, and 29%, respectively (Table 3.9). Corresponding nodule occupancy by another strain LN 1115b was 21, 24, and 18% and by the strain LN 7007 in its specific host ICC 5003 LN was 12, 18, and 29%, respectively. The identity of the streptomycin-resistant inoculant strains LN 707b, LN 1115b, and LN 7007 occupying the nodules was further confirmed by DNA finger printing, and 90–100% of the isolates retained the same ERIC-PCR banding pattern.

The cumulative means of different centres for 3 years trial are presented in Table 3.10. The host-specific *Mesorhizobium* strains LN 707b and LN 1115b recorded a seed yield of 1,563.11 and 1,509.53 kg/ha amounting to an increase of 35 and 33%, respectively, over the non-nodulating uninoculated check. The nodule occupancy was about 24–27% for different locations on overall mean basis of 3 years of the three locations.

Table 3.9 Competitiveness of chickpea rhizobia inoculated in compatible LN selections of chickpea based on ERIC-PCR pattern

Treatments	% Nodule occupied			
	Hisar	Sehore	Durgapura	Mean
ICC 4948 uninoculated	–	–	–	–
Inoculated LN 707b	22	27	18	22.3
Inoculated LN 1115b	20	18	23	20.3
HN uninoculated	–	–	–	–
ICC 5003 uninoculated	–	–	–	–
Inoculated LN 7007	12	28	18	19.3
HN uninoculated	–	–	–	–
ICC 4993 NN	0	0	0	0

Table 3.10 Cumulative data of 3 years of grain yield (kg/ha) of compatible LN selections of chickpea host inoculated with chickpea rhizobia conducted at three locations

Treatments	2003–2004	2004–2005	2005–2006	Mean	% Increase over control
ICC 4948 LN (Un ino)	1,594.66	1,384.50	812.00	1,263.72	20.07
ICC 4948 LN +LN 707b	1,914.33	1,765.00	1,010.00	1,563.11	35.38
ICC 4948 LN +LN 1115b	1,836.66	1,754.50	937.50	1,509.55	33.08
ICC 4948 HN (Un ino)	1,737.33	1,630.00	950.50	1,439.27	29.82
ICC 5003 LN (Un ino)	1,541.00	1,508.50	769.00	1,272.83	20.64
ICC 5003 LN +LN 7007	1,788.33	1,662.50	827.50	1,426.11	29.17
ICC 5003 HN (Un ino)	1,614.33	1,682.00	795.00	1,363.77	–
ICC 4993 NN (Un ino)	1,355.66	1,207.00	467.50	1,010.05	–

3.5 Future Thrust Areas and Conclusions

Trials conducted at various AICRP centres on pulses were aimed to evaluate symbiotic association for better N_2 fixation and productivity in legumes. The extent of response varied with rhizobial inoculant strain, with cultivar, under different agroecological zones, interaction with native rhizobial population, with other microbes, soil N status, cultural practices, and under different environmental conditions.

One of the important information gathered from these experiments was that the soil moisture plays very important role in rhizobial population dynamics in the soil. Sand dunes of Loharu/Badra region of Haryana state, where no or sparse nodulation was observed in chickpea, showed abundant nodulation with the availability of water in this region with the provision of sprinkler irrigation. Similarly, the locations near to a canal also showed better nodulation. Therefore, if proper moisture is provided than even native rhizobial population can multiply and better yields can be harvested. Under farmers' field conditions, the nitrogen fixation should be estimated at the harvest of crop and accordingly improvements for enhanced productivity should be suggested.

In the absence of proper nodule sterilization, contaminants appear and sometimes endophytic organisms having growth similar to rhizobia were picked up failing the plant infectivity test (Dudeja et al 2011). Further work on this aspect has been recently initiated at Hisar centre and a large number of organisms including non-rhizobial bacteria (*Micrococcus*, *Bacillus*, and other gram-negative bacteria) associated with chickpea and pea roots and nodules have been isolated. Identification and characterization of these organisms is in progress. During isolation few rhizobia-like colonies appear within 3–4 days, while others take 1 week to a month, indicating the presence of different rhizobial genera in a single nodule. Studies of such rhizobial strains are of great interest as *Mesorhizobium* from chickpea should show growth within 3–4 days that falls between *Rhizobium* and *Bradyrhizobium* growth pattern. Detailed studies of other nodule associated bacteria other than rhizobia are also important for enhancing crop productivity. Appropriate storage methods for isolated cultures needs to be investigated further as inactivation/loss of nodulation characters were frequently observed.

What is actually determining the success of rhizobia in soil is still unanswered? Attempts have been made to find out the various factors involved in the determination of success of rhizobia in the soil and enhancing crop productivity. Usually, survival is poor and competitiveness is low. *Mesorhizobium ciceri*, *M. mediterraneum*, *M. temperadae*, and *M. tianshanense* have been identified to nodulate chickpea (Dudeja and Singh 2008). Genetic features involved in the effectiveness and competitiveness of a strain and environmental effects on their expression need to be further characterized. Large numbers of chickpea biotypes, which do not form nodules but exist as endophytes (Sarita et al 2005), are reported in soil, performing functions other than nitrogen fixation. These organisms should be further characterized.

With the advent of advanced molecular techniques, it is possible to extract DNA directly from the soil and amplify specific rhizobia present in the soil. Density gradient gel electrophoresis (DGGE) can be used to identify bacteria even differing in single base pair. Recently, at Hisar centre the changes in rhizobial population due to the presence of chickpea host or non-host crop wheat were investigated using DGGE technique. Due to inoculation of rhizobia changes in the bacterial population, three culturable bacteria *Peusdomonas lubricans*, *Arthrobacter* sp., and *Limnobacter thiooxidans* in the adjoining area in soil were observed. Changes in a non-culturable *Rhizobium* sp. in chickpea rhizosphere and in a non-culturable *Pseudoaminobacter salicylatoxidans* were observed in wheat rhizosphere. Such studies to harvest the complete benefit of biofertilizers need to be explored in detail.

Quality aspect of the biofertilizers is of prime importance for better crop yield response. So far there is no proper full proof quality control for the biofertilizers. One can count the rhizobia-like colonies on the specified media plates, but that does not ensure the genetic characteristics of rhizobia suitable to a specific crop. Plant infection tests are time-consuming and not fool proof. Therefore, there is an emergent need to develop an appropriate rapid methodology for appropriate quality control.

References

Chandra R, Pareek RP (1985) Role of host genotype in effectiveness and competitiveness of chickpea (*Cicer arietinum* L.) *Rhizobium*. Trop Agric 62:90–94

Chaudhary P, Dudeja SS, Khurana AL (2001) Chickpea nodulation variants as a tool to detect the population diversity of chickpea rhizobia in soil. Physiol. Mol. Biol. Plants 7:47–54

Chaudhary P, Khurana AL, Dudeja SS (2002) Heterogeneity of rhizobia isolated from chickpea nodulation variants. Indian J. Microbiol. 42(3):195–199

Dudeja SS, Chaudhary P (2005) Fast chlorophyll fluorescence transient and nitrogen fixing ability of chickpea nodulation variants. Photosynthetica 43:253–259

Dudeja SS, Khurana AL (1999) Long term multiplication field evaluation of chickpea rhizobia in India. In: Proceedings of international symposium on long term fertilization trials as a basis for sustainable land use and quantification of matter cycles, Halle, UFZ-Bericht, Germany, 3–5 June, pp 277–280

Dudeja SS, Khurana AL (2001) Integrated management of N and P availability in chickpea through the use of *Rhizobium* and Phosphate solubilizers. In: Proceedings of XIV international plant nutrition colloquium, Hannover, Germany, July 27–Aug 03

Dudeja SS, Narula N (2008) Molecular diversity of root nodule forming bacteria. In: Khachatourians GG, Arora DK, Rajendran TP, Srivastava AK (eds) Agriculturally important microorganisms, vol II. Academic World International, Bhopal, India, pp 1–24

Dudeja SS, Singh P (2008) High and low nodulation in relation to molecular diversity of chickpea mesorhizobia in Indian soils. Arch Agron Soil Sci 54:109–120

Dudeja SS, Dahiya BS, Khurana AL, Rupela OP, Suddarshana MR, Thompson JA (1982) Behaviour of non-fixing chickpea genotype in pot tests. Intern Chickpea Newslett 7:19–20

Dudeja SS, Giri R, Saini R, Suneja-Madan P, Kothe E (2011) Interaction of endophytic microbes with legumes. J. Basic Microbiol. (in press)

Dudeja SS, Potdukhe SR, Namdeo SL, Datar VV, Vinod Kumar, Tilak KVBR, Khurana AL, Rupela OP (1997) Evaluation of some selected chickpea nodulation variants at five locations in India. In: Rupeala OP, Johansen C, Herridge DF (eds) Extending nitrogen fixation research to farmers' fields. Proceeding of an international workshop on managing legumes nitrogen fixation in cropping systems of Asia, ICRISAT Asia Center, 20–24 August, pp 261–276

Dudeja SS, Gupta SC, Majumdar VL, Chaudhary P (2007) Competitiveness of molecularly predominant and host specific chickpea rhizobia under field conditions. In: Vorisek K (ed) Practical solutions for managing optimum C and N content in agricultural soils IV international conference, Prague, 20–22 June, Czech University of Life Sciences and Crop Research Institute, Prague, pp 38–52

Dudeja SS, Nandwani R, Narula N (2009) Molecular diversity of mesorhizobia: particularly rhizobia infecting chickpea. In: Singh DP, Tomar VS, Behl RK, Upadhayaya SD, Bhale MS, Khare D (eds) Proceedings of an international conference on sustainable agriculture for food bioenergy and livelihood security, 14–16 Feb 2007, Agrobios International, Jodhpur, pp 336–347

Jarvis BDW, van Berkum P, Chen WX, Nour SM, Fernandez MP, Cleyet-Marel JC, Gillis M (1997) Transfer of *Rhizobium loti, Rhizobium huakuii, Rhizobium ciceri*, Rhizobium mediterraneum and Rhizobium tianshanense to Mesorhizobium gen. nov. Int J Syst Bacteriol 47:895–898

Khrurana AL, Dudeja SS (1997) Biological nitrogen fixation technology for pusles production in India. Technical Bulletin, Indian Institute of Pulses Research, Kanpur, pp 1–18

Khrurana AL, Dudeja SS, Singh M (1999) Increasing the efficiency of phosphatic fertilizer with phosphate solubilizing bacteria to improve soil fertility and chickpea (*Cicer arietinum*) productivity. In: Proceedings of international symposium on long term fertilization trials as a basis for sustainable land use and quantification of matter cycles, 3–5 June, Halle, UFZ-Bericht, Germany, pp 129–132

Khurana AL, Dudeja SS (1982) *Cicer-Rhizobium* interaction with different chickpea culitvars under field conditions. Zbl Mikrobiol 137:207–213

Khurana AL, Dudeja SS (2003) Current scenario and constraints in the use of biofertilizers in India. In: Marwaha SS, Arora JK (eds) Biotechnological strategies in agro-processing. Asiatech, New Delhi, pp 76–79

Khurana AL, Sharma PK, Dudeja SS (1991) Influence of host, moisture and native rhizobial population on nodule occupancy in chickpea (*Cicer arietinum*). Zbl Mikrobiol 146:137–142

Khurana AL, Namdeo SL, Patel BJ, Dudeja SS (1997) On-farm experiments on rhizobial inoculants: problems and possible solutions. In: Rupeala OP, Johansen C, Herridge DF (eds) Extending nitrogen fixation research to farmers' fields. Proceeding of managing legumes nitrogen fixation in cropping systems of Asia, ICRISAT Asia Center, 20–24 Aug 1996, pp 217–226

Khurana AL, Dudeja SS, Sheoran A (1998) Biological nitrogen fixation in chickpea for sustainable agriculture-prospects and limitations. In: El Basam N, Behl RK, Prochnow B (eds) Sustainable agriculture for food energy and industry: strategies towards achievments, vol 1. Science, New Delhi, pp 439–444

Nandwani R, Dudeja SS (2009) Molecular diversity of a native mesorhizobial population of nodulating Chickpea (*Cicer arietinum* L) in Indian soils. J Basic Microbiol 49:463–470

Pareek RP, Chandra R (2003) Chickpea microbiology and nitrogen fixation. In: Ali M, Kumar S, Singh NB (eds) Chickpea research in India. Indian Institute of Pulses Research, Kanpur, pp 167–193

Poinsot V, Bélanger E, Laberge S, Yang GP, Antoun H, Cloutier J, Treilhou M, Dénarié J, Promé JC, Debellé F (2001) Unusual methyl-branched α, ß-unsaturated acyl chain substitutions in the *Nod* factors of an Arctic *Rhizobium*, *Mesorhizobium* sp. Strain N33 (*Oxytropic arctobia*). J Bacteriol 183:3721–3728

Sarita, Sharma PK, Priefer UB, Prell J (2005) Direct amplification of rhizobial *nodC* sequences from soil total DNA and comparison to *nodC* diversity of root nodule isolates. FEMS Microbiol Ecol 54:1–11

Sheoran A, Khurana AL, Dudeja SS (1997) Nodulation competitiveness in *Rhizobium* chickpea nodulation variants symbiosis. Microbiol Res 152:407–412

Vincent JM (1970) A manual for the practical study of root nodule bacteria. Blackwell, Oxford

Weir BS (2010) The current taxonomy of rhizobia New Zealand rhizobia webstie. (http://www.rhizobia.co.nz/taxonomy/rhizobia.html)

Chapter 4
Phosphate-Solubilizing Microorganisms

Ramesh Chander Kuhad, Surender Singh, Lata, and Ajay Singh

4.1 Introduction

Phosphorus (P) is second major nutrient in crop productivity for its involvement in many essential processes such as cell division, development, photosynthesis, breakdown of sugar, nutrient transport within the plant, transfer of genetic characteristics from one generation to another and regulation of metabolic pathways (Tandon 1987; Armstrong 1988; Theodorou and Plaxton 1993). The maintenance of high level of soil phosphorus has been a major challenge to agricultural scientists, ecologists and farm managers because in most of the soils, phosphate is present in unavailable form due to complex formation with Ca^{2+}, Al^{3+}, Fe^{2+} or Mn^{2+} depending on soil pH and organic matter. The main problem of phosphorus in soil is its rapid fixation and the efficiency of P solubilization rarely exceeding 10–20%. The fixed forms of P in acidic soils are aluminium and iron phosphates while in neutral to alkaline soils as calcium phosphates. The manufacture of phosphatic fertilizers requires high-grade rock phosphate (RP) and sulphur which are getting depleted progressively and becoming costlier. The total world reserves of RP are estimated to be around 2,700 billion tons of which 80% are located in the USA, Russia, and Morocco. In India, RP deposits are estimated to be about 145 million tons but bulk of this is of poor quality and is unsuitable for manufacture of phosphatic fertilizers. Only 25% of the total P requirement is met through indigenous sources; hence, about 1.5 million tons of high grade RP is imported annually.

The concentration of total P in soil ranges from 0.01 to 0.2%, with an average approximately 0.05% (Barker 1984). But out of this only 20% of total P is available

R.C. Kuhad (✉)
Lignocellulose Biotechnology Laboratory, Department of Microbiology, University of Delhi, South Campus, New Delhi 110021, India
e-mail: kuhad85@gmail.com

S. Singh and Lata
Microbiology Division, Indian Agricultural Research Institute, New Delhi 110012, India

A. Singh
Lystek International Inc, Waterloo, Ontario, Canada N2J 3H8

A. Singh et al. (eds.), *Bioaugmentation, Biostimulation and Biocontrol*, Soil Biology 28,
DOI 10.1007/978-3-642-19769-7_4, © Springer-Verlag Berlin Heidelberg 2011

to plants. It makes about 0.2% of plant dry weight (Sahachtman et al. 1998), which indicates the importance of P availability in soil. Because of extreme reactivity and fixation of P in soil, it is unavailable for plant growth. Moreover, the applied phosphorus also gets fixed chemically with metal ions in the soil (Bagyaraj and Varma 1995; Holford 1997). Even if the total P is high, there is always a need to apply phosphatic fertilizers regularly, part of which gets fixed in soil and is unavailable to plants depending on soil pH and climatic conditions. Although various chemical transformations help in release of P immobilization, chemical fixation depletes soil of the available P. Therefore, it is important to develop technology for P solubilization for plant growth. Level of soluble P in soil can be increased either by using phosphate-solubilizing microorganisms (PSM) as bio-inoculants for solubilization of fixed soil P which can improve crop yields or application of phosphorus (P) rich compost. The use of PSM as bio-inoculants plays a vital role in maintaining soil nutrient status, structure and sustains the production base. It also reduces reliance on expensive imported phosphate; thereby a significant increase in yield was subsequently reported with inoculation of PSM in different crops.

Uptake of phosphorus by the plants is in the form of soluble orthophosphorus. In general, the availability of these ions to the plants is in the order of $H_2PO_4^{-1}$ > $H_2PO_4^{-2}$ > PO_4^{-3}. The availability also depends mainly on soil pH (Nath and Borah 1983). The soil microorganisms that are of diverse type (fungi, bacteria and actinomycetes) are responsible for solubilization of inorganic phosphates (Kapoor et al. 1989; Kucey et al. 1989; Holvorson et al. 1990; Illmer and Schinner 1992; Kapoor 1995) causing changes in pH of the soil microenvironment and by producing chelating substances that lead to the solubilization of inorganic phosphates. Iron (Fe) and aluminium (Al) at low pH and calcium (Ca) at high pH fix the available form of P into insoluble forms in soil (Dala 1973; Sanyal and De Datta 1991; Johnson and Loepper 2006; Rengel and Marschner 2005) which are not easily available.

4.2 Phosphate-Solubilizing Microorganisms

A large number of autotrophic and heterotrophic soil microorganisms have capacity to solubilize mineral phosphates. PSM are present in almost all the soils, although their number varies depending upon the soil and climatic conditions (Kucey et al. 1989). First time in 1903, tricalcium phosphate (TCP) solubilization was demonstrated by soil bacteria in liquid and in solid medium (Stalstrom 1903; Sackett et al. 1908). However, the extent of solubilization varies with the source of inorganic P and the microorganisms involved (Banik and Dey 1982). Phosphate-solubilizing fungi (PSF) and bacteria are known as effective organisms for phosphate solubilization (Reyes et al. 1999). Recent report suggests that PSM can be used in union with RP so that phosphorus in the RP can be made available in the soil for plant uptake (Jisha and Alagawadi 1996). Thus, PSM cause the release of nutrients into soil in naturally balanced proportion and exerts beneficial effects on plant development (Glick 1995).

The use of PSM as bio-inoculants plays a vital role in maintaining soil nutrient status and structure as it reduces reliance on expensive imported phosphatic fertilizers. Application of these bacteria along with RP resulted in increased availability of inorganic phosphate for plant utilization (Hebbara and Devi 1990; Rachewad et al. 1992; Jisha and Alagawadi 1996; Chen et al. 2006). Thus, the beneficial effect of inoculation on the availability of P to crops has led to the development of inoculum which is popularly known as phosphobacterin. The first evidence to show that inoculation of seedling with PSM increase uptake of P and crop yield was by Gerretsen (1948) on oat crop. Sundara Rao and Paul (1959) reported significant increase in the yield of barseem after inoculation with phosphobacterin. Since then, beneficial effects of inoculation with different PSM have been reported with different crops.

4.2.1 Major Groups of PSM

An extensive range of soil bacteria, actinomycetes, cyanobacteria and fungi belonging to the genera *Pseudomonas*, *Enterobacter*, *Bacillus*, *Rhizobium*, *Agrobacterium*, *Microccocus*, *Aereobacter*, *Erwinia*, *Streptomyces*, *Nocardia*, *Aspergillus*, *Penicillium*, *Trichoderma*, *Anabaena*, *Nostoc*, *Calothrix* and *Scytonema* that are able to solubilize various forms of precipitated P have been reported (Kucey et al. 1989; Roychoudhary and Kaushik 1989; Rodriguez and Fragaa 1999; Whitelaw 2000; Tran Thi Ngoc Son et al. 2006; Gulati et al. 2008; Sulbarán et al. 2009) and are generally considered to contribute a significant component of the total soil phosphatase activity (Richardson 1994).

In soil, phosphate-solubilizing bacteria (PSB) constitute 1–50% and fungi 0.5–0.1% of the total respective population (Banik and Dey 1982; Kucey 1983; Kucey et al. 1989; Chen et al. 2006). In general, fungal isolates exhibit greater P-solubilizing ability than bacteria in both liquid and solid media (Banik and Dey 1982; Gaur et al. 1973; Kucey 1983; Venkateswarlu et al. 1984). Fungi in soils are able to penetrate deep into soil more easily than bacteria, and hence may be more important to P solubilization in soils (Kucey 1983).

In addition, actinomycetes *Micromonospora*, *Nocardia* and *Streptomyces* have also been reported to solubilize phosphates. Yeast such as *Torula* sp. which is usually not present in soils has also been isolated from compost and characterized for solubilization of TCP and RP (Singh et al. 1980).

4.2.2 Phosphate-Solubilizing Bacteria

An extensive range of soil bacteria including actinomycetes both aerobic and anaerobic are able to solubilize various forms of insoluble inorganic phosphate compounds, such as TCP, dicalcium phosphate, hydroxyapatite and RP (Goldstein 1986; Rodriguez et al. 2000; Gulati et al. 2008; Sulbarán et al. 2009). The prominent genera involved in mineral phosphate solubilization are *Pseudomonas*, *Bacillus*, *Rhizobium*, *Burkholderia*, *Achromobacter*, *Agrobacterium*, *Microccocus*, *Aerobacter*, *Flavobacterium*, *Erwinia*, *Alcaligenes*, *Escherichia*, *Serratia* and *Xanthomonas* sp. (Li 1981; Datta et al. 1982; Venkateswarlu et al. 1984; Fernandez et al. 1984; Thomas and Shantaram 1986; Surange and Kumar 1993; Farhat et al. 2009).

The *Pseudomonas* sp. includes *P. striata*, *P aeruginosa*, *P. putida*, *P vermiculosam* and *P. fluroscence* (Sen and Paul 1957; Gaind and Gaur 1990a, b; Kumar et al. 2002; Buch et al. 2008). Population of PSB mainly depends on different soil properties such as physical, chemical properties, organic matter and P content (Kim et al. 1998).

Major *Bacillus* sp. known for their mineral phosphate-solubilizing abilities are *B. polymyxa*, *B. subtilis*, *B. brevis*, *B. circulans*, *B. megaterium*, *B. mesentricus*, *B. mycoides*, *B. pulvifaciens*, etc. (Sen and Paul 1958; Sundara Rao and Sinha 1963; Paul and Sundara Rao 1971; Rodriguez and Fragaa 1999; Swain and Ray 2009).

Most PSB were isolated from the rhizosphere of various plants and are known to be metabolically more active than those isolated from sources other than rhizosphere (Baya et al. 1981; Katznelson and Bose 1959; Vazquez et al. 2000). The P-solubilizing ability in bacteria is generally reduced or lost upon repeated subculturing but no such loss has been observed in the case of PSF (Kucey 1983; Sperber 1958).

P-solubilizing ability of PSB is affected by many physiological factors some of which are C and N source in medium, mineral source, pH, incubation temperature, aeration and incubation period (Gaind and Gaur 1990a, b; Vassilev and Vassileva 2003; Taiwo and Ogundiya 2008). Singh (1992) showed that PSB could solubilize 240 μg/ml and 5,300 μg/ml of TCP and mussoorie rock phosphate (MRP), respectively, after 9 weeks of incubation at 30°C under stationary conditions. Dave and Patel (1999) showed that *Pseudomonas* sp. release 27–163 mg of P_2O_5 at 37°C in 3 weeks under static conditions at cell concentrations of 6.6×10^7/ml in 100 ml of Pikovskaya's (PVK) medium containing TCP equivalent to 225 mg P_2O_5 (Dave and Patel 1999). Strains of *Pseudomonas* sp. are capable of releasing 162 μg/ml inorganic phosphate in the medium containing TCP (Santhi 1998). Strains of *Enterobacter* can release inorganic phosphate ranging from 83 to 551 μg/ml in medium containing hydroxyapatite (Kim et al. 1997). Kumar and his coworker (1999) observed that strains of *Acetobacter* released 142–431 μg/ml of inorganic phosphate from TCP. Illmer and Schinner (1992) reported that temperature 25–30°C is optimum for P solubilization for *Pseudomonas* sp. whereas fungi solubilize more P at a slightly higher temperature, i.e. 30–35°C (Gaur 1990). *B. polymyxa* was found to have a higher temperature range of 35–40°C for optimum solubilization (Gaur 1990). The maximum decrease in pH was observed on third and seventh day by *B. polymyxa* and *P. striata*, respectively, which increased on further incubation (Illmer and Schinner 1992). Recent report by Farhat et al. (2009) showed that the soluble phosphorus (P) concentration in optimized medium reached 967, 500, 595 and 326 mg/l from $CaHPO_4$, $Ca_3(PO_4)_2$, hydroxyapatite and RP within 72 h of incubation after inoculion with *Serratia marcescens*.

Actinomycetes as PSM are of special interest since these gram-positive filamentous sporulating bacteria can flourish in extreme environments (Jiang et al. 2005; Pathom-Aree et al. 2006) and also produce various plant growth-promoting substances (Fabre et al. 1988; Manulis et al. 1994; Ikeda 2003; Jain and Jain 2007). Many workers have isolated actinomycetes strains with phosphate-solubilizing ability (Ahmad and Jha 1968; Mahmoud et al. 1973; Ibrahim and Abdel-Aziz 1977; Banik and Dey 1983; Mba 1997; Hamdali et al. 2008). Mba (1994) isolated four actinomycetes from earthworm cast of *Pontoscolex corethrurus* which were able to solubilize up to 7 mg soluble P/g RP within one week of incubation. Interestingly, the RP-solubilizing power of these isolates was decreased with acidification of the medium and most of the isolates were able to hydrolyse carboxymethyl cellulose (CMC). Hamdali et al. (2008) isolated 55 actinomycete cultures, which were able to solubilize RP in synthetic minimum medium. Most of these isolates were from genera *Streptomyces* and *Micromonospora* and did not produce any organic acids but found to produce siderophores rendering the P available for plants.

Many workers have used a composite culture of bacteria and fungi to enhance the P solubilization but the efficiency of such mixed inoculants depends on the compatibility among microorganisms. Kundu and Gaur (1981) found that inoculation of *P. striata* and *Aspergillus awamori* together solubilized more phosphates than dual inoculation of *B. polymyxa* and *A. awamori*.

4.2.3 Phosphate-Solubilizing Fungi

A large number of fungi have also been reported to solubilize insoluble forms of phosphorus. These include *Aspergillus*, *Candida*, *Penicillium*, *Rhizopus*, *Cladosporium* and *Paecilomyces*, etc. (Agnihotri 1970; Gaind and Gaur 1991; Banik and Dey 1982; Singh et al. 1984; Venkateswarlu et al. 1984; Darmwal et al. 1989; Sheshadari et al. 2000). These PSF are well known for their ability to solubilize mineral phosphates owing to their ability to produce organic acids (Mattey 1992; Reddy et al. 2002; Bojinova et al. 2008; Richa et al. 2007; Xiao et al. 2009; Khan et al. 2010), and particularly some *Aspergillus* and *Penicillium* species have been tested by inoculating directly into the soil in order to solubilize RP (Kucey 1987; Vassilev et al. 1997). Inoculation of PSF and mycorrhizal fungi also improves the physico-chemical, biochemical and biological properties of RP-amended soil (Caravaca et al. 2004).

In another study, *Aspergillus tubingensis* and two isolates of *Aspergillus niger* have also shown the highest solubilization of RP under in vitro conditions (Reddy et al. 2002). Richa et al. (2007) tested *Aspergillus tubingensis* and *A. niger* for their efficacy to solubilize RP in RP-amended soils and they observed that available P along with organic carbon was significantly increased when compared to initial soil. Also the soil pH was lowered compared to initial pH of the soil. The improvement of physico-chemical and biochemical properties of RP-amended soil with the inoculation of A. *niger* and mycorrhizal fungi was also reported by Caravaca et al. (2004).

Similarly, El-Azouni (2008) tested the efficacy of *Aspergillus niger* and *Penicillium italicum* to solubilize TCP in vitro and they found the release of inorganic P was 490 and 275 μg P ml^{-1}, respectively, after one week incubation. Change in soluble P with time with respect to pH has been observed with *Paeciiomyces fusisporus* and other fungi (Goyal et al. 1982; Mishra et al. 1983). A nematofungus *Arthrobotrys oligospora* also has shown the ability to solubilize the phosphate rocks (Duponnois et al. 2006).

Arbuscular Mycorrhizae (AM) fungi are known to enhance P nutrition of plants especially in P-deficient soils by scavenging the available P due to the large surface area of their hyphae and by their high-affinity P uptake mechanisms (Hayman 1974; Moose 1980; Sanders and Tinker 1973). There are also reports of organic acid production by AM (Paul and Sundara Rao 1971) that could solubilize the insoluble mineral phosphates. As the soil phosphorus levels available to the plants increase, the amount of phosphorus also increases in the plant's tissues and thus the concomitant carbon drain from the plant by the AM fungi making symbiosis nonbeneficial to the plant (Grant et al. 2005).

4.2.4 Interaction of PSB with Other Microorganisms

The PSM when used with other plant growth-promoting rhizobacteria (PGPR) act synergistically to enhance crop yields (Saxena and Tilak 1994, 1997). Co-inoculation of *P. striata* and AM significantly increased the soybean yield and P uptake by plants over control. Dual inoculation of *Rhizobium* with PSM (Perveen et al. 2002) or arbuscular mycorrhizae (AM) fungi (Zaidi et al. 2003) has been shown to improve plant growth more than with their sole inoculation in P-deficient soils. Synergistic interactions on plant growth have been observed by co-inoculation of PSB with N_2 fixers such as *Azotobacter* (Kundu and Gaur 1984) and *Azospirillum* (Belimov et al. 1995) or with vesicular AM (Kim et al. 1998). Son et al. (2003, 2006) showed that co-inoculation of *Bradyrhizobium japonicum* and *Pseudomonas* sp. enhanced the number of nodules, dry weight of nodules, yield components, soil nutrient availability and uptake in soybean crop. Research workers have reported similar results with legume crops when PSBs are coinoculated with various N_2 fixing bacteria (Table 4.1).

4.3 Mechanisms of Phosphate Solubilization

Release of organic acids by PSM has been reported as a primary mechanism of phosphate solubilization (Hilda and Fraga 2000; Khiari and Parent 2005). Besides organic acids, the production of chelating substances (2-ketogluconic acid), humic

Table 4.1 Effect of co-inoculation of legumes with phosphate solubilizers and N_2 fixers

Crop	Rhizobia	Co-inoculating PS solubilizers	Plant responses to inoculation	References
Alfalfa	*R. meliloti*	*Pseudomonas*	Plant growth, nitrogenase activity, nodule number, total nodule weight and total plant nitrogen showed significant increase	Knight and Langston-Unkefer (1988)
Chickpea	*Mesorhizobim*	*Pseudomonas*	Marked increase in nodule weight and shoot biomass when coinoculated with *Mesorhizobium* and *Pseudomonas* in sterilized chillum jar conditions. In pot experiments, co-inoculation significantly increased root and shoot biomass	Sindhu et al. (2002a)
	Mesorhizobium	*Bacillus*	Dual inoculation significantly increased plant dry weight, nodulation, N content, protein content and seed yield, compared to single inoculation	Wani et al. (2007)
	Rhizobium	*Pseudomonas*, *Bacillus*	Significantly increased nodule weight, root and shoot biomass and total plant nitrogen	Parmar and Dadarwal (1999)
Clover	*R. leguminosarum bv.trifolii* 24	*Pseudomonas* sp.	Co-inoculation significantly increased shoot and nodule weight in comparison to plants inoculated with *R. leguminosarum* bv. *Trifolii*	Derylo and Skorupska (1993)
Common bean	*Rhizobium*	*A. brasilense*	Co-inoculation promoted root hair formation and an increase in secretion of the nod gene induced flavonoids resulting in greater number of nodules	Burdman et al. (1996)
Soybean	*B. japonicum*	*P. fluorescens*	Co-inoculation increased colonization of *B. japonicum* on soybean roots, nodule number and the acetylene reduction assay	Chebotar et al. (2001), Son et al. (2006)
Greengram	*Bradyrhizobium* sp. (*Vigna*)	*Bacillus*	Co-inoculation enhanced nodulation and growth of greengram	Sindhu et al. (2002b)
Wheat	*R. leguminismarum*	*Pseudomonas* sp.	Dual inoculation along with P fertilizer increase yield by 30–40%	Afzal and Asghari (2008)

substances, mineral acids (sulphuric acids), siderophores and proton extrusion mechanisms also play an important role (Kapoor et al. 1989; Kucey et al. 1989; Illmer et al. 1995).

Growth of PSM is generally accompanied by decrease in pH of the medium and soil (Singh et al. 1980; Mishra and Banger 1985; Darmwal et al. 1989; Stevenson 2005). Reduction in pH is due to the production of organic acids, which include citric, gluconic, fumaric, malic, oxalic, lactic, 2-ketogluconic, malonic acids, etc. (Banik and Dey 1981; Venkateswarlu et al. 1984; Illmer and Schinner 1992; Vassilev et al. 1996; Hwangbo et al. 2003; Patel et al. 2008). However, quantity and quality of organic acid produced is fully dependent on type of PSM. The highest solubilization of TCP and RP was reported by citric and fumaric acid (Gaur 1990). Calcium and magnesium in RP binds to the solubilized P and carbonate thereby neutralizing the organic acids, which play a vital role in solubilization.

The release of 2-ketogluconic acid by PSM was correlated with P solubilization due to its calcium chelating ability (Firsching 1969). Moreover, these compounds form stable organometallic complexes with Fe and Al and decrease precipitation of solubilized phosphates. Chelation of calcium, especially oxalic acid helps in solubilization of insoluble phosphates (Illmer and Schinner 1992). Besides acids production and chelators, bacterial exo-polysaccharides (EPS) also play an important role in mineral phosphate solubilization. Recently, Yi et al. (2008) observed the synergistic effects of EPS and organic acid on TCP solubilization, which varied with the origin and the concentration of EPS in medium. The increase of P solubilization brought by EPS is mainly attributed to the precipitation of EPS consequently resulting in greater phosphorus released from insoluble phosphate.

Humic substances are produced in soil as a result of organic matter decomposition and contain humic and fulvic acids as their main components. Humic acids are strong chelating agents which chelate Ca^{2+} and release H^+, and thus help in dissolution of insoluble phosphates (Gaur 1969; Pareek and Gaur 1973; Mishra et al. 1983; Banger et al. 1985; Singh and Amberger 1990). The functional group of these compounds, such as carboxylic, phenolic, hydroxylic and phenolic hydroxyl, forms stable complex with Ca, Fe and Al (Banger and Mishra 1990). They also check reprecipitation of the solubilized phosphates (Singh and Amberger 1990).

The sulphate-reducing bacteria and other heterotrophic organisms under anaerobic conditions release H_2S that solubilize ferric phosphate by forming insoluble sulphides of Fe and render soluble phosphate (Gaur 1990). Rhizospheric microorganisms release CO_2 forming carbonic acid (HCO_3), which lowers the pH, and thus help in P solubilization (Hayman 1975). The oxidation of inorganic sulphur and pyrite to sulphuric acid by sulphur oxidizing bacteria (*Thiobacillus*) lowers the soil pH and hence improves the availability of phosphates (Wainwright 1984; Kapoor et al. 1991) due to formation of $CaNO_3$ and $CaSO_4$. Certain rhizospheric microorganisms (*Rhizobium*, *Azotobacter*, *Pseudomonas*) synthesize iron-chelating compounds (siderophores), which in acidic soil remove *iron* from ferric phosphates and render phosphate for plants (Bossier et al. 1988; Suneja and Lakshminarayana 1999; Hamdali et al. 2008).

Proton extrusion mechanism is a probable mechanism of phosphate solubilization in some microorganisms, which can solubilize phosphates without the release of organic acids in the environment (Illmer et al. 1995). Release of H^+ accompanying respiration and assimilation has been observed in *Penicillium aurantiogriseum* and *Pseudomonas* sp. solubilizing hydroxyapatite (Illmer et al. 1995; Lin et al. 2006). The insoluble phosphates are solubilized at the cell surface of the microorganisms. The NH_4^+/H^+ exchange mechanism depends on the presence of NH_4^+ ions in the medium as uptake of NH_4^+ occurs with the concomitant release of H^+ in the environment (Roos and Luchner 1984). Quality and quantity of root exudates also alter the concentration of P in the soil solution due to presence of organic ligands in the root exudates (Hinsinger 2001).

4.4 Estimation and Enumeration of Phosphate-Solubilizing Microorganisms

PSMs can be isolated from different sources such as soil (Khanna et al. 1979; Gupta et al. 1986; Kapoor et al. 1989; Roychoudhary and Kaushik 1989), rhizosphere (Sardina et al. 1986; Thakkar et al. 1993; Singh and Kapoor 1994), root nodules (Chhonkar and Subba Rao 1967; Surange and Kumar 1993), compost (Kapoor et al. 1989; Thakkar et al. 1993), RPs (Gaur et al. 1973) and earthworm casts (Mba 1994).

Solubilization of tricalcium phosphate (TCP) in agar medium has been used as the initial criterion for isolation and enumeration of PSM. Stalstrom (1903) first demonstrated the solubilization of TCP by soil bacteria in solid and liquid medium. Organisms growing on such media are able to solubilize P and produce clear zone around colonies due to dissolution of calcium phosphate. Mineral P solubilizers have been routinely isolated and screened using PVK medium (Pikovskaya 1948) or Sperber medium (Sperber 1957) by plate assay method, looking for halo/clear zone around colonies of potential P solubilizers (Katznelson et al. 1962; Das 1963; Bardiya and Gaur 1974; Darmwal et al. 1989). The zone of clearance is due to solubilization of inorganic phosphate mainly due to the production of organic acids in the surrounding medium (Johnston 1952; Das 1963; Sethi and Subba Rao 1968; Pareek and Gaur 1973; Kundu and Gaur 1981; Kapoor et al. 1989; Halder et al. 1990; Singal et al. 1991; Yadav and Dadarwal 1997).

Although, microorganism capable of producing zone of clearance around its colonies in plate assay method is selected as a potential P solubilizer, however, several workers have reported that many isolates which did not produce any visible zone of clearance on agar plates but could solubilize various types of insoluble inorganic phosphate in liquid medium (Louw and Webley 1959; Das 1963). Thus, the existing plate assay fails where the halo zone is inconspicuous or absent. This may be because of the varying diffusion rates of different organic acids secreted by different microorganisms (Johnston 1952). On the other hand microorganisms, which showed P solubilization in laboratory medium, do not give any solubilization in soil. For example, Whitelaw et al. (1999) showed that, even though *Penicillium radicum* produced GA when grown in laboratory culture, the organic acid was not the primary mechanism for solubilization of precipitated P. Hence, there is need of suitable medium for quick and reliable plate assay for screening PSM. However, the plate assay can be regarded as generally suitable for isolation and preliminary characterization of PSM.

Gupta et al. (1994) gave a modified PVK medium formulated using different concentrations of bromophenol blue (BPB) dye for screening PSM. Results revealed that as the concentration of dye increased, the clarity and visibility of the yellow coloured halo/zone improved, and most appropriate dye concentration was found to be 2.4 mg/ml. There is no correlation between halo zone formation and quantity of inorganic phosphate solubilized (Ostwal and Bhide 1972; Arora and Gaur 1979). Indicator medium containing bromothymol blue (BTB) has also been developed to enhance chances of picking up efficient PSB (Krishnaraj 1996).

Advantage of using modified medium is that the incubation period required prior to selection of PSB is significantly reduced, i.e. minimum incubation period for PSB was 24 h than in original PVK medium that exceeds 7 days.

National Botanical Research Institute's phosphate growth medium (NBRIP), which is more efficient than PVK medium was developed for screening of PSM (Nautiyal 1999). NBRIP medium was comparable to PVK agar medium; however, in broth assay, NBRIP medium consistently resulted in a 3-fold increase in P solubilization. The rate of phosphate solubilization was increased with increased concentrations of glucose (Nautiyal 1999). Also, glucose, xylose and sucrose are reported to be good carbon sources for PSF (Ahmad and Jha 1968), while for bacteria glucose, galactose and sucrose were found to be effective. However, the solubilization potential of microorganisms varies with different carbon sources depending on the type of insoluble phosphate (Ahmad and Jha 1968; Thakkar et al. 1993). In anaerobic conditions, P release is affected by the physiological state of cells and carbon sources (acetic, propionic and butyric acid) as observed under stationary conditions (Rustrain et al. 1997). The utilization of glucose directly correlates with drop in pH, which was reflected as mineral phosphate solubilization due to conversion of glucose to GA (Goldstein and Liu 1987; Liu et al. 1992). It was observed that 60 mM GA produced results in release of 0.1 mM inorganic phosphate. The amount of acid liberated by PSB roughly is more than 5 per cent of the carbohydrate consumed (Banik and Dey 1982). Phosphate solubilization ability was improved with $(NH_4)_2SO_4$ at a lower concentration and was 27.1 percent more effective than KNO_3 as observed by Nautiyal (1999). *P. striata* (PS-9) has been found to utilize urea, asparagine, ammonium sulphate, ammonium nitrate, potassium nitrate and calcium nitrate for solubilization of RP (Gaur 1990). Sodium nitrate and peptone were found to be better substrate for P solubilization by *B. megaterium* var. phosphaticum. *P. striata* solubilized the higher amount of insoluble phosphates among bacteria, and quantities of RP solubilized were much less than TCP and hydroxyapatite (Arora and Gaur 1979). However, incorporation of bacteria with RP resulted in increased availability of inorganic phosphate for plant utilization (Hebbara and Devi 1990; Rachewad et al. 1992; Jisha and Alagawadi 1996). But, in plate assay, for efficient P solubilizers, tricalcium is used instead of MRP. As TCP gets metabolized faster, it is normally used for screening studies compared to RPs (Singh and Kapoor 1994). Also, TCP solubilization is higher as compared to aluminium phosphate and ferric phosphate (Narsian et al. 1993). Microbial solubilization of RP is influenced by the physical and chemical properties of the RP and the microorganisms involved (Garg et al. 1989; Kapoor et al. 1989).

Bacteria prefer to grow in neutral to alkaline (pH 7 to 8) reaction for maximum solubilization (Wani et al. 1979) and fungi do better at slightly acidic to neutral pH (Gaur 1990). A direct correlation was obtained between decrease in pH and increase in available P of the culture media in certain cases (Sperber 1958; Agnihotri 1970; Liu et al. 1992) but others have contradictory reports that solubilization is not always proportional to the decline in pH (Mehta and Bhide 1970; Krishnaraj 1996; Asea et al. 1988; Parks et al. 1990). The form in which inorganic phosphate exists also changes according to the soil pH. Below pH 6.0, most inorganic phosphate is present as monovalent H_2PO_4 species. The plant uptake is also high at the pH range of 5.0–6.0, which indicates that P is primarily taken up as monovalent form (Furihata et al. 1992). It has been noted that PSB could solubilize both RP and dicalcium phosphate in unbuffered media but failed to solubilize RP in buffered media. The organic acid secreted by these bacteria was 20–50 times less than that required to solubilize P from alkaline soil (Gyaneshwar et al. 1998).

4.5 PSM and Yield of Crops

PSM inoculation can increase crop yields up to 70% (Verma 1993). The increase in yield is mainly attributed by the increased availability of soluble phosphorus on inoculation with PSMs, which enhance the plant growth by improving biological nitrogen fixation (Kucey et al. 1989).

Sundara Rao et al. (1963) reported increase in yields and phosphate uptake in tomato and wheat with phosphobacterin and a strain of *Bacillus megaterium*. Kundu and Gaur (1980) reported increased potato yield with phosphobacteria in potato. Also, increase in wheat yield with inoculation of PSM-*Azotobacter chroococcum* is reported. Similarly, increase in yield was observed in various crops such as chickpea, rice, etc. (Alagawadi and Gaur 1988; Kavimadan and Gaur 1971; Monod et al. 1989).

Similarly, several authors have reported increased yield of wheat (Whitelaw et al. 1997; Omar 1998), onion (Vassilev et al. 1997), alfalfa (Rodríguez et al. 1999), rice (Khan et al. 2008), maize (Richa et al. 2007) and soybean (Abd-Alla and Omar 2001) through simple inoculation of PSF. Recently, El-Azouni (2008) observed that the dual inoculation of PSF (*A. niger* and *P. italicum*) significantly increased dry matter and yield of soybean plants compared to the control TCP-amended soil in pot experiment along with significant increase in the percentage of N and P content of the plant.

4.6 Future Prospects

PSMs are an integral component of soil microbial community and play an important role in P cycle in soil rendering the unavailable P to plants. These PSM have enormous potential for making use of fixed P in the soil particularly in soils with low P availability in tropical and subtropical developing countries. The mechanism of phosphate solubilization by microorganisms has been studied in detail but the P solubilization is a complex phenomenon affected by many factors, such as PSM used, nutritional status of soil and environmental factors. Moreover, the stability of the PSMs after inoculation in soil is also important in P solubilization to benefit crop growth. Therefore, it needs further studies to understand the characteristics and mechanisms of phosphate solubilization by PSM. To conclude, the efforts should be made to identify, screen and characterize more PSM for their ultimate application under field conditions.

Acknowledgement The authors acknowledge the help from Mr. Harsh Kuhad, Amity University, Noida, India, during the preparation of this chapter.

References

Abd-Alla MH, Omar SA (2001) Survival of rhizobia/bradyrhizobia and rock-phosphate-solubilizing fungus *Aspergillus niger* on various carriers from some agroindustrial wastes and their effects on nodulation and growth of faba bean and soybean. J Plant Nutr 24:261–272

Afzal A, Asghari B (2008) *Rhizobium* and phosphate solubilizing bacteria improve the yield and phosphorus uptake in wheat (*Triticum aestivum*). Int J Agric Biol 10:85–88

Agnihotri VP (1970) Solubilization of insoluble phosphates by some fungi isolated from nursery seed beds. Can J Microbiol 16:877–880

Ahmad N, Jha KK (1968) Solubilization of rock phosphate by microorganisms isolated from Bihar solis. J Gen Appl Microbiol 14:89–95

Alagawadi AR, Gaur AC (1988) Associative effect of *Rhizobium* and phosphate solubilizing bacteria on the yield and nutrient uptake of chickpea. Plant Soil 105:241–246

Armstrong DL (1988) Role of phosphorus in plants. In: Armstrong DL (ed) Better crops with plant food. Potash and Phosphate Institute, Atlanta, USA, pp 4–5

Arora D, Gaur AC (1979) Microbial solublization of different inorganic phosphates. Indian J Expt Biol 11:1258–1261

Asea PEA, Kucey RMN, Stewart JWB (1988) Inorganic phosphate solubilization by two *Penicillium* species in solution culture and soil. Soil Biol Biochem 20:459–464

Bagyaraj DJ, Varma A (1995) Interaction between arbuscular mycorrhizal fungi and plants: their importance in sustainable agriculture in arid and semiarid tropics. Adv Microbiol Ecol 14:119–142

Banger KC, Mishra MM (1990) Solubilization of insoluble phosphates by humic acid. Int J Trop Agric 8:209–213

Banger KC, Yadav KS, Mishra MM (1985) Transformation of rock phosphate during composting and the effect of humic acid. Plant Soil 85:259–266

Banik S, Dey BK (1981) Phosphate solubilizing microorganisms of a lateritic soil. I. Solubilization of inorganic phosphates and production of organic acids by microorganisms isolated in sucrose calcium phosphate agar plates. Zbt Bakt Abt II 136:478–486

Banik S, Dey BK (1982) Available phosphate content of an alluvial soil as influence by inoculation of some isolated phosphate solubilizing microorganisms. Plant Soil 69:353–364

Banik S, Dey BK (1983) Alluvial soil microorganisms capable of utilizing insoluble aluminium phosphate as a sole source of phosphorus. Zbt Microbiol 138:437–442

Bardiya MC, Gaur AC (1974) Isolation and screening of microorganisms dissolving low grade rock phosphate. Folia Microbiol 19:386–389

Barker SA (1984) Soil nutrient bioavailability. Wiley, New York

Baya AM, Robert SB, Ramos CA (1981) Vitamin production in relation to phosphate solubilization by soil bacteria. Soil Biol Biochem 13:527–532

Belimov AA, Kojemiakov AP, Chuvarliyeva CV (1995) Interaction between barley and mixed cultures of nitrogen fixing and phosphate-solubilizing bacteria. Plant Soil 173:29–37

Bojinova D, Velkova R, Ivanova R (2008) Solubilization of morocco phosphate by *Aspergillus niger*. Biores Technol 99:7348–7353

Bossier P, Hofte M, Verstraee W (1988) Ecological significance of siderophores in soil. Adv Microbiol Ecol 10:385–414

Buch A, Archana G, Naresh KG (2008) Metabolic channelling of glucose towards gluconate in phosphate solubilizing *Pseudomonas aeruginosa* P4 under phosphorus deWciency. Res Microbiol 159:635–642

Burdman S, Volpin H, Kigel J, Kapulnik Y, Okon Y (1996) Promotion of nod gene inducers and nodulation in common bean (*Phaseolus vulgaris*) root inoculated with *Azospirillum brasilense*. Appl Environ Microbiol 62:3030–3033

Caravaca F, Alguacil MM, Azcon R, Diaz G, Roldan A (2004) Comparing the effectiveness of mycorrhizal inoculation and amendment with sugar beet, rock phosphate and *Aspergillus niger* to enhance field performance of the leguminous shrub *Dorycnium pentaphyllum* L. Appl Soil Ecol 25:169–180

Chebotar VK, Asis CA Jr, Akao S (2001) Production of growth promoting substances and high colonization ability of rhizobacteria enhance the nitrogen fixation of soybean when coinoculated with *Bradyrhizobium japonicum*. Biol Fertil Soils 34:427–432

Chen YP, Rekha PD, Arun AB, Shen FT, Lai WA, Young CC (2006) Phosphate solubilizing bacteria from subtropical soil and their tricalcium phosphate solubilizing abilities. Appl Soil Ecol 34:33–41

Chhonkar PK, Subba Rao NS (1967) Phosphate solubilization by fungi associated with legume root nodules. Can J Microbiol 13:749–753

Dala RC (1973) Characterization of soil inorganic phosphorus. Agrochimica 17:385–396

Darmwal NS, Singh RB, Rai R (1989) Isolation of phosphate solubilizers from different sources. Curr Sci 58:570–571

Das AC (1963) Utilization of insoluble phosphates by soil fungi. J Indian Soc Soil Sci 11:203–207

Datta M, Banik S, Gupta RK (1982) Studies on the efficacy of a phytohormone producing phosphate solubilizing *Bacillus firmus* in augementing paddy yield in acid soils of Nagaland. Plant Soil 69:365–373

Dave A, Patel HH (1999) Inorganic phosphate solubilizing soil *Pseudomonas*. Indian J Microbiol 39:161–164

Derylo M, Skorupska A (1993) Enhancement of symbiotic nitrogen fixation by vitamin-secreting fluorescent *Pseudomonas*. Plant Soil 154:211–217

Duponnois R, Kisa M, Plenchette C (2006) Phosphate solubilizing potential of the nematofungus *Arthrobotrys oligospora*. J Plant Nutrit Soil Sci 169:280–282

El-Azouni IM (2008) Effect of phosphate solubilizing fungi on growth and nutrient uptake of Soybean (*Glycine max* L.) plants. J Appl Sci Res 4:592–598

Fabre B, Armau E, Etienne G, Legendre F, Tiraby G (1988) A simple screening method for insecticidal substances from Actinomycetes. J Antibiot 41:212–219

Farhat MB, Farhat A, Bejar W, Kammoun R, Bouchaala K, Fourati A, Antoun H, Bejar S, Chouayekh H (2009) Characterization of the mineral phosphate solubilizing activity of *Serratia marcescens* CTM 50650 isolated from the phosphate mine of Gafsa. Arch Microbiol 191: 815–824

Fernandez HM, Carpena AO, Dadahiya C (1984) Solubilization of mineral phosphorus in calcareous soils by *Bacillus cereus* in green house experiments. Agrociencia 43:235–245

Firsching FH (1969) Chelates in analytical chemistry. Dekker, New York, pp 4–12

Furihata T, Suzuki M, Sakuri H (1992) Kinetic characterization of two phosphate uptake systems with different affinities in suspension cultures *Cahtaranthus roseus* protoplasts. Plant Cell Physiol 33:1151–1157

Gaind S, Gaur AC (1990a) Influence of temperature on the efficiency of phosphate solubilizing microorganisms. Indian J Microbiol 30:305–310

Gaind S, Gaur AC (1990b) Shelf life of phosphate solubilizing inoculants as influenced by type of carrier, high temperature and low moisture. Can J Microbiol 36:846–849

Gaind S, Gaur AC (1991) Thermotolerant phosphate solubilizing microorganisms and their interaction with mungbean. Plant Soil 133:141–149

Garg N, Mishra MM, Garg KL (1989) In: Mukerji KG, Singh VP, Garg KL (eds) Frontiers in applied microbiology, vol III. Rastogi, Meerut, pp 263–271

Gaur AC (1969) Studies on the availability of phosphate in soils as influenced by humic acid. Agrochimica 14:62–65

Gaur AC (1990) Phosphate solubilizing microoganisms as biofertilizers. Omega Scientific, New Delhi, India

Gaur AC, Madan M, Ostwal KP (1973) Solubilization of phosphatic compounds by native microflora of rock phosphates. Indian J Expt Biol 11:427–429

Gerretsen FC (1948) The influence of microorganisms of the phosphate intake by the plants. Plant Soil 1:51–81

Glick BR (1995) The enhancement of plant growth by free living bacteria. Can J Microbiol 32:145–148

Goldstein AH (1986) Bacterial phosphate solubilization: historical perspective and future prospects. Am J Altern Agric 1:57–65

Goldstein AH, Liu ST (1987) Molecular cloning and regulation of a mineral phosphate solubilizing gene from *Erwinia herbicola*. Biotechnology 5:72–74

Goyal S, Phogat VK, Mishra MM, Yadva KS (1982) Transformation of phosphorus during solubilization of rock phosphate by *Aspergillus awamori*. Indian J Microbiol 22:136–138

Grant C, Bitman S, Montreal M, Plenchette C, Morel C (2005) Soil and fertilizer phosphorus: effects on plant supply and mycorrhizal development. Can J Plant Sci 85:3–14

Gulati A, Rahi P, Vyas P (2008) Characterization of phosphate solubilizing fluorescent *Pseudomonads* from the rhizosphere of seabuckthorn growing in the cold deserts of Himalayas. Curr Microbiol 56:73–79

Gupta RD, Bhardwaj KKR, Marwah BC, Tripathi BR (1986) Occurrence of phosphate dissolving bacteria in some soils of North-West Himalayas under varying biosequences and climosequence. J Indian Soc Soil Sci 34:498–504

Gupta R, Singal R, Shankar A, Kuhad RC, Saxena RK (1994) Modified plate assay for screening phosphate solubilizing microorganisms. J Gen Appl Microbiol 40:255–260

Gyaneshwar P, Kumar GN, Parekh LJ (1998) Effect of buffering on the phosphate solubilization ability of microorganisms. World J Microbiol Biotechnol 14:669–673

Halder AK, Mishra AK, Bhattacharya P, Chakraborty PK (1990) Solubilization of rock phosphate by *Rhizobium* and *Bradyrhizobium*. J Gen Appl Microbiol 36:81–92

Hamdali H, Bouizgarne B, Hafidi M, Lebrihi A, Virolle MJ, Ouhdouch Y (2008) Screening for rock phosphate solubilizing actinomycetes from Moroccan phosphate mines. Appl Soil Ecol 38:12–19

Hayman DS (1974) Plant growth responses to vesicular-arbuscular mycorrhiza VI. Effects of light and temperature. New Phytol 73:71–80

Hayman DS (1975) Phosphorus cycling by soil microorganisms and plant roots. In: Walker N (ed) Soil microbiology. Butterworths, London, pp 67–91

Hebbara M, Devi SL (1990) Effect of phosphorus solubilizing bacteria (PSB) on phosphorus availability to groundnut from rock phosphate. Curr Res 19:56–57

Hilda R, Fraga R (2000) Phosphate solubilizing bacteria and their role in plant growth promotion. Biotech Adv 17:319–359

Hinsinger P (2001) Bioavailability of soil inorganic P in the rhizosphere as affected by root induced chemical changes: a review. Plant Soil 237:173–195

Holford ICR (1997) Soil phosphorus its measurement and its uptake by plants. Aust J Soil Res 35:227–239

Holvorson HO, Kenyan A, Kornberg HL (1990) Utilization of calcium phosphates for microbial growth at alkaline pH. Soil Biol Biochem 22:887–890

Hwangbo H, Park RD, Kim YW, Rim YS, Park KH, Kim TH, Suh JS, Kim KY (2003) 2-ketogluconic acid production and phosphate solubilization by *Enterobacter intermedium*. Curr Microbiol 47:87–92

Ibrahim AN, Abdel-Aziz IM (1977) Solubilization of rock phosphate by *Streptomycetes*. Agrokem Alajtan 26:424–434

Ikeda T (2003) Pharmacological effects of ivermectin, an antiparasitic agent for intestinal strongyloidiasis: its mode of action and clinical efficacy. Nippon Yakurigaku Zasshi 122:527–538

Illmer P, Schinner F (1992) Solubilization of inorganic phosphates by microorganisms isolated from forest soils. Soil Biol Biochem 24:389–395

Illmer P, Babato A, Schinner F (1995) Solubilizattion of hardly soluble $AlPO_4$ with PSM. Soil Biol Biochem 27:265–270

Jain PK, Jain PC (2007) Isolation, characterization and antifungal activity of *Streptomyces sampsonii* GS 1322. Indian J Exp Biol 45:203–206

Jiang Y, Li WJ, Xu P, Tang SK, Xu LH (2005) Study on diversity of actinomycetes salt and alkaline environments. Wei Sheng Wu Xue Bae 46:191–195

Jisha MS, Alagawadi AR (1996) Nutrient uptake and yield of sorghum (*Sorghum biocolor* L. Moenck) inoculated with phosphate solubilzing bacteria and cellulolytic fungus in a cotton stalk amended vertisol. Microbiol Res 151:1–5

Johnson SE, Loepper RH (2006) Role of organic acids in phosphate mobilization from iron oxide. Soil Sci Soc Am J 70:222–234

Johnston HW (1952) The solubilization of phosphate: I. The action of various organic compounds on dicalcium and tricalcium phosphate. N Z J Sci Technol 33:436–444

Kapoor KK (1995) Phosphate solubilization through soil microorganisms. In: Behl KK, Khurana AL, Dogra RC (eds) Plant microbe interactions in sustainable agriculture. CCS Haryana Agricultural University, Hisar and MMB, New Delhi, pp 46–61

Kapoor KK, Mishra MM, Kukreja K (1989) Phosphate solubilization by soil microorganisms: a review. Indian J Microbiol 29:119–127

Kapoor KK, Mishra MM, Malik RS, Banger KC (1991) Solubilization of MRP by use of pyrite and *Thiobacilli*. Environ Ecol 9:635–641

Katznelson H, Bose B (1959) Metabolic activity and phosphate dissolving capability of bacterial isolates from wheat roots in the rhizosphere and non rhizosphere soil. Can J Microbiol 5:79–85

Katznelson H, Peterson EA, Rouat JW (1962) Phosphate dissolving microorganisms on seed and in the root zone of plants. Can J Bot 40:1181–1186

Kavimadan SK, Gaur AC (1971) Effect of seed inoculation with *Pseudomonas* sp. on phosphate uptake and yield of maize. Curr Sci 40:439–440

Khan SA, Hamayun M, Yoon H, Kim H, Suh S, Hwang S, Kim J, Lee I, Choo Y, Yoon U, Kong W, Lee BM, Kim J (2008) Plant growth promotion and *Penicillium citrinum*. BMC Microbiol 8:231

Khan MS, Zaidi A, Ahemad M, Oves M, Wani PA (2010) Plant growth promotion by phosphate solubilizing fungi: current perspective. Arch Agron Soil Sci 56(1):73–98

Khanna SS, Chaudhary ML, Bathla RN (1979) Influence of moisture, FYM and pyrites on the solubilization of rock phosphate in calcareous soils of Haryana. Bull Indian Soc Soil Sci 12:545–549

Khiari L, Parent LE (2005) Phosphorus transformations in acid light-textured soils treated with dry swine manure. Can J Soil Sci 85:75–87

Kim KY, McDonald GA, Jordan D (1997) Solubilization of hydroxyapatite by *Enerobacer agglomerans* and cloned *Escherichia coil* in culture medium. Biol Fertil Soils 24:347–352

Kim KY, Jordan D, McDonald GA (1998) Effect of phosphate-solubilizing bacteria and vesicular-arbuscular mycorrhizae on tomato growth and soil microbial activity. Biol Fert Soils 26:79–87

Knight TJ, Langston-Unkefer PJ (1988) Enhancement of symbiotic nitrogen fixation by a toxin releasing plant pathogen. Science 242:951–954

Krishnaraj PU (1996) Genetic characterization of mineral phosphate solubilization in *Pseudomonas* sp. Ph.D. Thesis, IARI, New Delhi.

Kucey RMN (1983) Phosphate solubilizing bacteria and fungi in various cultivated and virgin Alberta soils. Can J Soil Sci 63:671–678

Kucey RMN (1987) Increased phosphorus uptake by wheat and field beans inoculated with a phosphate solubilizing *Penicillium bilaii* strain and with vesicular-arbuscular mycorrhzal fungi. Appl Environ Microbiol 55:2699–2703

Kucey RMN, Janzen HH, Leggett ME (1989) Microbially mediated increases in plant available phosphorus. Adv Agron 42:199–221

Kumar NR, Arasu VT, Gunasekaran P (2002) Genotyping of antifungal compounds producing plant growth-promoting rhizobacteria, *Pseduomonas fluorescens*. Curr Sci 82:1463–1466

Kundu BS, Gaur AC (1980) Effects of N2 fixing and PSM as single and composite inoculant on cotton. Indian J Microbiol 20:225–229

Kundu BS, Gaur AC (1981) Effect of single and composite cultures on rock phosphate solubilization. Haryana Agric Univ J Res 11:559–562

Kundu BS, Gaur AC (1984) Rice response to inoculation with N2-fixing and P-solubilizing microorganisms. Plant Soil 79:227–234

Li SG (1981) Studies on phosphorite decomposing microorganisms. J Soil Sci 5:33–35

Lin TF, Huang HI, Shen FT, Young CC (2006) The protons of gluconic acid are the major factor responsible for the dissolution of tricalcium phosphate by *Burkholderia cepacia* CC-Al74. Biores Technol 97:957–960

Liu ST, Lee LY, Jai CY, Hung CH, Chang YS, Wolfram JH, Rogers R, Goldstein AH (1992) Cloning of an *Erwinia herbicola* gene necessary for gluconic acid production and enhanced mineral phosphate solubilization in *E. coli* HB 101: nucleotide sequence and probable involvment in biosynthesis of the co-enzyme pyrroloquinoline quinone. J Bacteriol 174:5814–5819

Louw HA, Webley DM (1959) A study of soil bacteria dissolving certain phosphate fertilizers and related compounds. J Appl Bacteriol 22:227–233

Mahmoud SAZ, Abdel-Hafez AM, El-Sawy M, Hanafy EA (1973) Phytin-hydrolysing bacteria in soils of Egypt. Zbt Bakt Abt II 128:528–531

Monod SPI, Gupta DN, Chavan AS (1989) Enhancement of phosphate availability and phosphorus uptake in rice by phosphate solubilizing culture. J Maharashtra Agric Univ 14:178–181

Manulis S, Shafrir H, Epstein E, Lichter A, Barash I (1994) Biosynthesis of indole-3-acetic acid via the indole-3-acetamide pathway in *Streptomyces* spp. Microbiology 140:1045–1050

Mattey M (1992) The production of organic acids. Rev Biotechnol 12:87–132

Mba CC (1994) Rock phosphate solubilizing and cellulolytic actinomycetes isolates of earthworm casts. Environ Manage 18:257–261

Mba CC (1997) Rock phosphate solubilizing *Streptosporangium* isolates from casts of tropical earthworms. Soil Biol Biochem 29:381–385

Mehta YR, Bhide VP (1970) Solubilization of tricalcium phosphate by some soil fungi. Indian J Expt Biol 8:228–229

Mishra MM, Banger KC (1985) Phosphocompost as a phosphorus source in neutral and alkaline soils. In: Mishra MM, Kapoor KK (eds) Soil biology: proceedings soil biology symposium. CCS Haryana Agricultural University, Hisar, India, pp 139–147

Mishra MM, Phogat VK, Goyal S, Yadav KS (1983) Solubilization of phosphorus from Mussoorie rock phosphate by *Aspergillus awamori* and humic substances. Tropic Plant Sci Res 1:221–224

Moose B (1980) Vesicular-arbuscular mycorrhiza research for tropical agriculture. Research Bulletin 194, Hawaii Institute of Tropical Agriculture and Human Resources, Honolulu, HI, USA, University of Hawaii

Narsian V, Thakkar J, Patel HH (1993) Isolation and screening of phosphate solubilizing fungi. Indian J Microbiol 34:113–118

Nath AK, Borah DK (1983) A study on the release on native and applied fixed phosphate as affected by pH and moisture regime. Indian J Agric Chem 16:247–251

Nautiyal CS (1999) An efficient microbiological growth medium for screening phosphate solubilizing microoganisms. FEMS Microbiol Lett 170:265–270

Omar SA (1998) The role of rock-phosphate-solubilizing fungi and vesicular-arbusular-mycorrhiza (VAM) in growth of wheat plants fertilized with rock phosphate. World J Microbiol Biotech 14:211–218

Ostwal KP, Bhide VP (1972) Solubilization of tricalcium phosphate by soil *Pseudomonas*. Indian J Expt Biol 10:153–154

Pareek RP, Gaur AC (1973) Release of phosphate from tricalcium phosphate and rock phosphates by organic acids. Curr Sci 42:278–279

Parks EJ, Olson GJ, Brickman FE, Baladi F (1990) Characterization of high performance liquid chromatography (HPLC) of the solubilization of phosphorus in iron ore by a fungus. Indian J Microbiol 5:18–190

Parmar N, Dadarwal KR (1999) Stimulation of nitrogen fixation and induction of flavonoid like compounds by rhizobacteria. J Appl Microbiol 86:36–44

Patel DK, Archana G, Kumar GN (2008) Variation in the nature of organic acid secretion and mineral phosphate solubilization by *Citrobacter* sp. DHRSS in the presence of different sugars. Curr Microbiol 56:168–174

Pathom-Aree W, Stach JE, Ward AC, Horikoshi K, Bull AT, Goodfellow M (2006) Diversity of actinomycetes isolated from challenger deep sediment (10, 898 m) from the mariana trench. Extremophiles 10:181–189

Paul NB, Sundara Rao WVB (1971) Phosphate dissolving bacteria in rhizosphere of some cultivated legumes. Plant Soil 25:127–132

Perveen S, Khan MS, Zaidi A (2002) Effect of rhizospheric microorganisms on growth and yield of green gram (*Phaseolus radiatus*). Indian J Agric Sci 72:421–423

Pikovskaya RI (1948) Mobilization of phosphorus in soil in connection with the vital activity of some microbial species. Microbiologiya 7:362–370

Rachewad SN, Raut RS, Malewar GU, Ganure CK (1992) Effects of phosphate solubilizing biofertilizer on biomass production and uptake of phosphorus by sunflower. J Maharashtra Agric Univ 17:480–481

Roychoudhary P, Kaushik BD (1989) Solubilization of Mussoorie rock phosphate by cyanobacteria. Curr Sci 58:569–570

Reddy MS, Kumar S, Babita K, Reddy MS (2002) Biosolubilization of poorly soluble rock phosphates by *Aspergillus tubingensis* and *Aspergillus niger*. Biores Technol 84: 187–189

Rengel Z, Marschner P (2005) Nutrient availability and management in the rhizosphere: exploiting genotypic differences. New Phytol 168:305–312

Reyes I, Bernier L, Simard RR, Antoun H (1999) Effect of nitrogen source on the solubilization of different inorganic phosphates by an isolate of *Penicillium rugulosum* and two UV induced mutants. FEMS Micobiol Ecol 28:281–290

Richa G, Khosla B, Reddy MS (2007) Improvement of maize plant growth by phosphate solubilizing fungi in rock phosphate amended soils. World J Agric Sci 3:481–484

Richardson AE (1994) Soil micro-organisms and phosphate availability. In: Pankhurst CE, Double BM, Gupts VVSR, Grace PR (eds) Soil biota management in sustainable agriculture. CSIRO, Melbourne, Australia, pp 50–62

Rodriguez H, Fragaa R (1999) Phosphate solubilizing bacteria and their role in plant growth promotion. Biotech Adv 17:319–339

Rodriguez H, Gonzalez T, Selman G (2000) Expression of a mineral phosphate solubilizing gene from Erwinia herbicola in two rhizobacterial strains. J Biotechnol 84:155–161

Roos W, Luchner M (1984) Relationships between proton extrusion and fluxes of ammonium ions and organic acids in *Penicillium cyclopium*. J Gen Microbiol 130:1007–1014

Roychoudhary P, Kaushik BD (1989) Solubilization of Mussoorie rock phosphate by cynobacteria. Curr Sci 10:569–570

Rustrain E, Delgens JP, Maletta R (1997) P release by pure culture of *Acinetobacter* sp. Effect of growth stage with cell cultivated on various carbon sources. Lett Appl Environ Microbiol 24:144–148

Sackett WG, Palten AJ, Brown CW (1908) The solvent action of soil bacteria upon the insoluble phosphates of raw bone meal and natural rock phosphate. Zbl Bakt Abt II 20:688

Sahachtman DP, Reid RJ, Ayling SM (1998) Phosphate uptake by plants from soil to cell. Plant Physiol 116:447–453

Sanders FE, Tinker PB (1973) Phosphate flow into mycorrhizal roots. Pest Sci 4:385–395

Santhi V (1998) Mechanism of mineral phosphate solubilization and growth promotion by diverse bacteria. M.Sc. Thesis, UAS, Dharwad

Sanyal SK, De Datta SK (1991) Chemistry of phosphorus transformations in soil. Adv Soil Sci 16:1–120

Sardina MG, Boiardi JL, Erbola RJ (1986) Solubilization of phosphorus from low-grade minerals by microbial action. Biotechnol Lett 8:247–252

Saxena AK, Tilak KVBR (1994) Interaction among beneficial microorganisms. Indian J Microbiol 34:91–106

Saxena AK, Tilak KVBR (1997) Interaction of soil microorganisms with vesicular arbuscular mycorrhiza. In: Tiwari JP, Saxena G, Tiwari I, Mittal N, Chamola BP (eds) New approches in microbial ecology. Aditya Books, New Delhi, India, pp 231–250

Sen A, Paul NB (1957) Solubilization of phosphates by some common soil bacteria. Curr Sci 26:222–223

Sen A, Paul NB (1958) Occurrence of phosphobacteria in the glands of *Cassia occidentalis*. Indian J Agric Sci 28:21–29

Sethi RP, SubbaRao NS (1968) Solubilization of tricalcium phosphate and calcium phytate by soil fungi. J Gen Appl Microbiol 14:329–331

Sheshadari S, Kumarasamy R, Lakshminarasimham C, Ignacimuthu S (2000) Solubilization of inorganic phosphates by *Azospirillium halopraeferans*. Curr Sci 79:565–567

Sindhu SS, Suneja S, Goel AK, Parmar N, Dadarwal KR (2002a) Plant growth promoting effects of *Pseudomonas* sp. on coinoculation with *Mesorhizobium* sp. cicer strain under sterile and wilt sick soil conditions. Appl Soil Ecol 19:117–120

Sindhu SS, Gupta SK, Suneja S, Dadarwal KR (2002b) Enhancement of greengram nodulation and growth by *Bacillus* sp. Biol Plantarum 45:117–120

Singal R, Gupta R, Kuhad RC, Saxena RK (1991) Solubilization of inorganic phosphates by a Basidiomycetous fungus *Cyathus*. Indian J Microbiol 31:397–401

Singh S (1992) Solubilization of insoluble phosphates by bacteria. M.Sc. Thesis, CCS HAU, Hisar, India

Singh CP, Amberger A (1990) Humic substances in straw compost with rock phosphate. Biol Waste 31:165–174

Singh S, Kapoor KK (1994) Solubilization of insoluble phosphate isolated from different sources. Enviorn Ecol 12:51–55

Singh CP, Mishra MM, Yadav KS (1980) Solubilization of insoluble phosphates by thermophilic fungi. Ann Microbiol 131:289–296

Singh HP, Pareek RP, Singh TA (1984) Solubilization of rock phosphate solubilizers in broth. Curr Sci 53:1212–1213

Son TTN, Vu van Thu VU and Kobayashi H (2003) Effect of organic and bio fertilizer application on rice-soybean-rice cropping systems. In: Proceedings of the final workshop of JIRCAS Mekong Delta Project – "Development of new technologies and their practice for sustainable farming systems in the Mekong Delta", November 25–26, 2003, pp 65–81

Son TTN, Diep CN, Giang TTM (2006) Effect of bradyrhizobia and phosphate solubilizing bacteria application on soybean in rotational system in the mekong delta. Omonrice 14:48–57

Sperber JI (1957) Solution of mineral phosphates by soil bacteria. Nature 180:994–995

Sperber JI (1958) Solution of apatite by soil microorganisms producing organic acids. Aust J Agric Res 9:778–781

Stalstrom VA (1903) Beitrag Zur Kenntrusder einwinsking sterilizer and in garung befindlieher striffe any dil loslieshkeit der phosphorus are destrical cum phosphours. Zbt Bakt Abt II 11:724–732

Stevenson FJ (2005) Cycles of soil: carbon, nitrogen, phosphorus, sulfur, micronutrients. Wiley, New York

Sulbarán M, Pérez E, Ball MM, Bahsas A, Yarzábal LA (2009) Characterization of the mineral phosphate-solubilizing activity of *Pantoea agglomerans* MMB051 isolated from an iron-rich soil in southeastern Venezuela (Bolívar State). Curr Microbiol 58:378–383

Sundara Rao WVB and Paul NB (1959) Bacterization of phosphobacterin, radioisotopes, fertilizers and cowdung gas plant. ICAR Proc Ser 322–326

Sundara Rao WVB, Sinha MK (1963) Phosphate dissolving microorganisms in the soil and rhizosphere. Indian J Agric Sci 33:272–278

Sundara Rao WVB, Bajpai PD, Sharma JP, Subbiah BV (1963) Solubilization of phosphates by PSM using 32P as tracer and influence of seed bacterization on the uptake by the crop. J Indian Soc Sci 11:209–219

Suneja S, Lakshminarayana K (1999) Siderophore production of *Azotobacter*. In: Narula N (ed) *Azotobacter* in sustainable agriculture. CBS, New Delhi, India, pp 64–73

Surange S, Kumar N (1993) Phosphate solubilization under varying pH by *Rhizobium* from tree legumes. Indian J Exptl Biol 11:427–429

Swain MR, Ray RC (2009) Biocontrol and other beneficial activities of *Bacillus subtilis* isolated from cowdung microflora. Microbiol Res 164:121–130

Taiwo LB, Ogundiya M (2008) Microbial solubilization of Ogun rock phosphate in the laboratory and in soil. Afr J Microbiol Res 2:308–312

Tandon HLS (1987) Phosphorus research and agricultural production in India. Fertilizer Development and Consultation Organization, New Delhi, p 172

Thakkar J, Narsian V, Patel HH (1993) I. Inorganic P solubilization by certain soil bacteria. II. Solubilization of natural rock phosphate and pure insoluble inorganic P by Aspergillus awamori. Indian J Expt Biol 31:743–747

Theodorou ME, Plaxton WC (1993) Metabolic adaptations of plant respiration to nutritional phosphate deprivation. Plant Physiol 101:339–344

Thomas GV, Shantaram MV (1986) Solubilization of inorganic phosphates by bacteria from coconut plantation soils. J Plant Crops 14:42–48

Vassilev M, Vassileva N (2003) Biotechnological solubilization of rock phosphate on media containing agro-industrial wastes. Appl Microbiol Biotechnol 61:435–440

Vassilev N, Massimiliano F, Federico F (1996) Rock phosphate solubilization with gluconic acid produced by immobilized *Penicillium* variable P16. Biotech Tech 10:585–588

Vassilev N, Toro M, Vassileva M, Azcón R, Barea JM (1997) Rock phosphate solubilization by immobilized cells of *Enterobacter* sp. in fermentation and soil conditions. Biores Technol 61:29–32

Vazquez P, Holguin G, Puente EM, Lopez-Cortes A, Bashan Y (2000) Phosphate-solubilizing microorganisms associated with the rhizosphere of mangroves in a semiarid coastal lagoon. Biol Fertil Soils 30:5–6

Venkateswarlu B, Rao AY, Raina P (1984) Evaluation of phosphorus solubilization by microorganisms isolated from Aridisols. J Indian Soc Soil Sci 32:273–277

Verma LN (1993) Biofertiliser in agriculture. In: Thampan PK (ed) Organics in soil health and crop production. Peekay Tree Crops Development Foundation, Cochin, India, pp 152–183

Wainwright W (1984) Microbial sulphur oxidation in soil. Sci Prog 65:459–475

Wani PV, More BB, Patil PL (1979) Physiological studies on the activity of phosphorus solubilizing microorganisms. Indian J Microbiol 19:23–25

Wani PA, Khan MS, Zaidi A (2007) Synergistic effects of the inoculation with nitrogen fixing and phosphate-solubilizing rhizobacteria on the performance of field grown chickpea. J Plant Nutr Soil Sci 170:283–287

Whitelaw MA (2000) Growth promotion of plants inoculated with phosphate-solubilizing fungi. Adv Agron 69:99–151

Whitelaw MA, Harden TJ, Bender GL (1997) Plant growth promotion of wheat inoculated with *Penicillium radicum* sp. nov. Aust J Soil Res 35:291–300

Whitelaw MA, Harden TJ, Helyar KR (1999) Phosphate solubilization in solution culture by the soil fungus *Penicillium radicum*. Soil Biol Biochem 32:655–665

Xiao C, Chi R, He H, Qiu G, Wang D, Zhang W (2009) Isolation of phosphate-solubilizing fungi from phosphate mines and their effect on wheat seedling growth. Appl Biochem Biotechnol 159:330–342

Yadav KS, Dadarwal KR (1997) Phosphate solubilization and mobilization through soil microorganisms. In: Dadarwal KR (ed) Biotechnological approaches in soil microorganisms for sustainable crop production. Scientific Publishers, Jodhpur, India, pp 293–308

Yi Y, Huang W, Ying G (2008) Exopolysaccharide: a novel important factor in the microbial dissolution of tricalcium phosphate. World J Microbiol Biotechnol 24:1059–1065

Zaidi A, Khan MS, Amil M (2003) Interactive effect of rhizotrophic microorganisms on yield and nutrient uptake of chickpea (*Cicer arietinum* L.). Eur J Agron 19:15–21

Chapter 5
Bioaugmentation and Biovalourization of Agro-Food and Beverage Industry Effluents

Alok K. Pandey, Brijesh K. Mishra, Anju Arora, Surender Singh, Lata, and Ramesh C. Ray

5.1 Introduction

Intensive agriculture has resulted in increased productivity. Every year there is a record increase in food production and huge surplus of various crops, e.g., cereals and tuber crops such as potato, cassava, sweet potato, sugar beet and sugarcane, etc., are produced and processed for value addition. Most of the starchy crops such as cassava, sweet potato and potato are perishable (Ray and Ward 2006) and enormous infrastructure is needed to store large quantities of such crops. In developing countries, due to poor infrastructure, post-harvest losses account for 25–30% of the total production owing to spoilage by bacteria, fungi and insect attack and hence it is becoming imperative to process these crops into value-added commodities. The food and beverage processing industry is growing fast the world over. These industries generate lots of solid waste and effluents, which are rich in nutrients and able to support growth of variety of microorganisms (Thassitou and Arvanitoyannis 2001). These effluents (wastewaters) if disposed untreated, add to the pollution problem. In view of the extensive contamination of the environment by persistent and toxic chemical pollutants originating from industrial wastewaters, it is imperative to develop cost-effective and efficient methods for their remediation. At present, an Ajmer, Rajasthan-305206 international trend promoting pollution prevention through cleaner production, which is based on the 5R policy (Fig. 5.1) namely reduction, replacement, reuse, recovery and recycling, is emerging (Olguín et al. 2004). Bioaugmentation is the popular and attractive technology that utilizes the metabolic potential of microorganisms to clean up the environment (Watanabe 2001;

A.K. Pandey, A. Arora, S. Singh, and Lata (✉)
Division of Microbiology, Indian Agricultural Research Institute, New Delhi 110 012, India
e-mail: latarajat@yahoo.co.in

B.K. Mishra
National Research Centre on Seed Spices, Tabiji, Ajmer, Rajasthan, India

R.C. Ray
Central Tuber Crops Research Institute (Regional Centre), Dumduma Housing Board, Bhubaneswar 751019, India

A. Singh et al. (eds.), *Bioaugmentation, Biostimulation and Biocontrol*, Soil Biology 28,
DOI 10.1007/978-3-642-19769-7_5, © Springer-Verlag Berlin Heidelberg 2011

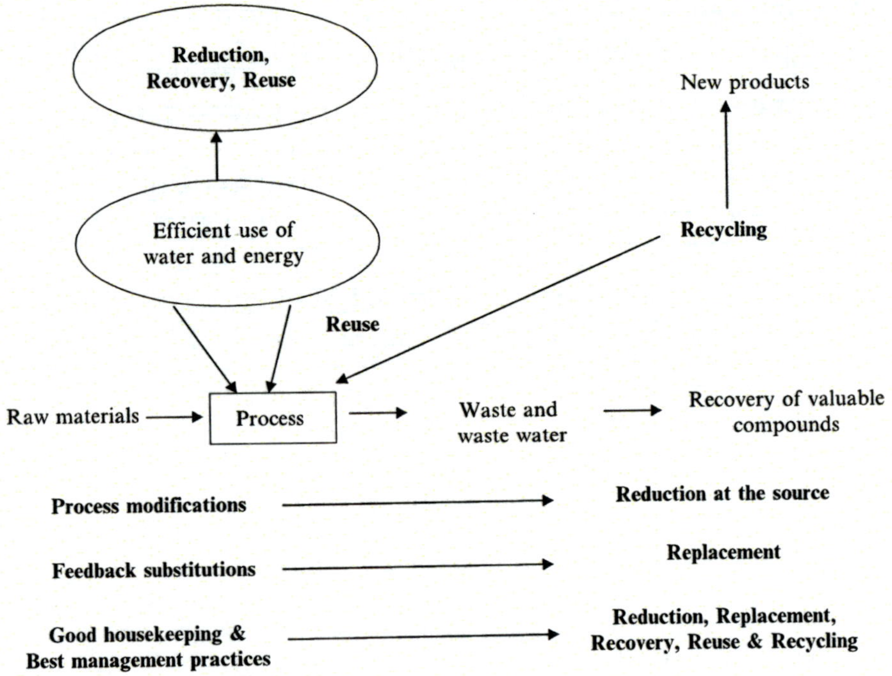

Fig. 5.1 The 5 R policy (Olguín et al. 2004)

Thassitou and Arvanitoyannis 2001). In the past, many types of treatment processes including physical, chemical and biological treatments have been recommended for wastewater treatment (Wilbey 2006), but they suffer from inherent shortcomings such as they are not fit for in situ application and may lead to formation of by-products, which further pose disposal problems. Furthermore, nearly all physical and chemical treatments are energy-intensive processes that cannot be afforded due to ecological fragility and sustainability. For organic wastewater disposal there is still a lot left to desire. It is because, in the past, treatment processes have ignored these wastes/effluents as potential feedstocks for useful microbial fermentations and product formation with notable exception of methane generation from municipal sludge. Most of the times, the waste treatment has been thought as a charge on society without a detailed examination of economical and ecological factors involved.

Whatever primary (physical separation) treatment process is used, microbially decomposed biomass is a major product in food processing industry wastes. It can be utilized as organic fertilizer and animal feed (Ward et al. 2008), as feedstock for anaerobic digestion to methane and production of bioethanol (Ward et al. 2006). An excellent example of commercial possibilities comes from potato starch processing industry. For example, the Symba process based on using mixed culture of *Candida utilis* and *Endomycopsis fibuligera* produced a high quality single cell protein (SCP) and simultaneously reduced biochemical oxygen demand (BOD) and chemical oxygen demand (COD) of potato processing wastes up to 90% (Skogman 1976).

5.2 Fermentation Methods

The main methods for bioaugmentation and biovalourization of food and beverage industry wastes are solid-state fermentation (SSF) processes, whereas effluents are mostly treated by submerged fermentation (SmF) processes.

5.2.1 Solid-State Fermentation

Solid-state processes involve growing the organisms on a substrate that is primarily insoluble such that there is essentially no free liquid. While research on industrial fermentation has chiefly focused on submerged processes, application of SSF is particularly relevant and applicable to the fermentative processing of agro-solid wastes with a view to production of SCP and other bioproducts such as bioethanol, organic acids, enzymes, etc.

Advantages of SSF include simplicity of the fermentation operation, the high capacity of the microbes to release concentrated enzymes during growth of substrate hydrolysis, the substrate penetrating capacity of fungal mycelium and the competitive advantage that fungi hold over bacteria in the low water-activity environment. SSF processes have been used for protein enrichment and improvement of digestibility of food industry solid wastes using substrates which include the following: filter-pressed cakes; for example, from vegetable recovery, grape marc, root crop wastes such as beet, cassava and sweet potato, and pulp of fruits and coconuts. SSF has likewise been used to upgrade feed quality of cellulosic wastes from tea manufacture, bagasse fractions from various processes, brewery-spent grains, corn stover, pollard and bran wastes, wheat and paddy straw, and hemp and cotton stalks. *Basidiomycetes* cultures of *Polyporus* strains cultivated on whole bagasse substrate in a solid-state system degrade the cellulose, hemicelluloses and lignin components of unwashed and untreated bagasse, while increasing the digestibility and protein content of the product and rendering it suitable for cattle feed (Ward et al. 2008).

5.2.2 Submerged Fermentation

Liquid wastes, typically generated in the aqueous washing or wastes liquid streams from food processing, may be used for SCP production, where the protein-rich biomass with or without wastes particulate matter originating from the plant material is recovered from the waste treatment process or for biomethane generation, with substantial reduction in BOD and COD. Examples of liquid wastes suitable for SCP enrichment and biomethane generation include: food processing effluents, processing filtrates and decantation liquid wastes, fruit and vegetable waste hydrolyzate, palm and olive oil mill effluents, canary effluents, pulp and peel extracts, waste coconut milk, fruit and sugarcane (vinasse) stillages, various pretreated hydrolyzate bagasse and other hydrolyzed cellulose wastes.

5.3 Types of Wastes Produced During Food and Beverage Processing

Food and beverage processing wastes include solid wastes in the form of pulp or effluents generated during processing of different agricultural produce. All processes generate waste by-products and effluents to a greater and lesser degree. The quality and quantity of wastes produced depend on the type of food and beverage being processed. There are large differences within sectors, and even from site to site therefore generalization is not only difficult, but could also be misleading. Fallow and Wheelock (1982) described wastes produced from industrial food processing into three kinds:

- Waste produced before storage on processing site
- Waste produced during storage on processing site
- Waste produced during processing

All these wastes whether solid or liquid, contain starch or sugars, which serve as primary carbon source during biological treatment processes. In case of starchy wastewater, success of the treatments depends on the capacity of the organism to catabolize starch, which has to be hydrolyzed to easily assimilable sugars, i.e., maltose and glucose. Three types of enzymes are involved in complete hydrolysis of starch to glucose viz., endo-amylase, exo-amylase and glucoamylase. Glucose produced after the action of amylolytic enzymes is assimilated. Therefore, the organism used should exhibit all the three enzyme components of amylolytic system or microbial consortia (mixed cultures) are to be used.

White rot fungi are versatile and robust organisms having enormous potential for oxidative bioremediation of a variety of toxic chemical pollutants such as phenols present in olive mill wastewater (OMWW) as well as brown colour toxic material produced from molasses-based wastewater (Thassitou and Arvanitoyannis 2001). They are capable of mineralizing a wide variety of toxic xenobiotics due to non-specific nature of their extracellular lignin mineralizing enzymes. In recent years, a lot of work has been done on the development and optimization of bioremediation processes using white rot fungi, with emphasis on the study of their enzyme systems involved in biodegradation of industrial pollutants (Asgher et al. 2008).

5.4 Bioaugmentation and Biovalourization of Solid Wastes

For economically feasible treatment of solid wastes, it has to be treated in situ. This can be achieved by SSF of these wastes and success depends on the ease of degradation of these wastes. A number of procedures using either peel or pulp from fruit, vegetable or tuber wastes have been used as fermentation substrate for amylolytic organisms to generate protein-enriched biomass as well as other value-added products (Ray et al. 2008a, b). The protein-enriched biomass can be used as animal feed. The earlier processes that used solid starchy wastes were operated in two steps mostly by using mixed cultures or extracted enzymes for initial hydrolysis of starch and subsequently fermentation of hydrolysate to value added products. Moreton (1978) described a process, which used enzymatically hydrolyzed potato wastes to grow *C. utilis* and it was proved to be an excellent medium to produce protein-enriched feed.

Similar processes have been evolved for utilizing solid wastes from other starchy crops like cassava and sweet potato. Cassava is one among the six main agricultural products (wheat, rice, corn, potato, oat meal and cassava) generated across the globe with an estimated production of 160 million tons per year worldwide (FAO 2005). Ofuya and Nwajiuba (1990) described a method by which cassava peel was readily degraded and utilized by a strain of *Rhizopus* growing in SSF. Maximal growth occurred at 45°C and was proportional to the degree of hydrolysis of the peel. The yield of dry mycelia biomass from the reducing sugars of the peel was 51%. After 72 h of fermentation, the peel contained 76% moisture, 6% cellulose, 7% hemi cellulose and 0.4% ash. The protein content increased from 5.6 to 16% thus bioconverting cassava into fungal biomass. SSF of cassava peels using a strain of *Rhizopus* spp. caused a drastic reduction (c.95%) in the HCN level of the peel and about 42% reduction in the soluble tannin content (Tweyongyere and Katongole 2002). In vitro digestibility of the fermented peel was 70.5% and the principal sugars of the fermented peel were xylose, mannose and galactose (Ofuya and Obilor 1994). In another study, mash prepared from cassava peels was fermented with *Saccharomyces cerevisiae* or *Candida tropicalis*. Temperature at 30°C, pH of 5.5 and moisture content at 30% were found to be the optimum for crude protein formation by these organisms (Anita and Mbongo 1994).

Yeast and yeast-like organisms were chosen for the SCP fermentation of cassava root, starch, cassava bagasse as well as waste water (Oliveira and Reis 2001). A mixed culture of *C. utilis* and *E. fibuligera* efficiently and rapidly utilized both starch and free sugars present in cassava starch factory effluents. After 28 h SmF, the protein content of the biomass was 22% (w/v) (Manilal et al. 1991).

Cassava and potato starch residues were used as substrate to produce the edible mushroom *Volvariella volvacea*. The best results were obtained in a medium containing 4.8% (w/v) declassified potato flour and 1.2% (w/v) cassava bagasse (Tonial et al. 2000).

Protein enrichment of sweet potato residue with amylolytic yeasts has also been attempted using SSF (Yang 1988). Sweet potato residues fermented with the yeast strain, *Saccharomyces* spp. IFO1426 could produce an animal feed product

Table 5.1 Microbes and substrates used in single-cell protein production studies

Microbes	Specific organisms	Substrates
Fungi	*Aspergillus niger*	Lemon pulp, banana waste, orange peel, cassava bagasse, corn cobs, sugar cane bagasse, rice straw
	Aspergillus terreus	Sugar cane bagasse, stillage
	Chaetomium cellulolyticum	Cellulosic wastes
	Fusarium graminearum	Starch hydrolyzate, vinasse, cereal fractions, cellulose hydrolyzate
	Geotrichum candidum	Orange peel, cassava waste water
	Sporotrichum pulverulantum	Bagasse, stalks, straws
	Trichoderma reesei	Beet pulp, cellulose, straws, bagasse, stillage
	Trichoderma viride	Cane sugar vinasse, orange peel, palm oil effluent
	Trichoderma aureoviride	Beet pulp
	Penicillium cyclopium	Whey
	Penicillium roqueforti	Citrus fruit peel, lemon pulp
	Pleurotus ostreatus	Cellulosic waste
	Aspergillus niger, *Rhizopus stolonifer*, *Neurospora sitophila*	Sweet potato bagasse
Yeasts	*Candida krusei*	Sorghum hydrolyzate, tea concentrate
	Candida tropicalis	Fruit waste, plant liquid waste, salad oil manufacturing wastewater
	Candida utilis	Banana waste, apple waste, pineapple cannery waste, cellulose and hemicellulose hydrolyzate, salad oil manufacturing wastewater, defatted rice polishing
	Saccharomyces cerevisiae	Stillage, molasses, sweet orange waste
	Saccharomyces fibuliger	Apple waste
	Saccharomyce lipolytica	Distillery waste
	Saccharomyces uvarum	Banana skin, vegetable plant extract, beet molasses
	Torula utilis	Plant origin liquid waste, olive oil wastewater
	Kluyveromyces fragilis	Whey
	Pichia pinus	Mango waste
	Rhodocyclus gelatinosus	Cassava starch
	Candida tropicalis, *Schwaaniomyces occidentalis*, *Torulopsis wickerham*, *Endomycopsis fibuligera*, and *Saccharomyces* spp.	Cassava waste water
	Saccharomyces cerevisiae	Sweet potato waste
Bacteria	*Bacillus subtilis*	Potato waste
	Brevibacterium flavum	Cane sugar vinasse
	Cellulomonas spp.	Agricultural wastes, potato waste, sugarcane bagasse, cotton debris
	Rhodopseudomonas gelatinous	Wheat bran
	Bacillus subtilis	Fruit waste
	Corynebacterium acetoacidophilum	Cane sugar vinasse

Fig. 5.2 (**a**) Cassava bagasse and (**b**) *Manipueira*

containing 16–21% protein. Yeasts such as *C. utilis*, *E. fibuligera*, *Pichia burtonii* and *Saccharomyces* spp. are used for SCP production (Yang 1988; Yang et al. 1993). Sweet potato distillery wastes are enriched with yeast protein and the feed is utilized for red carp (*Cyprinus carpiol*; Mokolensang et al. 2003).

In all these processes, production of SCP or protein-enriched biomass from solid starchy wastes proceeds in two steps, i.e., liquefaction and saccharification of starch (polysaccharide) to sugar (monosaccharide) and then utilization of sugar for production of biomass (Ward et al. 2008). Mixed cultures and enzymes have been used to perform these processes. In a further improvement of the process, Sukara and Doelle (1989) described a one-step process for the production of SCP and amyloglucosidase (which is a commercially important enzyme) using a newly isolated *Rhizopus* spp. This fungus was found to convert ground cassava tubers into SCP without pre-treatment due to its high amyloglucosidase production potential. Various microorganisms and agro-food wastes utilized in SCP production studies are given in Table 5.1 (Ward et al. 2008).

Table 5.2 Bioprocesses involving cassava bagasse[a]

Microorganism	Process	Application
Aspergillus niger LPB 21	SSF	Citric acid
A. niger NRRL 2001	SSF	Citric acid
A. niger CFTRI 30	SSF	Citric acid
Aureobasidium pullulans	SmF	Pullulan
Bacillus subtilis	SSF	α-amylase, pectinase
B. brevis	SSF	α-amylase
Candida lipolytica	SmF	Citric acid
Ceratocystis fumbriata	SSF	Aroma compounds
C. fumbriata	SSF	Aroma compounds
Kluyveromyces marxianus	SSF	Aroma compounds
Lactobacillus plantarum	SSF	Lactic acid
Lentinula edodes	SSF	Mushroom
Pleurotus sajor-caju	SSF	Mushroom
Rhizopus spp.	SSF	Biotransformation
R. arrahizus	SmF	Fumaric acid
R. ciricians	SmF	Fumaric acid
R. delemer	SmF	Fumaric acid
R. Formosa	SmF	Fumaric acid
R. oligosporus	SmF	Fumaric acid
R. oryzae	SmF	Fumaric acid
R.oryzae	SSF	Aroma compounds
Saccharomyces cerevisiae	SSF, SmF	Bio-ethanol

[a]*SSF*: solid-state fermentation, *SmF* submerged fermentation.
Source: Pandey et al. (2000), Ray and Moorthy (2007), Ray et al. (2008b), Ray and Kar (2009), Swain and Ray (2007), Swain et al. (2009)

Besides protein enrichment, the solid wastes of the starchy tubers like potato pulp and cassava bagasse (Fig. 5.2a) have been subjected to SSF with fungi, yeasts and bacteria for production of value added products like enzymes, lactic acid (Ray et al. 2008a) and bioplastics (Ghofar et al. 2005). Jyothi et al. (2005) have optimized a process for glutamic acid production from cassava bagasse by using *Brevibacterium divaricatum*. Ray and his co-workers have extensively studied the bioprocessing of cassava bagasse in to value-added products such as enzymes (Ray and Kar 2009; Swain et al. 2009), bioethanol, organic acid (Ray et al. 2008b) and microbial polysaccharide, pullulan (Ray and Moorthy 2007). The list of value-added products obtained by bioprocessing of cassava bagasse is given in Table 5.2.

5.5 Bioaugmentation and Biovalourization of Food Industry Effluents

5.5.1 *Starch and Flour*

Processing of different starchy crops for extraction of starch and processed food production entails the use of large volumes of water for unit operations like washing and peeling. Food industry effluents generated from such operations are characterized by high BOD, COD and high total suspended solids.

Washing of potatoes produces effluent rich in starch, soluble protein and sugars. This wastewater has a high concentration of suspended solids (on an average of 2.5% of the original potato solids) and a high BOD (Jin et al. 1998). Earlier procedures for treating these effluents mainly aimed at reducing the BOD values. The recent procedures utilize nutrients dissolved in these food industry effluents for production of useful by-products such as SCP and commercial enzymes. Symba process was one of the earliest procedures described in 1960 for treatment of effluents and getting useful SCP as by-products. Lemmel et al. (1979) optimized continuous production of *C. utilis* and *Saccharomycopsis fibuligera* on potato processing food industry effluents. The two cultures were propagated as a source of SCP in continuous, mixed, aerobic, single-stage cultivation on potato processing food industry effluents. This process was not as efficient as the *C. utilis* predominated the mixed cultures and amylase production by *S. fibuligera* appeared to be very less, thus limiting the efficiency of whole process. Other bacterial cultures were also tried for mixed culture growth on potato processing waste effluents. A batch process using mixed cultures of *Cellulomonas* spp. and *Bacillus* grown on waste from a factory producing potato crisps was developed. The process was operated at pH 7.0 and temperature 37°C at a dissolved oxygen saturation of 20%. The maximum percentage of protein (33.3%) in the biomass with 69.3% decrease in COD in 72-h incubation was reached by this biological treatment. It was also noticed that increased percentage of protein in dry matter coincides with reduction of starch in effluents (Rubio and Molina 1989). In a recent study, using concentrated media (25% solids) made from potato starch pre-hydrolyzed with malt flour and batch-fermented for 20 h at 26°C under aerobic conditions, *C. utilis* ATCC 9256 was the most efficient protein forming strain (Gelinas and Barrette 2007). Scaled-up at the 100 L level, the aerobic batch process was improved under fed-batch conditions with molasses supplementation. After drying, fermented starch contained 11–12% protein, including 7–8% yeast protein.

Malladi and Ingham (1993) developed thermophilic aerobic treatment of potato processing wastewater using indigenous bacteria which were later identified as different strains of *Bacillus* like *B. stearothermophilus*, *B. brevis*, *B. licheniformis*, *B. coagulans*, *B. acidocaldarus* and *Lactobacillus* spp. This thermophilic aerobic digestion had obvious advantages of being a faster method and effectively decreased BOD, total soluble solids and starch content of the wastewater in 96 h.

Amylolytic fungal strains have been tried for treating starch processing food industry effluents. This process holds commercial value by simultaneous production of fungal protein and glucoamylases which hold commercial value. In this simple, low cost, single-step process developed by Jin et al. (1999), the selected fungus *Rhizopus oligosporus* DAR2710 converted more than 95% of starch present in the effluent thereby generating 4.5–5.2 g of dry fungal biomass from 1l of starch processing waste water within 14 h at 45°C. The fungal biomass had 46% protein and was safe for human and animal consumption. In addition to fungal protein and glucoamylase, the batch process led to 95% reduction in COD. Many other promising strains of *Aspergillus* spp., i.e., *A. oryzae*, *Aspergillus niger* and *A. terreus* have been employed for reclamation of starchy effluents (Jin et al. 1998, 1999; Mishra et al. 2003). In a process described by Mishra et al. (2004), mixed culture of *A. niger* ITCC 2012 and *A. foetidus* MTCC 508 reduced about 90% of initial COD within 60 h of fermentation of effluents from the potato chips industry.

Cassava wastewater, "*manipueira*", (Fig. 5.2b) is a carbohydrate rich residue generated in large amounts during the production of cassava flour and starch. The production of 1 ton of cassava flour or starch generates 300 L of cassava waste; consequently, the treatment and disposal of cassava wastewater is a major concern for the cassava flour and starch industry (Ray et al. 2008b). Biological treatment of cassava wastewater has been investigated. For example, *Aspergillus oryzae* was used for the treatment of the cassava starch processing wastewater in a laboratory-based bioreactor study (Tung et al. 2004). In the typical pH range (pH 4–5) of the cassava wastewater, the formation of fungal biomass (up to 0.8 g/g COD) was achieved with lowering of BOD and COD. Use of cassava wastewater as feedstock in biotechnological process is a viable alternative which can contribute towards an increase in economic value of the residues. In different processes described by different workers, cassava wastewater has been used to produce surfactin by *Bacillus subtilis* (Nitschke and Pastore 2006), volatile compounds by *Geotrichum fragrans* (Damasceno et al. 2003), microbial polysaccharides such as pullulan by *Aureobasidium pullulans* (Ray and Moorthy 2007) and xanthan gum by *Xanthomonas campestris* (Selbmann et al. 2002).

5.5.2 Fruits and Vegetables

Market-based solid waste of vegetable origin is considered to be one of the important wastes which has the potential to generate energy due to its higher organic composition and easily biodegradable nature. Generation of H_2 from vegetable and fruit processing wastewater could substitute for non-renewable fossil fuels from energy security point of view (Sung 2004).

Vegetable residues can be processed by fermentation with lactic acid bacteria leading to a suitable transformation of low molecular materials like sugars into lactic acid. After lactic acid fermentation of carrot and grape, pomace the end product is rich in crude fibre, shows an acidic pH and can be used as a bread improver and for crude fibre enrichment of bakery goods (Mohan et al. 2009).

The wastes from fruit and vegetable processing industries generally contain large amounts of solid suspensions and a high BOD, COD, dissolved oxygen and total solids (Stabnikova et al. 2005). Indicative values for BOD, COD and suspended solids for the processing of some food wastes are summarized.

Fruit processing pomace wastes are commonly used as animal feed or as fertilizer developed through bioaugmented SSF process. Apple pomace is the main by-product resulted from pressing apples for juice or cider and it accounts for 25–35% of the mass of apple (Gullón et al. 2007). Dried apple pomace is considered as a potential food ingredient having high dietary fibre content (Carson et al. 1994; Sudha et al. 2007). Korkie et al. (2002) reported hydrolysis of complex polysaccharides present in grape pomace and its utilization for ethanol production using a potential yeast strain, *Pichia rhodanensis*. Cranberry pomace is a primary by-product of the traditional cranberry juice processing industry. Bioconversion of cranberry pomace can be achieved through its SSF by industrially beneficial fungi such as *Trichoderma*, *Penicillium* and *Rhizopus* (Zheng and Shetty 1998). These authors demonstrated that soil application of cranberry pomace waste augmented with *Trichoderma harzianum* inoculant can be used both for pest control as well as for enhancing pesticide degradation. In the case of polymeric dye pollution, a novel *Penicillium* spp. inoculant could be used.

Vegetable processing wastes have been biovalourized for protein enrichment by yeasts. For example, Chinese cabbage juice (Choi et al. 2002), waste brine generated from *kimchi* production (Choi and Park 1999), deproteinized leaf juices (Chanda and Chakrabatri 1996) and corn silage juice (Hang et al. 2003) can serve as nutrient source for yeast growth. Stabnikova et al. (2005) studied yeast cultivation using water extracts of cabbage, watermelon, a mixture of residual biomass of green salads and tropical fruits. These extracts contained from 1,420 to 8,900 mg/l of dissolved organic matter, and from 600 to 1,800 mg/l of nitrogen. pH of the extracts was in the range from 4.1 to 6.4. Biomass concentration of yeast, *S. cerevisiae* grown at 30°C for 96 h in the sterilized extracts without any nutrient supplements was from 6.4 to 8.2 g/l; content of protein was from 40 to 45% of dry biomass.

5.5.3 Coffee and Tea

Coffee pulp is the agro-industrial residue produced during the pulping of coffee berries to obtain coffee beans. In coffee-growing regions, coffee pulp is considered to be one of the most abundant agricultural wastes, as well as one of the hardest to handle. There have been many reports describing the composition, conservation, upgrading and utilization of coffee pulp. For example, Penaloza et al. (1985) studied the nutritive improvement of coffee pulp by using *Aspergillus niger* under SSF conditions. Likewise, Orozco et al. (2008) studied caffeine reduction in coffee pulp by *Streptomyces* strains. Brand et al. (2000) studied microbial detoxification of coffee husk by filamentous fungi such as *Rhizopus*, *Phanerochaete* and *Aspergillus* spp. In SSF, all the strains showed degradation of caffeine and tannins maximum up to 87 and 65%, respectively, at pH 6.0 and moisture 60% in 6 days. Coffee husk was also used as solid substrate for growth and flavor production by *Ceratocystis fimbriata* (Soares et al. 2000).

Tea wastes were also used as solid substrate for production of enzyme (glucoamylase) (Selvakumar et al. 1998) and gluconic acid (Sharma et al. 2008).

5.5.4 Olive Oil Mill

Olive mill wastes are produced by olive oil producing industries in Mediterranean countries. According to Food and Agricultural Organization (FAO 2006) statistics, 2.7 million tons of olive oil is produced annually worldwide, generating approximately 30 million tons of wastewater (Azbar et al. 2004) and substantial amount of olive pulp as solid waste. From the environmental point of view, OMWW is considered the most critical waste emitted by olive mills because of its high organic load and its chemical composition. The organic fraction contains large amounts of phenolic compounds which tend to polymerize into high-molecular-weight polymers that are difficult to degrade and have been found to be phytotoxic (Ayed et al. 2005; Crognale et al. 2006). Organic fraction also includes sugars, tannins, polyphenols, polyalcohols, pectins and lipids. Some of these substances (mainly sugars and polyalcohols) can be used as carbon and energy sources for microbial growth. Conventional biological wastewater treatments are ineffective for oil mill effluents treatment since phenolics possess antimicrobial activity (Ahmadi et al. 2006). Most of the studies have been focused on bioremediation as a means of reducing the polluting effect of OMWW and its biotransformation into valuable products.

In general, aerobic bacteria such as *Bacillus pumilis* (Ramos-Cormenzana et al. 1996) and *Azotobacter vinelandii* (Ehaliotis et al. 1999) appeared to be very effective in reducing the content of low-molecular-weight phenolics as well as phytotoxicity of OMWW. Many white rot fungi are more effective in degradation of low- and high-molecular-weight phenolics due to their inherent characteristic to produce lignolytic enzymes such as lignin peroxidases, manganese peroxidases and laccases (Giannoutsou et al. 2004; Sampedro et al. 2007). There are reports (Sanjust et al. 1991; Kalmis and Sargin 2004) on the cultivation of *Pleurotus* spp., *Pleurotus sajor-caju* and *P. cornucopiae* var. *citrinopileatus* using wheat straw moistened with mixtures containing 25 and 50% OMWW. The removal of total phenols from OMWW relative to the total organic load consumed indicates the highest capability for free as well as immobilized *Phanerochaete chrysosporium* (Garcia et al. 2004). OMWW can be detoxified through removal of organic matter, decreasing COD/BOD ratio by *P. chrysosporium* or *Trametes versicolor* in the presence of complex microbial consortia in combined aerobic/anaerobic systems for its reuse and biogas production on industrial scale (Dhouib et al. 2006). *Pycnoporus coccineus*, *P. sajor-caju*, *Coriolopsis polyzona* and *Lentinus tigrinus* are also very active in colour and COD removal of olive mill effluents at 50 and 75 g/l COD (Jaouani et al. 2003). At 100 g/l COD, only *P. coccineus* and *P. sajor-caju* are effective. *Panus trigrinus* CBS577.79 gives better COD reduction (60.9%), dephenolization (97.2%) and decolourization (75%) of olive mill effluents in bubbled column bioreactor as compared to stirred tank reactors due to possible occurrence of shear stress (D'Annibale et al. 2006). A better de-colourization of OMWW by *C. polyzona* has been reported (Jaouani et al. 2006) under lignin peroxidase induction conditions (5 mM veratryl alcohol addition) than when lignin peroxidase was repressed (100 mM Mn^{2+} addition). High levels of laccase have a detrimental effect on

OMWW decolourization concomitant to the formation of soluble polymeric aromatic compounds. However, high laccase activity produced by *Pleurotus* spp. in the growth medium reflected a close relationship between the amount of laccase produced and decrease in phenol content (Tsioulpas et al. 2002).

Different microbial enzymes, e.g., lipases, laccases, peroxidases and pectinases produced during fungal treatment of OMWW provide an opportunity for biotechnological valourization of residues (Crognale et al. 2006; D'Annibale et al. 2006). Some yeasts, e.g., *Geotrichum candidum* (Asses et al. 2003), *Saccharomyces* spp. (Giannoutsou et al. 2004) are also reported to reduce COD and phenolic content of OMWs. Some of these aerobic edible fungi, e.g., *Pleurotus* when employed at large scale in OMWW bioremediation, can be harvested to obtain fungal biomass after detoxification of phenolics by lignolytic enzymes (Laconi et al. 2007). Supplementation of wheat straw with 25% OMWW may be employed for commercial production of oyster mushroom (*Pleurotus ostreatus*) (Kalmis and Sargin 2004).

The production of microbial biopolymers, e.g., xanthan gum by *Xanthomonas compestris* (López et al. 2001) and metal binding microbial polysaccharide by *Paenibacillus jamilae* (Morillo et al. 2007, 2009) have also been proposed by way of microbial valourization process to produce biodegradable plastics. *Azotobacter chroococcum* strain H23, when grown on diluted OMWW provides a yield of 6.2 g polyhydroxyalkanoates per litre of culture medium (Pozo et al. 2002). Additionally, the high content of organic matter makes OMWs an interesting alternative resource to produce biofuel (Li et al. 2007).

5.5.5 *Palm Oil Mill*

Palm oil milling process can be categorized into a dry and a wet (standard) process. The wet process of palm oil milling is the most common and typical way of extracting palm oil (Wu et al. 2009). It is estimated that for each ton of crude palm oil that is produced, 5–7.5 ton of water are required, and more than 50% of this water ends up as palm oil mill effluent (POME) (Ahmad et al. 2003). Raw POME is a colloidal suspension containing 95–96% water, 0.6–0.7% oil and 4–5% total solids. Included in the total solids are 2–4% suspended solids, which are mainly constituted of debris from palm fruit mesocarp generated from three main sources, i.e. sterilizer condensate, separator sludge and hydrocyclone wastewater (Borja and Banks 1994; Khalid and Wan Mustafa 1992; Ma 2000). If the untreated effluent is discharged into water courses, it is certain to cause considerable environmental problems (Davis and Reilly 1980) due to its high BOD (25,000 mg/l), COD (53,630 mg/l), oil and grease (8,370 mg/l), total solids (43,635 mg/l) and suspended solids (19,020 mg/l) (Ma 1995).

The high compositions and concentrations of carbohydrate, protein, nitrogenous compounds, lipids and minerals in POME (Habib et al. 1997) render it possible to reuse the effluent for biotechnological means (Table 5.3) (Wu 2009). Preliminary investigations on enzymatically hydrolyzed substrates from POME have indeed

Table 5.3 Various products or metabolites produced in bioprocesses during the reuse of palm oil mill effluent (POME) or its derivatives as substrates (Wu 2009; modified)

Product	Microorganism	Fermentation medium based on POME
Penicillin	*Penicillium chrysogenum* FR2284	50% (v/v) concentrated POME $+KH_2PO4+ (NH_4)_2SO_4$
Bioinsecticide	*Bacillus thuringiensis* H-14	Raw POME
Acetone-Butanol-Ethanol (ABE)	*Clostridium acetobutylicum* NCIMB 13357	Particulate fraction of raw POME
ABE	*Clostridium acetobutylicum* NCIMB 13357	Particulate fraction of raw POME
Polyhydroxyalkanoates (PHA)	*Ralstonia eutropha* ATCC 17699	Concentrated organic acids from the anaerobically digested POME (100 g/l of total acids with acetic: propionic=3:1)
PHA	Mixed cultures	High concentration of POME with 490 COD/N ratio (g COD/g N) and 160 COD/P ratio (g COD/g P)
Organic acids	Mixed cultures	POME+palm oil sludge in the ratio of 1:1
Citric acid	*Aspergillus niger* (A103)	2% (w/w) POME+4% (w/w) wheat flour +4% (w/w) glucose with no added ammonium nitrate (optimized medium)
Itaconic acid	*Aspergillus terreus* IMI 282743	Retentate of POME
Cellulase (CMCase)	Mixed culture (1:1) of *Aspergillus niger* and *Trichoderma harzianum*	50% (v/v) raw POME
Cellulase (CMCase)	*Myceliophthora thermophila*	50% (v/v) raw POME
Cellulase (FPase)	*Penicillium* (P1-EFB)	1% (w/w) POME sludge
Lignin peroxidase	*Penicillium* (P1-EFB)	1% (w/w) POME sludge
Lipase	*Clostridium aurantibutyricum* ATCC 17777	Model medium for raw POME
Xylanase	Isolate SO1	10% (v/v) supernatant of POME+ another nine different types of supporting
Protease	*Aspergillus terreus* IMI 282743	75% (v/v) retentate of POME
Hydrogen	Thermophilic microflora	Raw POME

demonstrated the possibility of such substrates supporting the growth of *Candida tropicalis* (Wang et al. 1981). On the other hand, Barker and Worgan (1981) noted that unhydrolyzed POME could support good growth of *A. oryzae* in the presence of an added inorganic nitrogen source. Their results also revealed that celluloses, polyphenols and nitrogenous compounds were the least biodegradable of the substrate constituents. When palm oil effluent was treated using the fungus *Trichoderma viride*, COD of the palm oil effluent was reduced by 95% after 10–14 days of fermentation and the resulting fungal biomass was highly enriched with protein (Karim and Kamil 1989).

5.5.6 *Sugar Refining*

Fermentation processes using sugarcane molasses yield large volumes of dark brown and highly toxic molasses-based effluents (MBE) that contains considerable amounts of organic compounds. Although most of the organic matter of MBE is removed by means of conventional biodegradation treatments, the removal of dark colour due to the presence of melanoidin-type high-molecular-weight compounds is only marginal (Vahabzadeh et al. 2004). White rot fungi are, however, capable of catalyzing degradation of numerous recalcitrant organic compounds often present in MBE (Fu and Viraraghavan 2001). The colour removal ability of *P. chrysosporium* is correlated to the activity of ligninolytic enzymes lignin peroxidase (LiP) and manganese peroxidase (MnP). Increased expression of laccase genes in *Trametes* spp. I-62 (*lcc1* and *lcc2*) upon exposure to MBE accompanied by enhanced colour removal, suggested the involvement of laccase in the melanoidins metabolism (D'Souza et al. 2006). Molasses spent wash (MSW) or digested spent wash (DSW) or alcohol distillery wastewater (WAD) is another wastewater from molasses-based alcohol distilleries (Chopra et al. 2004). The brown colour of MSW is due to the presence of melanoidin pigments, which are highly recalcitrant to biodegradation. These pigments, polycyclic aromatic hydrocarbons (PAHs) like benzo(*a*) pyrenes and phenols are the causes of its toxicity (Raghukumar et al. 2004). Several species of white rot fungi have been reported to remove about 70–80% of the colour present in MSW-based effluents (Raghukumar 2002; D'Souza et al. 2006). Immobilized mycelia of *P. coccineus* on polyurethane foam removed nearly twofold higher WAD colour and threefold higher total phenol content than did free mycelia (Chairattanamanokorn et al. 2005). Decolourization of MSW using free and immobilized mycelia of *Flavodon flavus* is accompanied by simultaneous detoxification and decrease in PAH contents of the MSW possibly via the action of glucose oxidase, accompanied by the production of H_2O_2 that acts as a bleaching agent in the process. Decolourization and COD reduction (52%) of DSW by *Coriolus versicolor* is dependent on the carbon source and addition of organic/inorganic nitrogen has no enhancing effect on decolourization and COD reduction (Chopra et al. 2004).

5.5.7 *Fermented Beverage*

The fermented beverage industry is divided into three main categories: brewing, distilling and wine manufacture. Each of these industries produces liquid waste with many common characteristics, such as high BODs and CODs, but differs in the concentration of the organic compounds that determine the biological treatment that will be selected. The difficulty in dealing with fermentation wastewaters is in the flows and loads of the waste. Since the fermentation industry's wastewater contains high concentrations of tannins, phenols and organic acid, anaerobic treatment results in higher performance (Thassitou and Arvanitoyannis 2001). Mayer (1991) attempted to compare aerobic with anaerobic treatment of the wastewaters in a German brewery. Anaerobic treatment achieved 91% COD reduction at loading rates up to 20 g COD/L/day, whereas the aerobic treatment resulted in a 76% reduction at a loading rate of 69 g COD/L/day. In order to optimize the conditions of anaerobic treatment, Suzuki et al. (1997) conducted several experiments for the optimization of acidity and temperature of highly concentrated brewery wastewater by applying the upflow anaerobic sludge blanket. These experiments showed that the optimal conditions for the particular treatment were 40°C and pH 5–6. The amount and load of distillery waste varies according to the raw materials used. For example, the biological load for molasses is three times that of raisins (Stroo 1989). Fitzgibbon et al. (1995) studied the biological treatment of spent wash from molasses distilleries. Analysis of raw spent wash showed it to be a recalcitrant waste, with a high COD of 85,170 mg/L and containing inhibitory phenolic compounds such as gallic and vanillic acid. The fungi *G. candidum*, *C. versicolor*, *P. chrysosporium* and *Mycelia sterilia* were screened for their ability to decolourize spent wash and to reduce the COD level. A 10-day pre-treatment at 300°C resulted in reducing the COD by 53.17% and total phenols by 47.82%, enabling other remediating organisms to grow. *C. versicolor* immobilized in a packed-bed reactor reduced the COD of spent wash by a further 50.3%, giving an overall reduction in COD of 77% to 15,780 mg/l. Benito et al. (1997) conducted laboratory batch tests to examine the ability of the white rot fungus *Trametes vercicolor* to treat molasses-based distillery wastewater. Likewise, *Geotrichum* spp. when applied for dehydration of distillery waste water of *shochu*, an alcoholic drink prepared from sweet potato reduced COD and BOD level up to 90% (Yoshi et al. 2001).

In winery, the treatment methods are based on principles similar to the previous fermentation industries. Zhang et al. (2008) investigated the production of fungal biomass protein (FBP) in treatment of winery wastewater using microfungi. Three fungal strains, *T. viride* WEBL0702, *Aspergillus niger* WEBL0901 and *A. oryzae* EBL0401, were selected in terms of microbial capability for FBP production and COD reduction. *T. viride* appeared to be the best strain for FBP production due to high productivity and less nitrogen requirement. However, one of the main problems in winery waste treatment is the presence of vinasse, which needs to be treated biologically for 4–8 days in order to reduce by 90% the COD.

5.6 Conclusions

For disposal of food and beverage industry effluents, focus has been changed from mere pollution control to make the remediation processes economically feasible. Investigation of several sectors of the food industry (starch and flour, fruits and vegetables, olive and palm oil, fermentation, etc), confirmed the usefulness and potential of bioaugmentation of food waste. However, these processes have to be applied on large scale. With an objective to generate value added by-products such as enzymes or protein-rich biomass, which will defray the costs of the bioaugmentation processes. Many researchers have employed microorganisms for production of enzymes from food industry effluents. There are reports of microbial plastics production by starchy wastes and OMWW fermentation. There is a vast potential for development of a suitable bioremediation and biovalourization technology for safe disposal and recycling of food industry effluents.

References

Ahmad AL, Ismail S, Bhatia S (2003) Water recycling from palm oil mill effluent (POME) using membrane technology. Desalination 157:87–95

Ahmadi M, Vahabzadeh F, Bonakdarpour B, Mehranian M, Mofarrah E (2006) Phenolic removal in olive oil mill wastewater using luffa-immobilized *Phanerochaete chrysosporium*. World J Microbiol Biotechnol 22:119–127

Anita SP, Mbongo PM (1994) Utilization of cassava peels as substrate for crude protein formation. Plant Food Hum Nutr 46:345–351

Asgher M, Bhatti HN, Ashraf M, Legge RL (2008) Recent developments in biodegradation of industrial pollutants by white rot fungi and their enzyme system. Biodegradation 19: 771–783

Asses N, Marouani L, Hamdi M (2003) Scale down and optimization of olive mill wastewaters decolorization by *Geotrichum candidum*. Bioprocess Eng 22:503–507

Ayed L, Assas N, Sayadi S, Hamdi M (2005) Involvement of lignin peroxidase in the decolourization of black olive mill wastewaters by *Geotrichum candidum*. Lett Appl Microbiol 40:7–11

Azbar N, Bayram A, Filibeli A, Muezzinoglu A, Sengul F, Ozer A (2004) A review of waste management options in olive oil production. Crit Rev Environ Sci Technol 34:209–247

Barker TW, Worgan JT (1981) The utilization of palm oil processing effluents as substrates for microbial protein production by the fungus *Aspergillus oryzae*. Eur J Appl Microbiol Biotechnol 11:234–240

Benito GG, Miranda MP, de los Santos DR (1997) Decolourization of wastewater from an alcoholic fermentation process with *Trametes Vercicolor*. Biores Technol 61:1–33

Borja R, Banks CJ (1994) Anaerobic digestion of palm oil mill effluent using an up-flow anaerobic sludge blanket (UASB) reactor. Biomass Bioenergy 6:381–389

Brand D, Pandey A, Roussos S, Soccol CR (2000) Biological detoxification of coffee husk by filamentous fungi using a solid state fermentation system. Enz Microb Technol 27:127–133

Carson KJ, Collins JL, Penfield MP (1994) Unrefined, dried apple pomace as a potential food ingredient. J Food Sci 59:1213–1215

Chairattanamanokorn P, Imai T, Kondo R, Sekine M, Higuchi T, Ukita M (2005) Decolorization of alcohol distillery wastewater by thermotolerant white rot fungi. Appl Biochem Microbiol 41:662–667

Chanda S, Chakrabatri S (1996) Plant origin liquid waste: a resource for single cell protein production by yeast. Biores Technol 57:51–54

Choi MH, Park YH (1999) Growth of *Pichia guilliermondii* A9, an osmotolerant yeast, in waste brine generated from kimchi production. Biores Technol 70:231–236

Choi MH, Ji GE, Koh KH, Ryu YW, Jo DH, Park YH (2002) Use of waste Chinese cabbage as a substrate for yeast biomass production. Biores Technol 83:251–253

Chopra P, Singh D, Verma V, Puniya AK (2004) Bioremediation of melanoidin containing digested spent wash from cane molasses distillery with white rot fungus *Coriolus versicolor*. Indian J Microbiol 44:197–200

Crognale S, D'Annibale A, Federici F, Fenice M, Quaratino D, Petruccioli M (2006) Olive oil mill wastewater valorisation by fungi. J Chem Technol Biotechnol 81:1547–1555

D'Annibale A, Quaratino D, Federici F, Fenice M (2006) Effect of agitation and aeration on the reduction of pollutant load of olive mill wastewater by the white-rot fungus Panus tigrinus. Biochem Eng J 29:243–249

D'Souza DT, Tiwari R, Sah AK, Raghukumar C (2006) Enhanced production of laccase by a marine fungus during treatment of colored effluents and synthetic dyes. Enz Microb Technol 38:504–511

Damasceno S, Cereda MP, Pastone GM, Oliveira JG (2003) Production of volatile compounds by *Geotrichum fragrans* using cassava waste water as substrate. Process Biochem 39: 411–414

Davis JB, Reilly PJA (1980) Palm oil mill effluent: a summary of treatment methods. Oleagineux 35:323–330

Dhouib A, Ellouz M, Aloui F, Sayadi S (2006) Effect of bioaugmentation of activated sludge on olive mill wastewater detoxification with white-rot fungi. Lett Appl Microbiol 42:405–411

Ehaliotis C, Papadopoulou K, Kotsou M, Mari I, Balis C (1999) Adaptation and population dynamics of *Azotobacter vinelandii* during aerobic biological treatment of olive-mill wastewater. FEMS Microbiol Ecol 30:301–311

Fallow SJ, Wheelock JV (1982) Byproducts from the U.K. food system. Conserv Recycl 5: 163–172

FAO (2005) Food agricultural organization statistical database. Food and Agriculture Organization of the United Nations, Rome. Available at: http://www.fao.org

FAO (Food and Agriculture Organisation) (2006) FAOSTAT database. http://faostat.fao.org/

FitzGibbon FJ, Nigam P, Singh D, Marchant R (1995) Biological treatment of distillery waste for pollution-remediation. J Basic Microbiol 35:293–301

Fu Y, Viraraghavan T (2001) Fungal decolourization of dye wastewater: a review. Biores Technol 79:251–262

Garcia IG, Pena PRJ, Venceslada JLB, Martin AM, Santos MAM, Gomez ER (2004) Removal of phenol compounds from olive mill wastewater using *Phanerochaete chrysosporium, Aspergillus niger, Aspergillus terreus* and *Geotrichum candidum*. Process Biochem 35:751–758

Gelinas P, Barrette J (2007) Protein enrichment of potato processing waste through yeast fermentation. Biores Technol 98:1138–1143

Ghofar A, Ogawa S, Kokugan T (2005) Production of L-lactic acid from fresh cassava roots slurried with Tofu liquid waste by *Streptococcus bovis*. J Biosc Bioeng 100:606–612

Giannoutsou EP, Meintanis C, Karagouni AD (2004) Identification of yeast strains isolated from a two-phase decanter system olive oil waste and investigation of their ability for its fermentation. Biores Technol 93:301–306

Gullón B, Falqué E, Alonso JL, Parajó JC (2007) Evaluation of apple pomace as a raw material for alternative applications in food industries. Food Technol Biotechnol 45:426–433

Habib MAB, Yusoff FM, Phang SM, Ang KJ, Mohamed S (1997) Nutritional values of chironomid larvae grown in palm oil mill effluent and algal culture. Aquaculture 158:95–105

Hang YD, Woodams EE, Hang LE (2003) Utilization of corn silage juice by *Klyuveromyces marxianus*. Biores Technol 86:305–307

Jaouani A, Sayadi S, Vanthournhout M, Penninckx MJ (2003) Potent fungi for decolorization of olive oil mill wastewater. Enz Microb Technol 33:802–809

Jaouani A, Tabka MG, Penninckx MJ (2006) Lignin modifying enzymes of *Coriolopsis polyzona* and their role in olive oil mill wastewaters decolourisation. Chemosphere 62:1421–1430

Jin B, van Leeuwen HJ, Patel B, Yu Q (1998) Utilization of starch processing food industry effluents for production of microbial biomass protein and fungal α-amylase by *Aspergillus oryzae*. Biores Technol 66:201–206

Jin B, van Leeuwen HJ, Patel B, Doelle HW, Yu Q (1999) Production of fungal protein and glucoamylase by *Rhizopus oligosporus* from starch processing food industry effluents. Process Biochem 34:59–65

Jyothi AN, Sasikiran K, Nambisan B, Balagopalan C (2005) Optimisation of glutamic acid production from cassava starch factory residues using *Brevibacterium divaricatum*. Process Biochem 40:3576–3579

Kalmis E, Sargin S (2004) Cultivation of two *Pleurotus* species on wheat straw substrates containing olive mill wastewater. Int Biodet Biodeg 53:43–47

Karim MIA, Kamil AQA (1989) Biological treatment of palm oil mill effluent using *Trichoderma viridae*. Biol Waste 27:143–152

Khalid AR, Wan Mustafa WA (1992) External benefits of environmental regulation: resource recovery and the utilization of effluents. Environmentalist 12:277–285

Korkie LJ, Janse BJH, Viljoen-Bloom M (2002) Utilizing grape pomace for ethanol production. S Afr J Enol Vitic 23:31–36

Laconi S, Molle G, Cabiddu A, Pompei R (2007) Bioremediation of olive oil mill wastewater and production of microbial biomass. Biodegradation 18:559–566

Lemmel SA, Heimsch RC, Edwards LL (1979) Optimizing the continuous production of *Candida utilis* and *Saccharomycopsis fibuligera* on potato processing food industry effluents. Appl Environ Microbiol 37:227–232

Li A, Antizar-Ladislao B, Khraisheh MAM (2007) Bioconversion of municipal solid waste to glucose for bio-ethanol production. Bioprocess Biosyst Eng 30:189–196

López MJ, Moreno J, Ramos-Cormenzana A (2001) *Xanthomonas campestris* strain selection for xanthan production from olive mill wastewaters. Water Res 35:828–1830

Ma AN (1995) A novel treatment for palm oil mill effluent. Palm Oil Res Instit Malays 29:201–212

Ma AN (2000) Environmental management for the palm oil industry. Palm Oil Dev 30:1–9

Malladi S, Ingham SC (1993) Thermophilic aerobic treatment of potato-processing food industry effluents. World J Microbiol Biotechnol 9:45–49

Manilal VB, Narayanan CS, Balagopalan C (1991) Cassava starch effluent treatment with concomitant SCP production. World J Microbiol Biotechnol 7:185–190

Mayer ES (1991) Waste treatment experiments at the Gabriel Sedlmayr Spaten. Franziskaner-Braeu K.-G.a.A. Brauwetl. (Ger.) 131:2346

Mishra BK, Arora A, Lata (2003) Use of amylolytic cultures of *Aspergillus* for treatment of food industry effluents from potato processing industry. Indian J Microbiol 43:57–59

Mishra BK, Arora A, Lata (2004) Optimization of a biological process for treating potato chips industry food industry effluents using a mixed culture of *Aspergillus foetidus* and *Aspergillus niger*. Biores Technol 94:9–12

Mohan VS, Mohanakrishna G, Kannaiah Goud R, Sarma PN (2009) Acidogenic fermentation of vegetable based market waste to harness biohydrogen with simultaneous stabilization. Biores Technol 100:3061–3068

Mokolensang JF, Yamasaki S, Onone Y (2003) Utilization of sweet potato distillery by products as feedstuff for red carp *Cyprinus carpiol*. J World Aquac Soc 34(4):512–517

Moreton RS (1978) Growth of *Candida utilis* on enzymatically hydrolyzed potato waste. J Appl Bacteriol 44:373–382

Morillo JA, Guerra Del Aguila V, Aguilera M, Ramos-Cormenzana A, Monteoliva-Sanchez M (2007) Production and characterization of exopolysaccharide produced by *Paenibacillus jamilae* grown on olive mill-wastewaters. World J Microbiol Biotechnol 23:1705–1710

Morillo JA, Antizar-Ladislao B, Monteoliva-Sanchez M, Ramos-Cormenzana A, Russell NJ (2009) Bioremediation and biovalorization of olive-mill wastes. Appl Microbiol Biotechnol 82:25–39

Nitschke M, Pastore GM (2006) Production and properties of a surfactant obtained from *Bacillus subtilis* grown on cassava waste water. Biores Technol 97:336–341

Ofuya CO, Nwajiuba CJ (1990) Microbial degradation and utilization of cassava peel. World J Microbiol Biotechnol 6:144–148

Ofuya CO, Obilor SN (1994) The effects of solid state fermentation on the toxic components of cassava peel. Process Biochem 29:25–28

Olguín EJ, Sánchez G, Mercado G (2004) Cleaner production and environmentally sound biotechnology for the prevention of upstream nutrient pollution in the Mexican coast of the Gulf of México. Ocean Coast Manage 47:61–70

Oliveira MA, Reis EM (2001) Biological treatment of waste water from cassava meal industry. Environ Res 85:177

Orozco AL, Perez MI, Guevara M, Rodriguez J, Hernadez M, Gonzalez-Ville FG, Polvillo O, Arios ME (2008) Biotechnological enhancement of coffee pulp residues by solid state fermentation with *Streptomyces*. Py-GC/MS analysis. J Analyt Appl Pyrol 81:247–252

Pandey A, Soccol CR, Nigam P, Vanete T, Soccol VT, Vandenberghe LPS, Mohan R (2000) Biotechnological potential of agro-industrial residues. II: cassava bagasse. Biores Technol 74:81–87

Penaloza W, Molina MR, Brenes RG, Bressani R (1985) Solid state fermentation: an alternative to improve the nutritive value of coffee pulp. Appl Environ Microbiol 49:388–393

Pozo C, Martínez-Toledo MV, Rodelas B, González-López J (2002) Effects of culture conditions on the production of polyhydroxyalkanoates by *Azotobacter chroococcum* H23 in media containing a high concentration of alpechin (wastewater from olive oil mills) as primary carbon source. Biotechnol 7:25–131

Raghukumar C (2002) Bioremediation of colored pollutants by terrestrial versus facultative marine fungi. In: Hyde KD (ed) *Fungi in marine environments*, Fungal Diversity Research Series 7. Fungal Diversity Press, Hong Kong, pp 317–344

Raghukumar C, Mohandass C, Kamat S, Shailaja MS (2004) Simultaneous detoxification and decolorization of molasses spent wash by the immobilized white-rot fungus *Flavodon flavus* isolated from a marine habitat. Enz Microb Technol 35:197–202

Ramos-Cormenzana A, Juárez-Jiménez B, Garcia-Pareja MP (1996) Antimicrobial activity of olive mill waste-waters (alpechin) and biotransformed olive oil mill wastewater. Int Biodet Biodeg 38:283–290

Ray RC, Kar S (2009) Statistical optimization of α-amylase production by *Bacillus brevis* MTCC 7521 in solid state fermentation using cassava bagasse. Biol Sec Cell Mol Biol. doi:10.2478/s11756-009-0160-1

Ray RC, Moorthy SN (2007) Exopolysaccharide production from cassava starch residue by *Aureobasidium pullulans* strain MTCC 1991. J Sci Ind Res 66:252–255

Ray RC, Ward OP (2006) Post-harvest microbial biotechnology of topical root and tuber crops. In: Ray RC, Ward OP (eds) Microbial biotechnology in horticulture, vol 1. Science Publishers, New Hampshire, USA, pp 345–396

Ray RC, Shetty K, Ward OP (2008a) Solid-state fermentation and value-added utilization of horticultural processing wastes. In: Ray RC, Ward OP (eds) Microbial biotechnology in horticulture, vol 3. Science Publishers, New Hampshire, USA, pp 231–272

Ray RC, Mohapatra S, Panda S, Kar S (2008b) Solid substrate fermentation of cassava fibrous residue for production of α- amylase, lactic acid and ethanol. J Environ Biol 29(1/2):111–115

Rubio MC, Molina OE (1989) Treatment of potato waste effluents with bacterial protein production. Biol Waste 29:221–228

Sampedro I, Marinari S, D'Annibale A, Grego S, Ocampo JA, García-Romera I (2007) Organic matter evolution and partial detoxification in two-phase olive mill waste colonized by white-rot fungi. Int Biodet Biodeg 60:116–125

Sanjust E, Pompei R, Resciggno A, Augusto R, Ballero M (1991) Olive milling wastewater as medium for growth of four *Pleurotus* species. Appl Biochem Biotechnol 31:223–235

Selbmann L, Crognale S, Petruccioli M (2002) Exopolysaccharide production from *Sclerotium glucanicum* NRRL 3006 and *Botryosphaeria rhodina* DABAC-P 82 on raw and hydrolyzed starchy materials. Lett Appl Microb 34:51–55

Selvakumar P, Ashakumary L, Pandey A (1998) Biosynthesis of glucoamylase from *Aspergillus niger* by solid state fermentation using tea waste as the basis of a solid substrate. Biores Technol 65:83–85

Sharma A, Vivekanand V, Singh RP (2008) Solid state fermentation for gluconic acid production from sugarcane molasses by *Aspergillus niger* ARNU-4 employing tea waste as the novel product. Biores Technol 99:3444–3450
Skogman H (1976) Production of *symba* yeast from potato wastes. In: Birch GC, Parker KJ, Worgan JT (eds) Food from waste. Applied Science, London, pp 167–179
Soares M, Christen P, Pandey A, Soccol CR (2000) Fruity flavor production by *Ceratocystis fimbriata* grown on coffee husk in solid state fermentation. Process Biochem 35:857–861
Stabnikova O, Wang JY, Ding HB, Tay JH (2005) Biotransformation of vegetable and fruit processing wastes into yeast biomass enriched with selenium. Biores Technol 96:747–751
Stroo HF (1989) Bioremediation of hydrocarbon-contaminated soil solids using liquid/solids contract reactors. In: Proceedings of the 10th national conference: superfund '89. HMRCI, Washington, DC, pp 331–337
Sudha ML, Baskaran V, Leelavathi K (2007) Apple pomace as a source of dietary fibre and polyfenols and its effect on the rheological characteristics and cake making. Food Chem 104:686–692
Sukara E, Doelle HW (1989) A one-step process for the production of single-cell protein and amyloglucosidase. Appl Microbiol Biotechnol 30:135–140
Sung S (2004) Final technical report biohydrogen production from renewable organic wastes. Iowa State University, Ames, USA
Suzuki H, Yoneyama Y, Tanaka T (1997) Acidification during anaerobic treatment of brewery wastewaters. Water Sci Technol 35:265
Swain MR, Ray RC (2007) Alpha-amylase production by *Bacillus subtilis* CM3 in solid state fermentation using cassava fibrous residue. J Basic Microbiol 47(5):417–425
Swain MR, Kar S, Ray RC (2009) Exo-polygalacturonase production by *Bacillus subtilis* CM5 in solid state fermentation using cassava bagassae. Braz J Microbiol 40:636–648
Thassitou PK, Arvanitoyannis IS (2001) Bioremediation: a novel approach to food waste management. Trends Food Sci Technol 12:185–196
Tonial TM, Pandey A, Chiarello MD, Soccol CR (2000) Cultivation of *Volvariella volvaceae* to produce biomass from potato and cassava processing residues by submerged fermentation. Indian J Microbiol 40:35–40
Tsioulpas A, Dimou D, Iconomou D, Aggelis G (2002) Phenolic removal in olive oil mill wastewater by strains of *Pleurotus* spp in respect to their phenol oxidase (laccase) activity. Biores Technol 84:251–257
Tung TQ, Miyata N, Iwahor K (2004) Growth of *Aspergillus oryzae* during treatment of cassava starch processing wastewater with high content of suspended solids. J Biosci Bioeng 97:329–335
Tweyongyere R, Katongole I (2002) Cyanogenic potential of cassava peels and their detoxification for utilization as livestock feeding. Vet Hum Toxicol 44:366–369
Vahabzadeh F, Mehranian M, Saatari AR (2004) Color removal ability of *Phanerochaete chrysosporium* in relation to lignin peroxidase and manganese peroxidase produced in molasses wastewater. World J Microbiol Biotechnol 20:859–864
Wang CW, Chong CN, Rahim B (1981) Growth of SCP on hydrolysate of palm oil sludge. In: Proceedings of the first ASEAN workshop on the technology of animal feed production utilizing food waste materials. Bandung, Indonesia, pp 1–13
Ward OP, Singh A, Ray RC (2006) Production of renewable energy from agricultural and horticultural substrates and wastes. In: Ray RC, Ward OP (eds) Microbial biotechnology in horticulture, vol 1. Science, New Hampshire, USA, pp 517–558
Ward OP, Singh A, Ray RC (2008) Single-cell protein from horticultural and food processing wastes. In: Ray RC, Ward OP (eds) Microbial biotechnology in horticulture, vol 3. Science, New Hampshire, USA, pp 273–298
Watanabe K (2001) Microorganisms relevant to bioremediation. Curr Opin Biotechnol 12:237–241

Wilbey RA (2006) Water and waste treatment. In: Brennan JG (ed) Food processing handbook. Wiley-VCH, KGaA, Weinheim, Germany

Wu TY, Mohammad AW, Md. Jahim J, Anuar N (2009) A holistic approach to managing palm oil mill effluent (POME): Biotechnological advances in the sustainable reuse of POME. Biotechnol Adv 27:40–52

Yang SS (1988) Protein enrichment of sweet potato residue with amylolytic yeasts by solid state fermentation. Biotechnol Bioeng 32:886–890

Yang SS, Jang HD, Liew CM, Du Preez JC (1993) Protein enrichment of sweet potato residues by solid-state cultivation with mono- and co-cultures of amylolytic fungi. World J Microbiol Biotechnol 9:258–264

Yoshi H, Furuta T, Ikeda M, Ito T, Iefuji H, Linko P (2001) Characterization of the cellulose-binding ability of *Geotrichum* sp. MIII cells and its application to dehydration of the distilled waste of sweet potato shochu. Biosci Biotechnol Biochem 65:2187–2199

Zhang ZY, Jin B, Bai ZH, Wang XY (2008) Production of fungal biomass protein using micro-fungi from winery wastewater treatment. Biores Technol 99:3871–3876

Zheng Z, Shetty K (1998) Cranberry processing waste for solid state fungal inoculant production. Process Biochem 33:323–329

Chapter 6
Composting of Lignocellulosic Waste Material for Soil Amendment

Ramesh Chander Kuhad, Piyush Chandna, Lata, and Ajay Singh

6.1 Introduction

Biomass can be defined as any organic matter that is available on a renewable basis, including dedicated energy crops, trees, agricultural food, feed crop residues, aquatic plants, wood residues, animal wastes, and other waste materials (Kamm et al. 2006). Biomass has been recognized as a major renewable energy source of the world which supplements declining fossil fuel resources (Malherbe and Cloete 2002; Ozeimen and Karaosmanoglu 2005; Jefferson 2006). Biomass also acts as a major carbon sink as 60–87 billion tons of carbon can be stored in forests through plant photosynthesis (IEA/IPCC 2007). This accounts for 50% of photosynthates on the Earth, out of which 60% of the total plant biomass is produced annually on earth (Kartha and Larson 2000; Pérez et al. 2002; Nakasaki et al. 1985a; Nakasaki and Akiyama 1988). It is estimated that annually up to 1.7–2.0 $\times$ 10^{11} tons of biomass is produced on earth. However, only 6 $\times$ 10^9 tons of biomass is currently utilized for food and non-food applications (Zoebelin 2001).

As per the FAO (http://www.fao.org), 446 million dry tons of crop residues and 377 million dry tons of perennial crops have been generated every year. Global crop residues alone were estimated at about 4 billion mg and cereals crop residue alone contribute 3 billion mg per annum for lignocellulosic residues (Lal 2008). In addition to this, 87 million dry tons of animal manures, process residues, and other residues were generated which can be recycled (Table 6.1).

Lignocelluloses are the building blocks of all plants, composed of two linear polymers, cellulose and hemicellulose and a nonlinear, three-dimensional polymer

R.C. Kuhad (✉) and P. Chandna
Department of Microbiology, University of Delhi South Campus, Benito Juarez Road, New Delhi 110 021, India
e-mail: kuhad85@gmail.com

Lata
Microbiology Division, Indian Agricultural Research Institute, New Delhi 110 012, India

A. Singh
Lystek International Inc., Waterloo, Ontario, N2J 3H8, Canada

A. Singh et al. (eds.), *Bioaugmentation, Biostimulation and Biocontrol*, Soil Biology 28, DOI 10.1007/978-3-642-19769-7_6, © Springer-Verlag Berlin Heidelberg 2011

Table 6.1 Worldwide production of lignocellulosic residue

Source	Production (million tons) year 2008/2009
Wheat	658.0
Coarse grain	1088.6
Cereal	2191.9
Rice	445.3
Soybeans	220.8
Cottonseed	44.4
Rapeseed	48.0
Groundnuts (unshelled)	35.5
Sunflower	27.8
Palm kernels	10.8
Copra	5.3
Sugar	168.0

Source: FAO (http://www.fao.org)

lignin (Perez et al. 2002). Cellulose is a non-branched water-insoluble polysaccharide consisting of several hundreds to tens of thousands of glucose units linked through β-1,4-linkages and is usually arranged in microcrystalline structures. Hemicellulose is a polymeric material, lower in molecular weight than cellulose, and consisting of C6-sugars (glucose, mannose, and galactose) and C5-sugars (mainly arabinose and xylose). Lignin is a highly cross-linked polymer made from substituted phenylpropene units. Lignocellulosic materials mainly consist of 38–50% of cellulose, 23–32% hemicellulose, and 15–25% lignin (Deobald and Crawford 1987). Apart from these primary polymers, plants contain other structural polymers (5–13%) such as waxes and proteins.

A significant amount of the lignocellulose waste is often disposed of by biomass burning worldwide (Levine 1996). In China alone, more than 100 million tons of crop straw is burned every year. Thus all the lignocellulosic biomass is generally not available for composting/recycling because of their competing use in other commercial or agricultural practices. The biological conversion of this waste into value-added product as compost and its use as soil amendment may be the most sought out options with multiple benefits (Singh et al. 2006; Kausar et al. 2010). The complexity of degraded plant materials and the quality of the final product may depend upon the type of lignocellulosic biomass and its treatment.

6.2 Composting Methods

Composting is bio-oxidation of solid heterogeneous organic substrates through a biological degradation process resulting in a stabilized high nutrient product for soil amendment. Compost consists of microbial cells, plant and animal residues at various stages of decomposition, stable humus synthesized from the residues of the microorganisms and from their carbonized compounds (Nelson and Sommers 1996).

The composts prepared from different organic wastes differ in their quality and stability, which further depends upon the composition of raw material used for the compost production (Gaur and Singh 1995; Ranalli et al. 2001; Smith 2009).

Prior to 1970, there were very few composting operations except for those producing compost for the mushroom industry. Much of the early work was carried out in the USA where there are now over 3,500 green waste composting facilities. Similarly, Austria, Canada, Germany, and the Netherlands are well advanced in terms of green waste and source-separated organics composting. Various methods of composting are described below:

(a) *In-vessel composting*: In-vessel composting occurs within a contained vessel, enabling the operator to maintain closer control over the process in comparison with other composting methods. There are several types of in-vessel composting reactors: vertical plug-flow, horizontal plug-flow, and agitated bin.
(b) *Windrow composting*: In this method, solid waste is arranged in long rows and covered to allow decomposition. The material is turned over repeatedly by mechanical means. Old wooden pallets are an excellent size for a compost holding unit. This method is more suitable for Lignocellulosic waste composting.
(c) *Aerated pile composting*: Waste is arranged in piles and forced aeration is used to supply extra air through perforated pipes, which are buried inside the compost pile. The aerated pile process achieves substantially faster composting rates through improved aeration design.
(d) *Continuous-feed composting*: It uses a reactor that permits control of the environmental parameters. Reactor is like an industrial fermenter, where composting is complete within 2–4 days. This method is a very fast one and expensive and mainly applicable for municipal waste.
(e) *Vermicomposting*: Vermicomposting is a simple process of composting, in which certain species of earthworms are used to enhance the process of waste conversion and produce a better end product. Vermicomposting differs from general composting in several ways; it is a mesophilic process, utilizing microorganisms and earthworms that are active at 10–32°C and can reduce the volume of lignocellulosic biomass by 40–60%.

The capital costs of aerated static pile or windrow configuration may be lower than in-vessel composting configurations, but costs increase markedly when odor control system is required (http://www.epa.gov/owm/mtb/mtbfact.htm). Highly mechanized in-vessel systems are often more costly to construct and less flexible in their ability to adapt to changing properties of biosolids and bulking agent feedstock, but tend to be less labor intensive and have smaller footprint of the system (Fig. 6.1).

Capital costs of in-vessel systems range from $33,000 to $83,000 per dry metric ton per day of processing capacity. A typical aerated static pile facility costs approximately $33,000 per dry metric ton per day of processing capacity (USEPA 2002). Typical operation and maintenance (O&M) costs for in-vessel systems range from $150 to $225 per dry ton per day. Aerated static pile O&M

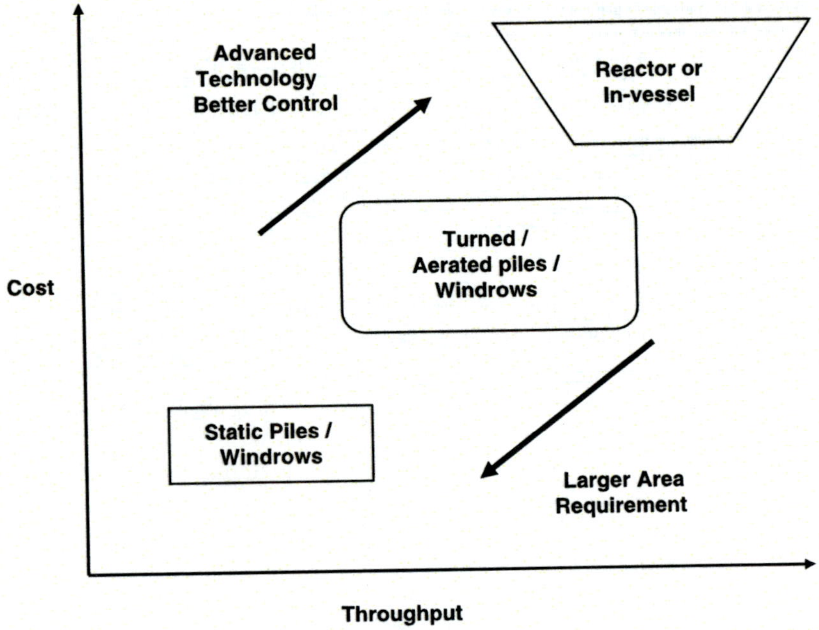

Fig. 6.1 A techno-economic comparison of different composting methods

costs average $150 per dry ton per day. Costs for windrow systems fall between the costs for in-vessel and aerated static pile. The selling price for compost ranges from $10 to $20 per ton. Some municipal facilities allow landscapers and homeowners to pick up compost for little or no charge.

6.3 Composting Process

Composting phases depend on the nature and the amount of the organic matter being composted. The efficiency and decomposition process is determined by the degree of aeration and agitation, size of compost pile as well as its ingredients which determine the activity of decomposer microbes (Biddlestone and Gray 1985). The composting process proceeds through three phases: the mesophilic phase, the thermophilic phase, and the cooling and maturation phase (Fig. 6.2).

Mainly three general categories of microorganisms (bacteria, actinomycetes, and fungi) populate the composting process, which decompose the organic material into humus-rich compost. Microbes in compost provide several benefits such as disease suppression, improvement in nutrient retention and mineralization in soil, improved soil structure, decomposition of toxic chemicals, production of plant growth-promoting compounds, and improvement in the crop quality (Hargreaves et al. 2008). The changes in temperature and the availability of substrates to bacteria

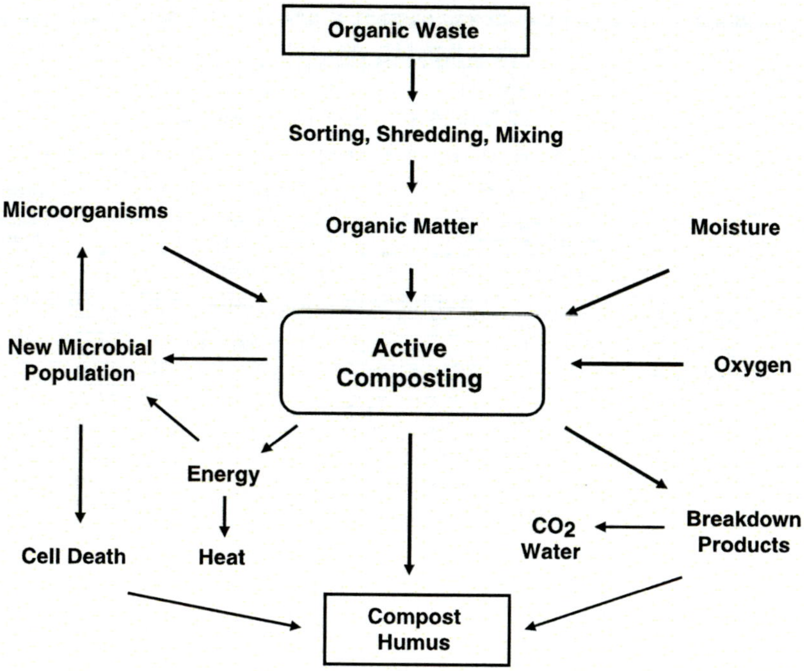

Fig. 6.2 Flow diagram of a typical composting process

seemed to mainly determine the composition of bacterial members at different stages of composting (Cahyani et al. 2003).

6.3.1 Microbial Community

Extensive studies are available on the population of bacteria, actinomycetes and fungi during composting, and the quantitative analysis of microorganisms in compost by using culture-based methods (Dees and Ghiorse 2001; Kuroda et al. 2004; Saludes et al. 2008). *Aspergillus* and *Penicillium* are the predominant mesophilic fungal genera in composting process (Thambirajah et al. 1995; van Heerden et al. 2002). Various mesophilic bacterial species have been isolated which belong to diverse families such as Alcaligenaceae, Alteromonadaceae, Burkholderiaceae, Bradyrhizobiaceae, Caryophyanaceae, Caulobacteraceae, Corynebacteriaceae, Clostridiaceae, Comamonadaceae, Corynebacteriaceae, Enterobacteriacea, Flavobacteriaceae, Flexibacteraceae, Hyphomicrobiaceae, Intrasporangiaceae, Methylobacteriaceae, Microbacteriaceae, Micrococcaceae, Moraxellaceae, Nocardiaceae, Nocardiopsaceae, Paenibacillaceae, Phyllobacteriaceae, Propionibacteriacea, Pseudomonadaceae, Pseudonocardiaceae, Rhodobacteraceae, Sphingobacteriaceae, Staphylococcaceae, and Xanthomonadaceae. Table 6.2 shows common microorganisms associated with composting process.

Table 6.2 Common microbial species associated with composting process

Microorganism	Mesophilic (20–45°C)	Thermophilic (45–70°C)
Bacteria	*Aerobacter aerogenes*	*Aneurinibacillus* sp.
	Alcaligenes denitrificans	*Barevibacillus* sp.
	Bacillus licheniformis	*Bacillus stearothermophilus*
	Cellulomonas folia	*Bacillus macerans*
	Corynebacterium	*Bacillus schlegelii*
	Nitrosospira sp.	*Bacillus thermodenitrificans*
	Nitrosomonas sp.	*Bacillus pallidus*
	Pseudomonas aeruginosa	*Hydrogenobacter* sp.
	Proteus vulgaris	*Rhodothermus marinus*
	Rhodococcus sp.	*Saccharococcus thermophilus*
	Serratia marcescens	*Thermus thermophilus*
Actinomycetes	*Actinoplanes* sp.	*Streptomyces thermofuscus*
	Nocardia brasiliensis	*Saccharomonospora* sp.
	Micromonospora parva	*Streptomyces thermovulgaris*
	Micromonospora vulgaris	*Thermomonospora glaucus*
	Pseudonocardia	*Thermomonospora fusca*
	Streptomyces violaceoruber	*Thermomonospora viridis*
	Streptomyces rectus	*Thermomonospora curvata*
Filamentous fungi	*Aspergillus niger*	*Absidia corymbifera*
	Aspergillus terreus	*Aspergillus fumigatus*
	Cladosporium cladosporioides	*Chaetomium thermophile*
	Fusarium solani	*Humicola insolens*
	Fusarium monoliforme	*Mycelia sterila*
	Geotrichum candidum	*Paecilomyces variotii*
	Mucor racemosus	*Rhizomucor pusillus*
	Penicillium digitatum	*Sporotrichum thermophile*
	Rhizopus nigricans	*Taleromyces thermophilus*
	Trichderma koningii	*Thermomyces lanuginosus*
Yeasts	*Candida tropicalis*	
	Candida krusei	
	Candida parapsilosis	
	Pichia sp.	
	Rhodotorula rubra	
	Saccharomyces sp.	

Various microbes were isolated from the thermophilic phase belonging to the families of Micromonosporaceae, Streptomycetaceae, Thermoactinomyceteaceae, Thermonomomonosporaceae, Streptosporangiaceae, etc. (Beffa et al. 1996a, b). Fungal population generally decreases from 10^6 CFU/g to around 10^3 CFU per g of compost as the temperature rises above 60°C with their elimination above 64°C (Thambirajah et al. 1995). As the temperature in compost fell below 60°C, both mesophilic and thermophilic fungi begin to re-colonize the substrate. Among the mesophilic fungi, a few lignin-degrading Basidiomycota including *Coprinus* sp., *Panaeolus* sp., *Corticium coronilla*, *Trametes* sp., and *Phanerochaete* sp. were isolated from compost at the cooling and maturation phases or from mature compost (Granit et al. 2007). Wide ranges of bacteria were isolated from different compost environments, including species of *Pseudomonas*, *Klebsiella*, and *Bacillus*

(Nakasaki et al. 1985a, b; Strom 1985a, b; Falcon et al. 1987). Typical bacteria of the thermophilic phase are species of *Bacillus*, e.g., *B. subtilis*, *B. cereus*, *B. licheniformis*, and *B. circulans*.

6.3.2 *Succession of Microflora During Composting*

In composting process, the distribution of specific microbial species and diversity undergoes changes in response to the temperature and availability of the type of substrate. The composting is generally characterized by a short mesophilic period at the beginning where mesophilic microbes predominate and it also decides the onset of thermophilic phase

The accumulation of heat is the critical limiting factor pertaining to microbial decomposition during thermophilic phase. Aeration systems can be used to dissipate excess heat so as to optimize stabilization. Heating is essential to enable the development of a thermophilic population of microorganisms, which are capable of degrading the more recalcitrant compounds like lignin and xenobiotics, and to kill pathogens and weed seeds (Boulter et al. 2000). As the temperature increases to over 40°C, thermophilic microbes take over and become responsible for the degradation process. This phase is characterized by increased temperature ranging from 45 to 70°C, and the growth and activity of thermolabile microbes is inhibited (Stetter 1998).

The microbial diversity of lignolytic microbes can be expected during the thermogenic phase because degradation and mineralization of complex organic matter take place during this phase (Nakasaki et al. 1985b). A temperature of 55°C for at least 3 days has been found adequate for the reduction of pathogen in aerated static pile systems (USEPA 2002). According to Stentiford (1996), temperatures higher than 55°C favor sanitation, 45–55°C biodegradation, and 35–45°C microbial diversity. When the temperature exceeds 50°C, microbial activity decreases dramatically but after the compost has cooled mesophilic bacteria and actinomycetes again dominate (McKinley and Vestal 1985; Strom 1985a). Actinomycetes play an important role in the later stages of composting and particularly in the degradation of relatively complex and recalcitrant compounds such as cellulose and lignin (Ryckeboer et al. 2003a, b). Anaerobic bacteria found during composting are generally highly cellulolytic and thus may play a significant role in the degradation of macromolecules. The majority of the mesophilic anaerobic bacteria in composting are facultative, while under thermophilic conditions more obligate anaerobic bacteria are found (Atkinson et al. 1996).

Among the composting techniques reported, the hyperthermal composting method reported by Kanazawa et al. (2003) is unique. It involves maintaining a relatively high temperature of 80°C during the process over a long period using a specific seed compost and aeration device. The compost prepared at thermophilic temperature is more suitable for mushroom production because it is free from the pathogen which affects the mushroom quality. After the subsequent decrease in temperature, the curing period (cooling and maturation) starts and may last for a long time. The amount of readily available nutrients becomes a limiting factor that causes a decline in microbial activity and heat output.

Microbes perform their essential function with the help of the enzymes they produce. Bioaugmentation with efficient lignocellulolytic microbes may improve and/or accelerate the composting process. Moreover, inoculation with cellulolytic fungi may increase the process of decomposition further. Rapid bioconversion of agroresidues into highly nutrient-enriched compost can be achieved by using efficient lignocellulolytic fungi such as *Trichoderma, Aspergillus awamori, Polyporous versicolor, Penicillium funiculosum, Phanerochaete chrysosporium*, etc. (Gaind et al. 2005; Lata et al. 2009; Kausar et al. 2010; Neklyudov et al. 2006; Gaind and Nain 2007) by supplementing with lignocellulosic residues with nitrogen-rich wastes of plant animal origin is necessary for desirable composting. Similarly, Pandey et al. (2009) has evaluated the effect of a hyperlignocellulolytic fungal consortium and different nitrogen amendments on paddy straw composting and observed the changes in terms of physico-chemical and biological parameters (Vuorinen 2000; Tuomela et al. 2000). Enrichment of composted material with nutrients, plant growth-promoting regulators such as hormones (indole acetic acid and giberellic acid), and microorganisms (*Azotobacter chroococum, Aspergillus awamori, Pseudomonas fluorescens*, and P-solubilizing microbes) produces further value-added organic fertilizers with higher contents of N and P for improved crop yields (Ahmad et al. 2007).

6.3.3 Process Kinetics

To determine waste biodegradability and generate a useful measure for the loss of organic matter during composting, it is necessary to determine process kinetics using the data obtained by an experimental study under controlled conditions (Hamoda and Abu Qdais 1998). Modeling composting processes is a prerequisite to realize the process control of composting. The growth rates of microorganisms and the use of Monod equation to simulate the composting process have been considered (Agamuthu 2000). Haug (1993) emphasized on the thermodynamic and kinetic changes taking place during composting procedure. Bari and Koenig (2000) stressed on kinetics analysis of forced aeration and studied composting processes under different aeration modes.

The degradation of organic matter as a function of time follows first-order kinetics expressed as:

$$\mathrm{d(OM)/dt} = -k\mathrm{T.OM},$$

where "OM" is the quantity of biodegradable volatile solids at any time of the composting process in kilogram, "t" is time in days, and "kT" is the reaction rate constant.

A proper design and operation of the composting reactor is necessary to guarantee a good compost quality and reduced emissions, and reduces the bulk density of the waste. The main function of the composting reactor will be the realization of optimal environmental conditions for the development of microbial population. The kinetic model to be developed should be able to predict the process rate in relation to the (actual) composition of the waste and (actual) conditions to which this waste is exposed in the reactor.

6.3.4 Pathogen Suppression

Enteric pathogens such as bacteria, viruses, helminthes, and protozoa are of concern as they can pose significant health hazard to both humans and animals. During composting, pathogen suppression is accomplished by several processes including competition between indigenous microbes and pathogens, antagonistic relationship between organisms, action of antibiotics produced by certain fungi and actinomycetes, natural die-off of the enteric pathogens in the non-ideal compost environment, toxic byproducts such as gaseous ammonia, nutrient depletion, and thermal conditions (Wichuk and McCartney 2007).

Time–temperature regulation and guidelines in North America, applicable to composting operations, are presented in Table 6.3. A minimum temperature of 55°C should be maintained for a period of 3 consecutive days in different composting methods except for the windrow, where temperature greater than 55°C should be maintained for at least 15 days with a minimum of five turnings during a high-temperature period.

In composting, the microbial community follows a predictable succession pattern resulting in the re-colonization of compost with metabolically active mesophilic and thermophilic populations that can be suppressive toward plant pathogens (Nakasaki et al. 1996). The pathogenic organisms and parasites present in the raw organic material disappear during the composting process at the elevated temperatures. Gram-negative bacteria and actinomycetes are more prevalent in suppressive compost, and effective bacterial antagonists in compost include *B. subtilis* and *Enterobacter* sp. (Atkinson et al. 1997).

Among the bacteria, *Salmonella, Shigella, Escherichia coli, Enterobacter, Yersinia, Streptococci,* and *Klebsiella* can emerge and cause infections among compost handlers and agricultural users (Strauch 1996) if temperature is not up to

Table 6.3 Time–temperature criteria for compost pathogen reduction

Composting technology	Time–temperature requirement as per the US EPA
Windrow	Temperature >55°C for 15 days or longer; during the >55°C period, there should be a minimum of five turnings of the windrow
Aerated static pile	Temperature >55°C for a period of 3 consecutive days
Reactor (in-vessel)	Temperature >55°C for a period of 3 consecutive days

60°C during thermophilic phase. Biological characteristics of disease suppression can involve a combination of mechanisms including competition for nutrients, antibiosis, production of extracellular hydrolytic enzymes, bioactive compounds, hyper- and myco-parasitism, predation, and host-mediated induction of resistance (Whipps 1997; Hoitink et al. 1997; Lucas 1998). *Bacillus subtilis* has been observed to act as an agent of biological control against several plant pathogens (Li et al. 1998; Walker et al. 1998). When *B. subtilis* was inoculated (10^7–10^8 cells/g wet compost) into compost products made from various organic wastes, in vitro suppressive effect was observed against the plant pathogens *Fusarium oxysporum*, *P. ultimum*, *Verticillium dahliae*, *Pyricularia oryzae*, and *Rhizoctonia solani* (Phae and Shoda 1990).

Strauch (1987) considered fecal streptococci to be more useful indicators of the disinfection processes in sewage sludge composts as compared to the fecal coliforms. Fecal streptococci are more resistant to environmental factors than fecal coliforms. Monitoring of the pathogenic fungal population in compost is as important as pathogenic bacteria to determine its quality and suitability in field application (Peters et al. 2000).

6.4 Factors Affecting the Composting Process

Various factors that affect the composting process and the final compost product quality are described in this section.

6.4.1 Carbon:Nitrogen Ratio

The microorganism involved in composting process requires a source of carbon to provide energy and multiplication of new cells, as well as nitrogen source for building the cell proteins. Lignocellulosic crop residues are mainly organic matter of which carbon is the chief element (48–58%). Based on the assumption that organic matter contains 58% organic carbon, a conversion of 1.724 has been proposed for the conversion of organic matter into its carbon content (Nelson and Sommers 1996). Table 6.4 shows C:N ratio of various plant waste materials. Since, high initial C:N ratio causes a slower start up of the process and results in longer than usual composting time, and low initial C:N ratio results in high emission of NH_3 (Tiquia et al. 2000), ideal C:N ratio is desired in composting process to maintain high protein/nitrogen levels to facilitate optimum and rapid degradation. In addition, phosphorus and other trace minerals are required by microorganisms for optimum activity. Chemical analysis of the microorganisms revealed that on an average they contained 50% C, 5% N, and 0.25–1% P on a dry weight basis.

Table 6.4 C:N ratio of various waste materials used in composting

Substrate	Material	C:N ratio
High carbon	Wood	700
	Sawdust	500
	Paper	170
	Straw	80
	Corn stalks	60
	Leaves	60
	Rice hulls	121
	Sugarcane residue	50
	Newspaper	175
	Cardboard	350
High nitrogen	Alfalfa	13
	Kitchen waste	15
	Green clover	16
	Mature clover	23
	Grass clippings	19
	Mustard	26
	Soybean meal	5
	Fruits and vegetable waste	35
	Peanut shells	35
	Garden waste	30
	Weeds	30
	Coffee grounds	20
	Seaweed	19
	Cow manure	20
	Poultry manure	10
	Horse manure	25
	Municipal wastewater sludge	8

Most of the lignocelluloses residues have wide C:N ratio varying from 35 to 325:1 (Hue and Liu 1995). Thus there is a need of supplementing nitrogen during composting of lignocellulosic biomass to bring the C:N ratio of lignocellulosic substrates in a desirable range. The crop residues may be amended with nitrogen-rich chemical fertilizers such as urea to bring the optimum C:N ratio of 25 to 35:1. Alternative organic waste material rich in nitrogen such as poultry droppings/manure, farmyard manure, dry blood, fish meal, and oil seed cakes such as soybean meal, neem, castor and jatropha have also been used effectively (Gaind et al. 2009; Pandey et al. 2009).

6.4.2 Particle Size

Small particles have much more surface area and can be degraded much faster, and also result in a decrease in air space with less porosity, thereby reducing the aeration and affecting the degradation adversely. Haug (1993) suggested that for a particle larger than 1 mm, oxygen diffusion would limit the decomposition. Decomposition and microbial activity would be rapid near the surfaces as oxygen diffusion is very high. Particle size also affects moisture retention as well as free air space and porosity of the compost mixture (Naylor 1996).

6.4.3 Moisture Content

An adequate amount of moisture is essential for microbial activity and is an important factor to be controlled during composting. Moisture in compost comes either from the initial feedstock or the metabolic water produced by microbial action. The optimal moisture content in composting is in the range between 50 and 60%. Bacterial metabolic activity is severely inhibited when the moisture content drops below 40%. Fungi have a lower moisture threshold and could grow well at 30–40% moisture. The oxygen uptake during composting reduced drastically at moisture levels below 30%. At water potentials below −20 kPa (about 60% moisture), bacteria progressively failed to colonize the compost mass (Miller et al. 1985). Total liquid content may be used as a guide, rather than the water content, given by:

$$\%\ \text{Liquid} = \frac{100 \times (\%\,\text{moisture} \pm \%\,\text{liquid})}{(100 - \%\,\text{ash})}.$$

Water used to moisten the compost pile should have a neutral pH and that may also help to reduce the acidity of the compost. The finished compost is usually neutral (with a pH between 7.1 and 7.5). Moisture content also influences the structural and thermal properties of the material, as well as the rate of biodegradation and metabolic process.

6.4.4 Oxygen

Oxygen is required by the microorganisms for oxidizing various organic molecules present in the composting mass (Richard and Walker 1999; Diaz et al. 2002; Liang et al. 2003). During aerobic composting 1 g of organic matter releases about 25 kJ of heat energy, which is enough to vaporize 10.2 g of water coupled with losses due to aeration resulting in water loss during composting (Finstein et al. 1986). A minimum oxygen concentration of 5% within the pore space of the composting pile is necessary to maintain aerobic conditions. Aeration has multiple function of supplying O_2 to support aerobic metabolism, controlling temperature, and removing CO_2 and other gases. While insufficient aeration promotes the formation of anaerobic zones and the generation of foul odors, excessive aeration limits microbial activity as a result of reduced moisture and associated cooling (Brodie et al. 2000; Hao et al. 2001).

6.4.5 Temperature

Another critical parameter influencing the rate of composting and the quality of product is temperature, which is also a fundamental factor affecting the rate and net outcome of chemical and biochemical reactions of the microorganisms (Michel et al. 1996). Variations in temperature affect the various phases of composting and kinetics of growth rate constant, diffusion coefficient, and hydrolysis rate constant. The metabolic heat trapped in an organic pile of sufficient size can elevate the temperatures in the pile from the ambient to 70–80°C within few days. Depending upon the phase of composting, turning of the pile is required to maintain different temperatures during various phases of composting such as mesophilic (25–45°C), thermophilic (>45°C), and cooling and maturation (again mesophilic phase) which ultimately affects the final product quality.

6.5 Compost Product Maturity and Quality

The maturity and quality of compost can be evaluated by using physical, chemical, or biological methods (Wu et al. 2000; Wu and Ma 2001). The maturity of compost can be determined by analytical tests, such as the content of inorganic nutrients such as total nitrogen, phosphorus, magnesium, calcium and nitrogen, salt content, electrical conductivity, and pH. Nutrient balance is very much dependent on the type of feed materials being processed. For evaluating compost stability, several indexes and methods have been proposed (Itävaara et al. 2002). Compost maturity is associated with plant-growth potential or phytotoxicity, whereas stability is related to the compost's microbial activity.

Various parameters that are monitored to evaluate the maturity of composts include enzymatic and biological activities, germination tests, calorimetry, thermogravimetry, respiration, and spectroscopic determination of humification (Provenzano et al. 2001; Domeizel et al. 2004; Castaldi et al. 2005; Chang et al. 2006). Concentration of carbon is reduced due to the evolution of CO_2 during degradation of organic matter while that of nitrogen is increased resulting in the reduction of carbon to nitrogen ratio (C:N) at the end of composting. Thus the C:N ratio is frequently used as an index of compost maturity. The C:N ratio below 20 is indicative of acceptable compost maturity. Chanyasak and Kubota (1981) established a water-soluble organic-C/organic-N ratio of 5–6 as an essential indicator of compost maturity.

Phytotoxicity is one of the most important criteria for evaluating the suitability of compost for agricultural purposes (Brewer and Sullivan 2003; Cooperband et al. 2003). Phytotoxicity is mainly caused by increased solubility of heavy metals or the production of phytotoxic substances such as ammonia, ethylene oxide, and organic acids (Jimenez and Garcia 1989). The decomposition of the organic matter during composting involves the disappearance of phytotoxic substances, such as low molecular weight organic acids (Pascual et al. 1997).

Germination index (GI) of seed is a common technique used to determine compost's maturity and toxicity, and the GI of 50% has been used as an indicator of phytotoxin-free compost (Tiquia et al. 1996). Quantitative detection of organic materials (e.g., humic acid) using ^{13}C-NMR spectra and ultraviolet spectroscopy at different steps of composting can also provide a good measure of maturity (Genevini et al. 2002; Laor and Avnimelech 2002).

Many researchers have proposed indices of maturity based on the monitoring of humic substances, humic acids, and fulvic acids (Lopez et al. 2002). Humic acid is generally considered to be more stable than fulvic acid and is associated with increasing the soil buffering capacity. Humification index is calculated from the ratio of humic acid and fulvic acid. The threshold value of humification index 0.5 in well-stabilized compost was reported (Ciavatta et al. 1990). As the compost matures, the humic material in compost tends to increase and is capable of binding many metals, thus decreasing their availability (Deportes and Benoit-Guyod 1995). In general, immature compost contains high levels of fulvic acids and low levels of humic acids.

Respirometric techniques can also provide accurate information on the microbial activity of a compost sample (Kalamdhad et al. 2008). Respirometric parameters that were used for the evaluation of stability are the specific oxygen uptake rate (SOUR), the oxygen demand (OD_{20}) and the dry-specific oxygen uptake rate (DSOUR) tests. Typical values of specific oxygen uptake rate vary from 4.0 to 17.5 mg O_2 $h^{-1}g^{-1}$ compost solids (Moreira et al. 2008).

Measurement of activity of some key enzymes involved in decomposition process may also indicate the progress of composting process. Important hydrolytic enzymes that were involved in the composting process include cellulases, hemicellulases, proteases, lipases, phosphatases, and arylsulphatases. High levels of protease, lipase, and cellulase activities have been detected throughout the active phase of composting (Cunha-Queda et al. 2002; Mondini et al. 2004). Oxidoreductases such as superoxide dismutase, catalase and protease activities reflect the intensity of microbial activity such as respiration and biodegradation (Garcia-Gil et al. 2000).

Other methods studied include the fertilizer stability index that can be evaluated according to the CO_2 evolution rate (Q), in mg CO_2-C g^{-1} organic matter day^{-1}, as being very stable ($Q < 2$), stable ($Q < 4$), or unstable ($Q > 4$) (USDA). Researchers have used ATP as a measure of microbial biomass, which can be a useful analytical method for compost maturity in animal waste compost (Tseng et al. 1995). Quinone profile method examines the microbial community structure of various compost products and soils have produced encouraging results (Hiraishi 2000). Fatty acid methyl esters (FAMES) can be effectively used over 50 mol% as the maturity index for cattle manure and poultry manure composts (Kato et al. 2005; Kato and Miura 2008). Color, odor, temperature, inorganic nitrogen, and cation exchange capacity are other indices of maturity (Zmora-Nahum et al. 2005).

6.6 Compost Application in Agriculture and Horticulture

Compost application improves the physical, chemical, and biological properties of soils. The compost adds airspace to the soil, improves structure, slows down crust formation, reduces erosion, and enhances hydraulic properties (Aggelides and Londra 2000). Increased organic matter would also increase soil porosity, aeration and fertilizer retention, and creates a better environment for healthy plant root growth. Soil organic matter is essential for maintaining soil quality by improving the biological, physical, and chemical soil conditions. Soil organic matter consists of a variety of simple and complex carbon compounds and thus provides food for a variety of organisms. Soil organisms can affect plant growth directly by binding atmospheric N_2 by free or symbiotically living bacteria, by mobilization of N, P and water by mycorrhizal fungi, through antagonism and predation of pathogens, by induction of resistance of plants against pathogens, and aspecific competition with pathogens.

A major goal of composting is the removal of pathogens. Although much emphasis is often put on the potential survival of some plant, human or animal pathogens during compositing, the majority of pathogens are promptly and completely killed during the heat phase of composting (Wichuk and McCartney 2007). For unrestricted compost use, it is generally accepted that risk is sufficiently minimized only when pathogens are not detectable in the finished product.

Compost enriches the soil with microbes and slowly released nutrients. Soil microorganisms govern the numerous nutrient cycling reactions in soils. Phosphorus flux through the microbial biomass is faster in organic soils, and more phosphorus would bind in the microbial biomass. Phosphorus consumption is directly proportional to microbial consortium growth rates as microbes use phosphorus to produce DNA, RNA, and ATP. These effects are clearly identified just after compost application in a rather compacted and heavy and sandy soil (Aggelides and Londra 2000; Celik et al. 2004). An active and diverse soil microflora and fauna is not only considered advantageous for healthy soil but also for the suppression of plant pathogens. Compost amendment to soil also has positive implication in suppressing soil-borne diseases (Hoitink and Boehm 1999; Martin 2003) including not only root pathogens (primarily through competition and parasitism), but also shoot pathogens (through induced resistance). Compost application has a strong stimulating effect on the parasitic microbial community including known antagonists as fluorescent pseudomonads and *Trichoderma* species (Termorshuizen et al. 2004).

Addition of compost to soil increases the soil's phosphorus, potassium, nitrogen, and organic carbon content. Biosolids compost also plays a role in bioremediation of hazardous sites and pollution prevention. Compost has proven effective in degrading or altering many types of contaminants, such as wood-preservatives, solvents, heavy metals, pesticides, petroleum products, and explosives.

When compost is added to the soil, carbon is sequestered for a long time in the compost fraction that is resistant to decomposition. Carbon sequestration is important to mitigate the greenhouse effect. In addition, more labile organic compounds from fresh organic matter can be protected by hydrophobic humic substances originating from compost. It has been estimated that the application of composted household biowaste leads to a net reduction of 57 g CO_2 per kg biowaste, while biowaste incineration results in a net reduction of only 25 g CO_2 per kg (Termorshuizen et al. 2004). On the other hand, landfill disposal of biowaste results in significant methane emissions, a strong greenhouse gas that contributes to breakdown of the ozone layer.

There is an increasingly accepted general view that genetically modified organisms (GMO) or their products introduced into the environment should be degradable and should disappear after a limited period of time. Due to the risk of possible horizontal gene transfer, disposal methods for GMOs need to address destruction of both the organism and the genetic material. The conditions created in a properly managed composting process environment may help in destroying GMOs and their genes, thereby reducing the risk of the spread of genetic material (Singh et al. 2006). When considering composting as a potential method for the disposal of GMOs, the establishment of controlled conditions providing an essentially homogenous environment appears to be an important requirement.

6.7 Potential Risks Associated with Composting

As discussed above, there are many benefits of composting including waste management through recovery of useful organic matter for use as soil amendment and reduction of waste that is disposed to landfill and incinerated. On the other hand, there may be certain occupational and health risks associated with the composting process and compost application such as odors, bioaerosol (organic dust containing bacterial and fungal spores) generation, emission of volatile organic compounds (VOCs), and potential pathway from use on land for contaminants to enter food chain (Domingo and Nadal 2009). Because of their potential toxicity, metals such as arsenic, cadmium, chromium, lead, mercury and nickel, pesticides, and organochlorinated compounds such as PCDD/Fs and PCBs, PAHs and halogenated hydrocarbons are among the potent chemical contaminants of concern.

An important problem in the composting facilities is the odor due to the VOC emissions, beginning with the arrival of fresh material. Under anaerobic conditions, reduced sulfur compounds of intense odor are generated, whereas aerobic degradation process generates emissions of alcohol, ketones, esters, and organic acids. The systemic toxic effects of VOCs are renal, hematological, neurological and hepatic alterations, as well as mucosal irritations.

For biological risks, it is necessary to consider two kinds of microorganisms: (1) the pathogenic agents present in the fresh organic material that are susceptible to elimination during the composting process, and (2) the microorganisms that are developed during the active composting process, which also play an important role in the degradation of the organic matter (mesophilic fungi and thermophilic bacteria). Occupational biorisks are those derived from the presence of organisms, and/or substances generated by them in occupational atmospheres, which can cause adverse effects (infectious, allergenic, toxic, and carcinogenic) on the workers (Schlosser and Huyard 2008). *Aspergillus fumigatus* is a well-known opportunistic pathogenic fungus in nose and throat, susceptible of causing infections. Other fungi such as *Aspergillus flavus*, *Stachybotrys atra*, and different species of *Fusarium* and *Penicillium* produce mycotoxins. It is known that the endotoxins produced by the gram-negative bacteria cause fever and respiratory problems, as well as gastrointestinal disturbances and diarrheas (Domingo and Nadal 2009).

6.8 Conclusions

Composting is the controlled decomposition of organic matter resulting in a stabilized product that can be used as soil amendment/conditioner in agriculture. If organic matter is targeted for agricultural use, the major reason for composting is the inactivation of pathogens of the material. Stabilized organic matter in compost improves the physical, chemical, and biological properties of soils, by improving structure, soil porosity, aeration, and fertilizer retention, which creates a better environment for healthy plant root growth. The diversion of biodegradable waste from landfill is of key importance in developing a sustainable waste strategy (Domingo and Nadal 2009). The selective collection and the recycling of biosolids and the organic fraction of the municipal waste are essential factors for the success of a modern policy of global waste management.

However, the compost derived from the organic fraction of municipal solid waste may contain metals, persistent organic pollutants, as well as microbial toxins, whose exposure in certain scenarios might mean health risks for the general population. From a point of view of occupational health, the results of epidemiological studies are the best evidence to establish a correlation between the potential risks and the current situations. Rigorous periodic analytical quality controls accompanied by the corresponding evaluations of risk of the chemical and biological components are highly recommended.

References

Agamuthu P (2000) Kinetic evaluation of composting of agricultural wastes. Environ Technol 21:185–92

Aggelides SM, Londra PA (2000) Effects of compost produced from town wastes and sewage sludge on the physical properties of a loamy and a clay soil. Bioresour Technol 71:253–259

Ahmad R, Jilani G, Arshad M, Zahir ZA, Khalid A (2007) Bio-conversion of organic wastes for their recycling in agriculture: an overview of perspectives and prospects. Ann Microbiol 57:471–479

Atkinson CF, Jones DD, Gauthier JJ (1996) Biodegradabilities and microbial activities during composting of municipal solid waste in bench-scale reactors. Compost Sci Utilize 4:14–23

Atkinson CF, Jones DD, Gauthier JJ (1997) Microbial activities during composting of pulp and paper-mill primary solids. World J Microbiol Biotechnol 13:519–525

Bari QH, Koenig A (2000) Kinetic analysis of forced aeration composting-II. Application of multiplayer analysis for the prediction of biological degradation. Waste Manage Res 18:313–320

Beffa T, Blanc M, Lyon P, Vogt G, Aragno M (1996a) Isolation of *Thermus* strains from hot composts (60ºC to 80ºC). Appl Environ Microbiol 62:1723–1727

Beffa T, Blanc M, Marilley L, Fisher JL, Lyon PF, Aragno M (1996b) Taxonomic and metabolic microbial diversity during composting. In: de Bertoldi M, Sequi P, Lemmes B, Papi T (eds) The science of composting, vol 1. Chapman and Hall, London, pp 149–161

Biddlestone AJ, Gray KR (1985) Composting. In: Moo-Young M (ed) Comprehensive biotechnology: the principles, applications, and regulations of biotechnology in industry, agriculture, and medicine. Oxford, Pergamon, pp 1059–1070

Boulter JI, Boland GJ, Trevors JT (2000) Compost: a study of the development process and end-product potential for suppression of turfgrass disease. World J Microbiol Biotechnol 16:115–134

Brewer LJ, Sullivan DM (2003) Maturity and stability evaluation of composted yard trimmings. Compost Sci Utilize 11:96–112

Brodie HL, Carr LE, Condon P (2000) A comparison of static pile and turned windrow methods for poultry litter compost production. Compost Sci Utilize 8:178–189

Cahyani VR, Matsuya K, Asakawa S, Kimura M (2003) Succession and phylogenetic composition of bacteria community responsible for the composting process rice straw estimated by PCR-DGGE analysis. Soil Sci Plant Nutr 49:619–630

Castaldi P, Alberti G, Merella R, Melis P (2005) Study of the organic matter evolution during municipal solid waste composting aimed at identifying suitable parameters for the evaluation of compost maturity. Waste Manag 25:209–213

Celik I, Ortas I, Kilic S (2004) Effects of compost, mycorrhiza, manure and fertilizer on some physical properties of a chromoxererrt soil. Soil Till Res 78:59–67

Chang JI, Tsai JJ, Wu KH (2006) Composting of vegetable waste. Waste Manage Res 24:354–362

Chanyasak V, Kubota H (1981) Carbon/organic nitrogen ratio in water extract as measure of compost degradation. J Ferment Technol 59:215–219

Ciavatta C, Govi M, Vittori Antisari L, Sequi P (1990) Characterization of humified compounds by extraction and fractionation on solid polyvinilpyrrolidone. J Chromatogr 509:141–146

Cooperband LR, Stone AG, Fryda MR, Ravet JL (2003) Relating compost measures of stability and maturity to plant growth. Compost Sci Utilize 11:113–124

Cunha-Queda AC, Vallini G, Agnolucci M, Coelho CA, Campos L, de Sousa RB (2002) Microbiological and chemical characterisation of composts at different levels of maturity, with evaluation of phytotoxicity and enzymatic activities. In: Insam H, Riddech N, Klammer S (eds) Microbiology of composting. Springer, Berlin, Germany, pp 346–355

Dees PM, Ghiorse WC (2001) Microbial diversity in hot synthetic compost as revealed by PCR-amplified rRNA sequences from cultivated isolates and extracted DNA. FEMS Microbiol Ecol 35:207–216

Deobald LA, Crawford DL (1987) Activities of cellulase and other extracellular enzymes during lignin solubilization by *Streptomyces viridosporus*. Appl Microbiol Biotechnol 26: 158–163

Deportes J, Benoit-Guyod ZD (1995) Hazard to man and the environment posed by the use of urban waste compost: a review. Sci Total Environ 172:197–222

Diaz MJ, Madejon E, Lopez F, Lopez R, Cabrera F (2002) Optimization of the rate vinasse/grape marc for co-composting process. Process Biochem 37:1143–1150

Domeizel M, Khalil A, Prudent P (2004) UV spectroscopy: a tool for monitoring humification and for proposing an index of the maturity of compost. Bioresour Technol 94:177–184

Domingo JL, Nadal M (2009) Domestic waste composting facilities: a review of human health risks. Environ Int 35:382–389

Falcon MA, Corominas E, Perez ML, Perestelo F (1987) Aerobic bacterial populations and environmental factors involved in the composting of agricultural and forest wastes of the Canary Islands. Biol Waste 20:89–99

Finstein MS, Miller FC, Strom PF (1986) Waste treatment composting as a controlled system. In: Schonborn W (ed) Biotechnology, vol 8, Biodegradation. VCH, Weinheim, Germany, pp 363–398

Gaind S, Pandey AK, Lata N (2005) Biodegradation study of crop residues as affected by exogenous inorganic nitrogen and fungal inoculants. J Basic Microbiol 45:301–311

Gaind S, Nain L (2007) Chemical and biological properties of wheat soil in response to paddy straw incorporation and its biodegradation by fungal inoculants. Biodegradation 18:495–503

Gaind S, Nain L, Patel VB (2009) Quality evaluation of co-composted wheat straw, poultry droppings and oil seed cakes. Biodegradation 20:307–317

Garcia-Gil JC, Plaza C, Soler-Rovira P, Polo A (2000) Long-term effects of municipal solid waste compost application on soil enzyme activities and microbial biomass. Soil Biol Biochem 32:1907–1913

Gaur AC, Singh G (1995) Recycling of rural and urban wastes through conventional and vermi-composting. In: Tandon HLS (ed) Recycling of crop, animal, human and industrial wastes in agriculture. Fertilizer Development and Consultation Organization, New Delhi, pp 31–49

Genevini P, Adani F, Veeken AHM, Nierop KGJ, Scaglia B, Dijkema C (2002) Qualitative modifications of humic acid-like and core-humic acidlike during high-rate composting of pig faces amended with wheat straw. Soil Sci Plant Nutr 48:143–150

Granit T, Chen Y, Hadar Y (2007) Humic acid bleaching by white-rot fungi isolated from biosolids compost. Soil Biol Biochem 39:1040–1046

Hamoda MF, Abu Qdais HA (1998) Evaluation of municipal solid waste composting kinetics. Resour Conservat Recycl 23:209–223

Hao X, Chang C, Larney FJ, Travis GR (2001) Greenhouse gas emissions during cattle feedlot manure composting. J Environ Qual 30:376–386

Hargreaves JC, Adl MS, Warman PR (2008) A review of the use of composted municipal solid waste in agriculture. Agric Ecosyst Environ 123:1–14

Haug RT (1993) The practical handbook of compost engineering. Lewis, Boca Raton, pp 385–436

Hiraishi A (2000) Effects of weather and other environmental factors on the performance of a composting system using flowerpots for disposal of household biowaste. Seibutsu kogaku 78:301–310

Hoitink HAJ, Stone AG, Han DY (1997) Suppression of plant disease by composts. HortScience 32:184–187

Hoitink HAJ, Boehm MJ (1999) Biocontrol within the context of soil microbial communities: a substrate-dependent phenomenon. Annu Rev Phytopathol 37:427–446

Hue NV, Liu J (1995) Predicting compost stability. Compost Sci Utilize 3:8–15

IPCC (2007) Climate change 2007: Intergovernmental Panel on Climate Change (IPCC) fourth assessment report (AR4). Cambridge University Press, Cambridge, UK

Itävaara M, Venelampi O, Vikman M, Kapanen A (2002) Compost maturity: problems associated with testing. In: Insam H (ed) Microbiology of composting. Springer, Heidelberg, Germany, pp 373–382

Jefferson M (2006) Sustainable energy development: performance and prospects. Renew Energ 31:571–582

Jimenez IE, Garcia VP (1989) Evaluation of city refuse compost maturity: a review. Biol Waste 27:115–142

Kalamdhad AS, Pasha M, Kazmi AA (2008) Stability evaluation of compost by respiration techniques in a rotary drum composter. Resour Conservat Recycl 52:829–834

Kamm B, Gruber PR, Kamm M (2006) Biorefineries: industrial processes and products. VCH, Weinheim

Kanazawa S, Yamamura T, Yanagida H, Kuramoto H (2003) New production technique of biohazard-free compost by the hyper-thermal and aerobic fermentation method. Soil Microorganisms 58:105–114

Kartha S, Larson ED (2000) Bioenergy primer: modernized biomass energy for sustainable development, energy and atmosphere programme. United Nations Development Programme, New York

Kato K, Miura N (2008) Effect of matured compost as a bulking and inoculating agent on the microbial community and maturity of cattle manure compost. Bioresour Technol 99:3372–3380

Kato K, Miura N, Tabuchi H, Nioh I (2005) Evaluation of maturity of poultry manure compost by phospholipid fatty acid analysis. Biol Fertil Soils 41:399–410

Kausar H, Sariah M, Saud HM, Alam MZ, Ismail MR (2010) Development of compatible lignocellulolytic fungal consortium for rapid composting of rice straw. Int Biodeterior Biodegradation 64:594–600

Kuroda K, Hanajima D, Fukumoto Y, Suzuki K, Kawamoto S, Shima J, Haga K (2004) Isolation of thermophilic ammonium-tolerant bacterium and its application to reduce ammonia emission during composting of animal wastes. Biosci Biotechnol Biochem 68:286–292

Lal R (2008) Crop residues as soil amendments and feedstock for bioethanol production. Waste Manag 28:747–758

Laor Y, Avnimelech Y (2002) Fractionation of compost-derived dissolved organic matter by occulation process. Org Geochem 33:257–263

Lata, Gaind S, Pandey AK, Grover M, Surya Kalyani S, Vasudevan V (2009) Practical applications of bioactive compost in organic agriculture, agriculturally important microorganisms Vol. II, Practical applications of bioactive compost in organic agriculture

Levine JS (1996) Biomass burning and global change. In: Levine JS (ed) Remote sensing and inventory development and biomass burning in Africa, vol 1. The MIT Press, Cambridge, USA, p 35

Liang C, Das KC, McClendon RW (2003) The influence of temperature and moisture contents regimes on the aerobic microbial activity of a biosolids composting blend. Bioresour Technol 86:131–137

Li H, White D, Lamza KA, Berger F, Liefert C (1998) Biological control of *Botrytis*, *Phytophthora* and *Pythium* by *Bacillus subtilis* Cot1 and CL27 of micropropagated plants in high humidity fogging glasshouses. Plant Cell Tissue Organ Cult 52:109–112

Lopez MJ, Elorrieta MA, Vargas-Garcia MC, Suarez-Estrella F, Moreno J (2002) The e.ect of aeration on the biotransformation of lignocellulosic wastes by white-rot fungi. Bioresour Technol 81:123–129

Lucas JA (1998) Plant pathology and plant pathogens. Blackwell, Oxford, p 247

Malherbe S, Cloete TE (2002) Lignocellulose biodegradation: fundamentals and applications. Rev Environ Sci Biotechnol 1:105–114

Martin FN (2003) Development of alternative strategies for management of soilborne pathogens currently controlled with methyl bromide. Annu Rev Phytopathol 41:325–350

McKinley VL, Vestal JR (1985) Physical and chemical correlates of microbial activity and biomass in composting municipal sewage sludge. Appl Environ Microbiol 50:1395–1403

Michel FC, Forney LJ, Huang AJF, Drew S, Czuprendski M, Lindeberg JD, Reddy CA (1996) Effects of turning frequency, leaves to grass mix ratio and windrow vs. pile configuration on the composting of yard trimmings. Compost Sci Utilize 4:126–143

Miller RL, Keularts JLW, Street JRWE, Pound RW, Shane W (1985) Control of turfgrass pests. Ohio Cooperative Extension Service Leaflet 187. The Ohio State University, Columbus

Mondini C, Fornasier F, Sinicco T (2004) Enzymatic activity as a parameter for the characterization of the composting process. Soil Biol Biochem 36:1587–1594

Moreira R, Sousa JP, Canhoto C (2008) Biological testing of digested sewage sludge and derived composts. Bioresour Technol 99:8382–8389

Nakasaki K, Kubo M, Kubota H (1996) Production of functional compost which can suppress phytopathogenic fungi of lawn grass by inoculating *Bacillus subtilis* into grass clippings. In: de Bertoldi M, Bert P, Tiziano P (eds) The science of composting. Blackie Academic and Professional, London, pp 87–95

Nakasaki K, Sasaki M, Shoda M, Kubota H (1985a) Change in microbial numbers during thermophilic composting of sewage sludge with reference to CO2 evolution rate. Appl Environ Microbiol 49:37–41

Nakasaki K, Sasaki M, Shoda M, Kubota H (1985b) Characteristics of mesophilic bacteria isolated during thermophilic composting of sewage sludge. Appl Environ Microbiol 49:42–45

Nakasaki K, Akiyama T (1988) Effects of seeding on thermophilic composting of household organic waste. J Ferment Technol 66:37–42

Naylor LM (1996) Composting. Environ Sci Pollut Ser 18:193–269

Neklyudov AD, Fedotov GN, Ivankin AN (2006) Aerobic processing of organic waste into composts. Appl Biochem Microbiol 42:341–353

Nelson DW, Sommers LE (1996) Total carbon, organic carbon, and organic matter. In: Sparks DL (ed) Methods of soil analysis, Part 3. Chemical methods. SSSA Book Series no 5. SSSA-ASA, Madison, WI

Ozeimen D, Karaosmanoglu F (2005) Production and characterization of bio-oil and biochar from rapeseed cake. Renew Energ 29:779–787

Pandey AK, Gaind S, Ali A, Nain L (2009) Effect of bioaugmentation and nitrogen supplementation on composting of paddy straw. Biodegradation 20:293–306

Pascual JA, Ayuso M, Garcia C, Hernández T (1997) Characterization of urban wastes according to fertility and phytotoxicity parameters. Waste Manag Res 15:103–115

Pérez J, Muñoz-Dorado J, de la Rubia T, Martinez J (2002) Biodegradation and biological treatments of cellulose, hemicellulose, and lignin: an overview. Int Microbiol 5:53–63

Peters S, Koschinsky S, Schwieger F, Tebbe CC (2000) Succession of microbial communities during hot composting as detected by PCR-single-strand-conformation polymorphism-based genetic profiles of small-subunit rRNA genes. Appl Environ Microbiol 66:930–936

Phae CG, Shoda M (1990) Expression of the suppressive effect of *Bacillus subtilis* on phytopathogens in inoculated composts. J Ferment Bioeng 6:409–414

Provenzano MR, de Oliveira SC, Santiago Silva MR, Senesi N (2001) Assessment of maturity degree of composts from domestic solid wastes by fluorescence and Fourier transform infrared spectroscopies. J Agric Food Chem 49:5874–5879

Ranalli G, Bottura G, Taddei P, Garavani M, Marchetti P, Sorlini C (2001) Composting of solid sludge residues from agricultural and food residues. Bioindicators of monitoring and compost maturity. J Environ Sci Health A 36:415–436

Richard TL, Walker PL (1999) Oxygen and temperature kinetics of aerobic solid state biodegradation. In: Bidlingmaier W, de Bertoldi M, Diaz LF, Papadimitriou EK (eds) Proceedings of the international conference ORBIT 99, "Biological treatment of waste and the environment", Rhombos-Verlag, Berlin, pp 85–91

Ryckeboer J, Mergaert J, Vaes K, Klammer S, De Clerq D, Coosemans J, Insam H, Swings J (2003a) A survey of bacteria and fungi occurring during composting and self-heating processes. Ann Microbiol 53:349–410

Ryckeboer J, Mergaert J, Coosemans J, Deprins K, Swings J (2003b) Microbiological aspects of biowaste during composting in a monitored compost bin. J Appl Microbiol 94:127–137

Saludes RB, Iwabuchi K, Miyatake F, Abe Y, Honda Y (2008) Characterization of dairy cattle manure/wallboard paper compost mixture. Bioresour Technol 99:7285–7290

Schlosser O, Huyard A (2008) Bioaerosols in composting plants: occupational exposure and health. Environ Risque Sante 7:37–45

Singh A, Billingsley K, Ward O (2006) Composting: a potentially safe process for disposal of genetically modified organisms. Crit Rev Biotechnol 26:1–16

Smith SR (2009) A critical review of the bioavailability and impacts of heavy metals in municipal solid waste composts compared to sewage sludge. Environ Int 35:142–156

Strauch D (1987) Microbiological specifications of disinfected compost. In: De Bertoldi M, Ferranti MP, L'Hermite P, Zucconi F (eds) Compost: production, quality and use. Elsevier, London, pp 210–229
Strauch D (1996) Occurrence of microorganisms pathogenic for man and animals in source separated biowaste and compost importance, control, limits and epidemiology. In: De Bertoldi M, Sequi P, Lemmes B, Papi T (eds) The sciences of composting. Blackie Academic and Professional, Glasgow, UK, pp 224–232
Stetter KO (1998) Hyperthermophiles: isolation, classification, and properties. In: Horikoshi K, Grant WD (eds) Extremophiles: microbial life in extreme environments. Wiley-Liss, New York, pp 1–24
Stentiford EI (1996) Composting control: principles and practice. In: de Bertoldi M, Sequi P, Lemmes B, Papai T (eds) The sciences of composting. Blakie Academic and Professional, Glasgow, UK, pp 49–59
Strom PE (1985a) Effect of temperature on bacterial species diversity in thermophilic solid-waste composting. Appl Environ Microbiol 50:899–905
Strom PE (1985b) Identification of thermophilic bacteria in solid-waste composting. Appl Environ Microbiol 50:906–913
Termorshuizen AJ, Moolenaar SW, Veeken AHM, Blok WJ (2004) The value of compost. Rev Environ Sci Biotechnol 3:343–347
Thambirajah JJ, Zukali MD, Hashim MA (1995) Microbiological and biochemical changes during composting of palm empty fruit-bunches. Effect of nitrogen supplementation on the substrate. Bioresour Technol 52:133–144
Tiquia SM, Tam NFY, Hodgkiss IJ (1996) Microbial activities during composting of spent pig-manure sawdust litter at different moisture contents. Bioresour Technol 55:201–206
Tiquia SM, Richard TL, Honeyman MS (2000) Effects of windrow turning and seasonal temperatures on composting hog manure from hoop structures. Environ Technol 21:1037–1046
Tseng DY, Chalmers JJ, Tuovinen OH, Hoitink HAJ (1995) Characterization of a bench-scale system for studying the biodegradation of organic solid wastes. Biotechnol Prog 11:443–451
Tuomela M, Vikman M, Hatakka A, Itavaara M (2000) Biodegradation of lignin in a compost environment: a review. Bioresour Technol 72:169–183
USEPA (2002) Use of composting for biosolids management. Biosolids Technology Fact Sheet. EPA 832F-02-024, September 2002, Washington, DC, USA
Van Heerden I, Cronje C, Swart SH, Kotze JM (2002) Microbial, chemical and physical aspects of citrus waste composting. Bioresour Technol 81:71–76
Vuorinen AH (2000) Effect of bulking agent on acid and alkaline phosphomonoesterase and b-D-glucosidase activities during manure composting. Bioresour Technol 75:113–138
Walker R, Powell AA, Seddon B (1998) *Bacillus* isolates from the spermosphere of peas and dwarf French beans with antifungal activity against *Botrytis cinerea* and *Pythium* species. J Appl Microbiol 84:791–801
Whipps JM (1997) Developments in the biological control of soil-borne plant pathogens. Adv Bot Res 26:11–34
Wichuk KM, McCartney D (2007) A review of the effectiveness of current time temperature regulations on pathogen inactivation during composting. J Environ Eng Sci 6:573–586
Wu L, Ma LQ (2001) Effects of sample storage on biosolids compost stability and maturity evaluation. J Environ Qual 30:222–228
Wu L, Ma LQ, Martinez GA (2000) Comparison of methods for evaluating stability and maturity of biosolids compost. J Environ Qual 29:424–429
Zmora-Nahum S, Markovitch O, Tarchitzky J, Chen Y (2005) Dissolved organic carbon (DOC) as a parameter of compost maturity. Soil Biol Biochem 37:2109–2116
Zoebelin H (2001) Dictionary of renewable resources. VCH, Weinheim

Chapter 7
Bioaugmentation for In Situ Soil Remediation: How to Ensure the Success of Such a Process

Thierry Lebeau

7.1 Introduction

This chapter aims at giving an up-to-date insight of in situ soil bioaugmentation, one of the most considered bioremediation methods. Biological soil remediation is a rather environmental sustainable solution for the treatment of contaminated soils. However, it is not sufficiently breaking through since depositing of contaminated soil at landfill sites is stimulated and/or legally permitted in most countries. Indeed, the destiny of polluted soil is still badly defined by laws, e.g., in the European Economic Community (EEC). An EEC soil directive is still in discussion and its acceptance will undoubtedly open a huge market for soil bioremediation. In his report for the Directorate General Research of the European commission, Vijgen (2002) concluded: "Looking at the available numbers, the prospects for any kind of treatment technology look almost non-existent. Only 10–25% of all excavated contaminated soils are treated. Assuming a simple equal division according to the three main techniques (thermal, physical/chemical and biological) it would appear that only a small part of the soil (~3–8%) excavated from remediation sites has any real chance of going to biological treatment", bioaugmentation being near 0%. This same author stated that, "despite economically and ecologically positive aspects of bioremediation along with the possibility to reuse the soil after the treatment, this technology suffers form a lack of reliability".

After giving an accurate definition of soil bioaugmentation, the following chapters make an inventory of the biotic and abiotic soil parameters (including sediment) exerting some effects on soil bioaugmentation. Thereafter, some recommendations are suggested to increase the reliability of this technology in spite of the continuous changes of the environmental conditions. The selection and implementation of the

---

T. Lebeau (✉)
Dpt Génie Biologique, Equipe Dépollution Biologique des Sols (EDBS), EA 3991 LVBE, University of Haute-Alsace, 29 rue de Herrlisheim, BP 50568, 68008 Colmar, France
and
UMR LPGN 6112 CNRS, Université de Nantes, 2 rue de la Houssinière, BP 92208, 44322 Nantes cedex 3, France
e-mail: thierry.lebeau@uha.fr; thierry.lebeau@univ-nantes.fr

A. Singh et al. (eds.), *Bioaugmentation, Biostimulation and Biocontrol*, Soil Biology 28,
DOI 10.1007/978-3-642-19769-7_7,

microbial inocula, the relevance of plant–microorganism association in the aim at better controlling this in situ bioremediation process, and the need in gaining default values involving the development of new monitoring methods are especially tackled. The emerging ecological engineering concept which combines process management and ecology is then presented as it could help us in increasing the reliability of bioremediation technologies, in particular bioaugmentation. Lastly, bioaugmentation performances are compared with bioattenuation and biostimulation.

7.2 How to Define and When to Use Bioaugmentation?

In their reviews, Gentry et al. (2004) and Singer et al. (2005) reminded that bioaugmentation is not new and is already being practiced in agriculture since the 1800s for growing legumes with symbiotic *Rhizobium* spp. or *Bradyrhizobium* spp. and later with free living bacteria such as *Azotobacter* and plant growth promoting rhizobacteria (e.g., *Azospirillum* spp.). Bioaugmentation also served for fighting plant pathogens and for food preparation (milk fermentation, beer, etc.).

In the last two decades, bioaugmentation focussed on soil remediation (Vogel 1996), but about half of the papers were published in the last five years. The consultation of the ISI Web of Knowledge database returned about 4,000,450 and less than 100 articles in response to respectively "soil+bioremediation", "soil +bioaugmentation", and "in situ+soil+bioaugmentation". The first papers in response to "in situ+soil+remediation" were published in 1989 (Portier et al. 1989).

Several similar definitions of bioaugmentation applied to soil pollution were proposed. The following two definitions can be kept in mind: (a) definition given by El Fantroussi and Agathos (2005), i.e., "the technique for improvement of the capacity of a contaminated matrix (soil or other biotope) to remove pollution by the introduction of specific competent strains or consortia of microorganisms" and (b) a broadened one suggested by Dejonghe et al. (2001): "this approach corresponds to increasing the metabolic capabilities of the microbiota present in the soil…". In that respect, bioaugmentation corresponds to an increase in the gene pool and, thus, the genetic diversity of that site.

Bioaugmentation thus encompasses the inoculation of single cells or consortia, microorganisms harboring degradation genes to be transferred to indigenous microflora, and rhizospheric microorganisms or not. Rhizoremediation, a bioaugmentation technique, was proposed by Kuiper et al. (2004) when rhizospheric microorganisms associated with plants are involved in remediation issues. The authors referred only to microorganisms inoculated in soil, the stimulation of the indigenous microorganisms of the rhizosphere being an alternative. Therefore, I suggest using either the terms rhizobioaugmentation or rhizobiostimulation.

Several authors considered that bioaugmentation should be applied when the bioattenuation and biostimulation have failed (Forsyth et al. 1995; Vogel 1996; Iwamoto and Nasu 2001; El Fantroussi and Agathos 2005; Mrozik and Piotrowska-Seget 2010). The main reasons are put forward: (a) low or nondetectable number of degrading microbes regarding the pollutant to be degraded, (b) compounds or mixtures of compounds requiring several metabolic pathways operating simultaneously with sometimes metabolic intermediates whose toxicity toward indigenous microbes may be high, and (c) some polluted areas requiring long microbial adaptation period of time justifying soil bioaugmentation.

The low cost of biological methods also argues for bioaugmentation. Costs (€t^{-1} of soil) vary between (Khan et al. 2004): 10–20 (phytoremediation), 30–60 (landfarming), and 15–70 (bioventing) for biological methods against 25–250 (soil flushing), 80–330 (solidification/stabilization), and 200–700 (incineration) for physiochemical solutions. The cost for in situ soil bioaugmentation is higher on average than other bioremediation methods without being however accurately estimated at present.

7.3 Which Parameters Control Inoculated Microorganisms?

Main parameters to be considered in bioremediation technologies were rewieved by Boopathy (2000) and Hazen and Stahl (2006), among others. Table 7.1 gathers the major abiotic and biotic parameters with an emphasis on bioaugmentation.

7.3.1 Abiotic Parameters

7.3.1.1 Temperature, pH, and Redox Potential

Bioremediation is influenced by temperature in the range 5–30°C (Diels and Lookman 2007) meaning that in temperate countries, in situ bioremediation is less effective, indeed ineffective during winter.

pH also acts strongly upon bioremediation performances and on the length of the bioremediation efficiency in the range 5–8, which are usual values encountered in soil (Diels and Lookman 2007; Grundmann et al. 2007). The most critical factor affecting atrazine degradation by two different microbial consortia was the soil pH and its organic matter content (Goux et al. 2003). Atrazine degradation was immediately effective at pH >7. While only one consortium could degrade atrazine at pH 6.1, bioaugmentation was ineffective at pH 5.7 (possible interaction with organic matter). The structure of the bacterial community showed that the poor degradation rate was not due to the lack of degrader survival but rather to an inhibition of atrazine catabolism. pH for phenol and TCE degradation was shown to vary depending on whether microbial cells were free or immobilized (Chen et al. 2007). The degradation was observed at pH >8 for immobilized cells, whereas pH between 6.7 and 10 did not affect the degradation with the free cells, most probably at the result of the mass transfer limitation that prevents physicochemical changes in the matrix immobilizing cells in spite of continuous modification in the surrounding environment.

Aerobic conditions are important for the soil cleaning-up. The use of O_2 as final electron acceptor is indeed the most energetically favorable reaction and is

Table 7.1 Major factors affecting bioaugmentation (adapted from Boopathy 2000 and Hazen and Stahl 2006)

Main factors	Some comments
Microbial factors: Survival and growth (amount of biomass) • Enzyme induction and activity • Metabolic activity • Production of toxic metabolites from degradation compounds • Microbial diversity • Microbial interaction (competition, mutualism, symbiosis, etc.) with indigenous populations • Mutation and horizontal gene transfer regarding GMO and wild type strains	• Monoxygenases and dioxygenases: two of the primary enzymes employed by aerobic organisms during organic contaminants degradation • Siderophores: molecules employed for the complexation of metals
Environmental factors: • pH • Temperature • Moisture content • Eh • Availability of electron acceptors • Availability of nutrients (carbon and energy source, mineral nutrients) • Toxic molecules	• Differentiation of surface soils and groundwater sediments by the organic matter level • Close relationship between soil humidity and microbial transport
Growth substrates: • Amount and bioavailability of substrates and contaminants (when they serve as substrates) • Limiting factors (C, N, etc.) • Preference for other substrates than contaminants	• Importance of organic matter as a source of substrates, a sorbant for pollutants and a microniche for inoculated microorganisms • Versatile effect of organic matter and minerals on pollutant degradation depending on the microbial metabolic traits (active metabolism vs. cometabolism). Some substrates possibly more easily metabolizable than contaminants depending on the bioavailability of rhizospheric exudates, Phophorous, etc.
Physicochemical bioavailability of pollutants: • Equilibrium and irreversible sorption • Solubility/miscibility in/with water • Complexation (ex: siderophores) • Mass transfer limitations • Toxicity of contaminants	• Decrease of organic contaminants bioavailability in the course of the time resulting from (a) chemical oxidation reaction incorporating contaminant into natural organic matter, (b) slow diffusion in very small pores and absorption into organic matter, (c) formation of semi-rigid films around non aqueous phase liquids (NAPL) with a high resistance toward NAPL-water mass transfer • Rate at which microbial cells can convert contaminants during bioremediation as a result of: (a) contaminant transfer rate to the cell and (b) contaminant uptake and metabolism
Biological aerobic vs. anaerobic process: • Oxidation/reduction potential • Availability of electron donor/acceptor	• Role of macrophytes in supplying O_2 in anoxic environments thanks to aerenchyma

likely to stimulate a number of biodegradation reactions. Empirically under aerobic conditions, a yield of 0.05–0.6 mol biomass mol carbon^{-1} can be obtained. Under anaerobic conditions, the yield falls to 0.04–0.083 mol carbon^{-1} (Diels and Lookman 2007). However, different bioremediation mechanisms requiring different final electron acceptors depend on contaminant characteristics. As a result of the highly reduced state of petroleum hydrocarbons, the preferred and most thermodynamically relevant terminal electron acceptor for microbial process is O_2. On the contrary, degradation of chlorinated solvents, depending on the degree of halogenation, is different from that of petroleum hydrocarbons and other oxidized chemicals, and the preferred redox condition is anaerobiosis (negative correlation between number of H_2 and Eh) (Diels and Lookman 2007). In hydromorphic soils or sediments where Eh is very low (<200 mV), the only one solution in maintaining aerobic conditions in the vicinity of inoculated microorganisms consists in using macrophytes as extensively reviewed by Stottmeister et al. (2003). Due to their specific tissue, aerenchym, a film of 1–4 mm thickness maintains redox gradients ranging from about −250 mV as frequently measured in reduced rhizospheres to about +500 mV directly on the root surface. Oxygen release rates are highest at $-250\,\text{mV} < \text{Eh} < -150\,\text{mV}$. Oxygen supply both protects the roots from toxic components in the anoxic, usually extremely reduced rhizosphere, and allows aerobic heterotrophic microorganisms to both grow and quickly degrade organic compounds.

7.3.1.2 Nutrients Origin, Availability, and Effect on Microbial Transport

Nutrients at inoculated microorganism's disposal are those accessible in soil for a more or less long period, i.e., organic matter as well as minerals in more or less available forms. Additional nutrients can be supplied to soil at the time of bioaugmentation in the aim at avoiding nutritional limiting factors. Ratios such as C/N in soil as well as C/S and C/P must be similar to those of microorganisms, i.e., 20, 200, and 300, respectively.

Organic matter is a source of substrate, a sorbant for pollutants and a microniche material for inoculated microorganisms. In their review, Mrozik and Piotrowska-Seget (2010) emphasized that organic matter is one of the most important soil parameters influencing the effectiveness of bioaugmentation as it plays a crucial role in bioavailability of pollutants and impairs the survival of inoculated strains. Nonetheless, nutrients in soil or sediment (from supply or not) may bring about opposite effects on microorganisms used in bioaugmentation depending on their metabolic traits and ecological considerations. Accordingly the pollutant degradation will be partial or total (mineralization). In case of cometabolism, the additional source of nutrients will be used for the microbial growth and nonspecific enzymes, mainly synthesized for cellular detoxification, and will be able to partly and fortuitously degrade pollutants. Metabolites from the compound degradation are most of the time observed in the culture medium and can be more toxic than the parental molecules as shown for 3,4-dichloroaniline, the main metabolite of diuron herbicide as a final degradation product whose toxicity is slightly below 100 times that of diuron (Tixier et al. 2002). On the contrary, active metabolism is characterized by a total degradation (mineralization) of pollutants by specific enzymes. However, mineralization can be impaired by a more easily metabolizable source of nutrients. Regarding the biodegradation of the glyphosate herbicide, some microorganisms were shown to be unable to mineralize glyphosate in the presence of phosphate in the culture media (Balthazor and Hallas 1986).

Another factor limiting bioaugmentation is the transport of selected microorganisms to the contaminated zones (Borges et al. 2008; Priestley et al. 2006). In particular, the characterization of the physiological responses of the inoculated microorganisms to starvation is fundamental to anticipate the success or failure of such a technique. It was shown by Borges et al. (2008) that the efficiency of bacterial transport through soils might be potentially increased by nitrogen starvation as a consequence of a significant reduction of the adhesion of inoculated bacteria to soil particles due to an alteration of the cell-surface hydrophobicity and cell adhesion to soil particles by bacterial strains previously characterized as able to use benzene, toluene, or xilenes as carbon and energy sources. Starvation also reduces the cell size potentially increasing the microbial transport (Caccavo et al. 1996; Lappin-Scott and Costerton 1992). However, the cell transport in soil was shown to be also governed by the dilution rate and other nutrients than nitrogen, e.g., succinate (Priestley et al. 2006). At a specific dilution rate, cells from nitrate-limited cultures were retained more strongly than cells from RDX-limited or succinate-limited cultures. In their review, Gentry et al. (2004) reported other techniques such as the use of adhesion-deficient strains with the experiment of Streger et al. (2002). A culture was passed several times through sterile sediment. While quite all strains were initially retained in the sediment, only 39% were retained after 27 passes as a result of higher hydrophobicity of the microbial cell surface. While these techniques are potentially able to enhance the transport of inoculants, the degrading capabilities must be evaluated. Conversely, the microbial transport can be reduced when means of growing inoculated biomass are used. Indeed, slimes and extracellular polymers may form, as a result of biofilm formation (Diels and Lookman 2007).

7.3.1.3 Soil Humidity and Hydraulic Regime

Soil humidity indirectly impacts on the soil aeration. Indeed, one liter of water (saturated in O_2) contains about 8 mg O_2 as compared to 300 mg for one liter of air suggesting sustaining accurate soil drainage (Diels and Lookman 2007).

In addition to the soil humidity, hydraulic properties of the soil on the microbial transport must be considered. While bioremedial feasibility studies often focus on soil chemical properties, more consideration should be given to the physical and hydraulic properties of the soil as well (Kinsall et al. 2000). Heterogeneity in hydraulic properties of porous media can limit microbial dispersion and resultant microbial activities. The soil texture, rather than porosity, was the most significant factor controlling the hydraulic properties and consequently the microbial transport. These findings demonstrate how apparent homogeneity in media properties does not equate with homogeneity in flow or transport of solutes and colloids. Microbial transport is directly related to the frequency of irrigation and length of the intervals between irrigation periods, making these variables important factors to consider when applying bioaugmentation through downward percolating water (Mehmannavaz et al. 2001). Accordingly other parameters measured after bacterial bioaugmentation are water infiltration, moisture loss, and surface hardness of inoculated soils.

7.3.1.4 Pollutant Accessibility and Availability for Microorganisms: A Key Issue

Although abundant literature focuses on the degradative performances of microorganisms, pollutant bioaccessibility and pollutant bioavilability are insufficiently taken into consideration, not to mention that the definition of these two words most often vary from one author to another. An explicit definition was given by Semple et al. (2004): "a bioavailable compound [was defined] as that is freely available to cross an organism's cellular membrane from the medium the organism inhabits at a given time" whereas a bioaccessible compound corresponds to "what is available to cross an organism's cellular membrane from the environment, if the organism has access to the chemical". These same authors estimate that until now most routine chemical techniques most often estimate the bioaccessible fraction, which must be taken in consideration as a matter of priority in bioremediation purposes. The concept of bioaccessibility was illustrated by Johnsen et al. (2005) in their review on polyaromatic hydrocarbons (PAH) bioremediation. They point out the fact that a large fraction of the PAH-degrading bacteria in soil is expected to be physically separated from the PAH sources. In soil, as opposed to well-mixed aqueous systems, the substrate consumption leads much faster to mass transfer-limited conditions as the number of cells increases. Therefore, PAH-degrading populations in soil are probably mostly not growing, but they are in a pseudostationary phase.

Regarding organic carbon (OC) compounds, Kd, Koc, and Kow are commonly used to determine the compound partition between particulate and dissolve phases indirectly giving some information about bioavailability and bioaccessibility. Kd is defined as the ratio between the particulate and the dissolved phases and represents a global value of several types of sorption processes. For nonpolar chemicals only sorption to natural organic matter is assumed to be relevant and will contribute to the concentration in pore water. Koc parameter is then defined as a function of the fraction of organic carbon (foc):

$$Koc = \frac{Kd}{foc},$$

where

$$foc = \frac{\text{mass of } OC}{\text{total mass of sorbent}}.$$

Accordingly Koc allows comparing soils with different OC contents. In the aim at being more accurate, OC composition should be taken into account (Pignatello 1998). In the example of pesticides (OC and metals), Table 7.2 shows the parameters governing their sorption in soils. OC of a soil or sediment plays the same role as an immiscible solvent with water, leading to use Kow, i.e., octanol–water partition since it replicates fairly well the partitioning between soil and the soil solution. The value of Koc can indeed be fitted using Kow as follows:

$$\log Koc = a \log Kow + b$$

In soil the relation suggested by Bohn et al. (2001) was the following:

$$\log Koc = -0.99 \log Kow - 0.34$$

These partition ratios were used to calculate some global indicators taken into account the bioavailability and bioaccessibility of organic pollutants in sediments. With the aim at determining whether it is possible to set environmental quality standards (EQS) for sediments from EQS defined for surface waters in the Directive of the European parliament, Dueri et al. (2008) studied, from several experimental data, the relationships between the concentration in water, pore water, and sediments for different families of organic contaminants. They showed that even though in some specific cases there is a coupling between water column – used to define EQS – and sediments, this coupling was rather the exception. Conversely, the dissolved pore water concentration (interstitial water) is related to the sediment toxicity. The EQS developed for water could be then applied to pore water and these authors suggested to calculate sediment EQS from water EQS. Unfortunately, the partitioning coefficient strongly depends on sediment characteristics such as the physical (texture, material, and temperature), chemical (organic matter content, black carbon, pH, redox, etc.), and biological (e.g., bioturbation) properties introducing an important uncertainty in the definition of sediment EQS. Therefore, the direct measurement of pore water concentration is regarded as a better option. Since bioavailability and toxicity seem to be related to the dissolved contaminant fraction, the most suitable method would be to measure directly in pore water. Different pore water sampling techniques are available,

Table 7.2 Parameters acting on soil pesticides

Organic compounds		Mineral compounds
Apolar	Polar	
Clay	Soil humidity	pH
Organic matter	Cation exchange capacity	Eh
Particle size	pH	Fine soil particles
Specific surface	Solubility	Organic matter
Solvant nature	Organic matter	Oxydes and hydroxydes (Fe, Mn, Al)
		Microorganisms

Adapted from Kabata-Pendias (2000) and Delle Site (2000)

mostly relying of passive sampling, but are not yet a standard practice for all environmental laboratories. In case that pore water sampling is not feasible, the partitioning approach can be used to calculate the EQS of sediment, as total sediment concentrations, but an additional partitioning term should be considered for the contaminants binding to soot carbon. Nevertheless, for other contaminants such as pesticides, the level of uncertainty introduced by the method is still between 2 and 3 orders of magnitude, which are not adequate for setting EQS.

Regarding metals, different approaches are able to determine their bioavailability and their bioaccessibility. The bioavailability of metals from soils is considered with respect to a series of single-extraction methods, including the use of ethylenediaminetetraacetic acid (EDTA), acetic acid, diethylenetriaminepentaacetic acid (DTPA), ammonium nitrate, calcium chloride, and sodium nitrate (Dean 2010). These authors also underlined in their review two alternate approaches for assessing the environmental health risk to humans by undertaking in vitro gastrointestinal extraction also known as the physiologically based extraction test. The free ionic metal concentration in the soil solution is a suited indicator of what is extracted by plants corresponding to biodisponibility (Checkai et al. 1987; Csillag et al. 1999). The soil solution recovery by centrifugation is an alternate means of estimating this availability (Csillag et al. 1999; Zhang et al. 2001). Bioavailability is driven by soil pH, CEC, and OM (Kayser et al. 2000).

7.3.2 *Biotic Parameters*

7.3.2.1 Protozoan Grazing, Microbial Competition and Earthworm Effect

After their inoculation, microorganisms must face protozoan grazing and harsh competition with indigenous microorganisms (Gentry et al. 2004 and van veen et al. 1997). Most of the time microbial population declines after a few weeks as shown with *Pseudomonas* sp. ADP that scarcely survived (Moran et al. 2006). As a result, residual simazine removals were the same in bioaugmented and not bioaugmented microcosms after 28 days. It was calculated from data reported in the literature that the decline in cell numbers of *Pseudomonas fluorescens* could vary between 0.2 and 1.0 log unit in 10 days (Van veen 1997). The pollutant level can also impair the microbial colonization as shown with *Sphingomonas chlorophenolica* strain RA2 in a soil contaminated by pentachlorophenol (Colores and Schmidt 1999). Regarding plant growth promoting rhizobacteria (PGPR), competition for limiting resources between introduced and indigenous microorganisms is the most important factor determining PGPR survival (Strigul and Kravchenko 2006). These authors also stated that the most effective PGPR inoculation was expected in organic and mineral-poor soils or stressed soils, when the development of indigenous microflora was inhibited. Another important factor for PGPR survival was compatibility between the composition of the host plant root exudates and the ability of the PGPR to utilize these compounds.

Earthworm bioturbation also acts upon inoculated microorganisms as shown by Monard et al. (2008) in their original study on atrazine-degrading bacteria and atrazine mineralization. Digestion by earthworms significantly impacted atrazine mineralization in bioaugmented soils. Regarding the two atrazine degraders tested, *Pseudomonas* sp. (strain ADP) survived better than *Chelatobacter heintzii* within the 10 days of experiment, although the latter was still metabolically active and able to mineralize atrazine. A positive "burrow-lining" effect on the atrazine gene (atzA) sequence copy number was observed in soil whether bioaugmented with *Chelatobacter heintzii* or not, indicating that burrow-linings form a specific 'hot spot' for atrazine degraders, including indigenous bacteria.

7.3.2.2 Characteristics of Microorganisms Selected for Bioaugmentation

In addition to abiotic and biotic characteristics of environment acting on bioaugmentation functioning, autecological properties of introduced microorgnanisms that play a crucial role in the success or failure of such a technique are thought to be important for in contaminated soils. Regarding PAH degradation by bacteria, Johnsen et al. (2005) identified efficient biofilm formation, cell-surface hydrophobicity, surfactant production, motility, and chemotaxis. The application of bacteria that exhibit chemotaxis toward pollutants has received less attention. Cells displaying chemotaxis can sense chemicals such as those adsorbed to soil particles in a particular niche and swim toward them (Singh et al. 2006). Minor differences in the physical properties of bacteria (shape and cell-wall type) can also lead to major differences in transport behavior at the field scale (Becker et al. 2003).

7.4 How to Make Sure of Bioaugmentation Success in Uncontrolled Soil Environments?

7.4.1 Microbial Inoculum Choice: A Crucial Step

7.4.1.1 Microbial Origin and Selection

Microbial selection (Table 7.3) must be undertaken basing at the same time on performances toward pollutants (i.e., tolerance of high concentrations of contaminants and high degradation rate or change in chemical forms for metals), fast growing, survival, and activity of selected microorganisms in a wide range of environmental conditions (Mrozik and Piotrowska-Seget 2010). The selected microorganisms must be well characterized and stored in a culture collection where their catabolic ability must be preserved (Thompson et al. 2005).

Unfortunately, the point regarding the cell survival, and more generally the microbial ecology issues, is often neglected (Thompson et al. 2005; Vogel and

Table 7.3 Some recommendations for the selection of microbial inocula for soil bioaugmentation

Item	Recommendations
Microbial strategy and performance towards pollutants	• Microorganisms exhibiting chemotaxis towards pollutants and secreting polymers: biofilm formation on the surface of hydrocarbons are especially well suited for the treatment of recalcitrant or slow-degrading compounds • Tolerance to wide range of pollutants and high degradation rate for organic compounds or change in chemical forms for metals
Microbial metabolism	• Aerobic vs. anaerobic microorganisms: aerobic microorganisms are more efficient but the choice depends also on the pollutant characteristics (e.g., degradation of low- and high-chlorinated compounds require aerobic and anaerobic conditions respectively) • Active metabolism vs. cometabolism. Microorganisms with active metabolism do not generate undesirable metabolites • Order in which soil nutrients, including pollutants, are used by inoculated microorganisms. Microorganisms that first used other nutrients than pollutants should be avoided • Superbugs: resilient to environment stresses, harboring catabolically superior pollutant-degrading enzymes. Unfortunately, they are most of the time human pathogens precluding field implementation
Microbial growth strategy and growth medium	• Fast (r-) or slow (K-) growing microorganisms: K-strategists are more suitable to survive long term in a habitat, slowly and continuously degrading a contaminant • Selected microorganisms should be able to grow on poor media
Microbial origin of inocula	• Origin: hot springs, pristine places, deep sea, deep underground, deep intestine, sites with industrial activity, river bank soils, sludges • Microorganisms should be ubiquitous and abundant • Selection of microorganisms from the soil to be cleaned up is sometimes more relevant than selection of exogenous microorganisms
Microorganisms-plants association	• Positive effects of plants on microorganisms: survival, higher growth rate and stability of the microbial activity irrespective of the environmental conditions, higher pollutant degradation • Macrophytes are able to support the aerobic microbial growth and pollutant degradation in anaerobic environments such as sediment
Pure microbial strains vs. consortia	• Consortia are more efficient than pure strains in terms of global survival and whole degradation of complex molecules and mixture of pollutants (mutalism and syntrophic effects) • "Natural" are better than "artificial" consortia in supporting harsh conditions but remediation performances are lower than superbug assemblages and are less engineered inocula

Walter 2001) leading to numerous failures that could impair the credibility of such a technique. As underlined by Thompson et al. (2005), there is still a lack of knowledge regarding conditions that suit the best to microorganisms in their natural conditions or habitat. The development of tools to determine in situ these conditions is in progress and must be spread outside research laboratory once their use in routine will be effective. It was also observed that bioremediation efficacy is more likely to rely on the selectivity and specialization of added microorganisms rather than on nutrient load (Hamdi et al. 2007).

First in the selection procedure, it should be kept in mind that bacteria become dominant with depth as the number of actinomycetes and fungi decreases (Boopathy 2000). This fact should be taken into consideration in the microbial selection scheme according to the depth of the area to be cleaned. In their review, Mrozik and Piotrowska-Seget (2010) reported that most experiments dealing with bioaugmentation were carried out using gram-negative bacteria belonging to the genus *Achromobacter*, *Alcaligenes*, *Flavobacterium*, *Pseudomonas*, and *Sphingobium*,. A focus should be also directed to gram-positive bacteria: *Bacillus*, *Mycobacterium*, and *Rhodococcus*. Regarding fungi, *Absidia*, *Aspergillus*, *Mucor*, *Penicillium*, and *Verticillium* should be considered. Among them, 'superbugs', i.e., "microorganisms highly resilient to environmental stresses, harboring catabolically superior pollutant-degrading enzymes", should be selected as a matter of priority (Singer et al. 2005). Unfortunately, these authors also reported that a 'superbug' is often closely related to a human pathogen (Singer et al. 2005) precluding its field implementation (e.g., *Pseudomonas aeruginosa*). Thompson et al. (2005) suggested that preselected bacteria must be identified before being used in bioaugmentation purposes. A subset of the superbug population, denoted as heirloom microorganisms, are 'superbugs' that are maintained and handed down from generation to generation within a research group. These heirloom superbugs are usually easily cultured, fast growing, thoroughly characterized, and accessible worldwide through bioresource centers such as ATCC, BCCM, CBM, and DSMZ.

Up to now, attention has been focused, regarding strain selection, on various unique sites as hot springs and pristine places. Deep sea, deep underground, and deep intestine should also be explored as suggested by Verstraete et al. (2007). In addition to these "natural habitat", a number of sites altered by industrial actions, often unwanted, are now to be earmarked as resources of microbial diversity. For example, soil or sludge from river banks is relevant because of unspecified group of bacteria as the result of the presence of different and time varying pollutants (Dejonghe et al. 2001). In this way, not only the maintenance of microbial culture collections can be justified, but just as well the preservation of special sites. These authors began to suggest a strategy in microbial resource management: (a) who is there by using the genomic methods, (b) who is doing that with 16S rRNA probing in combination with microautoradiography, stable isotope probing, and microarrays, (c) who is doing what with whom by learning about exchanges between different groups of organisms, particularly about types of "trading" of electron donors and/or acceptors. But several information are still missing: (a) minimum differential in delta G – a measure for the change of a system's free energy – to switch to another partner or to another metabolism, (b) importance of the cross-feeding between species and the prevalent transport mechanism between the partners (e.g., mutualism and syntrophy) (review by Orphan 2009), and the need for physical cell-to-cell contact, and (c) architectural configuration of cells in flocs, granules, and biofilms as a response to both substrate supply and sharing between microbial population with some consequences on the transport of microorganisms and hence specific genetic exchange, biodiversity, and coordinate functioning.

When a microbial screening is undertaken from environmental samples, missing some relevant strains is the main fear of microbiologists. Regarding this point, some basic information on microbial ecology was reported by Verstraete et al. (2007). First, Beijerinck axioma was reminded by these authors, i.e., "every microorganism is everywhere". Accordingly specific inoculations would be useless as shown in many examples. But there are as many examples where inoculation is shown to be essential. Verstraete and his colleagues attempted to explain these contradictory examples by taking into consideration the Darwin-based niche assembly theory supplemented by the neutral theory of Hubbell leading them to conclude that microbial ecologists "should give particular attention to a continuous influx of new species into their systems. Microbial communities must be continuously challenged by new arrivals". In addition, according to the Hubbell theory, high number of species is bound to be of little abundance with only some 10% species that has abundance of more than 1%. Accordingly plenty of rare but useful species are not taken into account by the current microbial ecological methods and screening techniques. According to these authors, another principle must be kept in mind: Pareto principle stating that 20% of microorganisms cover for about 80% of the energy flux involved. The last principle concerns the Cooperative Community Continuum (CCC) considering the continuous evolvement of microbial communities and metacommunity basing on the relative proportions of species making them rather than on the absolute abundance of each species. In this principle, general parameters such as overall growth yield and rate, maintenance metabolism rate, etc. should be considered. Verstraete et al. (2007) derived some conclusions from these principles: (a) stable microbial communities indicate small biodiversity reservoir that limits the influx of new microbial species. It can be seen positively in terms of maintenance of the remediation performances and negatively because of the low adaptability potential facing the possible changes of the environmental conditions, (b) making bioaugmentation processes flexible by maintaining both predominant species and other species that are at standby and await their chance to become more active in the CCC suggesting to perform the microbial selection on microbial communities more than on a list of species, and (c) considering the microbial growth strategy. Some microbial species are fast-growing microorganisms (r-strategy also called zymogenous microbes) and act to a minor extent "offensive", while most of them are slow-growing (K-strategy also called autochtonous microbes) microorganisms and act as "defensive". If a strain for bioaugmentation was required that could survive long term in a habitat, slowly and continuously degrading a contaminant, a K strategist would be most suitable. Thompson et al. (2005) suggested that strains selected from populations that have low abundance, or are transient within the source or target habitat, are less likely to persist as inocula, when compared with those that are spatially and temporally ubiquitous. In addition, they must tolerate co-contaminants (metals and organic compounds) and be active, i.e., be able to degrade components and/or to modify the metal speciation. Other approaches of strains selection that have received less attention are the following: (a) microorganisms exhibiting chemotaxis toward pollutants and (b) microorganisms that secrete polymers and form biofilms on the surface of hydrocarbons are especially well suited for the treatment of recalcitrant or slow-degrading compounds.

In a majority of cases, the selected strains ("superbugs") are isolated through enrichment culture that relies on their ability to grow rapidly on synthetic culture media in defined laboratory conditions (Singer et al. 2005). Unfortunately, the composition of these media is most of the time far from the reality and some strains are wrongly selected. For example, in a study devoted to Cd biosorption (Lebeau et al. 2002) the best performing microorganism in a rich synthetic medium showed relatively low performance when it was cultivated in a soil extract medium ("poor" medium), supposed to be closer to the soil composition than the synthetic medium. According to Thompson et al. (2005), strains selected by selective enrichment are not typical or representative of indigenous communities in the target habitat and could equally be derived from transient populations.

Following their isolation from environmental sites, the selected microorganisms (pure cultures, natural, or artificial consortia) are cultivated under laboratory conditions and returned or not to the same soil. Bento et al. (2003) concluded that the best approach was the bioaugmentation performed by inoculating microorganisms preselected from their own environment. Belotte et al. (2003) estimated that isolates were 50% fitter when reintroduced into their home sites and that fitness diminished exponentially with distance from their origin. Since Heinaru et al. (2005) found that contaminants present in studied microcosm determined the presence and activity of specific microorganisms, a second possibility is the selection of appropriate microorganisms from sites with similar contaminants that are present in soils to be cleaned up.

The basic role of plants in the soil microbial functioning should be systematically taken into consideration in the microbial selection schemes. Plant-assisted bioaugmentation is indeed a way to stabilize the mitigation process regarding both microbial survival and activity even in environments subject to variable conditions. The related selection procedure is called rhizo-directed strain selection (Kuiper et al. 2004). In contrast to the bulk soil characterized by its oligotrophy, the rhizospheric soil is supplied with nutrients exudated by plants (Gentry et al. 2004) ensuring a continuous flux of substrates to microorganisms. Microorganisms are able to grow on one or more root exudates within the rhizosphere stimulating and maintaining the pollutant-degrading enzymes induction (Singer et al. 2005). Kuiper et al. (2001) suggested a dual selection: population dominance in the rhizosphere (target habitat) and the ability to degrade target contaminant. The importance of nutrients in determining inocula performance in the rhizosphere at least is not surprising regarding organic acids, as they are significant constituents of root exudates (Goddard et al. 2001). Only organic acids' utilization characteristics, not the presence of flagella and lipopolysaccharides, were found to be able to distinguish transient to stable rhizospheric populations.

7.4.1.2 Considering the Type of Pollutants in the Microbial Selection Scheme

Mechanisms where microorganisms benefit and derive energy for growth from contaminant transformation (i.e., active metabolism) are generally preferable over fortuitous cometabolic process (Rittmann et al. 2006) as they may generate undesirable metabolites as already mentioned above (Sect. 3.1.2). One concern regarding bioremediation purposes should be to guarantee the complete mineralization of the targeted pollutants, not their partial degradation, i.e., dissipation in metabolites of possibly higher toxicity. Yet official regulation most of the time only demands that the pollutant(s) concentration is below a threshold, not that the pollutant is completely mineralized. In this prospect, isotopic approaches are being explored for field validation of contaminant degradation (Löffler and Edwards 2006) and will most probably be used at establishing the future regulation regarding soil.

As already underlined (Sect. 3.1.1), overall aerobically degradated contaminants should be preferred as a result of higher performance, but it depends also on the molecule characteristics. In case of polychlorobiphenyls (PCBs), halogenation degree must be taken into account. Molecules with less than six atoms of chlorine can be aerobically degraded, whereas anaerobic dechlorination is necessary for more than six atoms and goes on with aerobic degradation (Smith et al. 2007). For pollutants of low accessibility such as PAH and PCBs – not very water-soluble pollutants – it is necessary to use strains able to produce surfactants to make these pollutants more accessible (Dua et al. 2002; Johnsen et al. 2005; Mrozik and Piotrowska-Seget 2010), but the result varies with the soil aging. The facility with which hydrocarbons can be removed from soils varies inversely with aging of soil samples as a result of weathering (Trindade et al. 2005). Contradictory results were shown however for PCB degradation when surfactants are used as they may alter the bacterial community implicated in PCB degradation resulting in a decrease of the performance (Chavez et al. 2006). Regarding PCBs, only anionic surfactants play a role in their degradation, although nonionic surfactants washed more of the PCBs from contaminated soil (up to 89%) (Singh et al. 2007). In fact PCBs have lower affinity for the interior of anionic rather than nonionic micelles and this may have promoted release of the PCBs from the micelles, bringing them in contact with the degrading bacteria. Some interesting results were reported with synthetic surfactants such as Tween 80, sodium dodecyl sulfonate, and sodium dodecylbenzene sulfonate. Tween 80 enhanced PCB degradation in aerobic (Liz et al. 2009) and anaerobic conditions (Field and Sierra-Alvarez 2008). Nonetheless, their ecotoxicity was already reported as well as their sometimes chemical instability, their tendency in trapping PCB in the micelles impeding the PCB degradation, and finally their high costs (Xia et al. 2009). Surfactant-producing strains as an alternative were shown to be ubiquitous (Ohtsubo et al. 2004). Biosurfactants are highly specific with a low toxicity and they are easily degradable (Xia et al. 2009). Soil can be thus bioaugmented by either cocultures – one being for the production of surfactants and the other for the PCB degradation – or single cell inoculum with, e.g., *Rhodococcus erythropolis* Z6, which both produces surfactants and degrades PCB up to penta-PCBs (Petric et al. 2007).

Compared to organic pollutants, the in situ remediation of toxic metals in porous matrices (soil and sediment) requires a specific approach since only the chemical form of metals varies. Microbial immobilization of metals or on the contrary their mobilization (e.g., by solubilization or complexation) combined with their extraction by plants can be undertaken.

Nonpoint source contaminations, i.e., moderate metal concentrations, but wide contaminated surfaces typically concern many agricultural soils as a result of repeated applications of both fertilizers and pesticides containing trace metals at various concentrations, along with atmospheric deposits. For example, about 1% of 11,400 French agricultural soil samples analyzed exceed the French limit values in case of sludge recycling for Pb, i.e., 100 mg kg^{-1} dw soil (Mench and Baize 2004). Metal extraction by plants is thus a relevant remediation technology named phytoextraction. Unfortunately, metal phytoextraction suffers from several limitations, the major limit steming from the slowness of the treatment (Baker et al. 2000) as a consequence of the low availability of metals at a given time. Reasonable period for remediation is indeed considered to be less than 5 years (Khan et al. 2000), while much more time is usually required to clean the soils (Baker et al. 2000; Dickinson and Pulford 2005). A promising alternative consists in optimizing the synergistic effect of plants and microorganisms (Glick 2003; Lebeau et al. 2008) by coupling phytoextraction with soil bioaugmentation, called also rhizoremediation (Kuiper et al. 2004). Overall, the uptake of metals by plants can be enhanced by two complementary means: (a) enhancement of the mobility of metals in porous matrices (soil and sediment), resulting in higher metal concentrations in plants, thanks to bioaugmentation with microorganisms producing surfactants (Herman et al. 1995; Mulligan et al. 1999, 2001), siderophores (Diels et al. 1999; Dubbin and Ander 2003), and organic acids (Di Simine et al. 1998; Majewska et al. 2007), and/or (b) enhancement of the plant biomass by associating plants with PGPR (Zhuang et al. 2007) and/or arbuscular mycorrhizal fungi (AMF) (Khan 2006).

An alternative consists in metal immobilization when the requisite time for the metal extraction is too long during which any agricultural activity must be suspended. Therefore, in order to maintain the agricultural activity, in situ metal immobilization is the only one alternative able to avoid any damage for living beings (Gadd and White 1993), even though metals remain in the soil. Soils can be bioaugmented by microorganisms exhibiting high affinity for metals (Volesky and Holan 1995) and can biosorb/precipitate both heavy and toxic metals by various mechanisms. Several experiments have proved that bioaugmentation with symbiotic microorganisms or not may reduce the accumulation of metals in plants (Jézéquel et al. 2005; Jézéquel and Lebeau 2008; Joner et al. 2000; Karagiannidis and Nikolaou 2000; Lovley and Lloyd 2000; Tonin et al. 2001).

Most matrices to be cleaned up are polycontaminated by several organic compounds or metals and often by both. Several microbial strains must be used either for the whole degradation of one or several molecules (mutualist and syntrophic effect) or because some strains are able to reduce the toxicity of some pollutants avoiding other microbial strains involved in the soil cleaning-up to be efficient. The need for several microorganisms was shown for the 2,4-D degradation (Mrozik and Piotrowska-Seget 2009). Bioaugmentation with *Pseudomonas* strain H1 and *Ralstonia eutropha* JMP134 increased 2-4-D degradation in the presence of Cd as compared with microcosm not inoculated or inoculated only with one strain. Intracellular sequestration of Cd by *Pseudomonas* reduced its toxicity for *Ralstonia*, thus improving 2,4-D degradation. Similarly, the dual bioaugmentation strategy for MTBE (methyl-tert-butyl ether) and TCE (trichloroethylene) remediation in the presence of heavy metals was studied by Fernandes et al. (2009). Cocultivating bacterial strains capable of resisting high concentrations of heavy metals (Cd^{2+} or Hg^{2+} or Pb^{2+}) and other able to degrade the common soil and groundwater pollutants MTBE or TCE showed degradation efficiencies well higher (49–182% higher) than those expected under the conditions employed. Regarding the whole degradation of molecules, it is shown for PCB that a mixture of aerobic and anaerobic microorganisms is necessary, suggesting either to sequentially make soil anaerobic for the dechlorination of molecules with more than six atoms of chlorine and then to shift them to aerobic conditions (e.g., H_2O_2 supply and soil turning over) or to create biphasic conditions in soils by using macrophytes releasing oxygen through aeremchyma (Stottmeister et al. 2003).

7.4.1.3 Single Cells vs. Consortia

Microorganisms tend to act synergistically with others (Verstraete et al. 2007). This is all the more true in the environment since very harsh conditions are encountered with most of the time polycontaminations as mentioned above. Thus, most effort should be devoted to the preservation and collection of novel consortia. In many cases, consortia were shown to be more effective than single strains by the fact that intermediates of a catabolic pathway of one strain may be further degraded by other strains possessing suitable catabolic pathway (Kuiper et al. 2004). A defined consortium (five bacteria and one fungus) was used for PAH degradation (anthracene, phenanthrene, and pyrene). Bacterial and fungal isolates from the consortium, when inoculated separately to the soil, were less effective in anthracene mineralization compared to the consortium (Jacques et al. 2008). Sometimes, one microorganism making consortium is used to modify the physical and chemical properties of the pollutant; the other one is chosen for its ability to degrade it as shown by Kumar et al. (2006) for the enhancement of oil degradation by coculture of hydrocarbon-degrading and biosurfactant-producing bacteria.

"Natural" and "artificial" microbial consortia can be used. In the former case, the global performance may be not the highest, but assemblages can generally sustain very harsh conditions as the result of a selection over a long time. Conversely, it could be attractive to create assemblage only comprising superbugs in this aim at obtaining the highest degradative performances. Unfortunately, as mentioned above (Sect. 4.1.1), ecological factors are also to be considered. In their review, Mrozik and Piotrowska-Seget (2010) reported successful experiences with consortia among which consortia made by mixing pure bacterial cultures were shown to be able to degrade, e.g., hydrocarbons such as diesel, crude oil, and engine oil (Ghazali et al. 2004). A consortium of four isolates was generated that proved to be 85% more effective at processing waste working metal fluids than undefined inocula from sewage (Van der Gast et al. 2003, 2004). Some microbial species that have exceptional survival ability improved the persistence of the poorly surviving transient isolates when they are coinoculated together (Goddard et al. 2001). Han et al. (2008) experimented with success the synergy between fungi and bacteria in the bioaugmented remediation of petroleum-contaminated soil. Van der Gast et al. (2004) demonstrated that the strategy of using tailor-made consortia, which links functionally the in situ microbial community structure with the contaminants revealed by chemical analysis, seems to be a promising avenue toward rational selection of effective inocula for bioremediation applications. In addition to their ability to accomplish bioremediation, it was shown that the use of a microbial formula tailored with selected native strains selected for multiple resistance to heavy metals among the native microbial community could take advantage with respect to sensitive strains when heavy metals are present by avoiding inhibition by heavy metals of the respiratory process (Alisi et al. 2009).

7.4.2 *Inoculation Techniques: How to Inoculate Soils?*

Different methods for soil bioaugmentation as well as some key factors involved are reported in Table 7.4.

7.4.2.1 Soil Priming or Activated Soil

"Priming effect" was first reported by Bingemann et al. (1953). Basically, priming is predisposing an isolate or population of microorganisms to future conditions in which they are designed to perform a function (Singer et al. 2005). Practically, priming consists in the addition to a soil to be clean up a same soil or not whose microflora is already adapted to the pollution in a certain proportion, e.g., primed soils represented 2% of the soil to be bioremediated (Lamberts et al. 2008). In this prospect, soil priming and activated soil are closely related. Activated soil is indeed a concept based on the cultivation of microbial biomass from a fraction of the contaminated soil for subsequent use as an inoculum for bioaugmentation of the same soil (Otte et al. 1994) or other matrices such as water (Bester and Schäfer 2009). The only one difference between these two techniques is the following: soil

Table 7.4 Different methods for soil bioaugmentation and factors to be considered

Bioaugmentation methods		Advantages	Drawbacks
Cell bioaugmentation	Priming and activated soils	• Consortia of indigenous microorganisms potentially more resilient to stress than a single isolate and able to ensure several remediation functions • Maintenance of primed consortia within their native soil, potentially enhancing their survival in the target soil • Inclusion of unculturable microorganisms possibly containing one or more highly competent pollutant degraders • Microbial efficiency of activated soil towards pollutants checked before soil bioaugmentation	• Unknown composition of the microbial inoculum • No certainty of the microbial efficiency toward the target pollutants with the priming method
	Plant-microorganism association (rhizoremediation)	• Regular delivery of plant exudates used as substrates for inoculated microorganisms • Rhizospheric area serving as specific ecological niche for rhizobacteria (microbial survival enhancement) • The only one method for metal extraction	• More complicated inoculum formation (microbial-seed coating)
	Culture medium for inoculum production	• High and reproducible cell concentrations using artificial culture medium • Higher survival rate of soil inoculated microorganisms previously cultivated in a medium with composition close to soil	• Microbial stress after inoculation in soil (oligotrophy) • Low cell concentration using culture medium with composition close to soil
	Inoculum formulation	• Higher survival and prolonged period of activity with immobilized cells during storage and after inoculation • Positive effect of nutrients formulated with inoculum (possibility in colocalizing nutrients and inocula by using immobilization technique)	• Not proper distribution of microorganisms in soil profile nor survival guaranteed by free cell culture inocula

(*continued*)

Table 7.4 (continued)

Bioaugmentation methods		Advantages	Drawbacks
	Inoculum size and reinoculatio	• Several successive inoculations with lower cell concentrations better than only one with high concentration	• Bioaugmentation cost closely related to the inoculum size and reinoculation frequency
Gene bioaugmentation	Gene introduction and plasmid transfer	• Gene introduction not dependent to the microbial survival • Introduction by, e.g., self-transmissible plasmid of remediation genes into indigenous microorganisms already adapted to survive and proliferate in the environment	• Legal constraints governing gene scattering

priming aims at directly selecting microbial degraders in the soil to be cleaned up by using, e.g., waste substrates, whereas "activated soil" lies on firstly the priming of a fraction of soil by the addition of the pollutant(s) in the aim at selecting relevant microorganisms and secondly at bioaugmenting the soil to be cleaned up by the activated soil. Priming and activated soil are expandable to solving the issue of co-contaminated soil. Activated soil contains degrader population able to eliminate the target pollutant as a result of the soil previously exposed to pollutants. Not surprisingly, it was shown that during activation, the biodiversity dropped dramatically (Beaulieu et al. 2000). Activated soil serves at the same time of the inoculant, carrier, and source of nutrients without extracting the degraders from the soil (Gentry et al. 2004).

2-, 3-, or 4-chlorobenzoate was degraded by activated soils (Gentry et al. 2004). A significant reduction in the time required for the degradation of pentachlorophenol (PCP) and PAHs in contaminated soil was achieved using activated soil as an inoculum (Otte et al. 1994). Once produced, the activated soil biomass use for PCP degradation was shown to remain active for 5 weeks at 20°C and for up to 3 months when kept at 4°C (Barbeau et al. 1997). Otte et al. (1994) showed that PCP mineralization rate increased from 70 mg L^{-1} per day when no PCP was added to the soil to 700 mg L^{-1} per day when PCP was added. The study of Johnsen et al. (2007) was, however, less conclusive where strong impact was observed on the PAH-degrading community of a PAH-polluted soil but resulted only in a marginal effect on PAH degradation as compared to nonprimed soil. Interesting is the lag phase which was shown to be shorter for PAH biodegradation (Lamberts et al. 2008). The difference between a long-term contaminated soil and an unpolluted soil used as inoculum also results in a range of molecules that are degraded as shown by Wang et al. (2009). Complete biodegradation of benzo(*a*)anthracene and benzo(*b*) fluoranthene only occurred with the HAP-contaminated soil.

Although "priming" and "activated soils" can be criticized as a "black box" method as does biostimulation, this approach is attractive from a practical perspective. It remains to be seen how effective different matrices respond to primin, e.g., clay-primed soil augmented into sandy target soil or high pH primed soil augmented into neutral or low pH target soil and to validate at field scale (Singer et al. 2005).

7.4.2.2 Gene Bioaugmentation

Recently, special attention has been focused on enhancing the biodegradative potential of microorganisms by transfer of packaged catabolic genes from one or more donor strains to indigenous microflora existing in contaminated areas (Mrozik and Piotrowska-Seget 2010). The aim of such a method not only aims at gaining of new degradative pathways but also at making possible the extent catabolic potential of microbial communities.

Some advantages can be put forward, by comparing this method with cell bioaugmentation where introduced microorganisms often do no survive following soil inoculation (Gentry et al. 2004). The main potential advantage of this method is the introduction of remediation genes into indigenous microorganisms already colonizing soil. The transfer of plasmids via conjugaison is the technology most studied in bioaugmentation studies where the long-term survival of the introduced host strain is not required.

7.4.2.3 Inoculum Formulation

Numerous studies use bacteria introduced in liquid culture stage, which does not guarantee proper distribution of bacteria in soil profile, their shelf life, and activity (Mrozik and Piotrowska-Seget 2010).

The most promising option in cleaning-up of contaminated sites seems to be using carrier materials that maintain sufficient activity of inoculants over a prolonged period after release (Cassidy et al. 1996; van veen et al. 1997) as they provide protective niche and possibly temporary nutrition. Using carrier for the storage of inocula avoids also any decrease in microbial survival. Many of the microbial inoculants all over the world are based on solid peat formulations, but other carriers among which cork compost was shown to be the best alternative carrier to peat for inoculant technology (Albareda et al. 2008). This substrate could maintain densities of rhizobia and PGPR similar or higher than those obtained with peat. Artificial immobilization in polysaccharidic matrices is a suited alternative. They imitate biofilms as natural immobilization of microorganisms in their own exoplysaccharides to endure environmental stresses (reviewed by Singh et al. 2006). Organisms within a biofilm can indeed withstand shear forces, nutrient deprivation, pH changes, and toxic molecules in a more pronounced manner. Biofilms also facilitate gene transfer of mobile elements (plasmids) between biofilm organisms.

While contaminants dissolved in aqueous medium can undergo biodegradation, hydrophobic pollutants that remain adsorbed in the nonaqueous phase liquid pose problems (see Sect. 3.1.4). Bacteria access these compounds by either dissolution in the aqueous medium or by direct adhesion thanks to biofilm formation. Microorganisms that form biofilms on the surface of hydrocarbons are especially well suited for the treatment of recalcitrant or slow-degrading compounds because of their high microbial biomass and ability to immobilize compounds by biosorption, bioaccumulation, and biomineralization. Once the cells are brought into contact with a surface, the mechanism of biofilm formation and surfactant production commences, which leads to enhanced bioavailability and biodegradation by either dissolution or direct adhesion of the compound–water interface, a process that is facilitated by biofilm formation. Microbial cell immobilization enhances the microbial survival and then the remediation capability by different means: (a) the slow release of cells from the immobilizing matrice can prolong the degrading activity (Mertens et al. 2006), and (b) immobilization avoids protozoan grazing (Bouchez et al. 2000; Matz and Kjelleberg 2005). For example, immobilized *Burkholderia cepacia* on corncob for the degradation of carbofuran survived through 30 days, while the free cells decreased continuously after 10 days (Plangklang and Reungsang 2009). The stability of the introduced *Arthrobacter protophormiae* cells used for p-nitrophenol degradation was enhanced upon immobilization (Labana et al. 2005).

Several carriers were tested for soil bioremediation, e.g., polyvinyl alcohol (Cunningham et al. 2004), chitin and chitosan (Gentili et al. 2006), vermiculite (Straube et al. 1999), sugarcane bagasse, corncob and corncob powder (Labana et al. 2005; Plangklang and Reungsang 2009), gellan gum (Moslemy et al. 2002), wheat straw (Zhang et al. 2008), organomineral carriers consisting of zeolite-clinoptilolite and humic acids (Dercova et al. 2006), and even earthworm egg capsules (Daane and Häggblom 1999).

As mentioned above (see Sect. 3.1.4), the close contact between microorganisms and pollutants is essential. Some experiments tested the addition of surfactants such as cyclodextrins and rhamnolipids (Berselli et al. 2004; Garon et al. 2002; Straube et al. 1999).

The microbial survival is strongly driven by the availability of nutrients in soil, sometimes requiring formulating inocula with nutrients. Indeed, soil oligotrophy most of the time prevents to greet additional microorganisms or to sustain their growth. Mass balance – available nutrients compared to the additional amount of nutrients necessary to get the desired biomass – should be always calculated to determine the capability of the inoculum to growth (Vogel and Walter 2001). Major difficulty lies in the choice of suited nutrients by taking into account both the microbial nutritional needs and the nutrients availability in soil (including pollutants). The microbial substrate preference was determined for bioremediation of diesel oil contamination by using the method of isotope-labeled CO_2 assimilation to measure the substrate preferences at a single cell level of phylogenetically defined microbial subgroups in bioaugmentation mixtures, based on combined analyses of microautoradiography and fluorescence in situ hybridization (Hesselsoe et al. 2008). Other methods are described in Sect. 7.4.4.

Inocula formulated with nutrients were already tested in bioaugmentation studies. Unfortunetaly, pollutants are sometimes used once nutrients are exhausted, particularly when inoculated microorganisms are metabolically active and when nutrients are metabolically easier to degrade than pollutants. As a result, pollutant degradation is delayed as shown by Kao et al. (2005) using acetate and glucose for PCP degradation (Wolicka et al. 2009) with the use of citrate and yeast extract. The bioremediation performances may even decrease as shown by Alvey and Crowley (1995): 100 mg kg^{-1} of atrazine were totaly degraded in 11 days whithout any nutrient supply, whereas less than 10% were degraded in case of glucose or citrate supply. Another explanation is the nonspecific use of nutrient by microbial population, which is not involved in the pollutant degradation as reported by Carmichael and Pfaender (1997) for PAH. Survival of inoculated microorganisms as well as their bioremediation performances can be most likely enhanced by the supply of carbon substrates colocalized with the inoculum in the immobilization matrix (Duquenne et al. 1999), which could at the same time selectively promote immobilized cell growth and not indigenous microbial population in soil.

When bioaugmentation is associated with the culture of plants, seed adhesion and inoculum incorporation into soil are the two possibilities. Ciccillo et al. (2002)

compared these two methods by using *Burkholderia ambifaria* MCI 7 inoculated in the rhizosphere of maize plants. When applied as a maize seed treatment, *B. ambifaria* MCI 7 promoted plant growth significantly, while opposite effect was observed when incorporated into soil. *B. ambifaria* MCI 7 affected the indigenous microflora of plant bioaugmentation treatment according to the application method: seed treatment reduced the bacterial diversity, whereas incorporation into soil showed the contrary.

7.4.2.4 Inoculum Size and Reinoculation

There is a high controversy about the benefit of inoculation size and reinoculation on the microbial survival and colonization of the soil. Some authors showed positive effects, while others observe marginal positive effects to the best.

Regarding the inoculum size Habe et al. (2001) showed that soil inoculation with 10^{12} CFU kg^{-1} of a 2,3-dichlorodibenzo-p-dioxine degrading strain allows reaching a 80% degradation rate against 46% with 10^{10} CFU kg^{-1}. Similarly, 100% of 50 mg kg^{-1} of fenithrotion was degraded in 4 days by 2×10^{12} CFU kg^{-1} inculum against 15 days with 2×10^{8} CFU kg^{-1} inoculum (Hong et al. 2007). Comeau et al. (1993) estimated that the time for the complete degradation of 2,4-dichlorophenoxyacetic acid by *Pseudomonas cepacia* decreased by 0.5 day by additional log of inoculum. Regarding atrazine, Rousseau et al. (2003) considered that inoculum size greater than 10^{10} CFU kg^{-1} was required to observe an increase of atrazine degradation. Opposite results were shown however (Cassidy et al. 1997; Dechesne et al. 2005); i.e., the increase of the inoculum size did not modify the colonization rate of *Pseudomonas*.

Reinoculation aims at throwing off balance the ecosystem for the benefit of the inoculum. Three successive inoculations compared to only one increased atrazine biodegradation by 13% (Lima et al. 2009) and from 35% up to 90% by reinoculating each 3 days at 10^{11} CFU kg^{-1} during 35 days instead of a sole inoculation (Newcombe and Crowley 1999). These authors used a 500-liter batch fermenter for its ability to deliver inoculum repeatedly to atrazine-contaminated soils via irrigation lines. Similarly, 100% of 2,3-dichlorodibenzo-p-dioxine was degraded in 14 days due to soil reinoculation (10^{12} CFU kg^{-1} each 2 days) against 25% for only one inoculation (Widada et al. 2002). No effect of repeated inoculation was, however, observed by Cassidy et al. (1997) with *Pseudomonas* sp. UG30 for pentachlorophenol degradation

Repeated inoculation, pre-adaptation of cultures, and addition of surfactants showed removal enhancement of polychlorodibenzo-p-dioxines (PCDD) by up to 10.3% upon repeated inoculation, but was not much affected by the addition of surfactant (Nam et al. 2005). Higher degradation rate of phenanthrene by *Arthrobacter* sp. was also observed after three inoculations, the degradation being stopped, however, below an induction threshold concentration for metabolic activity of phenanthrene-degrading bacteria as a consequence of sorbed phenanthrene on soil particles (Schwartz and Scow 2001). Another experiment conducted by Singer et al. (2000) showed on the contrary partial bioremediation of PCB-contaminated soil achieved by repeated applications of PCB-degrading bacteria and a surfactant applied 34 times. Not conclusive results were also reported by Bouchez et al. (2000), although it was sludge, not soil: first bioaugmentation induced a decrease in the pollution level, but inocula were rapidly eaten by protozoa and the second massive inoculation unbalanced the ecosystem with an overgrowth of protozoa.

7.4.2.5 Commercial Inocula

Some commercial inocula were tested in bioaugmentation studies but not for in situ soil bioaugmentation. Bio-Dechlor INOCULUM™ and KB-1™ consortia consisting of multiple *Dehalococcoides* strains together with other bacterial groups, e.g., acetogens, fermentors, and other PCE-to-*cis*-DCE dechlorinators were stably maintained for groundwater remediation purposes (Löffler and Edwards 2006). Biodegradation of six PCB congeners (nos 28, 52, 101, 138, 153, and 168) in transformer oil was evaluated under anoxic, oxic, or anoxic/oxic treatments using commercial consortia among which Synbron was able to degrade 76.0 and 91.3%, respectively, in oxic and anoxic conditions (Sobiecka et al. 2009). Reactor study regarding biodegradation of two commercial diesel fuels, i.e., Diesel and HiQ Diesel spiked to an agricultural soil, by a characterized commercial source of microorganisms and nutrients (Enzyveba) was studied (Di Toro et al. 2008). Enzyveba is a complex and stable consortium of prokaryotic and eukaryotic microorganisms patented and commercialized by Marcopolo Engineering SpA (Cuneo, Italy) as bioactivator for landfills, composting facilities, and wastewater treatment plants. Degradation of crude oil was experimented by comparing eight commercial inocula with natural inocula (activated sludge and tropical aquarium water) (Thouand et al. 1999). Only one commercial inoculum was able to degrade 18% of the crude oil vs. 16–25% for natural inocula. A commercial inoculum was tested for the degradation of diesel oil (Mariano et al. 2008). Some few commercial inocula were reported in the book by Roberts (1998) on remediation of petroleum contaminates soils.

7.4.2.6 Microorganism-assisted Plants and the Vice Versa

Overall, bioaugmentation-assisted plants can enhance directly the remediation performances as plants attract water with their root system, accumulating water-soluble pollutant in the rhizosphere and concluding directly or indirectly with the degradation or translocation of the pollutant. In particular, plants are essential in metal soil cleaning-up as they directly extract and accumulate metals.

Plants play also an indirect part in remediation processes as they sustain a high microbial activity in the rhizosphere due to the continuous flux of substrates released by plants. This important contribution to the degradation of pollutants was ascribed to the increase in the number and metabolic activity of microbes present in the rhizosphere of plants especially used during phytoremediation or of plants which are emerging as natural vegetation on a contaminated site (Kuiper et al. 2004; Romantschuk et al. 2000). For the remediation of organic compounds, plants are almost used as helper for the settlement of inoculated microorganisms since they marginally extract and degrade organic compounds. Some experiments indeed proved that pollutant degradation was almost due to microorganisms more than plants as shown for TNT where axenic plants were able to remove 32–38% of the TNT reaching 80–88% when plants were inoculated with *Pseudomonas putida* JLR11 (Van Dillewijn et al. 2007). Similarly, atrazine removal by *Phragmites australis* required 40 or 7 days when microorganisms were removed or not from rhizosphere (McKinlay and Kasperek 1999). The increase in the number and metabolic activity of microbes is most of the time the result of root exudates that serve as growth substrates. Benizri et al. (2007) showed that the addition of maize root mucilage moderately increased microbial C (+23%) due to high turnover rate of microorganisms consuming this substrate. However, the number of cultivable bacteria was enhanced by 450%. Exudates modified also the metabolic and the genetic structure of the rhizobacteria compared to bulk soil with an overall decrease of the microbial diversity. Counterexample was given however with maize mucilage amendment, which contributed only minor change in the atrazine-degrading community and even reduced the maximal percentage of atrazine mineralization (López-Gutiérrez et al. 2005). Since plant roots are initially small and do not produce large amounts of exudates, when first seeded, the addition of exogenous substrates may be needed to increase initial microbial concentrations at the start of phytoremediation projects (Sung et al. 2006). Plant roots were also suggested to be a substitute for the tilling of soil to improve aeration, thus increasing activity of aerobic populations (Romantschuk et al. 2000). In anoxic environments such as natural and constructed wetlands, macrophytes are able to transport oxygen through the plant right down to the deepest roots thanks to their aerenchyma (see Sect. 3.1.1) and to maintain oxic conditions. Except during winter, oxygen is continuously released in the vicinity of roots at a rate around 100 up to 200 μmol O_2 h^{-1} g^{-1} of root dry mass according to pH, Eh, temperature, and plant (biomass, species, and stage of plant development), thus supporting a continuous microbial activity (Stottmeister et al. 2003). The root system of plants can help to spread bacteria through soil as well and help to penetrate otherwise impermeable soil layers (Kuiper et al. 2004). Hyphae of mycorrhiza-forming fungi that increase the volume of the root-soil colonization are also more or less densely colonized by bacteria (Sarand et al. 1998), helping bacteria in colonizing deeper horizon.

A key question, however, lies in the fact that bacteria could preferentially metabolize exudates more than the less easily degradable pollutants. The low availability of carbon and energy source remains the growth-limiting factor for the bacteria in soils, thus ensuring efficient contaminant-degradation (Romantschuk et al. 2000). Overall, rhizosphere functioning is not easy to predict as the result of both numerous parameters involved in its functioning and some still lacking informations, e.g., full composition of most plant exudates is unknown. Additionally degradation heterogeneity may be observed (Juhanson et al. 2009), the highest degradation rate being observed in the upper soil layer that has the highest root density. Nonetheless, modeling approach to quantify the microbial biomass in the rhizosphere was already undertaken (Sung et al. 2006) as a necessity in engineering process.

This contribution of the rhizomicrobial population is referred to as rhizoremediation with some substantial development in rhizo-directed strain selection. Rhizoremediation has been widely developed for the remediation of soils contaminated by organic pollutants (Barac et al. 2004; van Aken et al. 2004), less for metals (Lebeau et al. 2008). These authors showed that on average bioaugmentation increased metals accumulated by shoots by a factor of about 2 (metal concentration) and 5 (amount) without any obvious differences between bacteria and fungi.

A future prospect would be to determinate the exact role of inoculated microorganisms on bioremediation, i.e., direct or indirect role via indigenous microflora. Most of the time, one cannot conclude. In a soil bioaugmented with indigenous microfungi and plant, nonylphenol contamination was reduced and resident microfungal communities were stimulated (Girlanda et al. 2009). Nonylphenol depletion following fungal inoculation correlated also with biostimulation of indigenous fungi, suggesting positive interactions between introduced and resident fungi and then an indirect effect of bioaugmentation.

7.4.2.7 Increasing Bacterial Transport

In order to increase bacterial transport over distances >1 m, low ionic strength solutions and the nonionic surfactant Tween 20 were shown to reduce bacterial adhesion to ultraclean surfaces such as glass and quartz porous media and then increase the covering distance (Li and Logan 1999). Foam and strains resistant to adhesion may also be applied (Franzetti et al. 2009; Wang and Mulligan 2004).

Soil humidity also affects the microbial transport (Fang and Logan 1999). Indeed in unsaturated conditions, bacteria preferentially accumulate at the air–water interface, whereas under water-saturated conditions, bacteria readily adhere to soil particle. These authors suggest using air sparging solution under a saturated porous medium to produce a mobile water interface that was hypothesized to facilitate bacterial transport.

Any substrate percolated in the soil and metabolizable by microorganisms also modifies the distribution of the inoculated microorganisms as shown by Dechesne et al. (2005) with *Pseudomonas putida*. The microscale distribution of introduced bacteria was modified toward a more widely dispersed distribution with consequently a higher probability of the introduced cells to be in contact with other components of the soil ecosystems (contaminant, other microorganisms in the aim at transferring conjugal genes).

7.4.2.8 Soil Survival and Colonization

To increase the probability of the initial establishment and the long-term efficacy of an inoculum in soil, some prerequisites should be carried out: (a) the microbial genome introduced should be competent to participate in the main energy flux processes, (b) the new population should not fill a metabolic niche that is already being utilized by the indigenous microbiota, (c) enough nutrients should be available to support the growth of an additional population, and (d) the pollutant level should be compatible with the microbial tolerance. Additionally, according to the concept of the carrying capacity of microbial communities, the added populations stabilize at 10^3 CFU g^{-1} soil and this concentration appears to be independent of the ecosystem used (Dejonghe et al. 2001).

The survival and effectiveness of a bioaugmentation strain in its target environment depend also on the physiological state of the inoculated organism (Cunliffe and Kertesz 2006). The way in producing the microbial inoculum is thus crucial along with its conditioning (Sect. 4.2). In particular, the culture medium used to produce the inoculum must be rich enough to get a high amount of biomass without being however highly different in composition than that of soil characterized by its oligotrophy.

Plants may then play a major part in the soil colonization: (a) thanks to the movement of roots through the soil, new niches are created enhancing the chance that certain bacteria can establish in the soil (Dejonghe et al. 2001) and the (b) ability to efficiently use major exudate components during root colonization was underlined, in particular organic acids more than sugar or amino acids (Kuiper et al. 2004).

These results suggest a systematic association of a suitable rhizosphere strain with a compatible plant allowing the settlement of the inoculum on the root together with the indigenous population, thereby enhancing the bioremediation process. It can even be appropriate to use mixed instead of monoplant cultures to promote microbial diversity (mixture of exudates) along with cleanup of multicontaminated sites. Kuiper et al. (2001) were the first to methodically select a strain indigenous to a host's plant rhizosphere. Narasimhan et al. (2003) used an engineered bacterium to utilize the predominant root exudates of *Arabidopsis* for remediation of PCB, providing a selective advantage for the strain when in the *Arabidopsis* rhizosphere. De Weert et al. (2004) suggested a novel method of enhancing competitive root-tip colonization with a *Pseudomonas fluorescens* strain through repetitive cycling of the strain and mutant derivatives on the plant roots in situ. The microorganisms able to rapidly adapt to the plant rhizosphere were selected.

7.4.3 A Lack of Data and Experimental Methodology for Better Understanding of Bioaugmentation Functioning

We must recognize that the understanding of the whole mechanisms driven by soil remediation still remains partial as a result of the numerous biotic and abiotic parameters that control the soil functioning along with the high heterogeneity of this matrix. Overall, soil to be cleaned-up can be shown as a "black box" or at best as a "gray box" where only some few basic data are available, e.g., pollutant concentration in soil and in plants, allowing to estimate the mitigation performances of the system (yield and rate) without being possible most of the time to explain why it works or not.

This partial understanding reveals analytical methods and protocols at the researcher's disposal that are not always specifically designed for soil (e.g., TTGE was elaborated for the detection of human genetic mutations and soil DNA extraction was inspired by protocols regarding simple matrices) and that give information sometimes not specific enough (e.g. microbial cell counts and metal bioavailability estimated by chemical extractions). Moreover, there is not yet a consensus in the scientific community on a basic set of analytical protocols to be applied in the aim at comparing the results of different studies on the same basis and that could be used by soil remediation practitioners. Eventually, most analytical methods are based on heavy protocols. Hence, we are still far away from the fully and mechanistic understanding of the functioning of such a matrix and consequently to be able to optimize bioaugmentation.

Fortunately, the recent development of new analytical methods and protocols will help us in the next decade in a better understanding of the soil functioning. In their review Desai et al. (2010) reported the recent innovation in genotypic profiling, ultrafast genome pyrosequencing, metagenomics, metatranscriptomics, metaproteomics, and metabolomics along with bioinformatics tools that have provided crucial insights of microbial communities and their mechanisms into bioremediation of environmental pollutants.

Diplock et al. (2009) compared different analytical techniques/parameters to assess which best predicted hydrocarbon bioremediation of field collected soils under laboratory conditions. At elevated concentrations, the rate of degradation was best described by respiration and the total hydrocarbon in soil. The number of bacterial degraders and heterotrophs as well as quantification of the bioavailable fraction allowed an estimation of how bioremediation would progress. In absence of other microbial data, biosensors proved a useful predictor of bioremediation. Similarly, Imfeld et al. (2009) reported the different methods and techniques for assessing and monitoring processes of organic chemicals removal in constructed wetlands.

Some specific data regarding more especially bioaugmentation are reported below, in particular taking into account inoculum survival as well as its activity and the effect of bioaugmentation on soil ecosystem.

7.4.4 Monitoring of Inoculated Microorganisms in Soil

Several informations are needed: structure, function, and activity of the soil microflora and more specific data relating to the microbial population inoculated in the soil. We are now in a better position to obtain a more comprehensive assessment of the composition and structure of microbial communities in the environment. Nonetheless, available methods (e.g., T-RFLP, and TTGE) face three major drawbacks preventing the comparison of the results between different studies in terms of the data available (Marzorati et al. 2008):

1. Several experimental settings (e.g., DNA extraction protocol, denaturing gradient, and voltage) are used most of the time preventing any comparison between studies and even laboratories
2. Limits in the data interpretation. Dejonghe et al. (2001) pointed out that fingerprinting analysis does not necessarily reveal the organisms largely involved in the mainstream energy flux of the ecosystem. It should be completed with metabolic mass balance studies. For these authors, the methods suited to quantify temporal variation of genomes controlling 80% of the energy flux are mRNA-based techniques and proteomics that both rely on the expression of catabolic genes in active organisms
3. The lack of a common way of interpretation of the fingerprinting patterns. Marzorati et al. (2008) suggest that the interpretation can be performed through range-weighed richness reflecting the carrying capacity of the system, dynamics reflecting the specific rate of species coming to significance, or functional organization (Fo) defined through a relation between the structure of a microbial community and its functionality. Fo was defined as the ability of the community to organize in an adequate distribution of dominant microorganisms and resilient ones a condition that should ensure the potentiality of counteracting the effect of sudden stress exposure. These three parameters plotted together are supposed to represent a visual ecological interpretation of the initial raw fingerprinting pattern. Stability of the functionality does not necessarily imply stability of community structures. Minority community members can become dominant in a short period following a significant perturbation

In bioaugmentation studies, fingerprinting analysis giving an overall view of the soil microflora must be associated with analysis specifically dedicated to the microbial population inoculated in the soil. Fluorescent and luminescent marker (gfp, lux, etc.) and reporter genes provide easily detectable phenotypes to microbial cells and are valuable tools for the study of microorganisms in the nature (Jansson et al. 2000; Jansson 2003). Although these tools are becoming widely adopted, they are still issues that remain to be solved, such as the dependence of the reporter output on the physiological status of the cell. Moreover, autofluorescence of soil may impair the signal of the markers. Fluorescent in situ hybridization (FISH) allows more direct assessments of microbial community composition and the identification of specific populations in situ along with direct measures of their relative abundance (Jansson 2003). FISH was used to detect simazine-degrading bacteria in soil samples (Martin et al. 2008). FISH with rRNA-targeted oligonuclotide probes gives access to the microbial activity, but the problem in utilizing FISH in studies of natural bacterial communities is its sensitivity. Standard FISH with mono-fluorescein isothiocyanate (FITC)-labeled probes indeed gives a strong signal only if cells are metabolically active and, hence, contain large number of rRNAs. HNPP-FISH using 3-hydroxy-*N*-2′-biphenyl-2-naphthalenecarboxamide phosphate ester (HNPP) and Fast Red TR enhances the fluorescence signal eightfold compared to FITC-FISH (Iwamoto and Nasu 2001). In situ PCR was another method reported by the same authors. Amplification and detection of target genes are carried out inside individual bacterial cells. This technique enables to detect individual functional genes present in single copy or low copy numbers in intact bacterial cells that cannot be detected by FISH. Although establishment of the microbial population in a soil is related to the presence of specific niches, few studies have focused on indigenous bacteria and their spatial relationship within various microhabitats. Combining fluorescence staining techniques with soil thin section technology allows one to obtain images of microorganisms in situ (Li et al. 2004). Soil texture and the procedures used for resin embedding are important factors affecting the quality of stained soil thin sections. Limitations are about the nonspecific binding of dyes to the soil matrix and the autofluorescence of some soil components. These authors added that FISH and confocal laser scanning microscopy techniques provide a new potential for microbial distribution studies.

In the aim at monitoring the amount and activity of inoculated cells, qPCR and RT-qPCR are useful methods, of which the common target for qualitative and quantitative assessment of a target bacterial population is the 16S rRNA gene. Application of these methods was recently applied to various bioaugmentation studies: activity and abundance of the crude-oil-degrading bacterium Nocardia sp. H17-1 during bioremediation of oil-contaminated soil (Baek et al. 2009), quantification of a green fluorescent protein-labeled, genetically engineered *Pseudomonas putida* strain during 2-chlorobenzoate degradation in soil (Wang et al. 2004), concentration of total bacteria and oxygenase genes involved in the biodegradation of aromatic compounds (i.e., toluene dioxygenase, ring hydroxylating monooxygenase, naphthalene dioxygenase, phenol hydroxylase, and biphenyl dioxygenase) (Dominguez et al. 2008), molecular characterization of microbial populations at two sites with differing reductive dechlorination abilities. qPCR quantification of specific dechlorinating species provided the most effective and direct prediction of community dechlorinating potential, compared to T-RFLP and RFLP analysis with clone sequencing (Rahm et al. 2006).

Microarray technology that recently emerged as a new innovative tool in microbial ecology has the potential to simultaneously determine the dynamics and/or activities of most, if not all, of the microbial populations in complex environments such as soils and sediments Gentry et al. (2006). In their review, these authors identified five categories of microarrays, based on the genes targeted by the array, that have been successfully applied to microbial ecology research:

1. Phylogenetic oligonucleotide arrays (POAs) mainly based on 16S rRNA gene to compare the relatedness of communities in different environments
2. Functional gene arrays (FGAs) designed for key functional genes that code for proteins catalyzing various biogeochemical processes and also providing information on the microbial populations controlling these processes
3. Community genome arrays (CGAs) containing the whole genomic DNA of cultured organisms, and they able to describe a community based on its relationship to these cultivated organisms

4. Metagenomic arrays (MGA) that are a potentially powerful technique because, unlike the other arrays, they contain probes produced directly from environmental DNA itself and can be applied with no prior sequence knowledge of the community
5. Whole-genome open reading frame (ORF) arrays (WGA) containing probes for all of the ORFs in one or multiple genomes. These arrays have traditionally been used for functional genomic analyses of individual organisms, but they can also be used for comparative genomic analyses or to investigate the interactions of multiple organisms at the transcriptional level

Microarray technology is, however, limited by its specificity as the result of the highly conserved nature of many genes and the vast amount of unknown sequence data in environmental samples making difficult to design and validate microarray probes that are specific to a given target sequence. Although oligonucleotide probes have many advantages for probe design, they are typically less sensitive than longer PCR-based or CGA. Probe quantitation is also limited given the potential variability in steps including DNA extraction, labeling, hybridization, and analysis. FGAs and CGAs can be quantitative within a range of concentrations. Nonetheless, it can be difficult to compare data between, and even within, microarray experiments due to the use of different analysis methods and variability in printing, labeling, and hybridization. Even scrupulously well conducted, differences in hybridization signals cannot be always correlated with changes in specific populations due to the large amount of unknown nucleic acid sequences in environmental samples.

7.4.5 Monitoring of Pollutant Degradation in Soil

In the aim at rigorously quantifying pollutant degradation, not only dissipation, a stoechiometric balance would ideally be achieved – disappearance of the metabolite and appearance of CO_2 – which is in fact impossible in open environment. In this prospect, stable isotope probing (SIP) recently allowed to determine exactly which organisms assimilate specific contaminants. Historically, Radajewski et al. (2000) described the feeding of aerobic soil bacteria with [^{13}C]-CH_4, followed by separation of [^{13}C]-DNA and [^{12}C]-DNA by density-gradient centrifugation. Unlike radioisotopes, the use of stable isotopes allows implementing experiments in open environments. Recent studies in this area have focused on, e.g., the reductive dechlorination of chlorinated solvents, the degradation of the fuel additive methyl *tert*-butyl ether, and the removal of long-term hydrocarbon contamination (Scow and Hicks 2005).

Microbial ecologists seized SIP methods in the aim at identifying microorganisms responsible for some environmental processes. They are thus able to directly observe organisms of interest in multispecies consortia performing a chemical transformation by associating SIP with other well-known techniques used in ecology such as phospholipid fatty acid (PLFA) and DNA analysis. According to

Manefield et al. (2004) who reviewed applications in the field of biodegradation, DNA and RNA-SIP and PLFA-SIP allow answering the question "Which organisms among the active members of the community assimilates this substrate?" To the question "Does this organism assimilate this substrate?" then FISH-SIMS (secondary ion mass spectrometry) and SSU (small-subunit rRNA)-IRMS (isotope ratio mass spectrometry) can be applied and are well suited to bioaugmentation studies. SSU-IRMS can detect tiny isotopic enrichments in lipids, which means that very little substrate needs to be added to the environmental sample or incorporated by the active organisms. This diminishes or virtually eliminates the potential for an enrichment bias, which is a major concern, particularly for DNA-SIP studies. These authors stress the choice of the substrate too, basing on natural concentrations and stable isotope abundances as well as the presence of structurally related molecules, which may complicate the analyses. When elemental flows are complex, as they will be for general substrates in mixed communities (e.g., CO_2), stable isotope signatures will become diluted. In such cases, these authors believe that PLFA-SIP, FISH-SIMS, and SSU-IRMS will find the greatest utility. When elemental flows are more restricted, as they are likely to be with the recalcitrant or xenobiotic compounds of interest in bioremediation studies, they suggest using DNA- or RNA-SIP.

More recent methods, of which some of them being at the researcher's disposal, were Neufeld et al. (2007) (a) isotope array offers a level of throughput not readily afforded by FISH-SIP techniques in case of environment that yields sufficient RNA for direct hybridization. The main advantages of the isotope array are the direct detection of labeled RNA from active organisms, without the potential to detect unlabeled background as can occur with DNA- and RNA-SIP, (b) Raman microscopy and its combination with FISH. Raman microscopy employs an excitation laser incident on samples to measure the vibrational energy of chemical bonds. The scattered laser light is captured by a charge-coupled device camera, and a spectrum is obtained that is able to discern different biological molecules in the cell. When tested on single microbial cells, the magnitude of peaks in the Raman spectra is independent of bacterial species tested and corresponds to the proportion of label molecules. As the optical resolution of Raman microscopy (ca. 1 μm) is similar to that of FISH, the two methods are conceivably complementary. Raman also provides information into which cellular compounds the labeled substrate was incorporated, and (c) multi-isotope imaging mass spectrometry (or nano SIMS) is conceptually similar to the approach of Raman confocal microscopy. It improves upon the earlier secondary ion mass spectrometry (SIMS) technology, which was originally coupled with FISH but at only low resolution (ca. 10–15 μm). The nano-SIMS technology analyzes the stable isotope content of single cells at a high resolution (50 nm) that exceeds a Raman microscope, exhibiting a stable isotope measurement precision of $\pm 1\%$.

Potential of SIP to study plant–microbe interactions was suggested and experimented by Prosser et al. (2006). Plants were pulse-labeled with $^{13}CO_2$, and ^{13}C-enriched rRNA was recovered from rhizosphere bacteria after $^{13}CO_2$ is converted into plant sugars by photosynthesis and root exudates are metabolized by the rhizosphere microorganisms. SIP methods might begin to elucidate the fine details of plant–microbe interactions and could be a means of identifying relevant plants to be associated with rhizospheric microorganisms – chosen for their efficiency in bioremediation – as large part of molecules making exudates are metabolized by these microorganisms.

Although many pieces of information regarding bioaugmentation are known at the laboratory scale, it is only within an environmental context, in which microorganisms are constantly exposed to multiple changing environmental stresses that there will be a full understanding of microbial adaptative resiliency (Hazen and Stahl 2006). Knowledge of the stress response in the environment will facilitate the control of bioremediation. For example, immediate changes in protein and mRNA structure are part of the first line of defense. The techniques reported above have been used in situ and will contribute to provide more direct evidence of metabolic state, kinetics, and stress status.

7.4.6 Experimental Design and Modeling of Bioaugmentation Processes

Most of the time, only a few parameters are studied at a time leading to the lack of some default values precluding to fully understand the ecosystem functioning such as soil. One can explain this by the heaviness of experiments to be implemented inciting researchers to simplify their experimental devices. Indeed, the traditional method of process optimization involves the study of one variable at a time, which requires a number of combinations of experiments that are time, cost, and labor intensive (Rao et al. 2008). Unfortunately, it sometimes prevents bioremediation to succeed.

Conventional experimental procedures involve altering of one factor at a time keeping all other factors constant, resulting in assessing the impact of those particular factors. These procedures are time-consuming, require more experimental sets, and are unable to provide mutual interacting information of the factors (Mohan et al. 2007).The Taguchi method of design of experiments is a simple statistical tool involving a system of tabulated designs (arrays) that allows a maximum number of main effects to be estimated in an unbiased (orthogonal) fashion with a minimum number of experimental runs (Mohan et al. 2007; Rao et al. 2008). Design of experiments helps gaining information about the optimized levels, by taking large number of variables: set of independent variables (factors), both controllable and uncontrollable (dynamic/noise), over a specific region of interest. While traditional experimental design focuses on the average process performance characteristics, this approach concentrates on the effect of variation on the process characteristics and makes the product/process performance insensitive to variation by proper design of parameters (Mohan et al. 2007).The proposed method facilitated systematic mathematical approach to understand the complex bioremediation process and the optimization of near optimum design parameters. This methodology has been widely applied in many industrial sectors; however, its application in biological sciences, in particular in wastewater treatment and bioremediation, has been limited. This methodology was applied to bioslurry phase remediation of chlorpyrifos contaminated soil (Mohan et al. 2007). Eight biotic and abiotic factors were evaluated (substrate loading rate, pH, dissolved O_2, soil/water ratio, temperature, soil microflora load, application of bioaugmentation, and humic substances concentration). Substrate loading rate showed significant influence on the bioremediation process.

Although lots of data are available at the laboratory scale, it is only within an environmental context, whereas in open ecosystems, microrganisms are constantly exposed to multiple changing environmental stresses. Indeed, the tremendous variety of pollutants types impose site-specific requirements, thus often preventing the informations developed at one site to from being used to design treatment strategies for other systems and pollution types. Therefore, predictions regarding microbial adaptative resiliency must be supplied by developing stress response systems as tools for effective and general process control (Hazen and Stahl 2006). Knowledge of the stress response in the environment will facilitate the control of bioremediation. Motility and chemotaxis are important traits underlying the ability of microorganisms to adapt to rapid physical and chemical environmental changes. Stresses encompass starvation, heat shock, cold shock, oxidative stress, O_2 deprivation, osmotic challenges, acid stress, sodium stress, SOS response to DNA damage, etc. Stress is also relative; e.g., temperature and pH are stressful for one species and optimal for the other.

Modeling is a mean at predicting bioremediation efficiency and then to avoid too many experiments, but they are, however, unable to fully describe the huge complexity of ecosystems. In particular, the future prospects should consider the role of plants in bioremediation as a microbial regulating factor. The microbial biomass in the rhizosphere was modeled by Sung et al. (2006). From the simulation, results showed that the influence of root exudates may be smaller than the influence of microbial degradation on contaminant dissipation due to indigenous substrate conversion or the application of exogenous substrates. Since plant roots are initially small and do no produce large quantities of exudates when first seeded, the addition of exogenous substrates may be needed to increase initial microbial concentrations at the start of phytoremediation projects.

7.4.7 Ecological Engineering Applied to Microbial Resource Management: Emergent Concept Combining Process Approach and Ecology

According to Vogel and Walter (2001) microbial ecology issues are among the most important in bioaugmentation approaches, although unfortunately, they are rarely addressed, e.g., taking into account indigenous populations, environmental parameters together with phenotypic characteristics of the strains, and procedures for introduction to determine activity, persistence, and performance of bioaugmented strains.

Ecological engineering was first defined in the 1960s (Odum 1962), considering energy flows, to describe "those cases in which the energy supplied by man is small relative to the natural sources, but sufficient to produce large effects in the resulting patterns and processes". By extension, ecological engineering corresponds to the design of sustainable ecosystems that integrate human society with its natural environment for the benefit of both (Mitsch and Jørgensen 2004). It includes both the restoration of ecosystems that have been substantially disturbed by human activities and the development of new sustainable ecosystems that have both human and ecological value. This implies the design, the development, and the maintenance of ecosystems by using engineering technologies based on ecological principles. Natural ecosystems or ideas based upon them can be used to reduce or eliminate pollution problems (Jorgensen and Mitsch 1989).

Microbial communities are a part of these open ecosystems. These assemblages are most of the time an asset in terms of genes and functionalities Verstraete (2007). Although microbial communities are constantly changing, the subtle equilibrium between the different populations making up the consortium is generally not broken resulting in a high resilient capacity of these systems. Only the proportions of the populations making consortia may change. A minority population can become the most active population during nutrient limitation (LaPara et al. 2002). Hence, microbial communities are most probably the basis of microbial resource management in the domain of environmental biotechnology.

According to Verstraete (2007), the challenge for microbial ecologists is to develop the soil metabolome by introducing relevant microorganisms to benefit from their actions within the system. Unfortunately, "it would be naive to believe that by simply picking the 'right' organisms or manipulating the right field parameter, bioaugmentation will suddenly become as reliable and predictable as engineered systems. Inevitably, as with most biological systems, there will always be an element of unpredictability" (Thompson et al. 2005). Because of the unpredictability of this technology as well as politic choices still giving permission for putting polluted soils in dump, practitioners are unfortunately not prone to widely use bioremediation among which bioaugmentation. (Vijgen 2002).

The question that arises is the following: can we imagine to fully controlling in situ bioaugmentation, which is the main goal in process engineering, in spite of the spatial and temporal variability of open ecosystems? As described in the previous sections, some prerequisites must be put together to make bioaugmentation a success (Table 7.4), i.e., keep in mind some microbial ecology concepts leading to concrete techniques. In particular, the use of microbial consortia instead of monoculture is more relevant as they can perform more complicated tasks and endure more changeable environments (Brenner et al. 2008). According to these authors, natural and artificial consortia can be used. The former can maintain homeostasis since members generally do not outcompete one another and do not exhaust the resources in their environments. When constructing synthetic microbial consortia, cell–cell communication must be considered first (Brenner et al. 2008), by exploiting intraspecific signals, e.g., bacterial quorum sensing (QS), interspecific cues involving multispecies QS, and cross-feeding by exchanging chemicals involved in metabolism and growth. Opposite to natural microbial communities, the long-term behavior and destiny of artificial consortia are unpredictable. Artificial immobilization is thus a relevant mean to stabilize growth and activity of these inocula (Labana et al; 2005; Plangklang and Reungsang 2009; Siripattanakul et al. 2009) as biofilms do for natural consortia. The artificial assemblages of engineering organisms are less likely to display the robustness associated with natural assemblages because the role of keystone species, the greater complementary of resources use, and the greater facilitation are not fully addressed (Goldman and Brown 2009). Similarly, Dejonghe et al. (2001) consider that increasing the degradative potential of a microbial community do not require a high gene diversity but the right competence under the given conditions.

Competence refers to the capability of important genes to be activated by the microbial community and thus related metabolic traits to participate in the energy flux of the system. Vogel and Walter (2001) claimed, however, that it is not practical to tailor consortia specifically to each habitat. Another constraint for the future of bioaugmentation is the cost-to-benefit ratio. Ecological engineering indeed considers the sustainability of treatments such as bioaugmentation in terms of environmental cost and long-term effect of the treatment. In other words, is it better to bioaugment the soil with exogenous microorganins or to stimulate indigenous ones? Finally, one can wonder how to do better than nature in designing ecosystems. Other important question to be solved lies in the scale to which ecosystem studies should be undertaken for anticipating the ecological engineering consequences. According to Odum and Odum (2003), microcosm is the relevant scale, whereas van der Gast et al. (2006) considered that microbial diversity is related to volume possibly preventing any full-scale prediction. An example of the implementation of the Hubbell Law reported by Verstraete et al. (2007) is that the number of different species increases with the size of the reactors (ten dominant species in 1 m^3 reactor; 50 in 1,000 m^3). Much more attention should be given to the rate of change of microbial communities with time. Factors such as cell residence time and occurrence of temporal or continuous stress factors should strongly influence this rate of change. Similarly, atrazine biodegradation was shown to be greatly influenced by the amount of soil and the microcosm implementation (closed microcosms with 5 g of soil vs. open microcosm with 160 g of soil) (Lima et al. 2009). Although both experiments were performed at a similar scale, this low scaling-up significantly modified the degradation performances, probably as a result of different mass transport limitations and spatial heterogeneity, among other variables. One can thus wonder whether the degradation performances change continuously with the scaling or whether some key experimental scales allow predicting degradation performances at much more higher scales. Future prospects should examine thoroughly on this questions.

To conclude, the future prospects for in situ bioaugmentation technologies should be to reconcile process engineering, based on the fully control of any system, with variable environment conditions. One solution, as above evocated, would consist in combining plants with microorganisms. Indeed, physicochemical conditions in the rhizosphere are less susceptible to change in the course of the time than in the bulk soil. Plants are thus relevant means of controlling the environmental conditions surrounding microorganisms used in pollutant degradation.

7.5 State of Practice

Overall, bioaugmentation technology still remains more explorable as a result of several constraints regarding contaminated land management policies in the different countries and being driven by technical and economical feasibility principles (Iwamoto and Nasu 2001, Ruberto et al. 2009, Sorvari et al. 2009). Some reasons are often evoked:

1. Risk-based land management with the adoption of the "fitness for use" principle
2. Tight time schedule of remedial actions tending to be the driver of decisions on contaminated sites regarding the remediation choice. Time-consuming in situ methods are most of the time a constraint
3. Sustainability of the remediation method in the absence of comprehensive data available on the overall life-cycle impacts of different remediation methods

Overall, environmental impacts of remedial options have rarely been taken into account in decision making. Prevailing soil excavation combined with disposal off-site was seen mainly as non-ecoefficient. Nonetheless being a quick method, soil excavation is also the most efficient way to remove risks from the site all the more since contamination problems are often identified only at a very late stage. One can notice that basically decisions are mainly controlled by legislation and administrative practices and strongly driven by time and money. Bioaugmentation confronts us as well with the problem of the microorganisms' dissemination.

In Japan, first field experiment with bioaugmentation was undertaken in 2000 under strict control of the Ministry of International Trade and Industry. Nonetheless, the complexity involved in the transport of soils and the restrictive legislation in some area makes on-site bioremediation the strategy of choice, and (d) acceptance of biological and in situ remediation methods is considered as a prerequisite for the realization of ecoefficiency. First field experiments were conducted about ten years ago.

Although some experiments were performed at the field scale, the definition of this term varies from one author to another. Overall, it means time outdoor experiments with field soil (disturbed or not) and concerns most of the time very small surfaces (a few square meters). One square meter field plots were used in p-nitrophenol experiments (Labana et al. 2005). First field-scale atrazine remediation study in the United States was performed in 2.5 m^2 treatment plots (Strong et al. 2000). Soil bioaugmentation with 2,4,6-trinitrotoluene (TNT) was studied in 300 m^2 plots containing disturbed soil (Van Dillewijn et al. 2007). Field experiments were carried out in order to test the effect of phytoremediation and bioaugmentation for remediation of semicoke in 50 m^2 plots (Truu et al. 2003) and oil shale chemical industry solid waste in 10 m^2 plots (Juhanson et al. 2009).

The largest soil surface treated by bioaugmentation – 7,000 m^2 – was reported by Zhang et al. (2008) with a method tested for petroleum and salt contaminated soil in which wheat straw was used to enhance salt leaching and subsequent petroleum degradation by a bacteria–fungi consortium of *Enterobacter cloacae* and *Cunninghamella echinulata*. On average, field-scale trials were shown to take three times as long to reach the same endpoint as the laboratory trial (Diplock et al. 2009), underlying the necessity to validate laboratory experiements at the field scale.

One can also observe that a few long-term experiments were implemented and that the length of the treatment varies from a few weeks (Hesselsoe et al. 2008) up to 3 years (Juhanson et al. 2009). According to Weston and Balba (2003), commercial companies argue that, as soon as tailor-made consortia become more and more commercially available, bioaugmentation will become the standard for rapid and precise cleanup of a variety of contaminated soil situations.

7.6 Environmental Impact of Bioaugmentation

The basic question to answer is: "Comparing pollutants and bioaugmentation, which shows the strongest effect on environment?" One can also note that bioremediation, among which bioaugmentation, focuses mainly on the pollutant dissipation without most of the time taking into account (a) its destiny while metabolites can be more toxic than the parental molecule as shown for, e.g., PCP (Zuzana et al. 2009) and diuron herbicide metabolites (Tixier et al. 2002), and (b) ecological impact of introduced microorganisms on indigenous microorganisms. In their review, Dejonghe et al. (2001) pointed out that the species diversity can be maintained if a site is capable of supporting similar number of individuals for each of the different species present in the community. If no, the microbial type that is best adapted to the most productive niche (task) will then become dominant in the community and diversity will be low. In that case, adverse effect of bioaugmentation may be observed. Until now, environmental impacts of remedial options have unfortunately rarely been taken into account in decision making and only recently in laboratory experiments.

Overall, it is not easy to conclude whether bioaugmentation reduces or not the negative environmental impact of pollutants as the result of a variety of situations and tests used for environmental assessment. Some of them are reported here. Ecotoxicological tests such as the germination and vigor index of maize plants seeded in lindane-contaminated soils were used (Benimeli et al. 2008). Lindane at different concentrations in soil did not affect the germination and vigor index of maize plants seeded, while the index value decreased when *Streptomyces* sp. M7 was inoculated, showing a better vigor index at the same time as lindane removal. Similarly, D'Annibale et al. (2005) observed a significant decrease in soil toxicity contaminated by aromatic hydrocarbons when bioaugmented by fungi using two different soil contact assays, including the *Lepidium sativum* L. germination test and the *Collembola* mortality test. Recently, application of DNA microarrays to ecotoxicogenomics links ecotoxicological effects of exposure with expression profiles of several thousand genes (Steinberg et al. 2008). Biofilms as they occupy a strategic phase between water and sediments in aquatic systems could be regarded as early warning systems for the detection of the effects of toxicants (Geiszinger et al. 2009). Unfortunately, such approach seems not feasible in soil.

The structure of the microbial populations is another relevant indicator. The response of a single-species test might indeed differ from the response of the same species in a whole community. Baxter et al. (2006) observed that bacterial communities from contaminated soils with low biodiversity are much more readily perturbed through interventions such as contamination events or bioaugmentation treatments. Soil inoculation with *Sphingobium yanoikuyae* B1 able to degrade a range of PAHs was not shown to cause extensive changes in the native bacterial community of either soil, as assessed by DGGE (Cunliffe and Kertesz 2006), but its presence led to an increase in the population level of two other species in the aged contaminated soil community and appeared to have an antagonistic affect on several members of the HAP-contaminated compost community, indicating niche competition. Same technique was used for assessing the environmental innocuousness of an inoculum production (activated soil) to be used in the bioaugmentation of creosote-contaminated soils (D'Amours et al. 2008). The results suggest that soil activation may have the potential to ensure the environmental safety. Similarly, Goux et al. (2003) observed low effect of bioaugmentation for atrazine remediation on the pattern of the indigenous soil microflora by using DGGE. Using TTGE analysis Stallwood et al. (2005) showed that 12 weeks after microcosms' biostimulation or bioaugmention with *Pseudomonas* strain ST41, *Pseudomonas* species were the dominant soil bacteria in both bioaugmented and biostimulated systems. For more accurate results, some studies combined several methods, e.g., composition of the community by T-RFLP, physiological profile in Biolog ECOplates, and ecotoxicity tests (*Vibrio fischeri*, *Daphnia magma*, and *Selenastrum capricornutum*) (Alisi et al. 2009). Concurrently, with the increase of metabolic activity at community level and the microbial load, the gradual decrease of the ecotoxicity was observed. Although the structural analysis of microbial populations provides plenty of informations, some precautions must be taken. Indeed, although some effects of bioaugmentation can be observed by studying the genetic structure of microbial populations, some effects occurring in the biological communities do not directly affect biomass or composition, but their metabolism and success in nature, and further, pollutants may not affect just a single biological compartment, but several of them (Geiszinger et al. 2009). Consequently, one challenge to be faced is then combining chemical and biological assessment strategies into a single one.

But above all there is a need in (a) available certified reference materials resembling natural soils to validate all these analytical methods and other in ecological risk assessment in the aim at scaling up the responses to the whole ecosystem, (b) analysis at the whole community since the response of a single-species test might differ from the response of the same species at a higher organization degree, and (c) standardizing the tests for community responses. Unfortunately, standardized laboratory tests are carried out under optimized conditions, whereas the real environment conditions are often harsher and more varied compared to conditions in the laboratory – several factors co-occur, produce complex responses, tolerance of organisms toward toxicants is not linear in natural environments, natural environments most often contain mixture of compounds, etc.

Regarding microorganism-assisted plant for enhanced metal extraction, one could be afraid that the increase of the metal availability allowed by inoculated microorganisms is higher than the plant accumulation ability (Barona et al. 2001) with a risk of contamination of the subsoil and groundwater. In their study, Di Gregorio et al. (2006) have observed a severe modification of the bacterial community structure of the soil, using DGGE, due to the vegetation with *B. juncea*. Conversely, EDTA slightly affects the bacterial community structure, with the exception of the simultaneous presence of *B. juncea* and PGPR *Sinorhizobium* sp. In contrast, pseudomonads inoculated in soils, contaminated or not with Cd, induced a shift in the microbial communities, as suggested by analyzing in situ catabolic potential (Duponnois et al. 2006). But these studies did not report the relationships between EDTA supply and bioaugmentation with metal lixiviation.

7.7 Bioattenuation, Biostimulation, and Bioaugmentation: Added Value of Bioaugmentation Compared to Other Techniques

Although bioaugmentation shows most of the time measurable positive effects on pollutant, i.e., dissipation of organic compounds and changes in the chemical form of metals, the questions that raise, comparing this method with bioattenuation and biostimulation, are the following: "which additional dissipation rate?", "which additional time-saving?", and at the same time "which additional costs?" and "which difference in environmental effect?"

Additionaly to these questions, Boopathy (2000) suggested: "Is the contaminant biodegradable?" "Is biodegradation occurring in the site naturally?", "Are environmental conditions appropriate for biodegradation?" Overall, one can suggest that bioaugmentation should be set up if only substantial enhancement of remediation performance is expected, though some studies suggested that bioaugmentation may be of great interest in starting up the perequisite metabolic process.

Some data showing additional performance of bioaugmentation compared to biostimulation and bioattenuation are reported in Table 7.5.

7.8 Concluding Remarks and Future Prospects

Although exhibiting most of the time higher performances than in situ soil bioattenuation and biostimulation, controlling bioaugmentation in environments subject to variable conditions still remains the limiting factor preventing the broadcasting of

Table 7.5 Additional performance of bioaugmentation compared to biostimulation and bioattenuation

	Pollutant	Biostimulation	Bioaugmentation	Variation in the degradation rate	Comments	References
A	Total petroleum hydroacarbon			−6% for light fractions (C-12-C-23); -1% for heavy fractions (C-23-C-40)		Bento et al. (2005)
C				+48% for light fractions (C-12-C-23); + 59% heavy (C-23-C-40) fractions		
A	Dichloroethene	N:P:K at a ratio 3:1:6	Synthetic consortium: *Pseudomonas aeruginosa*, *Acinetobacter haemolyticus*, *Acinetobacter* sp., *Achromobacter xylosoxidans*, *Bacillus subtilis*, *Bacillus cereus*, *Klebsiella* sp.	+3.4% (*cis*-DCE); +18.1% (*trans*-DCE)		Olaniran et al. (2006)
C				+11.6% (*cis*-DCE); +8.7% (*trans*-DCE)		
E				+14% (*cis*-DCE); +18% (*trans*-DCE)		
A C	Acrylonitrile ACN		*Rhodococcus* sp. AJ270	No effect	Community affected by ACN, added carbon, and *Rhodococcus* sp. AJ270	Baxter et al. (2006)
A	Anthracene (ANT), pyrene (PYR) and berizo[*a*]pyrene (B[a]P)	Sludge compost (SSC) and decaying rice straw (DRS).	PAH-contaminated soil containing PAH degraders	+15–33% (ANT), +13–63% (PYR), +17–21% (B[a]P)	Cost-effective alternative of DRS for hydrocarbon bioremediation: positive impact on soil microbial activity, PAH removal	Hamdi et al. 2007

					enhancement with no apparent changes in soil physicochemical properties	
C				+33% (ANT), +63% (PYR), +25% (B[a]P)	PAH-spiked soils chronically lethal to ostracod Heterocypris Incongruens, sensitivity of the microcrustaccan, inhibition of lettuce root elongation (strong phytotoxicity of SSC)	
A	4-chloroaniline (4CA)	Aniline	4CA-degrading *Klebsiella* sp. CA17	+33–38%	Bioattenuation: 10–38% of 4CA degradation	Tongarun et al. (2008)
C				+24–28%		
A	PAH	Ground corn cob	*Monilinia* sp. W5-2	+433%		Wu et al. (2008)
B				+1067%		
A	Trichloroethene (TCE)	Methanol, ethanol, acetate, lactacte mixture	Dechlorinating culture of *Dehalococcoides* strain KB-1	Dechlorination of TCE to ethane: +82%	Laboratory study and a pilot test	Hood et al. (2008)
E				+95%	*Dehalococcoides* abundance increased by 2 orders of magnitude following biostimulation and bioaugmentation	
C	Gas-oil	Nitrogen, phosphorus	*Psychrotolerant* strain (B-2-2)	+75%		Ruberto et al. (2003)
A	PAH and BaP	Ground rice hulls	*Pseudomonas aeruginosa* strain 64	+47% (HAP); +19% (BaP)		Straube et al. (2003)
F		Dried blood		+155% (HAP); +17% (BaP)		

(*continued*)

Table 7.5 (continued)

	Pollutant	Biostimulation	Bioaugmentation	Variation in the degradation rate	Comments	References
D	Atrazine	Citrate	1 inoculation of *Pseudomonas* sp. ADP 3 inoculations	+0–27% (+70% of 20 L ha^{-1}) +10–11% (+87–98%)	Higher atrazine degradation rate with 3 vs. only one inoculation	Lima et al. (2009)
A C E	1,2,3,4-tetrachlorodibenzo-p-dioxin (1,2,3,4-TeCDD)	Dechlorination stimulated by addition of electron donor and halogenated priming compounds	Mixed culture containing *Dehalococcoides* ethenogenes strain 195	Dechlorination of 1,2,3,4-TeCDD to 2-monochlorodibenzo-p-dioxin (2MCDD): No effect of treatments (similar degradation level for A, C, E)	Bioattenuation: 12 up to 24% of 1,2,3,4-TeCDD dechlorination At least an order of magnitude higher gene copy numbers of *Dehalococcoides*-specific 16S rRNA genes and the D. ethenogenes strain 195 dehalogenase gene, tceA in the bioaugmented than in the nonbioaugmented microcosms No effect of amendmenent except 1,2,3,4 TeCB on bacteria community	Ahn et al. 2008
A B C D	Hydrocarbon	N and P	*Acinetobacter* strain B-2-2	−15% +45% +30% −5%	Abiotic degradation and bioattenuation: 30 and 65%, respectively Initial inhibition of bacterial growth and reduction in hydrocarbon degradation by supplying high levels of N and P	Ruberto et al. 2003

A	PAH	Ground corn cob	*Monilinia* strain W5-2	+13%	Detoxification of soils in bioaugmented microcosms confirmed by genetic toxicity assay	Wu et al. 2008
B				+19%	Biostimulation with ground corn cob: Increasing of both number and abundance of indigenous aromatic hydrocarbons degraders and change in microbial community composition in soil	
C				+32%	Negligible effect of bioaugmentation with *Monilinia* strain W5-2 on indigenous microbial community	
B	Total petroleum hydrocarbons (TPH)		*Rhizopus* sp.	+36%		Mancera-Lopez et al. (2008)
B			*Penicillium funiculosum*	+30%		
B			*Aspergillus sydowii*	+17%		
A	PAH	$CuSO_4$, H_2O_2, mono-ammonium phosphate	Undefined microbial consortium selected from culture enriched with HAP-contaminated soil	Dibenzofuran degradation: +46% up to 61% (depending of biostimulation method)	Abiotic degradation and bioattenuation: 29 and 39%, respectively	Atanaga (2006)
F				No significant positive effect	Similar results for dibenzo-p-dioxine, fluoren-9-one, and benzo(a) anthracene	

(*continued*)

Table 7.5 (continued)

	Pollutant	Biostimulation	Bioaugmentation	Variation in the degradation rate	Comments	References
A	Benzo(*a*)pyrene (BaP)	Surfactant		+61%	Surfactant-enhanced biostimulation: better bioremediation option than natural attenuation alone, biostimulation without surfactant, and bioaugmentation, with or without an added surfactant	Talley et al. 2007

A Biostimulation vs. bioattenuation
B Bioaugmentation vs. biostimulation
C Bioaugmentation vs. bioattenuation
D Bioaugmentation + biostimulation vs. bioaugmentation
E Bioaugmentation + biostimulation vs. bioattenuation
F Bioaugmentation + biostimulation vs. biostimulation

such a technique. Until now the microbial selection still remains based on the ability of microorganisms to act on pollutants to be removed from soil without taking into account their ecological traits and their ability to reach pollutants in soil. Consequently low survival and colonization rates are sometimes observed, when microorganisms are inoculated in soil, along with low microbial activity towards pollutants.

Some future prospects for in situ soil bioaugmentation are sum up below. A more mechanistic approach shoud be adopted to determinate the exact role of inoculated microorganisms in the pollutant removal and further to be able to control with accuracy this remediation process even in environments subject to variable conditions. In this respect, the use of plants in association with bioaugmentation could be helpful in regulating the microbial activity. It would be also relevant to reconcile process engineering, based on the full control of any system, with variable environment conditions through the emerging ecological engineering approach. "In situ" soil bioaugmentation are most often performed in laboratory or outdoor conditions at smale scales and often means naturally contaminated soils or nonreconstructed soils. Long-term studies at field scale should be more widely experimented. One can indeed wonder whether the degradation performances change continuously with the scaling or whether some key experimental scales allow predicting degradation performances at much more higher scales. A set of basic indicators (physicochemical and biological) should be suggested as a result of a scientific consensus in the aim at laying down accurate diagnostics to further implement bioaugmentation with accuracy as well at being able to estimate the environmental impact of such a technique. Finally, putting at the user's disposal commercially formulated inocula for different kinds of contaminants and soils is a prerequisite for the further bioaugmentation dissemination, not to mention social acceptance to be checked.

References

Ahn YB, Liu F, Fennel DE, Hagglom MM (2008) Biostimulation and bioaugmentation to enhance dechlorination of polychlorinated dibenzo-p-dioxins in contaminated sediments. FEMS Microbial Ecol 66:271–281

Albareda M, Rodríguez-Navarro DN, Camacho M, Temprano FJ (2008) Alternatives to peat as a carrier for rhizobia inoculants: solid and liquid formulations. Soil Biol Biochem 40:2771–2779

Alisi C, Musella R, Tasso F, Ubaldi C, Manzo S, Cremisini C, Sprocati AR (2009) Bioremediation of diesel oil in a co-contaminated soil by bioaugmentation with a microbial formula tailored with native strains selected for heavy metals resistance. Sci Total Env 407:3024–3032

Alvey S, Crowley DE (1995) Influence of organic amendments on biodegradation of atrazine as a nitrogen source. J Environ Qual 24:1156–1162

Atanaga HI (2006) Biodegradation of polyacyclic aromatic hydrocarbons in contaminated soil by biostimulation and bioaugmentation in the presence of copper(II) ions. World J Microb Biot 22:1145–1153

Baek KH, Voon BD, Cho DH, Kim BH, Oh HM, Kim HS (2009) Monitoring bacterial population dynamics using real-time PCR during the bioremediation of crude-oil-contaminated soil. J Microbial Biotech 19:339–345

Baker AJM, McGrath SP, Reeves RD, Smith JAC (2000) Metal hyperaccumulator plants: a review of the ecology and physiology of a biological resource for phytoremeditaion of metal-polluted soils. In: Terry N, Bañuelos G (eds) Phytoremediation of contaminated soil and water. Lewis, Boca Raton, FL, pp 85–107

Balthazor TM, Hallas LE (1986) Glyphosate-degrading microorganism from industrial activated sludge. Appl Environ Microb 51:432–434

Barac T, Taghavi S, Borremans B, Provoost A, Oeyen L, Colpaert JV, Vangronsveld J, van der Lelie D (2004) Engineered endophytic bacteria improve phytoremediation of water-soluble, volatile organic pollutants. Nat Biotechnol 22:583–588

Barbeau C, Deschenes L, Karamanev D, Comeau Y, Samson R (1997) Bioremediation of pentachlorophenol-contaminated soil by bioaugmentation using activated soil. Appl Microbiot Biot 48:745–752

Barona A, Aranguiz I, Elias A (2001) Metal associations in soils before and after EDTA extractive decontamination: implications for the effectiveness of further clean-up procedures. Environ Pollut 113:79–85

Baxter J, Garton NJ, Cummings SP (2006) The impact of acrylonitrile and bioaugmentation on the biodegradation activity and bacterial community structure of a topsoil. Folia Microbiol 51:591–597

Beaulieu M, Becaert V, Deschenes L, Villemur R (2000) Evolution of bacterial diversity during enrichment of PCP-degrading activated soils. Microb Ecol 40:345–355

Becker MW, Metge DW, Collins SA, Shapiro AM, Harvey RW (2003) Bacterial transport experiments in fractured crystalline bedrock. Ground Water 41:682–689

Belotte D, Curien JB, Maclean RC, Bell G (2003) An experimental test of local adaptation in soil bacteria. Evolution 57:27–36

Benimeli CS, Fuentes MS, Abate CM, Amoroso MJ (2008) Bioremediation of lindane-contaminated soil by *Streptomyces* sp. M7 and its effects on Zea mays growth. Int Biodeter Biodegr 61:233–239

Benizri E, Nguyen C, Piutti S, Slezack-Deschaumes S, Philippot L (2007) Additions of maize root mucilage to soil changed the structure of the bacterial community. Soil Biol Biochemi 39:1230–1233

Bento FM, Camargo FAO, Okeke B (2003) Bioremediation of soil contaminated by diesel oil. Braz J Microbiol 34:65–68

Bento FM, Camargo FAO, Okeke B, William T (2005) Frankenberger Comparative bioremediation of soils contaminated with diesel oil by natural attenuation, biostimulation and bioaugmentation Bioresource Technol 96:1049–1055

Berselli S, Milone G, Canepa P, Di Gioia D, Fava F (2004) Effects of cyclodextrins, humic substances and rhamnolipids on the washing of a historically contaminated soil and on the aerobic bioremediation of the resulting effluents. Biotechnol Bioeng 88:111–120

Bester K, Schäfer D (2009) Activated soil filters (bio filters) for the elimination of xenobiotics (micro-pollutants) from storm- and waste waters. Water Res 43:2639–2646

Bingemann CW, Varner JE, Martin WP (1953) The effect of the addition of organic materials on the decomposition of an organic soil. Proc Soil Sci Soc Am 17:34–38

Bohn HL, McNeal BL, O'Connor GA (2001) Soil chemistry. Wiley, New York

Boopathy R (2000) Factors limiting bioremediation technologies. Bioresource Technol 74:63–67

Borges MT, Nascimento AG, Rocha UN, Totola MR (2008) Nitrogen starvation affects bacterial adhesion to soil. Braz J Microbiol 39:457–463

Bouchez T, Patureau D, Dabert BP, Juretschko S, Doré J, Delgenès P, Moletta R, Wagner M (2000) Ecological study of a bioaugmentation failure. Environ Microbiol 2:179–190

Brenner K, You L, Arnold FH (2008) Engineering microbial consortia: a new frontier in synthetic biology. Trends Biotechnol 26:483–489

Caccavo F, Ramsing NB, Costerton JW (1996) Morphological and metabolic responses to starvation by the dissimilatory metal-reducing bacterium Shewanella alga BrY. Appl Environ Microb 62:4678–4682

Carmichael LM, Pfaender FK (1997) The effect of inorganic and organic supplements on the microbial degradation of phenanthrene and pyrene in soils. Biodegradation 8:1–13

Cassidy MB, Lee H, Trevors JT (1996) Environmental applications of immobilized microbial cells: a review. J Ind Microbio Biot 16:79–101

Cassidy MB, Mullineers H, Lee H, Trevors JT (1997) Mineralization of pentachlorophenol in a contaminated soil by *Pseudomonas* sp UG30 cells encapsulated in *κ-carrageenan*. J Ind Microbiol Biot 19:43–48

Chavez FP, Gordillo F, Jerez CA (2006) Adaptive responses and cellular behaviour of biphenyl-degrading bacteria toward polychlorinated biphenyls. Biotechnol Adv 24:309–320

Checkai RT, Corey RB, Helmke PA (1987) Effect of ionic and complexed metal concentrations on plant uptake of cadmium and micronutrient cations from solution. Plant Soil 99:335–345

Chen YM, Lin TF, Huang C, Lin JC, Hsieh FM (2007) Degradation of phenol and TCE using suspended and chitosan-bead immobilized *Pseudomonas putida*. J Hazard Mater 148: 660–670

Ciccillo F, Fiore A, Bevivino A, Dalmastri C, Tabacchioni S, Chiarini L (2002) Effects of two different application methods of Burkholderia ambifaria MCI 7 on plant growth and rhizospheric bacterial diversity. Environ Microbiol 4:238–245

Colores GM, Schmidt SK (1999) Colonization of contaminated soil by an introduced bacterium: effects of initial pentachlorophenol levels on the survival of Sphingomonas chlorophenolica strain RA2. J Ind Microbio Biot 23:326–331

Comeau Y, Greer CW, Samson RR (1993) Role of inoculum preparation and density on the bioremediation of 2, 4-D-contaminated soil by bioaugmentation. Appl Microbiol Biot 38:681–687

Csillag J, Partay G, Lukacs A, Bujtas K, Nemeth T (1999) Extraction of soil solution for environmental analysis. Int J Environ Anal Chem 74:305–324

Cunliffe M, Kertesz MA (2006) dynamics and polycyclic aromatic hydrocarbon degradation in aged and freshly PAH-contaminated soils. Environ Pollut 144:228–237

Cunningham CJ, Ivshina IB, Lozinsky VI, Kuyukina MS, Philp JC (2004) Bioremediation of diesel-contaminated soil by microorganisms immobilised in polyvinyl alcohol. Int Bioteter Biodegr 54:167–174

D'Amours D, Samson R, Deschenes L (2008) Environmental innocuousness of the activation of a microbial consortium from creosote-contaminated soil in a slurry bioreactor. J Environ Eng Sci 7:581–595

D'Annibale A, Ricci M, Leonardi V, Quaratino D, Mincione E, Petruccioli M (2005) Degradation of aromatic hydrocarbons by white-rot fungi in a historically contaminated soil. Biotechnol Bioeng 90:723–731

Daane LL, Haggblom MM (1999) Earthworm egg capsules as vectors for the environmental introduction of biodegradative bacteria. Appl Environ Microb 65:2376–2381

de Weert S, Dekk'ers LC, Kuiper I, Bloemberg GV, Lugtenberg BJJ (2004) Generation of enhanced competitive root-tip-colonizing Pseudomonas bacteria through accelerated evolution. J Bacteriol 186:3153–3159

Dean JR (2010) Heavy metal bioavailability and bioaccessibility in soil. Meth Mol Biol 599:15–36

Dechesne A, Pallud C, Bertolla F, Grundmann GL (2005) Impact of the microscale distribution of a *Pseudomonas* strain introduced into soil on potential contacts with indigenous bacteria. Appl Environ Microb 71:8123–8131

Dejonghe W, Boon N, Seghers D, Top EM, Verstraete W (2001) Bioaugmentation of soils by increasing microbial richness: missing links. Environ Microb 3:649–657

Delle Site A (2000) Factors affecting sorption of organic compounds in natural sorbent/water systems and sorption coefficients for selected pollutants: a review. J Phys Chem Ref Data 30: 187–439

Dercova K, Sejakova Z, Skokanova M, Barancikova G, Makovnikova J (2006) Potential use of organomineral complex (OMC) for bioremediation of pentachlorophenol (PCP) in soil. Int Biodeter Biodegr 58:248–253

Desai C, Pathak H, Madamwar D (2010) Advances in molecular and "-omics" technologies to gauge microbial communities and bioremediation at xenobiotic/anthropogen contaminated sites. Biores Technol 101:1558–1569

Di Gregorio S, Barbafieri M, Lampis S, Sanangelantoni AM, Tassi E, Vallini G (2006) Combined application of Triton X-100 and *Sinorhizobium* sp. Pb002 inoculum for the improvement of lead phytoextraction by Brassica juncea in EDTA amended soil. Chemosphere 63:293–299

Di Simine CD, Sayer JA, Gadd GM (1998) Solubilization of zinc phosphate by a strain of *Pseudomonas fluorescens* isolated from a forest soil. Biol Fer Soils 28:87–94

Di Toro S, Zanaroli G, Varese GC, Marchisio VF, Fava F (2008) Role of Enzyveba in the aerobic bioremediation and detoxification of a soil freshly contaminated by two different diesel fuels. Int Biodeter Biodegrad 62:153–161

Dickinson NM, Pulford ID (2005) Cadmium phytoextraction using short-rotation coppice Salix: the evidence trail. Environ Int 31:609–613

Diels L, Lookman R (2007) Microbial systems for in-situ soil and groundwater remediation conference information. In: Marmiroli N, Samotokin B (eds) Advanced science and technology for biological decontamination of sites affected by chemical and radiological nuclear agents, earth and environmental sciences, vol 75. NATO Science series IV, Springer ISSN: 1568–1238 pp 61–77

Diels L, De Smet M, Hooyberghs L, Corbisier P (1999) Heavy metals bioremediation of soil. Mol Biotechnol 12:154–158

Diplock EE, Mardlin DP, Killham KS, Paton GI (2009) Predicting bioremediation of hydrocarbons: laboratory to field scale. Environ Pollut 157:1831–1840

Dominguez RF, da Silva MLB, McGuire TM, Adamson D, Newell CJ, Alvarez PJJ (2008) Aerobic bioremediation of chlorobenzene source-zone soil in flow-through columns: performance assessment using quantitative PCR. Biodegradation 19:545–553

Dua M, Singh A, Sethunathan N, Johri AK (2002) Biotechnology and bioremediation: successes and limitations. Appl Microbiol Biot 59:143–152

Dubbin WE, Ander EL (2003) Influence of microbial hydroxamate siderophores on Pb(II) desorption from α-FeOOH. Appl Geochem 18:1751–1756

Dueri S, Castro-JiménezJ CJMZ (2008) On the use of the partitioning approach to derive environmental quality standards (EQS) for persistent organic pollutants (POPs) in sediments: a review of existing data. Sci Total Environ 403:23–33

Duponnois R, Kisa M, Assigbetse K, Prin Y, Thioulouse J, Issartel M, Moulin P, Lepage M (2006) Fluorescent pseudomonads occuring in *Macrotermes subhyalinus* mound structures decrease Cd toxicity and improve its accumulation in sorghum plants. Sci Total Environ 370:391–400

Duquenne P, Chenu C, Richard G, Catroux G (1999) Effect of carbon source supply and its location on competition between inoculated and established bacterial strains in sterile soil microcosm. FEMS Microbiol Ecol 29:331–339

El Fantroussi S, Agathos SN (2005) Is bioaugmentation a feasible strategy for pollutant removal and site remediation? Curr Opin Microbiol 8:268–275

Fang Y, Logan BE (1999) Bacterial transport in gas-sparged porous medium. J Environ Eng-ASCE 125:668–673

Fernandes VC, Albergaria JT, Oliva-Teles T, Delerue-Matos C, De Marco P (2009) Dual augmentation for aerobic bioremediation of MTBE and TCE pollution in heavy metal-contaminated soil. Biodegradation 20:375–382

Field JA, Sierra-Alvarez R (2008) Microbial transformation and degradation of polychlorinated biphenyls. Environ Pollut 155:1–12

Forsyth JV, Tsao YM, Bleam RD (1995) Bioremediation: when is augmentation needed. In: Hinchee RE, Fredrickson J, Alleman BC (eds) Bioaugmentation for site remediation. Battelle, Columbus, pp 1–14

Franzetti P, Caredda C, Ruggeri L, La Colla E, Tamburini PM, Bestetti G (2009) Potential application of surface active compounds by *Gordonia* sp. strain BS29 in soil remediation technologies. Chemosphere 75:801–807

Gadd GM, White C (1993) Microbial treatment of metal pollution: a working biotechnology? Trends Biotechnol 11:353–359

Garon D, Krivobok S, Wouessidjewe D, Seigle-Murandi F (2002) Influence of surfactants on solubilization and fungal degradation of fluorene. Chemosphere 47:303–309

Geiszinger A, Bonnineau C, Faggiano L, Guasch H, Lopez-Doval JC, Proia L, Ricart M, Ricciardi F, Romani A, Rotter S, Munoz I, Schmitt-Jansen M, Sabater S (2009) The relevance of the community approach linking chemical and biological analyses in pollution assessment. TRAC Trend Anal Chem 28:619–626

Gentili AR, Cubitto MA, Ferrero M, Rodriguez MS (2006) Bioremediation of crude oil polluted seawater by a hydrocarbon-degrading bacterial strain immobilized on chitin and chitosan flakes. Int Biodeter Biodegradation 57:222–228

Gentry TJ, Christopher R, Ian P (2004) New approaches for bioaugmentation as a remediation technology. Crit Rev Env Sci Tec 34:447–494

Gentry TJ, Wickham GS, Schadt CW, He Z, Zhou J (2006) Microarray applications in microbial ecology research. Microbiol Ecol 52:159–175

Ghazali M, Rahman RNZA, Salleh AB, Basri M (2004) Biodegradation of hydrocarbons in soil by microbial consortium. Int Biodeterior Biodegrad 54:61–67

Girlanda M, Favero-Longo SE, Lazzari A, Segreto R, Perotto S, Siniscalco C (2009) Indigenous microfungi and plants reduce soil nonylphenol contamination and stimulate resident microfungal communities. Appl Microbiol Biot 82:359–370

Glick BR (2003) Phytoremediation: synergistic use of plants and bacteria to clean up the environment. Biotechnol Adv 21:383–393

Goddard VJ, Bailey MJ, Darrah P, Lilley AK, Thompson IP (2001) Monitoring temporal and spatial variation in rhizosphere bacterial population diversity: a community approach for the improved selection of rhizosphere competent bacteria. Plant Soil 232:181–193

Goldman RP, Brown SP (2009) Making sense of microbial consortia using ecology and evolution. Trends Biotechnol 27:3–4

Goux S, Shapir N, El Fantroussi S, Lelong S, Agathos SN, Pussemier L (2003) Long term maintenance of rapid atrazine degradation in soils inoculated with atrazine degraders. Water Air Soil Pollut Focus 3:131–142

Grundmann S, Fuß R, Schmid M, Laschinger M, Ruth B, Schulin R, Munch JC, Reiner Schroll R (2007) Application of microbial hot spots enhances pesticide degradation in soils. Chemosphere 68:511–517

Habe H, Ide K, Yotsumoto M, Tsuji H, Hirano H, Widada J, Yoshida T, Nojiri H, Omori T (2001) Preliminary examinations for applying a carbazole-degrader, *Pseudomonas* sp. strain CA10, to dioxin-contaminated soil remediation. Appl Microbiol Biotechnol 56:788–795

Hamdi H, Benzarti S, Manusadzianas L, Aoyama I, Jedidi N (2007) Bioaugmentation and blostimulation effects on PAH dissipation and soil ecotoxicity under controlled conditions. Soil Biol Biochem 39:1926–1935

Han H, Tang J, Jiang H, Zhang ML, Liu Z (2008) Synergy between fungi and bacteria in fungi-bacteria augmented remediation of petroleum-contaminated soil. Huanjing Kexue 29:189–195

Hazen TC, Stahl DA (2006) Using the stress response to monitor process control: pathways to more effective bioremediation. Curr Opin Biotech 17:285–290

Heinaru E, Merimaa M, Viggor S, Lehiste M, Leito I, Truu J, Heinaru A (2005) Biodegradation efficiency of functionally important populations selected for bioaugmentation in phenol- and oil-polluted area. FEMS Microbiol Ecol 51:363–373

Herman DC, Artiola JF, Miller RM (1995) Removal of cadmium, lead, and zinc from soil by a rhamnolipid biosurfactant. Environ Sci Technol 29:2280–2285

Hesselsoe M, Bjerring ML, Henriksen K, Loll P, Nielsen JL (2008) Method for measuring substrate preferences by individual members of microbial consortia proposed for bioaugmentation. Biodegradation 19:621–633

Hong Q, Zhang ZH, Hong Y, Li S (2007) A microcosm study on bioremediation of fenitrothion-contaminated soil using Burkholderia sp. FDS-1. Int Biodeter Biodegrad 59:55–61

Hood ED, Major DW, Quinn JW, Yoon WS, GavastarA EEA (2008) Demonstration of enhanced bioremediation in a TCE source area at Launch Complex 34, Cape Canaveral Air Force Station. Ground Wat Monitor Remed 28:98–107

Imfeld G, Braeckevelt M, Kuschk P, Richnow HH (2009) Monitoring and assessing processes of organic chemicals removal in constructed wetlands. Chemosphere 74:349–362

Iwamoto T, Nasu M (2001) Current bioremediation practice and perspective. J Biosci Bioeng 92:1–8

Jacques RJS, Okeke BC, Bento FM, Teixeira AS, Peralba MCR, Camargo FAO (2008) Microbial consortium bioaugmentation of a polycyclic aromatic hydrocarbons contaminated soil. Bioresource Technol 99:2637–2643

Jansson JK (2003) Marker and reporter genes: illuminating tools for environmental microbiologists. Curr Opin Microbiol 6:310–316

Jansson JK, Björklöf K, Elvang AM, Jorgensen KS (2000) Biomarkers for monitoring efficacy of bioremediation by microbial inoculants. Environ Pollut 107:217–223

Jézéquel K, Lebeau T (2008) Soil bioaugmentation by free and immobilized bacteria to reduce potentially phytoavailable cadmium. Bioresource Technol 99:690–698

Jézéquel K, Perrin J, Lebeau T (2005) Bioaugmentation with a *Bacillus* sp. to reduce the phytoavailable Cd of an agricultural soil: comparison of free and immobilized microbial inocula. Chemosphere 59:1323–1331

Johnsen AR, Wick LY, Harms H (2005) Principles of microbial PAH-degradation in soil. Environ Pollut 133:71–84

Johnsen AR, Schmidt S, Hybholt TK, Henriksen S, Jacobsen CS, Andersen O (2007) community of a PAH-polluted soil but marginal effect on PAH degradation when priming with bioremediated soil dominated by mycobacteria. Appl Environ Microbiol 73:1474–1480

Joner EJ, Briones R, Leyval C (2000) Metal-binding capacity of arbuscular mycorrhizal mycelium. Plant Soil 226:227–234

Jorgensen SE, Mitsch WJ (1980) Classification and examples of ecological engineering. In: Mitsch WJ, Jorgensen SE (eds) Ecological engineering: an introduction to ecotechnology. Wiley, New York, pp 13–19

Juan Antonio Zermeño-Eguia Liz, Janet Jan-Roblero, Javier Zavala-Díaz de la Serna, Arturo Vera-Ponce de León and César Hernández-Rodríguez (2009). Degradation of polychlorinated biphenyl (PCB) by a consortium obtained from a contaminated soil composed of Brevibacterium, Pandoraea and Ochrobactrum. World Journal of Microbiology and Biotechnology Volume 25, Number 1, 165–170

Juhanson J, Truu J, Heinaru E, Heinaru A (2009) Survival and catabolic performance of introduced Pseudomonas strains during phytoremediation and bioaugmentation field experiment. FEMS Microbiol Ecol 70:446–455

Kabata-Pendias A (2000) Trace elements in soils and plants. CRC, Boca Raton

Kao CM, Liu JK, Chen YL, Chai CT, Chen SC (2005) Factors affecting the biodegradation of PCP by Pseudomonas mendocina NSYSU. J Haz Mater 124:68–73

Karagiannidis N, Nikolaou N (2000) Influence of arbuscular mycorrhizae on heavy metal (Pb and Cd) uptake, growth, and chemical composition of Vitis vinifera L. (cv. Razaki). Am J Enol Vitic 51:269–275

Kayser A, Wenger K, Keller A, Attinger W, Felix HR, Gupta SK, Schulin E (2000) Enhancement of phytoextraction of Zn, Cd and Cu from calcareous soil: the use NTA and sulfur amendments. Environ Sci Technol 34:1778–1783

Khan AG (2006) Mycorrhizoremediation-an enhanced form of phytoremediation. J Zhejiang Univ-Sc B 7:503–514

Khan AG, Kuek C, Chaudhry TM, Khoo CS, Hayes WJ (2000) Role of plants, mycorrhizae and phytochelators in heavy metal contaminated land remediation. Chemosphere 41:197–207

Khan FI, Husain T, Hejazi R (2004) An overview and analysis of site remediation technologies. J Environ Manage 71:95–122

Kinsall BL, Wilson GV, Palumbo AV (2000) The effect of soil heterogeneity on the vadose zone transport of bacteria for bioaugmentation. In: Wickramanayake GB, Gavaskar AR, Alleman

BC, Magar VS (eds) Bioremediation and phytoremediation of chlorinated and recalcitrant compounds. Battelle, Columbus, California, pp 395–403

Kuiper I, Bloemberg GV, Lugtenberg BJJ (2001) Selection of a plant-bacterium pair as a novel tool for rhizostimulation of polycyclic aromatic hydrocarbon-degrading bacteria. Mole Plant Microbe Int 14:1197–1205

Kuiper I, Lagnedijk EL, Bloemberg GV, Lugtenberg BJJ (2004) Rhizoremediation: a beneficial plant-microbe interaction. Mole Plant Microbe Int 17:6–15

Kumar M, Leon V, De Sisto MA, Ilzins OA (2006) Enhancement of oil degradation by co-culture of hydrocarbon degrading and biosurfactant producing bacteria. Pol J Microbiol 55:139–146

Labana S, Pandey G, Paul D, Sharma NK, Basu A, Jain RK (2005) Pot and field studies on bioremediation of p-nitrophenol contaminated soil using arthrobacter protophormiae RKJ100. Environ Sci Technol 39:3330–3337

Lamberts RF, Johnsen AR, Andersen O, Christensen JH (2008) Univariate and multivariate characterization of heavy fuel oil weathering and biodegradation in soil. Environ Pollut 156:297–305

LaPara TM, Zakharova T, Nakatsu CH, Konopka A (2002) Functional and structural adaptations of bacterial communities growing on particulate substrates under stringent nutrient limitation. Microbiol Ecol 44:317–326

Lappin-Scott HM, Costerton JW (1992) Ultramicrobacteria and their biotechnological applications. Curr Opin Biotech 3:283–285

Lebeau T, Bagot D, Jézéquel K, Fabre B (2002) Cadmium biosorption by free and immobilised microorganisms cultivated in a liquid soil extract medium: effects of Cd, pH and techniques of culture. Sci Total Environ 291:73–83

Lebeau T, Braud A, Jézéquel K (2008) Performance of bioaugmentation-assisted phytoextraction applied to metal contaminated soils: a review. Environ Pollut 153:497–522

Li Q, Logan BE (1999) Enhancing bacterial transport for bioaugmentation of aquifers using low ionic strength solutions and surfactants. Water Res 33:1090–1100

Li Y, Dick WA, Tuovinen OH (2004) Fluorescence microscopy for visualization of soil microorganisms: a review. Soil Biol Biochem 39:301–311

Lima D, Viana P, André S, Chelinho S, Costa C, Ribeiro R, Sousa JP, Fialho AM, Viegas CA (2009) Evaluating a bioremediation tool for atrazine contaminated soils in open soil microcosms: the effectiveness of bioaugmentation and biostimulation approaches. Chemosphere 74:187–192

Liz JA, Jan-Roblero J, Zavala-Díaz de la Serna J, Vera-Ponce de León A, Hernández-Rodríguez C (2009) Degradation of polychlorinated biphenyl (PCB) by a consortium obtained from a contaminated soil composed of *Brevibacterium, Pandoraea* and *Ochrobactrum*. Chem Mater Sci 25:165–170

Löffler FE, Edwards EA (2006) Harnessing microbial activities for environmental cleanup. Curr Opin Biotech 17:274–284

López-Gutiérrez JC, Philippot L, Martin-Laurent F (2005) Impact of maize mucilage on atrazine mineralization and atzC abundance. Pest Manage Sci 61:838–844

Lovley DR, Lloyd JR (2000) Microbes with a mettle for bioremediation. Nat Biotechnol 18:600–601

Majewska M, Kurek E, Rogalski J (2007) Microbially mediated cadmium sorption/desorption processes in soil amended with sewage sludge. Chemosphere 67:724–730

Mancera-Lopez M, Esparza-Garcia F, Chavez-Gomez B, Rodriguez-Vazquez R, Saucedo-Castaneda G, Barrera-Cortes J (2008) Bioremediation of an aged hydrocarbon-contaminated soil by a combined system of biostimulation-bioaugmentation with filamentous fungi. Int Biodeter Biodegrad 61:151–160

Manefield M, Whiteley AS, Bailey MJ (2004) What can stable isotope probing do for bioremediation? Int Biodeter Biodegrad 54:163–166

Mariano AP, Bonotto DM, de Angelis DD, Contiero P (2008) Biodegradability of commercial and weathered diesel oils. Braz J Microbiol 39:133–142

Martin M, Gibello A, Lobo C, Nande M, Garbi C, Fajardo C, Barra-Caracciolo A, Grenni P, Martinez-Inigo MJ (2008) Application of fluorescence in situ hybridization technique to detect simazine-degrading bacteria in soil samples. Chemosphere 71:703–710

Marzorati M, Wittebolle L, Boon N, Daffonchio D, Verstraete W (2008) How to get more out of molecular fingerprints: practical tools for microbial ecology. Environ Microbiol 10:1571–1581

Matz C, Kjelleberg S (2005) Off the hook: how bacteria survive protozoan grazing. Trends Microbiol 13:302–307

McKinlay RG, Kasperek K (1999) Observations on decontamination of herbicide-polluted water by marsh plant systems. Water Res 33:505–511

Mehmannavaz R, Prasher SO, Ahmad D (2001) Effect of bioaugmentation on microbial transport, water infiltration, moisture loss, and surface hardness in pristine and contaminated soils. J Environ Sci Heal A 36:123–139

Mench M, Baize D (2004) contamination des sols et de nos aliments d'origine végétale par les éléments en traces Courrier de l'environnement de l'INRA n°52, septembre 2004 31–56

Mertens B, Boon N, Verstraete W (2006) Slow-release inoculation allows sustained biodegradation of gamma-hexachlorocyclohexane. Appl Environ Microb 72:622–627

Mitsch WJ, Jørgensen SE (2004) Ecological engineering and ecosystem restoration. Wiley, New York

Mohan SV, Sirisha K, Rao RS, Sarma PN (2007) Bioslurry phase remediation of chlorpyrifos contaminated soil: process evaluation and optimization by Taguchi design of experimental (DOE) methodology. Ecotox Environ Safe 68:252–262

Monard C, Martin-Laurent F, Vecchiato C, Francez AJ, Vandenkoornhuyse P, Binet F (2008) Combined effect of bioaugmentation and bioturbation on atrazine degradation in soil. Soil Biol Biochem 40:2253–2259

Moran AC, Muller A, Manzano M, Gonzalez B (2006) Simazine treatment history determines a significant herbicide degradation potential in soils that is not improved by bioaugmentation with Pseudomonas sp. ADP. J Appl Microbiol 101:26–35

Moslemy P, Neufeld RJ, Guiot SR (2002) Biodegradation of gasoline by gellan gum-encapsulated bacterial cells. Biotechnol Bioeng 80:175–184

Mrozik A, Piotrowska-Seget Z (2010) Bioaugmentation as a strategy for cleaning up of soils contaminated with aromatic compounds. Microbiol Res 165:363–375

Mulligan CN, Yong RN, Gibbs BF, James S, Bennett HPJ (1999) Metal removal from contaminated soils and sediments by biosurfactants surfactin. Environ Sci Technol 33:3812–3820

Mulligan CN, Yong RN, Gibbs BF (2001) Heavy metal removal from sediments by biosurfactants. J Haz Mater 85:111–125

Nam IH, Hong HB, Kim YM, Kim BH, Murugesan K, Chang YS (2005) Biological removal of polychlorinated dibenzo-p-dioxins from incinerator fly ash by Sphingomonas wittichii RW1. Water Res 39:4651–4660

Narasimhan K, Basheer C, Bajic VB, Swarup S (2003) Enhancement of plant–microbe interactions using a rhizosphere metabolomics-driven approach and its application in the removal of polychlorinated biphenyls. Plant Physiol 132:146–153

Neufeld JD, Wagner M, Murrell JC (2007) Who eats what, where and when? Isotope-labelling experiments are coming of age. ISME J 1:103–110

Newcombe DA, Crowley DE (1999) Bioremediation of atrazine-contaminated soil by repeated applications of atrazine-degrading bacteria. Appl Microbiol Biot 51:877–882

Odum HT (1962) Ecological tools and their use Man and the ecosystem. In: Waggoner PE, Ovington JD (eds) Conference of the suburban forest and ecology. Lockwood, New Zealand, pp 57–75

Odum HT, Odum B (2003) Concepts and methods of ecological engineering. Ecol Eng 20:339–361

Ohtsubo Y, Kudo T, Tsuda M, Nagata Y (2004) Strategies for bioremediation of polychlorinated biphenyls. Appl Microbiol Biot 65:250–258

Olaniran AO, Pillay D, Pillay B (2006) Biostimulation and bioaugmentation enhances aerobic biodegradation of dichloroethenes. Chemosphere 63:600–608

Orphan VJ (2009) Methods for unveiling cryptic microbial partenerships in nature. Curr Opin Microbiol 12:231–237

Otte MP, Gagnon J, Comeau Y, Matte N, Greer CW, Samson R (1994) Activation of an indifenous microbial consortium for bioaugmentation of pentachlorophenol/creosote contaminated soils. Appl Microbiol Biot 40:926–932

Petric I, Hrsak D, Fingler S, Voncina E, Cetkovic H, Bejonga KA, Udikovic KN (2007) Enrichment and characterization of PCB-degrading bacteria as potential seed cultures for bioremediation of contaminated soil. Food technol biotech 45:11–20

Pignatello JJ (1998) Soil organic matter as a nanoporous sorbent of organic pollutants. Adv Colloid Interface Sci 76–77:445–467

Plangklang P, Reungsang A (2009) Bioaugmentation of carbofuran residues in soil using Burkholderia cepacia PCL3 adsorbed on agricultural residues. Int Biodeter Biodegrad 63:515–522

Portier R, Bianchini M, Fujisaki K, Henry C, McMillin D (1989) Comparison of effective toxicant biotransformation by autochthonous microorganisms and commercially available cultures in the in situ reclamation of abandoned industrial sites. Schriftenr Ver Wasser Boden Lufthyg 80:273–292

Priestley JT, Coleman NV, Duxbury T (2006) Growth rate and nutrient limitation affect the transport of Rhodococcus sp strain DN22 through sand. Biodegradation 17:571–576

Prosser JI, Rangel-Castro JI, Killham K (2006) Studying plant-microbe interactions using stable isotope technologies. Curr Opin Biotech 17:98–102

Radajewski S, Ineson P, Parekh NR, Murrell JC (2000) Stable-isotope probing as a tool in microbial ecology. Nature 403:646–649

Rahm BG, Chauhan S, Holmes VF, Macbeth TW, Sorenson KSJ, Alvarez-Cohen L (2006) Molecular characterization of microbial populations at two sites with differing reductive dechlorination abilities. Biodegradation 17:523–534

Rao RS, Kumar CG, Prakasham RS, Hobbs PJ (2008) The Taguchi methodology as a statistical tool for biotechnological applications: a critical appraisal. Biotechnol J 3:510–523

Riser-Roberts E (1998) Remediation of petroleum contaminated soil: biological, physical, and chemical processes. CRC, Boca Raton

Rittmann BE, Hausner M, Löffler FE, Love NG, Muyzer G, Okabe S, Oerther DB, Peccia J, Raskin L, Wagner M (2006) A vista for microbial ecology and environmental biotechnology. Environ Sci Technol 40:1096–1103

Romantschuk M, Sarand I, Petanen T, Peltola R, Jonsson-Vihanne M, Koivula T, Yrjala K, Haahtela K (2000) Means to improve the effect of in situ bioremediation of contaminated soil: an overview of novel approaches. Environ Pollut 107:179–185

Rousseau M, Ruy S, Di Pietro L, Angulo-Jaramillo R (2003) Unsaturated hydraulic conductivity of structured soils from a kinematic wave approach. J Hydraul Res 42:83–91

Ruberto L, Vazquez SC, Mac Cormack WP (2003) Effectiveness of the natural bacterial flora, biostimulation and bioaugmentation on the bioremediation of a hydrocarbon contaminated Antarctic soil. Int Biodeter Biodegrad 52:115–125

Sarand I, Timonen S, Rajamäki M, Peltola R, Nurmiaho-Lassila EL, Koivula T, Haahtela K, Romantschuk M, Sen R (1998) Biofilms, degradative fluorescent Pseudomonads and bioremediation potential in the scots pine mycorrhizosphere exposed to petroleum derived hydrocarbons. FEMS Microb Ecol 27:115–126

Schwartz E, Scow KM (2001) Repeated inoculation as a strategy for the remediation of low concentrations of phenanthrene in soil. Biodegradation 12:201–207

Scow KM, Hicks KA (2005) Natural attenuation and enhanced bioremediation of organic contaminants in groundwater. Curr Opin Biotech 16:246–253

Semple KT, Doick KJ, Burauel P, Craven A, Jones KC, Harms H (2004) Defining bioavailability and bioaccessibility for the risk assessment and remediation of soils and sediment is complicated. Environ Sci Tech 38:209A–212A

Singer AC, Gilbert ES, Luepromchai E, Crowley DE (2000) Bioremediation of polychlorinated biphenyl-contaminated soil using carvone and surfactant-grown bacteria. Appl Microbiol Biot 54:838–843

Singer AC, van der Gast CJ, Thompson IP (2005) Perspectives and vision for strain selection in bioaugmentation. Trends Biotechnol 23:74–77
Singh R, Paul D, Jain R (2006) Biofilms: implications in bioremediation. Trends Microbiol 14:389–397
Singh A, van Hamme JD, Ward OP (2007) Surfactants in microbiology and biotechnology: part 2. Application aspects. Biotechnol Adv 25:99–121
Siripattanakul S, Wirojanagud W, McEvoy JM, Casey FXM, Khan E (2009) A feasibility study of immobilized and free mixed culture bioaugmentation for treating atrazine in infiltrate. J Haz Mater 168:1373–1379
Smith KE, Schwab AP, Banks MK (2007) Phytoremediation of polychlorinated biphenyl (PCB)-contaminated sediment: a greenhouse feasibility study. J Environ Qual 36:239–244
Sobiecka E, Cedzynska K, Bielski C, Antizar-Ladislao B (2009) Biological treatment of transformer oil using commercial mixtures of microorganisms. Int Biodeter Biodegrad 63: 328–333
Sorvari J, Antikainen R, Kosola ML, Hokkanen P, Haavisto T (2009) Eco-efficiency in contaminated land management in Finland: barriers and development needs. J Environ Manage 90:1715–1727
Stallwood B, Shears J, Williams PA, Hughes KA (2005) Low temperature bioremediation of oil-contaminated soil using biostimulation and bioaugmentation with a Pseudomonas sp. from maritime Antarctica. J Appl Microbiol 99:794–802
Steinberg CEW, Sturzenbaum SR, Menzel R (2008) Genes and environment: striking the fine balance between sophisticated biomonitoring and true functional environmental genomics. Sci Total Env 400:142–161
Stottmeister U, Wiener A, Kuschk P, Kappelmeyer U, Kastner M, Bederski O, Muller RA, Moormann H (2003) Effects of plants and microorganisms in constructed wetlands for wastewater treatment. Biotechnol Adv 22:93–117
Straube WL, Jones-Meehan J, Pritchard PH, Jones WR (1999) Bench-scale optimization of bioaugmentation strategies for treatment of soils contaminated with high molecular weight polyaromatic hydrocarbons. Resour Conser Recy 27:27–37
Straube WL, Nestler CC, Hansen LD, Ringleberg D, Pritchard PH, Jones-Meehan JJ (2003) Remediation of Polyaromatic Hydrocarbons (PAHs) through Landfarming with Biostimulation and Bioaugmentation Acta Biotechnologica 23:179–196
Streger SH, Vainberg S, Dong HL, Hartzinger PB (2002) Enhancing transport of Hydrogenophaga flava ENV735 for bioaugmentation of aquifers contaminated with methyl tert-butyl ether. Appl Environ Microbiol 68:5571–5579
Strigul NS, Kravchenko LV (2006) Mathematical modeling of PGPR inoculation into the rhizosphere. Environ Model Softw 21:1158–1171
Strong LC, McTavish H, Sadowsky MJ, Wackett LP (2000) Field-scale remediation of atrazine-contaminated soil using recombinant *Escherichia coli* expressing atrazine chlorohydrolase. Environ Microbiol 2:91–98
Sung K, Kim J, Munster CL, Corapcioglu MY, Park S, Drew MC, Chang YY (2006) A simple approach to modeling microbial biomass in the rhizosphere. Ecol Model 190:277–286
Talley JW, Cui XF, Berveiler P (2007) Distribution and bioremediation of PAHs among different particle classes found in a contaminated surface soil. In: Li S, Wang Y, Cao F, Huang P, Zhang Y (eds) Progress in environmental science and technology, Science Press, Beijing, vol 1. pp 1159–1164
Thompson IP, van der Gast CJ, Ciric L, Singer AC (2005) Bioaugmentation for bioremediation: the challenge of strain selection. Environ Microbiol 7:909–915
Thouand G, Bauda P, Oudot J, Kirsch G, Sutton C, Vidalie JF (1999) Laboratory evaluation of crude oil biodegradation with commercial or natural microbial inocula. Can J Microbiol 45: 106–115
Tixier C, Sancelme M, Aït-Aïssa S, Widehem P, Bonnemoy F, Cuer A, Truffaut N, Veschambre H (2002) Biotransformation of phenylurea herbicides by a soil bacterial strain, *Arthrobacter* sp.

N2: structure, ecotoxicity and fate of diuron metabolite with soil fungi. Chemosphere 46: 519–526

Tongarun R, Lueprомchai E, Vangnai AS (2008) Natural attenuation, biostimulation, and bioaugmentation in 4-chloroaniline-contaminated soil. Curr Microbiol 56:182–188

Tonin C, Vandenkoornhuyse P, Joner EJ, Straczek J, Leyval C (2001) Assessment of arbuscular mycorrhizal fungi diversity in the rhizosphere of Viola calaminaria and effect of these fungi on heavy metal uptake by clover. Mycorrhiza 10:161–168

Trindade PVO, Sobral LG, Rizzo ACL, Leite SGF, Soriano AU (2005) Bioremediation of a weathered and a recently oil-contaminated soils from Brazil: a comparison study. Chemosphere 58:515–522

Truu J, Talpsep E, Vedler E, Heinaru E, Heinaru A (2003) Enhanced biodegradation of oil shale chemical industry solid wastes by phytoremediation and bioaugmentation. Oil Shale 20: 421–428

Van Aken B, Yoon JM, Schnoor JL (2004) Biodegradation of nitro-substituted explosives TNT, RDX, and HMX by a phytosymbiotic *Methylobacterium* sp. associated with poplar tissues (*Populus deltoides* x *nigra* DN34). Appl Environ Microb 70:508–517

van der Gast CJ, Whiteley AS, Starkey M, Knowles CJ, Thompson IP (2003) Bioaugmentation strategies for remediating mixed chemical effluents. Biotechnol Prog 19:1156–1161

van der Gast CJ, Whiteley AS, Thompson IP (2004) Temporal dynamics and degradation activity of an bacterial inoculum for treating waste metal-working fluid. Environ Microbiol 6:254–263

van der Gast CJ, Jefferson B, Reid E, Robinson T, Bailey MJ, Judd SJ, Thompson IP (2006) Bacterial diversity is determined by volume in membrane bioreactors. Environ Microbiol 8:1048–1055

van Dillewijn P, Caballero A, Paz JA, Gonzalez-Perez MM, Oliva JM, Ramos JL (2007) Bioremediation of 2, 4, 6-trinitrotoluene under field conditions. Environ Sci Technol 41: 1378–1383

van Veen JA, van Overbeek LS, van Elsas JD (1997) Fate and activity of microorganisms introduced into soil. Microbiol Mol Biol Rev 61:121–135

Verstraete W (2007) Microbial ecology and environmental biotechnology. ISME J 1:4–8

Verstraete W, Wittelbolle L, Heylen K, Vanparys B, de Vos P, van de Wiele T, Boon N (2007) Microbial resource management: the road to go for environmental biotechnology. Eng Life Sci 7:117–126

Vijgen J (2002). Summary report on technical and legal barriers for the introduction of ex-situ biological soil treatment in the EU member states. EU Commission, Directorate General Research

Vogel TM (1996) Bioaugmentation as a soil bioremediation approach. Curr Opin Biotechnol 7:311–316

Vogel TM, Walter MV (2001) Bioaugmentation. In: Hurst CJ, Crawford RL, Knudsen GR, McInerney MJ, Stetzenbach LD (eds) Manual of environmental microbiology. American Society for Microbiology, Washington, pp 952–959

Volesky B, Holan ZR (1995) Biosorption of heavy metals. Biotechnol Prog 11:235–250

Wang G, Gentry TJ, Grass G, Josephson K, Rensing C, Pepper IL (2004) Real-time PCR quantification of a green fluorescent protein-labeled, genetically engineered Pseudomonas putida strain during 2-chlorobenzoate degradation in soil. FEMS Microbiology Letters 233: 307–314

Wang S, Mulligan CN (2004) An evaluation of surfactant foam technology in remediation of contaminated soil. Chemosphere 57:1079–1089

Wang QY, Zhou DM, Cang L (2009) Microbial and enzyme properties of apple orchard soil as affected by long-term application of copper fungicide. Soil Biol Biochem 41:1504–1509

Weston A, Balba T (2003) Bioaugmentation. Innov Technol Group 3:1–2

Widada J, Nojiri H, Yoshida T, Habe H, Omori T (2002) Enhanced degradation of carbazole and 2,3-dichlorodibenzo-p-dioxin in soils by Pseudomonas resinovorans strain CA10. Chemosphere 49:485–491

Wolicka D, Suszek A, Borkowski A, Bielecka A (2009) Application of aerobic microorganisms in bioremediation in situ of soil contaminated by petroleum products. Bioresource Technol 100: 3221–3227

Wu YC, Luo YM, Zou DX, Ni JZ, Liu WX, Teng Y, Li ZG (2008) Bioremediation of polycyclic aromatic hydrocarbons contaminated soil with *Monilinia* sp.: degradation and microbial community analysis. Biodegradation 19:247–257

Xia H, Chi X, Yan Z, Cheng W (2009) Enhancing plant uptake of polychlorinated biphenyls and cadmium using tea saponin. Biores Technol 100:4649–4653

Zhang H, Zhao FJ, Sun B, Davison W, McGrath S (2001) A new method to measure effective soil solution concentration predicts copper availability to plants. Environ Sci Technol 35: 2602–2607

Zhang K, Hua XF, Han HL, Wang J, Miao CC, Xu YY, Huang ZD, Zhang H, Yang JM, Jin WB, Liu YM, Liu Z (2008) Enhanced bioaugmentation of petroleum- and salt-contaminated soil using wheat straw. Chemosphere 73:1387–1392

Zhuang X, Chen J, Shim H, Bai Z (2007) New advances in plant growth-promoting rhizobacteria for bioremediation. Environ Int 33:406–413

Zuzana S, Katarina D, Livia T (2009) Biodegradation and ecotoxicity of soil contaminated by pentachlorophenol applying bioaugmentation and addition of sorbents. Word J Microb Biotechnol 25:243–252

Chapter 8
Bioremediation of Contaminated Soils: Effects of Bioaugmentation and Biostimulation on Enhancing Biodegradation of Oil Hydrocarbons

Iwona Zawierucha and Grzegorz Malina

8.1 Introduction

Modern societies still continue to rely primarily on the use of petroleum hydrocarbons to cover their energy demands. Despite recent technological advances, accidental spills of crude oil and its refined products occur on a frequent basis during routine operations such as extraction, transportation, storage, refining and distribution (Nikolopoulou et al. 2007). The release of oil hydrocarbons into the environment may pose severe environmental problems due to sustained contamination of air, water and soil (Scherr et al. 2007). Various physical and chemical processes have been employed for effective remediation of oil hydrocarbon contaminated soil (Khan et al. 2004; Malina 2007). However, most of these techniques are expensive, and the byproducts may cause secondary contamination of soil and water, resulting in the need for additional post-treatments (Liang et al. 2009). Moreover, they require continuous monitoring and control for optimum performance. In addition, they do not usually result in a complete destruction of the contaminants (Gouda et al. 2008). Biological methods, such as bioremediation, are considered to be relatively cost-effective and environmentally friendly (Hosokawa et al. 2009). Bioremediation is a treatment method that uses microbiological restoration potentials for decontamination of polluted sites (Scherr et al. 2007). It is a relatively simple practical approach for the complete mineralization of hydrocarbons to carbon dioxide and water under aerobic conditions (Vidali 2001; Sarkar et al. 2005). However, the rate of hydrocarbon biodegradation in soil is affected by several physicochemical properties of the soil and contaminants, as well as biological characteristics of indigenous microorganisms. These include the number

I. Zawierucha (✉)
Institute of Chemistry, Environmental Protection and Biotechnology, Jan Dlugosz University of Czestochowa, Armii Krajowej 13/15, 42-200 Czestochowa, Poland
e-mail: iwona_zawierucha@o2.pl

G. Malina
Department of Hydrogeology and Engineering Geology, AGH University of Science and Technology, Mickiewicza 30, 30-059 Cracow, Poland

A. Singh et al. (eds.), *Bioaugmentation, Biostimulation and Biocontrol*, Soil Biology 28,
DOI 10.1007/978-3-642-19769-7_8,

and species of microorganisms present, concentrations of hydrocarbons and environmental conditions (pH, temperature, nutrients, oxygen and moisture content) suitable for microbial degradation (Betancur-Galvis et al. 2006; Gouda et al. 2008; Leahy and Colwell 1990; Perfumo et al. 2007; Horel and Schiewer 2009). Two methods are usually considered to increase the activity of microorganisms, thus enhancing the biodegradation rates during bioremediation of soil contaminated with oil hydrocarbons: (a) bioaugmentation through the direct application of selected oil-degraders to the site, and (b) biostimulation involving the application of a proper agent to soil to enhance the activity of indigenous microorganisms (Odokuma and Dickson 2003; Perfumo et al. 2007; Malina and Zawierucha 2007).

Bioaugmentation is a promising and low-cost bioremediation method, in which effective bacterial isolates or microbial consortia capable of degrading oil hydrocarbons are introduced to the contaminated soil. Multiplied indigenous microflora are generally applied in this technique; however, inoculation of soil with exogenous or laboratory-modified bacterial cultures still arouses many reservations (Gentry et al. 2004; Fantroussi and Agathos 2005; Zawierucha and Malina 2006). Sometimes, the application of oil-degrading microorganisms may lead to a failure of bioaugmentation (Vogel 1996; Gentry et al. 2004). This is because the survival and degrading ability of microorganisms introduced to a contaminated site are highly dependent on environmental conditions (Vogel 1996). Thus, in many cases, potentially degrading strains isolated from one site are not necessarily applicable to the other site. Moreover, isolates, including genetically engineered microorganisms, that are efficient oil-degraders under laboratory conditions, are not necessarily effective in situ (Sayler and Ripp 2000). In addition, introducing alien species to soil is not easily acceptable by the public (Hosokawa et al. 2009).

Biostimulation relies on increasing the activity of indigenous bacteria by providing nutrients, oxygen, surfactants or water to the contaminated soil (Coulon and Delille 2003) or modifying the environmental conditions (e.g., temperature, pH, redox potential). It is considered that indigenous bacteria are best adapted to the environment of the treated site (Rahman et al. 2003). Biostimulation, however, does not always work well due to the scarcity of indigenous oil-degraders or very high contaminant concentrations (Ueno et al. 2007).

The major objective of this chapter is to provide various bioremediation strategies based on bioaugmentation and biostimulation for enhancing biodegradation rates of oil hydrocarbon contaminated land.

8.2 Bioaugmentation

Bioaugmentation can be realized in three ways: (a) the enrichment or isolation of indigenous microorganisms from the target site, their subsequent culturing and re-inoculation; (b) isolates or enrichments are not inoculated to the source of the original culture; and (c) with the use of constructed or force-mutated microorganisms (Vogel and Walter 2001).

Bioaugmentation that uses indigenous microorganisms to the sites (soil and water) to be remediated is defined as autochthonous bioaugmentation (Ueno et al. 2007). Isolated single strains or enriched cultures, which can be obtained "before" or "after" the contamination of the target sites, are inoculated into the sites to be remediated (Hosokawa et al. 2009). The use of indigenous microorganisms with adapted biochemical potentials was proved to be one of the most powerful tools for bioaugmentation (Devinny and Chang 2000). Bento et al. (2005) noted that the addition of a bacterial consortium previously isolated from the Long Beach soil degraded 73–75% of the light (C_{12}–C_{23}) and heavy (C_{23}–C_{40}) fractions of total petroleum hydrocarbons (TPH) present in the soil, while only 46–49% removal was obtained as a result of intrinsic biodegradation. Next, in field-scale microcosms, Gouda et al. (2008) observed that about 86% of kerosene was degraded upon bioaugmentation of clay with indigenous *Pseudomonas* sp. CK for 20 days, but only 80% in the case of intrinsic biodegradation of the natural microflora. Introducing naturally developed microbial consortia may be more effective in comparison with single strains isolated and applied as pure cultures (Mrozik and Piotrowska-Seget 2009). This is in agreement with the results of Mancera-Lopez et al. (2008), who noted that in the bioaugmented systems three indigenous fungi strains (*Rhizopus* sp., *Penicillium funiculosum* and *Aspergillus sydowii*) removed 47, 45 and 40% of TPH, respectively, and these were even higher by 29–36% with respect to the pure fungi strains. Thus, the most effective bioaugmentation performance may be approached by the use of multiplied indigenous microorganisms to increase their abundance in soil (Malina and Zawierucha 2007). In the study of Wu et al. (2008) microcosms were set up with a PAHs contaminated soil using bioaugmentation with indigenous filamentous fungus, *Monilinia* sp. W5-2. After 30 days of treatment, bioaugmented microcosms resulted in a 35% decrease in the total PAHs, while the control microcosms showed only a 3% decrease. Bioaugmented microcosms also revealed about 70 and 72% decreases in benzo[*a*]pyrene and anthracene, respectively, while the values for the control microcosms were much lower.

Our respirometry studies conducted to determine the effect of bioaugmentation for enhancing biodegradation in soils contaminated with oil hydrocarbons at a former military airport showed the highest biodegradation rates (estimated from the O_2 uptake and CO_2 production rates) for samples to which the bacterial inoculum containing 4.8×10^{15} CFU/ml was added (Zawierucha and Malina 2006). Enhanced biodegradation rates were in this case four times higher than intrinsic biodegradation rates. Moreover, when the indigenous bacterial consortium was applied, the increase of biodegradation rate was about 22–46% higher compared to the exogenous bacterial consortium. This could be explained by the autochthonous adaptation that allows microorganisms to be physiologically compatible with their habitat, as compared to transient autochthonous organisms that do not occupy a functional niche (Atlas and Bartha 1998). Ueno et al. (2007) also noticed that bioaugmentation capacity of isolated bacterial species in soil microcosms contaminated with diesel oil was much higher than that of exogenous *P. aeruginosa* strain WatG. Therefore, it will be more practical to apply bioaugmentation with bacteria isolated from the soil that is to be cleaned-up (Hosokawa et al. 2009).

Native populations present in contaminated sites are certainly adapted to the climatic, physicochemical and nutrient conditions. However, these communities frequently do not include species with the enzymatic abilities needed to allow bioremediation to start and/or to proceed at increased rates, thus resulting in long-time processes. Application to contaminated soils of exogenous microorganism with proven hydrocarbon-degrading abilities may solve this problem. Possible sources to obtain exogenous microorganisms are remediated or contaminated sites, commercial suppliers and genetic engineering (Diaz-Ramirez et al. 2008).

Biodegradation studies of 4CA in two soil types (loam soil-NN and sandy clay loam soil-CM) using exogenous 4CA-degrading *Klebsiella* sp. CA17 for bioaugmentation were carried out by Tongarun et al. (2008). Biodegradation of 4CA in soil-NN microcosms was substantially enhanced by bioaugmentation with 4CA-degrading *Klebseilla* sp. CA17. Compared to that of intrinsic biodegradation (40% of 4CA degradation), the total 4CA biodegradation at the end of bioaugmentation treatment period finally reached 70 ± 4%. In the case of soil-CM microcosms, total 4CA degradation was 44%, as compared to rather poor (10%) intrinsic biodegradation at the end of the treatment period. Moreover, bioaugmentation of a 4CA-degrading culture was successful in soil-NN, where the bioaugmented culture survived and maintained its population size, with a gradual increase, throughout the entire treatment, while it was eventually outnumbered by indigenous microorganisms in soil-CM. These results may indicate that the degree of 4CA biodegradation in soil microcosms depends not only on the site characteristics, for example, soil properties, but also on the characteristics of the indigenous microbial population, and the stability of population density of the bioaugmented culture, which definitely affected the efficiency of biodegradation.

The choice of chronically contaminated areas or the sites where land farming was applied for remediation for the screening of exogenous bacterial strains potentially useful in bioaugmentation seems to be an appropriate approach. Ruberto et al. (2003) observed that in soil with B-2-2, a psychrotolerant hydrocarbon-degrading *Acinetobacter* strain previously isolated from a chronically polluted river degraded 75% of hydrocarbons, whereas autochthonous bacterial communities were able to degrade important fractions of the gas oil only by 35%. Next, Jacques et al. (2008) evaluated the capacity of an exogenous microbial consortium (five bacteria: *Mycobacterium fortuitum, Bacillus cereus, Microbacterium* sp., *Gordonia polyisoprenivorans, Microbacteriaceae bacterium* – naphthalene-utilizing bacterium; and a fungus identified as *Fusarium oxysporum*) obtained from a petrochemical site treated by means of land farming, to degrade and mineralize anthracene, phenanthrene and pyrene present in soil at different concentrations. They noted that the microbial consortium mineralized on average 98% of the three PAHs present at different concentrations in the soil after 70 days. On the contrary, the autochthonous soil microbial population showed no substantial mineralization of the PAHs. Moreover, bacterial and fungal isolates from the consortium, when inoculated separately to the soil, were less effective in anthracene mineralization compared to that of the consortium. These results may indicate synergistic promotion of PAHs mineralization by mixtures of the monoculture isolates (the microbial consortium). Individual microorganisms can mineralize only a limited range of substrates; so assemblages of mixed populations with overall broad enzymatic capacities are required to increase the rate and extent of petroleum biodegradation (Farinazleen et al. 2004; Heinaru et al. 2005). Enhanced biodegradation of spiked anthracene (ANT), pyrene (PYR) and benzo[*a*]pyrene (B[*a*]P) in soil was also studied by Hamdi et al. (2007). In this case, bioaugmentation was carried out by mixing the previously treated aged PAH-contaminated soil containing PYR- and B[*a*]P-degraders with the experimental PAH-spiked soil. In the control samples, the PAH removal was the lowest revealing ANT, PYR and B[*a*]P dissipation rates of 63, 33 and 35%, respectively, after 120 days. In turn, in bioaugmented samples, the final degradation rates were higher than those observed in nonamended PAH-spiked soils, and they were above 96% for ANT and PYR and 60% for B[*a*]P. Therefore, the bioaugmentation with exogenous bacteria can be recommended in the case of more recalcitrant chemicals, or when the local microbial population is insufficient or inadequate (Mariano et al. 2007).

Another bioaugmentation approach involves the use of genetically engineered microorganisms (GEMs) with increased capacity to degrade and tolerate toxic compounds. Mutations and horizontal gene transfer using molecular biology are employed to improve the microbial degradation activity (Rodrigues et al. 2006). The aim of the application of genetically modified bacterial strains is to enhance the ability of newly generated strains to degrade a broader range of xenobiotics, and to increase the degradation effectiveness in comparison with "wild" (natural) strains (Mrozik and Piotrowska-Seget 2009). Filonov et al. (2005) studied the effectiveness of a genetically tagged, plasmid-containing, naphthalene-degrading strain *Pseudomonas putida* KT2442 (pNF142: TnMod-OTc) in the experimental soil contaminated with naphthalene. They noted that the concentration of naphthalene in the experimental soil block, into which laboratory naphthalene-degrading strain KT2442 was introduced, decreased from 2.0 to 0.2 mg/g, whereas in the control soil block, which contained only indigenous naphthalene degraders, the concentration decreased only to 0.6 mg/g over the same time period (20 days). Moreover, 20 days after introducing the strain KT2442, the number of bacterial cells increased from 10^5 to 6×10^6 CFU/g of soil, amounting to 90% of the total population of naphthalene degraders. Mishra et al. (2004) also noted that the TPH level in the microcosm soil bioaugmented with a recombinant *Acinetobacter baumannii* pJES strain was reduced by 39.6% at the end of a 90 days treatment, while in the untreated soil this reduction was only by 6.9%. Next, Massa et al. (2009) compared two bacterial strains, the natural isolate *Arthrobacter* sp. FG1 and the engineered strain *Pseudomonas putida* PaW340/pDH5, for their efficiency in degrading 4-chlorobenzoic acid (4-CBA) in a slurry phase system. The recombinant strain was obtained by cloning the *Arthrobacter* sp. FG1 dehalogenase encoding genes in *P. putida* PaW340. The 4-CBA-grown engineered strain appeared to be significantly more efficient in the 4-CBA degradation than the "wild" *Arthrobacter* in soil slurries regardless of the presence or absence of indigenous bacteria, which did not affect biodegradation. On the other hand, Lima et al. (2009) examined the efficacy of bioaugmentation with rifampicin-resistant mutant of *Pseudomonas* sp. ADP for bioremediation of an atrazine-contaminated land. They observed that, for a more moderate level of soil contamination (ca. 7 g of atrazine per g of soil), bioaugmentation using one single inoculation with *P.* ADP could be sufficient for successful treatment. The atrazine removal was of 99% after 8 days of the treatment. Therefore, the use of GEMs has been suggested to improve or accelerate the remediation of sites polluted with xenobiotics (Filonov et al. 2005; Lima et al. 2009).

The novel approach, the so-called immobilized bioaugmentation based on delivering microbial cultures in an immobilized form, is applied to achieve more complete and/or more rapid degradation of hydrocarbons in soil. Such treatment offers the protection of inoculated microorganisms from sub-optimal environmental conditions (improper pH, presence of toxic substances, etc.), and it reduces their competition with indigenous microflora. Moreover, immobilization increases the biological stability of cells, including plasmids (Cunningham et al. 2004). For immobilization, both natural and synthetic materials are used. The first group includes dextran, agar, agarose, alginate, chitosan polyacrylamides, and κ-carrageenan, while the second includes poly(carbamoyl)sulphonate, polyacrylamide and polyvinyl alcohol (Mrozik and Piotrowska-Seget 2009). Immobilized matrix may also act as a bulking agent in contaminated soil, facilitating the transfer of oxygen crucial for rapid hydrocarbon mineralization (Cunningham et al. 2004; Liang et al. 2009). Results obtained by Oh et al. (2003) indicate that the inoculum immobilized on diatomaceous earth could be very effective for retaining microbial cells in association with the sand contaminated with crude oil. Cunningham et al. (2004) examined the potential of immobilized hydrocarbon-degrading microorganisms for the cleanup of diesel-contaminated soil, and compared it with the liquid-culture bioaugmentation. Using polyvinyl alcohol (PVA) cryogelation as an entrapment technique, they noted that bioaugmentation with a liquid enrichment culture reduced by 36.7% of oil and grease contents after 32 days of treatment, while the immobilized system resulted in the 48.1% reduction. Moreover, the reductions of diesel in these bioaugmentation systems were about 25.3–36.7% higher as compared to the non-amended (control) treatment pile. Next, Liang et al. (2009) explored the role of bio-carriers, such as activated carbon and zeolites, in immobilizing indigenous hydrocarbon-degrading bacteria and enhancing biodegradation in crude oil contaminated soils. They observed high microbial colonization of both zeolites and activated carbon. Microbial biomass reached concentrations of 10^{10} cells/g of activated carbon and 10^{6} cells/g of zeolites, indicating that the first carrier was better for the enrichment of bacteria. Total microbial and dehydrogenase activity were 12 and 3 times higher, respectively, in activated carbon than in zeolite. Moreover, the activated carbon bio-carrier enhanced the biodegradation of crude oil resulting in the removal of 48.9%, in comparison to the intrinsic biodegradation (13.0%), and liquid-culture bioaugmentation (free-living bacteria) – 37.4%. According to the authors, the bio-carrier improved the mass transfer of oxygen and nutrients, as well as the water-holding capacity of the soil, which were the limiting factors for biodegradation of crude oil.

The successful soil bioaugmentation requires the knowledge of not only types and contents of contaminants but also microbial strains or their consortia that are suitable for biodegradation. The selection of a culture appropriate for bioaugmentation should take into consideration the following features of microorganisms: fast growth rate, ease of cultivation, resistance to high contaminant concentrations and ability to survive in a wide range of environmental conditions (Mrozik and Piotrowska-Seget 2009). Moreover, for designing an optimum bioaugmentation method, it is necessary to evaluate the fractions of bioavailable contaminants to determine the required concentrations of a degrading inoculum to be added (Zawierucha and Malina 2006).

8.3 Biostimulation

Biostimulation is a technique that relies on increasing the activity of the indigenous bacteria by adjusting the factors that may limit their activity, mainly oxygen and nutrients. The main aim of biostimulation is to provide bacterial communities with a favorable environment, in which they can effectively degrade contaminants (Mohan et al. 2006; Ueno et al. 2007).

8.3.1 Biostimulation by Oxygen Supply

Bioremediation of hydrocarbon contaminated soils under aerobic conditions is preferable to improve degradation yields, given that the most common microbial degraders are aerobic (Menendez-Vega et al. 2007). Oxygen is supplied to soil to stimulate microbial activity and enhance aerobic biodegradation rates in the case when O_2 is considered as a limiting factor. Commonly used oxygen supply techniques may include tilling, forced aeration and chemical methods (Atlas 1991; Brown and Crosbie 1994; Riser-Roberts 1998). Tilling is recommended as a physical method to accelerate biodegradation during land farming, but it is effective only for top soils. Forced aeration techniques, including injection of aerated water, air or pure oxygen, are commonly used for enhancing biodegradation in soils and groundwater contaminated with petroleum hydrocarbons (Brown and Crosbie 1994; Riser-Roberts 1998). Chemical methods involve addition of alternative oxygen sources, such as oxygen-releasing compounds ORC®, or agents such as potassium permanganate ($KMnO_4$), hydrogen peroxide (H_2O_2) and ozone (O_3) (Riser-Roberts 1998).

In our study, we tested the effectiveness of diverse sources of oxygen (aerated water, aqueous solutions of H_2O_2 and $KMnO_4$) to enhance biodegradation of oil hydrocarbons in soil at a former military airport (Malina and Zawierucha 2007). Based on respirometric tests the highest CO_2 production rates (71–97% higher compared to a control) were achieved when the aqueous solution of $KMnO_4$ in concentration of 20 g/L was applied. On the other hand, on average, only 15% increase of CO_2 production rates was observed when the aqueous solutions of H_2O_2 were used, whereas aerated water did not cause any improvement of the biodegradation rates (addition of aerated water resulted in a decrease of CO_2 production rates as compared to a control). Most probably, the amount of added water led to excessive soil moisture that could reduce, in fact, the air-filled porosity and, consequently the oxygen contents in soil (Malina 1999).

Potassium permanganate is known to readily oxidize alkene carbon–carbon double bonds (Wolfe and Ingold 1981; Walton et al. 1992). Brown et al. (2003) observed the concentration reduction of benzo[*a*]pyrene (72.1%), pyrene (64.2%), phenanthrene (56.2%) and anthracene (53.8%) in soil, after 30 min of oxidation using 160 mM $KMnO_4$. They suggested that permanganate oxidation could be applied in remediation technology for soils contaminated with oil hydrocarbons. While hydrocarbons are not completely mineralized by permanganate oxidation reactions, their structure is altered by polar functional groups providing vast improvements in aqueous solubility, increased bioavailability for microorganisms, thus biodegradation enhancement.

Hydrogen peroxide is known in environmental applications as a chemical oxidizing agent, disinfectant and source of oxygen (Hamby 1996; Olexsey and Parker 2006; Goi et al. 2006). Aerobic biodegradation of hydrocarbons in soil can benefit from the presence of oxygen released during the H_2O_2 decomposition (Sturman et al 1995). On the other hand, it can be toxic to microorganisms (Riser-Roberts 1998) as high contents of H_2O_2 (100–200 mg/l) can inhibit bacterial metabolism (Huling et al. 1990). However, Tsai and Kao (2009) noted that 43 and 47% of TPH were removed from soil using 15 and 30% aqueous solution of H_2O_2, which corresponded to the H_2O_2 concentrations of 150 and 300 mg/L, respectively, after 40 h of treatment, while the TPH removal using 1% aqueous solutions of H_2O_2 was only of 1.1%. These results indicate that the TPH oxidation can be enhanced by higher H_2O_2 concentration. Moreover, Goi et al. (2006) noted that the efficiency of H_2O_2 application was strongly dependent on the soil matrix. Treatment of shale and transformer oils adsorbed on peat (a model of organic-rich soil) resulted in lower degree of oil removal, and required more H_2O_2 than the treatment of oil in sand matrix representing the mineral part of soil.

Aerated water can be an effective O_2 carrier for aerobic biodegradation of oil hydrocarbons in soil, and it may facilitate the transport of substrates to bacterial cells (Malina 2007). For example, Liu et al. (2001) found a 50% increase of phenanthrene biodegradation when increasing the soil water content to 200% of the soil field capacity. But on the other hand, at higher water contents, near or over saturation, all the pores are filled with water, which limits the oxygen transfer that determines the activity of aerobic microorganisms, thus biodegradation of contaminants may be hampered (Ramirez et al. 2009), and which could actually be the case in our experiment (Malina and Zawierucha 2007).

8.3.2 Biostimulation by Nutrients Supply

Additional nutrients (mainly nitrogen and phosphorus) introduced into contaminated soil in the form of organic and/or inorganic fertilizers may enhance intrinsic biodegradation of petroleum hydrocarbons by improving the C:N:P ratio (Sarkar et al. 2005). In theory, approximately 150 mg of nitrogen and 30 mg of phosphorus are utilized in the conversion of 1 g of hydrocarbon to cell materials (Rosenberg and Ron 1996). Based on this, the optimal C:N:P mole-ratio recommended for enhancing hydrocarbon removal is 100:10:1 (Malina 1999). However, as the soil environment is very complex and heterogeneous, the effectiveness of nutrient sources tends to be affected by the soil physicochemical properties (Malina 2007). We compared various sources of nutrients (N, P) with different C:N:P ratios for enhancing biodegradation of oil hydrocarbons in soil (Zawierucha et al 2008). The highest enhanced biodegradation rates (2–26 times higher than intrinsic biodegradation rates) were observed at the C:N:P ratio of 100:10:5. Moreover, the best results were achieved when the combination of $(NH_4)_2SO_4$ and Na_2HPO_4 was used, as the enhanced biodegradation rates were 120–1,556% higher compared to the intrinsic biodegradation rates. Ubochi et al. (2006) examined the potential of biostimulation treatment options in oil contaminated soil with different contents of the NPK fertilizer in soil. They noted that the application of 60 g NPK fertilizer was the best treatment option with the removal of 50.5% of crude oil while in the control (no nutrient addition) the removal was only of 29.5%. However, the effectiveness of biostimulation depends not only on the proper C:N:P ratio but also on the type of soil. Aspray et al. (2008) observed that for the sandy gravel and silty clay soils contaminated with petroleum hydrocarbons, both O_2 consumption and CO_2 production showed enhanced microbial activity when amended with NH_4NO_3, whereas, these results differed for the sandy loam soil. In this soil amended with nitrogen, inhibition of respiration was observed. Moreover, the form of nutrients (especially nitrogen) supply plays a role in effective fertilization (Chaillan et al. 2006). The amendment with nitrogen (particularly using inorganic fertilizers) can have no effect or, when applied at high concentrations, even deleterious effects (Bento et al. 2005; Walworth et al. 2007). Inorganic nitrogen fertilizers composed of nitrate and ammonium salts increase the salt concentration of soil pore water, lowering the soil osmotic potential and, thus inhibiting the microbial activity (Walworth et al. 2007). In addition, Sarkar et al. (2005) found that the microbial population in the fertilizer-amended soils dropped appreciably, suggesting a toxic effect due to fertilizer-induced acidity and/or NH_3 overdosing. Therefore, the fertilizers must be precisely dosed taking the local environmental conditions into consideration. The effectiveness of hydrocarbon biodegradation is not proportional to the nutrient concentrations and over-fertilization may inhibit decomposition of less biodegradable compounds (Chaillan et al. 2006). It is also recommended that the C:N:P ratio should be calculated on the basis of the concentration of saturated hydrocarbons, degradation of which is most sensitive to the level of nutrients in soil (Chaineau et al. 2005).

8.3.3 Biostimulation by Surfactants Supply

A critical aspect of bioremediation of oil-contaminated soils is the availability of contaminants for microorganisms limited by their water solubility (Menendez-Vega et al. 2007). This problem can be solved using natural and synthetic surfactants (Lai et al. 2009). Hydrocarbon-degrading microorganisms produce a variety of surface-active natural agents (the so-called biosurfactants) that improve bioavailability. However, synthetic surfactants may still be required when the contaminants are highly hydrophobic, and/or firmly sorbed in clay particles or soil organic matter (Menendez-Vega et al. 2007). Surfactants contain both hydrophobic and hydrophilic fractions and are useful in reducing the interfacial tension between hydrocarbons and soil water, thereby improving the water solubility of hydrophobic substances (Urum et al. 2006; Zawierucha et al. 2007). Consequently, surfactants increase bioavailability of hydrocarbons to microorganisms and in turn their biodegradation (Lee et al. 2005). In our study on the effect of surfactant (Tween 80) on biodegradation of oil hydrocarbon and microbial activity in soil, the highest O_2 consumption and CO_2 production rates, as well as dehydrogenase activities were observed at the surfactant dose of 1% (v/v) (Zawierucha et al. 2007). In this case, the O_2 consumption and CO_2 production rates were 115 and 49% higher, respectively, while the dehydrogenase activity increased 98%, as compared to a control (no addition of the surfactant). These results indicate that surfactants can improve biodegradation effectiveness in the soil contaminated with oil hydrocarbons, which was also postulated by Bento et al. (2005), Rous et al. (1994) and Xie (2003). The potential application of biosurfactants, surfactin (SF) and rhamnolipid (RL), for enhanced diesel biodegradation was investigated by Whang et al. (2009). They observed that, compared to the control treatment (no biosurfactants added), application of RL or SF resulted in diesel emulsification, and therefore enhanced biodegradation. Lai et al. (2009) compared the effectiveness of biosurfactants with that of synthetic surfactants in heavily oil-polluted soil. They found that biosurfactants exhibited much higher TPH removal efficiency than the synthetic ones. By using rhamnolipids, surfactin, Tween 80, and Triton X-100 in the concentration of 0.2% (w/w), the TPH removal for the soil contaminated with 3,000 mg TPH/kg dry soil was of 23, 14, 6, and 4%, respectively, and it increased to 63, 62, 40, and 35%, respectively, for the soil contaminated with 9,000 mg TPH/kg dry soil. Moreover, the biosurfactants-enhanced TPH removal efficiency did not vary significantly with the contact time. These results indicate the superior performance of biosurfactants over synthetic surfactants in terms of mobilization of oil pollutants from the contaminated soil, and thus confirm their use as biostimulation agents for bioremediation. Biosurfactants are preferred to chemical surfactants, as they have lower toxicity and shorter persistence in the environment (Nievas et al. 2008). The potential advantages of biosurfactants for enhancing bioremediation of hydrocarbon contaminated soils also include their unusual structural diversity that may lead to unique properties, as well as their biodegradability. Moreover, biosurfactants could easily be produced from renewable resources via microbial fermentation, making it an additional advantage over chemically synthetic surfactants (Mulligan 2005).

8.4 Conclusions

Technologies employing biological treatments are developing worldwide over the last decade, and are viewed as ready-made approaches for bioremediation of petroleum contaminated sites. This chapter presented the potential of oil hydrocarbons biodegradation enhancement in contaminated soils by applying various bioaugmentation and biostimulation methods. Based on our extensive review of literature and practical experience, bioremediation is not a panacea for all problems associated with oil hydrocarbon contaminated soil. The successful bioremediation requires the strategies tailored for the site-specific environmental parameters of both contaminated soils and contaminants. The key parameters to select the appropriate bioremediation strategy includes number and activity of microorganisms; types, concentrations and bioavailability of contaminants; oxygen and nutrients supply, and characteristics of soil. In the case of bioaugmentation, an additional database needs to be established, containing abilities of microorganisms to degrade oil hydrocarbons, together with their proliferation in the respective ecosystems, as well as their cellular resistance to xenobiotics and adaptation potentials to environmental conditions.

Although both bioaugmentation and biostimulation alone appeared to be effective in enhancing intrinsic biodegradation of oil hydrocarbons in soil, the simultaneous action of these techniques seems to improve the biodegradation rates more efficiently.

References

Aspray T, Gluszek A, Carvalho D (2008) Effect of nitrogen amendment on respiration and respiratory quotient (RQ) in three hydrocarbon contaminated soils of different type. Chemosphere 72:947–951

Atlas RM (1991) Microbial hydrocarbon degradation-bioremediation of oil spills. J Chem Tech Biotechnol 52:149–156

Atlas RM, Bartha R (1998) Microbial ecology: fundamentals and applications. Benjamin/Cummings, Menlo Park, CA

Bento FM, Camargo FAO, Okeke BC, Frankenberger WT (2005) Comparative bioremediation of soils contaminated with diesel oil by natural attenuation, biostimulation and bioaugmentation. Bioresour Technol 96:1049–1055

Betancur-Galvis LA, Alvarez-Bernal D, Ramos-Valdivia AC, Dendooven L (2006) Bioremediation of polycyclic aromatic hydrocarbon-contaminated saline-alkaline soils of the former Lake Texcoco. Chemosphere 62:1749–1760

Brown RA, Crosbie JR (1994) Oxygen sources for in situ bioremediation. In: Flathman PE, Jerger DE, Jurgen HE (eds) Bioremediation: field experience. CRC, LLC Boca Raton, FL, pp 311–332

Brown GS, Barton LL, Thomson BM (2003) Permanganate oxidation of sorbed polycyclic aromatic hydrocarbons. Waste Manage 23:737–740

Chaillan F, Chaineau CH, Point V, Saliot A, Oudot J (2006) Factors inhibiting bioremediation of soil contaminated with weathered oils and drill cuttings. Environ Pollut 144:255–265

Chaineau CH, Rougeux G, Yepremian C, Oudot J (2005) Effects of nutrient concentration on the biodegradation of crude oil and associated microbial populations in the soil. Soil Biol Biochem 37:1490–1497

Coulon F, Delille D (2003) Effects of biostimulation on growth of indigenous bacteria in sub-Antarctic soil contaminated with oil hydrocarbons. Oil Gas Sci Technol 58:469–479

Cunningham CJ, Ivshina IB, Lozinsky VI, Kuyukina MS, Philp JC (2004) Bioremediation of diesel-contaminated soil by microorganisms immobilised in polyvinyl alcohol. Int Biodeterior Biodegradation 54:167–174

Devinny J, Chang SH (2000) Bioaugmentation for soil bioremediation. In: Wise DL, Trantolo D (eds) Bioremediation of contaminated soils. Dekker, New York, pp 465–488

Diaz-Ramirez IJ, Escalante-Espinosa E, Favela-Torres E, Gutierrez-Rojas M, Ramirez-Saad H (2008) Design of bacterial defined mixed cultures for biodegradation of specific crude oil fractions, using population dynamics analysis by DGGE. Int Biodeterior Biodegradation 62:21–30

Fantroussi SE, Agathos SN (2005) Is bioaugmentation a feasible strategy for pollutant removal and site remediation? Curr Opin Microbiol 8:268–275

Farinazleen MG, Raja Noor ZA, Abu BS, Mahiran B (2004) Biodegradation of hydrocarbons in soil by microbial consortium. Int Biodeterior Biodegradation 54:61–67

Filonov AE, Akhmetov LI, Puntus IF, Esikova TZ, Gafarov AB, IzmalkovaTY SSL, Kosheleva IA, Boronin AM (2005) The construction and monitoring of genetically tagged, plasmid-containing, naphthalene degrading strains in soil. Microbiology 74:453–458

Gentry TJ, Rensing C, Pepper IL (2004) New approaches for bioaugmentation as a remediation technology. Crit Rev Environ Sci Tech 34:447–494

Goi A, Kulik N, Trapido M (2006) Combined chemical and biological treatment of oil contaminated soil. Chemosphere 63:1754–1763

Gouda MK, Omar SH, Nour Eldin HM, Checkroud ZA (2008) Bioremediation of kerosene II: a case study in contaminated clay (Laboratory and field: scale microcosms). World J Microbiol Biotechnol 24:1451–1460

Hamby DM (1996) Site remediation techniques supporting environmental restoration activities: a review. Sci Total Environ 191:203–224

Hamdi H, Benzarti S, Manusadzianas L, Aoyama I, Jedidi N (2007) Bioaugmentation and biostimulation effects on PAH dissipation and soil ecotoxicity under controlled conditions. Soil Biol Biochem 39:1926–1935

Heinaru E, Merimaa M, Viggor S, Lehiste M, Leito I, Truu J, Heinaru A (2005) Biodegradation efficiency of functionally important populations selected for bioaugmentation in phenol- and oil-polluted area. FEMS Microbiol Ecol 51:363–373

Horel A, Schiewer S (2009) Investigation of the physical and chemical parameters affecting biodegradation of diesel and synthetic diesel fuel contaminating Alaskan soils. Cold Reg Sci Technol 58:113–119

Hosokawa R, Nagai M, Morikawa M, Okuyama H (2009) Autochthonous bioaugmentation and its possible application to oil spills. World J Microbiol Biotechnol 25:1519–1528

Huling SG, Bledsoe BE, White MV (1990) Enhanced bioremediation utilizing hydrogen peroxide as a supplemental source of oxygen: a laboratory and field study. US EPA RS Kerr Environ Res Lab, Ada, OK, Rep EPA/600/2-90/006. http://nepis.epa.gov/Exe/ZyPURL.cgi?Dockey=20007M9S.txt

Jacques RJS, Okeke BC, Bento FM, Teixeira AS, Peralba MCR, Camargo FAO (2008) Microbial consortium bioaugmentation of a polycyclic aromatic hydrocarbons contaminated soil. Bioresour Technol 99:2637–2643

Khan FI, Husain T, Hejazi R (2004) An overview and analysis of site remediation technologies. J Environ Manage 71:95–122

Lai CC, Huang YC, Wei YH, Chang JS (2009) Biosurfactant-enhanced removal of total petroleum hydrocarbons from contaminated soil. J Hazard Mater 167:609–614

Leahy JG, Colwell RR (1990) Microbial degradation of hydrocarbons in the environment. Microbiol Rev 54:305–315

Lee M, Kim MK, Kwon M, Park BD, Kim MH, Goodfellow M, Lee S (2005) Effect of the synthesized mycolic acid on the biodegradation of diesel oil by *Gordonia nitida* strain LE31. J Biosci Bioeng 100:429–436

Liang Y, Zhang X, Dai D, Li G (2009) Porous biocarrier-enhanced biodegradation of crude oil contaminated soil. Int Biodeterior Biodegradation 63:80–87

Lima D, Viana P, Andre S, Chelinho S, Costa C, Ribeiro R, Sousa JP, Fialho AM, Viegas CA (2009) Evaluating a bioremediation tool for atrazine contaminated soils in open soil microcosms: the effectiveness of bioaugmentation and biostimulation approaches. Chemosphere 74:187–192

Liu B, Banks M, Schwab P (2001) Effects of soil water content on biodegradation of phenanthrene in a mixture of organic contaminants. Soil Sediment Contam 10:633–658

Malina G (1999) The bioventing of unsaturated zone contaminated with oil compounds. Monograph 66, Wydawnictwo Politechniki Częstochowskiej, Częstochowa

Malina G (2007) Risk reduction of soil and groundwater at contaminated areas. Monograph 132, Wydawnictwo Politechniki Częstochowskiej, Częstochowa

Malina G, Zawierucha I (2007) Potential of bioaugmentation and biostimulation for enhancing intrinsic biodegradation in oil hydrocarbon-contaminated soil. Biorem J 11:141–147

Mancera-Lopez ME, Esparza-Garcia F, Chavez-Gomez B, Rodriguez-Vazquez R, Saucedo-Castaneda G, Barrera-Cortes J (2008) Bioremediation of an aged hydrocarbon-contaminated soil by a combined system of biostimulation-bioaugmentation with filamentous fungi. Int Biodeterior Biodegradation 61:151–160

Mariano AP, Kataoka AP, Angelis D, Bonotto DM (2007) Laboratory study on the bioremediation of diesel oil contaminated soil from a petrol station. Braz J Microbiol 38:346–353

Massa V, Infantino A, Radice F, Orlandi V, Tavecchio F, Giudiuci R, Conti F, Urbini G, Guardo D, Barbieri P (2009) Efficiency of natural and engineered bacterial strains in the degradation of 4-chlorobenzoic acid in soil slurry. Int Biodeterior Biodegradation 63:112–125

Menendez-Vega D, Gallego JLR, Pelaez AI, de Cordoba GF, Moreno J, Munoz D, Sanchez J (2007) Engineered in situ bioremediation of soil and groundwater polluted with weathered hydrocarbons. Eur J Soil Biol 43:310–321

Mishra S, Sarma PM, Lal B (2004) Crude oil degradation efficiency of a recombinant *Acinetobacter baumannii* strain and its survival in crude oil-contaminated soil microcosm. FEMS Microbiol Lett 235:323–331

Mohan SV, Kisa T, Ohkuma T, Kanaly RA, Shimizu Y (2006) Bioremediation technologies for treatment of PAH-contaminated soil and strategies to enhance process efficiency. Rev Environ Sci Biotechnol 5:347–374

Mrozik A, Piotrowska-Seget Z (2009) Bioaugmentation as a strategy for cleaning up of soils contaminated with aromatic compounds. Microbiol Res. doi:10.1016/j.micres.2009.08.001 Microbiol Res 165:363–375

Mulligan CN (2005) Environmental applications for biosurfactants. Environ Pollut 133:183–198

Nievas ML, Commendatore MG, Esteves JL, Bucala V (2008) Biodegradation pattern of hydrocarbons from a fuel oil-type complex residue by an emulsifier-producing microbial consortium. J Hazard Mater 154:96–104

Nikolopoulou M, Pasadakis N, Kalogerakis N (2007) Enhanced bioremediation of crude oil utilizing lipophilic fertilizers. Desalination 211:286–295

Odokuma LO, Dickson AA (2003) Bioremediation of a crude oil polluted tropical rain forest soil. Global J Environ Sci 2:29–40

Oh YS, Sim DS, Kim SJ (2003) Effectiveness of bioremediation on oil-contaminated sand in intertidal zone. J Microbiol Biotechnol 13:437–443

Olexsey RA, Parker RA (2006) Current and future in situ treatment techniques for the remediation of hazardous substances in soil, sediments, and groundwater. In: Twardowska I, Allen HE, Häggblom MM, Stefaniak S (eds) Viable methods of soil and water pollution: monitoring, protection and remediation. Earth and Environmental Sciences, 69, pp 211–219

Perfumo A, Banat IM, Marchant R, Vezzulli L (2007) Thermally enhanced approaches for bioremediation of hydrocarbon-contaminated soils. Chemosphere 66:179–184

Rahman KSM, Rahman TJ, Kourkoutas Y, Petsas I, Marchant R, Banat IM (2003) Enhanced bioremediation of *n*-alkane in petroleum sludge using bacterial consortium amended with rhamnolipid and micronutrients. Bioresour Technol 90:159–168

Ramirez ME, Zapien B, Zegarra HG, Rojas NG, Fernandez LC (2009) Assessment of hydrocarbon biodegradability in clayed and weathered polluted soils. Int Biodeterior Biodegradation 63:347–353

Riser-Roberts E (1998) Remediation of petroleum contaminated soil: biological, physical, and chemical processes. CRC, LLC Boca Raton, Florida

Rodrigues JLM, Kachel A, Aiello MR, Quensen JF, Maltseva OV, Tsoi TV, Tiedje JM (2006) Degradation of Aroclor 1242 dechlorination products in sediments by *Burkholderia xenovorans* LB400(ohb) and *Rhodococcus* sp. strain RHA1(fcb). Appl Environ Microbiol 72:2476–2482

Rosenberg E, Ron EZ (1996) Bioremediation of petroleum contamination. In: Crawford RL, Crawford DL (eds) Bioremediation: principles and applications. Cambridge University Press, Cambridge, pp 100–124

Rous JD, Sabatini DA, Suflita JM, Harwell JH (1994) Influence of surfactants on microbial degradation of organic compounds. Crit Rev Environ Sci Tech 24:325–370

Ruberto L, Vazquez SC, Cormack WPM (2003) Effectiveness of the natural bacterial flora, biostimulation and bioaugmentation on the bioremediation of a hydrocarbon contaminated Antarctic soil. Int Biodeterior Biodegradation 52:115–125

Sarkar D, Ferguson M, Datta R, Birnbaum S (2005) Bioremediation of petroleum hydrocarbons in contaminated soils: comparison of biosolids addition, carbon supplementation and monitored natural attenuation. Environ Pollut 136:187–195

Sayler GS, Ripp S (2000) Field application of genetically engineered microorganisms for bioremediation processes. Curr Opin Biotechnol 11:286–289

Scherr K, Aichberger H, Braun R, Loibner AP (2007) Influence of soil fractions on microbial degradation behaviour of mineral hydrocarbons. Eur J Soil Biol 43:341–350

Sturman PJ, Stewart PS, Cunningham AB, Bouwer EJ, Wolfram JH (1995) Engineering scale-up of in situ bioremediation processes: a review. J Contam Hydrol 19:171–203

Tongarun R, Luepromchai E, Vangnai AS (2008) Natural attenuation, biostimulation and bioaugmentation in 4-chloroaniline-contaminated soil. Curr Microbiol 56:182–188

Tsai TT, Kao CM (2009) Treatment of petroleum-hydrocarbon contaminated soils using hydrogen peroxide oxidation catalyzed by waste basic oxygen furnace slag. J Hazard Mater 170:466–472

Ubochi KC, Ibekwe VI, Ezeji EU (2006) Effect of inorganic fertilizer on microbial utilization of hydrocarbons on oil contaminated soil. Afr J Biotechnol 5:1584–1587

Ueno A, Ito Y, Yumoto I, Okuyama H (2007) Isolation and characterization of bacteria from soil contaminated with diesel oil and the possible use of these in autochthonous bioaugmentation. World J Microbiol Biotechnol 23:1739–1745

Urum K, Grigson S, Pekdemir T, McMenamy S (2006) A comparison of the efficiency of different surfactants for removal of crude oil from contaminated soils. Chemosphere 62:1403–1410

Vidali M (2001) Bioremediation: an overview. Pure Appl Chem 73:1163–1172

Vogel TM (1996) Bioaugmentation as a soil bioremediation approach. Curr Opin Biotechnol 7:311–316

Vogel TM, Walter MV (2001) Bioaugmentation. In: Hurst CJ, Crawford RL, Garland JL, Lipson DA, Mills AL (eds) Manual of environmental microbiology. American Society for Microbiology, Washington, DC, pp 952–959

Walton J, Labine P, Reidies A (1992) The chemistry of permanganate in degradative oxidations. In: Eckenfelder WW, Bowers AR, Roth JA (eds) Chemical oxidation technologies for the nineties. Technomic Publishing Company, New York, pp 205–221

Walworth J, Pond A, Snape I, Rayner J, Ferguson S, Harvey P (2007) Nitrogen requirements for maximizing petroleum bioremediation in a sub-Antarctic soil. Cold Reg Sci Technol 48:84–91

Whang LM, Liu PWG, Ma CC, Cheng SS (2009) Application of rhamnolipid and surfactin for enhanced diesel biodegradation: effects of pH and ammonium addition. J Hazard Mater 164:1045–1050

Wolfe S, Ingold CF (1981) Oxidation of olefins by potassium permanganate, mechanism of α-ketol formation. J Am Chem Soc 103:938–939

Wu Y, Luo Y, Zou D, Ni J, Liu W, Teng Y, Li Z (2008) Bioremediation of polycyclic aromatic hydrocarbons contaminated soil with *Monilinia* sp.: degradation and microbial community analysis. Biodegradation 19:247–257

Xie W (2003) Effect of bioventing remediation by surfactant on in-situ oil-contaminated soil. Taiwan, China (M.Sc. Dissertation. NYUST) http://ethesys.yuntech.edu.tw/ETD-db/ETD-search/view_etd?URN=etd-0823104-163027

Zawierucha I, Malina G (2006) Bioaugmentation as a method of biodegradation enhancement in oil hydrocarbons contaminated soil. Ecohydrol Hydrobiol 6:163–169

Zawierucha I, Szewczyk A, Malina G (2007) Surfactant-enhanced natural attenuation of oil hydrocarbons in soil: the effect of surfactant dose on biodegradation and dehydrogenases activity. In: Goliński P, Zabawa S (eds) Reclamation and revitalization of demoted areas. PZITS, Poznań, pp 237–246

Zawierucha I, Szewczyk A, Malina G (2008) Biostimulation with nutrients as a method of biodegradation enhancement in oil hydrocarbons contaminated soil. In: Malina G (ed) Reclamation and revitalization of demoted areas. PZITS, Poznań, pp 99–110

Chapter 9
Biosurfactants for Soil Biology

Poonam Mudgil

9.1 Introduction

Biosurfactants are amphiphilic molecules that are produced by microorganisms (bio=biological origin) and are surface active agents (surfactant). They reduce the surface and interfacial tensions and act as solubility enhancer and emulsifying agents at low concentrations. Interest in biosurfactants emanates from their potential for wide applicability and advantages over chemical surfactants (Banat et al. 2000; Desai and Banat 1997; Rodrigues et al. 2006; Singh et al. 2007). Unlike chemical surfactants, biosurfactants are non-toxic, biodegradable, and biocompatible. Additionally, they have high specificity and selectivity, and can be synthesised using renewable sources. Functional groups of biosurfactants render them with additional properties such as antibacterial, antiviral, and antifungal. Rhamnolipids and surfactin are the most studied biosurfactants. They were first reported in literature in 1949 (Jarvis and Johnson 1949) and 1968 (Arima et al. 1968), respectively. Earlier interest in these molecules was due to their antibiotic and surfactant activities. Later it was found that microbial cells grown in the presence of hydrocarbon substrates produce biosurfactants suggesting their role in the treatment of oil-spillage and enhanced oil recovery. Furthermore, it has been reported that biosurfactants have applications in soil remediation and can be used for removing heavy metals from soil. They have also been shown to have applications in the fields of food, pharmaceuticals and medicine.

The molecular weights of biosurfactants range between 500 and 1,500 Da (Lang and Wagner 1987). They have a range of physicochemical properties, but essentially all of them are amphiphilic having a hydrophilic and a hydrophobic part in the molecule. Most of the biosurfactants are neutral or anionic (negatively charged), only few are cationic (positively charged); the reason might be that cationic surfactants are generally toxic in nature. The hydrophobic or non-polar part of the biosurfactant

P. Mudgil
School of Natural Sciences, University of Western Sydney, Locked Bag 1797, Penrith South DC, NSW 1797, Australia
e-mail: poonam_mudgil@yahoo.com.au

A. Singh et al. (eds.), *Bioaugmentation, Biostimulation and Biocontrol*, Soil Biology 28, DOI 10.1007/978-3-642-19769-7_9, © Springer-Verlag Berlin Heidelberg 2011

molecule is generally made up of fatty acid or hydroxyl fatty acid with a size ranging from C8 to C18. The hydrophilic or polar part can be made up of carbohydrate, peptide, cyclic peptide, carboxylic acid, alcohol or phosphate group. Biosurfactants are classified into various groups based on their chemical composition and microbial origin. The main groups of biosurfactants include glycolipids, lipopeptide, phospholipids, fatty acids, polymeric surfactants and particulate surfactants (Desai and Banat 1997).

Amphiphilic nature of the biosurfactant molecules allows them to partition at the interfaces and reduces surface (air–liquid) and interfacial (liquid–liquid) tension. Biosurfactants can reduce the surface tension of water from 72 to 30 mN/m or less (Guerra-Santos et al. 1984; Lang and Wullbrandt 1999; Robert et al. 1989). In case of liquid–liquid interfaces such as oil–water interface, the presence of a biosurfactant reduces interfacial tension and repulsive forces between the two interfaces allowing them to mix well. They have been found to reduce the interfacial tension to as low as 1 mN/m (Parra et al. 1989). The ability of a biosurfactant to reduce the surface or interfacial tension increases with its increasing concentration until it reaches a critical concentration, also known as critical micelle concentration (CMC). There is no further reduction in surface or interfacial tension above CMC. At CMC, the biosurfactant molecules self-assemble to form structures such as vesicles, bilayers or micelles. CMC formation is dependent on pH, ionic strength and temperature. Biosurfactants exhibit low CMC and the values range from 0.15 to 30 mg/L (Desai and Banat 1997). Low values of CMC for biosurfactants make them better and efficient surfactants than their synthetic counterparts.

9.2 Applications of Biosurfactants in Soil Biology

Biosurfactants have great potential in soil biology because they are biodegradable, have low toxicity, are effective in solubilising and degrading insoluble compounds in soil and can be economically produced using cheap and renewable resources. Use of biosurfactants to accelerate removal of soil contaminants has gained increasing attention recently (Christofi and Ivshina 2002; Mulligan 2005, 2009; Singh et al. 2007). Biosurfactants can be effectively used for removing hydrocarbon contaminants (Table 9.1) and heavy metals from the contaminated soils (Table 9.2).

9.3 Use of Biosurfactants in Removal of Hydrocarbons from Soil

Hydrocarbon pollutants in soil are of major concern to environment and human health. Oil spillage, discharge of oily waste and oil leakage are the main contributors to the problem. Many physical, chemical and biological methods have been proposed for treating the soil having hydrocarbon contaminants. Bioremediation, i.e., use of biological organisms or their products to remove or degrade toxic contaminants into

Table 9.1 Removal of hydrocarbons from soil using biosurfactants

Type of hydrocarbon	Type of biosurfactant	Type of soil	Removal efficiency	Reference
Phenanthrene	Rhamnolipid	Sand, artificially contaminated	85%	Gu and Chang (2001)
		Unsaturated soil (sand 14%, silt 38%, clay 48%) spiked with phenanthrene	20–30%	Chang et al. (2009)
		Sandy loam soil (sand 55%, clay 15%, silt 30%) spiked with phenanthrene	17%	Shin et al. (2006)
PAHs (phenanthrene, anthracene, pyrene)	Rhamnolipid	Artificially contaminated soil	74% for phenanthrene, 45% for anthracene, 69% for pyrene	Franzetti et al. (2009)
	Bioemulsan	Artificially contaminated soil	32% for phenanthrene, 19% for anthracene, 26% for pyrene	Franzetti et al. (2009)
Hexadecane	Monorhamnolipid	Sand, artificially contaminated	80%	Bai et al. (1997)
		Sand, artificially contaminated	58%	Bai et al. (1998)
Total petroleum hydrocarbons	Rhamnolipid	Sandy loam soil, heavily contaminated	63%	Lai et al. (2009)
	Surfactin	Sandy loam soil, heavily contaminated	62%	Lai et al. (2009)
Crude oil	Rhamnolipid	Crude oil contaminated soil	80%	Urum et al. (2003); Urum and Pekdemir (2004)
		Artificially contaminated soil	51%	Franzetti et al. (2009)
	Glycolipid biosurfactant	Model soil (50% sand, 30% clay, 20% peat) contaminated with crude oil	65–82%	Kuyukina et al. (2005)
	Bioemulsan	Artificially contaminated soil	33%	Franzetti et al. (2009)

a less toxic form, is one of the eco-friendly solutions. Because of their highly hydrophobic nature, hydrocarbon contaminants in soil have very low water solubility. They are mostly adsorbed on soil particles and therefore have poor bioavailability for bioremediation. Use of surfactants can increase their bioavailability and transfer them from soil particles to the aqueous phase of soil where they can be biodegraded

Table 9.2 Removal of heavy metals from soil using biosurfactants

Type of heavy metal	Type of biosurfactant	Type of soil	Removal efficiency	Reference
Cadmium	Rhamnolipid	Feldspar	96%	Asci et al. (2008a)
		Artificially contaminated soil	92%	Juwarkar et al. (2007)
		Kaolin	71.9%	Asci et al. (2007)
		Artificially contaminated soil	61.7%	Mulligan and Wang (2006)
		Artificially contaminated soil	35%	Franzetti et al. (2009)
	Bioemulsan	Artificially contaminated soil	19%	Franzetti et al. (2009)
Lead	Rhamnolipid	Artificially contaminated soil	88%	Juwarkar et al. (2007)
		Artificially contaminated soil	52%	Franzetti et al. (2009)
		Sandy loam	43%	Herman et al. (1995)
	Bioemulsan	Artificially contaminated soil	47%	Franzetti et al. (2009)
Arsenic	Rhamnolipid	Mine tailings (sandy soil)	7%	Wang and Mulligan (2009a)
Zinc	Rhamnolipid	Na-feldspar	98.9	Asci et al. (2008b)
		Artificially contaminated soil	87%	Franzetti et al. (2009)
		10% sand, 70% silt, 20% clay	18%	Mulligan et al. (2001)
		Soil sediment	13%	Dahrazma and Mulligan (2007)
	Sophorolipid	Sandy soil	100%	Mulligan et al. (1999a)
		10% sand, 70% silt, 20% clay	60%	Mulligan et al. (2001)
	Surfactin	10% sand, 70% silt, 20% clay	6%	Mulligan et al. (2001)
		10% silt, 90% sand	6%	Mulligan et al. (1999b)
	Bioemulsan	Artificially contaminated soil	31%	Franzetti et al. (2009)
Copper	Rhamnolipid	10% sand, 70% silt, 20% clay	65%	Mulligan et al. (2001)
		Artificially contaminated soil	48%	Franzetti et al. (2009)
		Construction site soil	46%	Mulligan et al. (2007)
		Soil sediment	37%	Dahrazma and Mulligan (2007)
	Sophorolipid	10% sand, 70% silt, 20% clay	25%	Mulligan et al. (2001)
	Surfactin	Sandy soil	70%	Mulligan et al. (1999a)
		10% silt, 90% sand	25%	Mulligan et al. (1999b)
		10% sand, 70% silt, 20% clay	15%	Mulligan et al. (2001)
	Bioemulsan	Artificially contaminated soil	17%	Franzetti et al. (2009)
Nickel	Rhamnolipid	Artificially contaminated soil	64%	Franzetti et al. (2009)
		Artificially contaminated soil	51%	Mulligan and Wang (2006)
		Soil sediment	27%	Dahrazma and Mulligan (2007)
	Bioemulsan	Artificially contaminated soil	25%	Franzetti et al. (2009)

by microorganisms or removed by soil washing. Synthetic surfactants used for this purpose are toxic and resist biodegradation while biosurfactants have clear advantage over them due to their low toxicity, higher surface activity, biocompatibility and biodegradability. Soil-washing method to remove hydrocarbon contaminants without damaging soil structure is a fast and effective method (Table 9.1). Mobilisation and solubilisation are two mechanisms by which hydrocarbons are removed by biosurfactants in soil washing. Mobilisation occurs below the CMC where lowering of interfacial tension causes displacement and dispersion of hydrocarbons. Solubilisation occurs above the CMC where hydrocarbon gets associated with micelles of the surfactants and gets removed in the washing step.

9.3.1 Removal of Polycyclic Aromatic Hydrocarbons

Polycyclic aromatic hydrocarbons (PAHs) are considered as priority pollutants due to their high toxicity and carcinogenic properties. They get into the soil from disposal of coal, petroleum and chemical wastes. Biodegradation and removal are important processes in bioremediation of PAHs, but sorption of PAH to soil inhibits its mobilisation and limits its bioremediation. Biosurfactants have been used effectively to enhance mobilisation, desorption and biodegradation of PAHs. Biosurfactants facilitate desorption of PAHs by lowering surface tension and solubilising them in aqueous phase of soil where they can be biodegraded or removed. Rhamnolipids are efficient in removing PAHs from contaminated soils (Table 9.1). In soil-washing experiments, Franzetti et al. (2009) showed that rhamnolipids removed 74% of phenanthrene, 45% of anthracene and 69% of pyrene from an artificially contaminated soil. Another type of biosurfactants, BS29 bioemulsans, produced by *Gordonia* sp. was also effective as soil-washing agents removing 32% of phenanthrene, 19% of anthracene and 26% of pyrene (Franzetti et al. 2009).

Phenanthrene is quite widespread in the environment and used as a model to study remediation of PAHs. Many biosurfactants are effective in enhancing solubility, bioavailability and biodegradation of phenanthrene. The biosurfactant, rhamnolipid, enhances phenanthrene solubility and removal efficiency from unsaturated soil. It may also promote microbial growth in the soil-water system that helps in biodegradation of phenanthrene (Chang et al. 2009). The biosurfactant, alasan, produced by *Acinetobacter radioresistens* enhances the solubilisation and biodegradation of phenanthrene and other PAHs (Barkay et al. 1999). The biosurfactants, sophorolipids, produced by *Candida bombicola* augment bioavailability of phenanthrene enhancing its degradation by biodegrading bacteria in a sandy silt soil contaminated with phenanthrene (Schippers et al. 2000).

Rhamnolipids have been extensively used for studying the effect of biosurfactants on the removal of phenanthrene from the contaminated soils (Table 9.1). Using phenanthrene spiked sand, Gu and Chang (2001) found that rhamnolipid increased the rate of mass transfer of phenanthrene from sorbed soil to the aqueous phase of the contaminated soil. They showed that use of biosurfactant coupled with a bioluminescent bacterium could be developed as a biosensor for detecting toxicity of phenanthrene in contaminated soils. The effect of pH on solubilisation of phenanthrene by rhamnolipid was studied by Shin et al. (2006) in a sandy loam soil spiked with phenanthrene. In soil-packed columns, the percentage removal of phenanthrene was 17.3 and 9.5% at pH 5 and 7, respectively. The observed highest solubility at pH 5 suggested that adjusting pH could enhance the solubility of phenanthrene. Pei et al. (2009) showed that rhamnolipid inhibited sorption of phenanthrene to soil. They suggested that the biosurfactant used for bioremediation should be added frequently in soil because it can itself get sorbed and biodegraded in soil reducing the overall efficiency of bioremediation. Ochoa-Loza et al. (2007) suggested that soil type governs the sorption of rhamnolipids determining its amount in aqueous phase available for bioremediation. Soils with low aluminosilicates and iron oxides exhibit low sorption of rhamnolipid while soils with high iron content may not be suitable for bioremediation by rhamnolipids. This information is helpful in predicting the feasibility of use of biosurfactant as remediation option for a particular soil type.

9.3.2 Removal of Aliphatic Hydrocarbons

There are many reports indicating the use of biosurfactants to facilitate and enhance bioremediation of soil contaminated with various aliphatic hydrocarbons such as tetradecane, pentadecane, hexadecane, octadecane and pristane. Oberbremer et al. (1990) showed that sophorolipids doubled the rate of hydrocarbon degradation in a model system containing 10% soil and 1.35% hydrocarbon mixture of tetradecane, pentadecane, hexadecane, pristane and an aromatic hydrocarbon, naphthalene. In the presence of the biosurfactants such as trehalose lipids, sophorolipids, cellubiose lipids and rhamnose, 93–99% of the hydrocarbon mixture was degraded within 71–79 h, whereas in their absence 81% was degraded in 114 h. Jain et al. (1992) showed that biosurfactants produced by *Pseudomonas aeruginosa* UG2 enhanced the biodegradation of a hydrocarbon mixture in a silt loam soil. The biosurfactants, having a CMC value of 18–19 mg/L, at the concentration of 100 mg/kg soil, significantly enhanced the degradation of tetradecane, hexadecane and pristane at 20°C over a 2-month incubation period. Pristane is a highly branched recalcitrant molecule in soil and in bioremediation experiments; Franzetti et al. (2009) showed that bioemulsans, produced by *Gordonia* sp., were effective in increasing biodegradation and decreasing the residual concentration of pristane in an artificially contaminated soil. Bai et al. (1997) used rhamnolipids to remove up to 80% of residual hydrocarbon (hexadecane) from sand columns (Table 9.1). They found that mobilisation was the primary mechanism of removal while solubilisation was insignificant. In further studies (Bai et al. 1998) they reported that the presence of cations such as Na^+ and Mg^{2+} improved the solubilisation of hexadecane. Increasing concentration of the cations and lowering pH were associated with a reduction in the interfacial tension that helped in solubilisation. Perfumo et al. (2007) reported that the removal of hexadecane with the help of biosurfactants was enhanced by increasing the temperature to 60°C as compared with the room temperature of 18°C. They suggested that thermally enhanced bioremediation can be used to complement the biosolubilisation with biosurfactants. Zhang and Miller (1992) showed the use of rhamnolipid to enhance the dispersion and biodegradation of octadecane. Rhamnolipid (300 mg/L) increased the mineralisation of octadecane by 20% but the dispersion of octadecane was affected by pH. This might be due to the influence of pH on the structure of rhamnolipids, which changes from lamellar sheet to vesicles and micelles due to changes in pH (Champion et al. 1995; Ishigami et al. 1987). When the pH increases from 5.5 to 8.0, there is repulsion between the negatively charged head groups. This makes a bigger head diameter changing the structure from lamellar to vesicles and to micelles.

Some biosurfactants have been shown to remove both aliphatic as well as aromatic hydrocarbons from the contaminated soil. For example, rhamnolipid from *P. aeruginosa* UG2 removed both aliphatic and aromatic hydrocarbons effectively from a sandy loam soil, which was artificially contaminated with these hydrocarbons (Scheibenbogen et al. 1994). The extent of removal was dependent on the type of hydrocarbon removed and concentration of surfactant used.

9.3.3 Removal of Petrochemical Mixtures

Most of the studies on the effect of biosurfactants on the removal of petroleum hydrocarbons have been carried out using pure compounds. The information on the effect of biosurfactants on removal and biodegradation of complex petrochemical mixtures is rather limited (Mulligan 2005).There are few reports on the use of biosurfactants in treating soil contaminated with petrochemical mixtures such as crude oil, gasoline and diesel fuel. Biosurfactants help by either removing these from soil or solubilising these prior to biodegradation. Application of different biosurfactants to remove oil from a sandy loam soil collected from an oil-contaminated site in an oil-refinery plant was studied by Lai et al. (2009). Using a screening protocol to assess the efficiency of different surfactants to remove total petroleum hydrocarbons from the contaminated soils, they showed that biosurfactants had better removing efficiency than synthetic surfactants. At a concentration of 0.2%, rhamnolipids and surfactin showed removal efficiencies of 23 and 14% for slightly contaminated soil, and 63 and 62% for highly contaminated soils, respectively. On the other hand, synthetic surfactants, Tween 80 and TitonX-100, showed only 6 and 4% of removal efficiencies for slightly contaminated soil, and 40 and 35% for highly contaminated soils, respectively. Removal efficiency increased with increasing concentration of biosurfactants but was independent of time of contact.

9.3.3.1 Crude Oil

Crude oil, which is a complex mixture of many aliphatic and aromatic hydrocarbons, can be removed from contaminated soils by soil washing with biosurfactants (Table 9.1). Using laboratory-contaminated soil, Urum et al. (2003) suggested that temperature and concentration of biosurfactant are the most important parameters governing removal of crude oil from contaminated soils. They showed almost 80% of removal of oil from crude oil contaminated soil using rhamnolipids. In further studies, they evaluated the ability of different biosurfactant solutions (aescin, lecithin, rhamnolipid, saponin and tannin) to remove crude oil from contaminated soil making measurements of surface tension, foaming and emulsification ability, sorption to soil and solubilisation (Urum and Pekdemir 2004). Removal of oil by biosurfactants was due to mobilisation caused by the reduction in surface tension, while solubilisation and emulsification had negligible effects. In general, the biosurfactants with low CMC and high soil sorption value had stronger ability to remove oil. The non-ionic biosurfactants produced by *Rhodococcus ruber* has been shown to remove 65–82% of crude oil by mobilisation in soil washing from model soil-packed columns heavily contaminated with crude oil (Kuyukina et al. 2005). The biosurfactants were found to be more effective than the chemical surfactant, Tween 60. The biosurfactants, BS29 bioemulsans, produced by *Gordonia* sp. has been shown to remove 33% of crude oil from an artificially contaminated soil in soil-washing experiments, although rhamnolipids were found to be more effective in its removal efficiency (51%) (Franzetti et al. 2009).

Biodegradation of crude oil products in soil is often limited by their low water solubility. Exogenous addition of biosurfactants enhances the bioavailability of hydrocarbons and facilitates biodegradation by indigenous microbial population. Abalos et al. (2004) showed that rhamnolipids from *P. aeruginosa* AT 10 enhanced the bioavailability and biodegradation of crude oil by a microbial consortium. Addition of rhamnolipid enhanced the biodegradation of total petroleum hydrocarbon from 32 to 61%. Biodegradation of isoprenoids from the aliphatic fraction, and alkylated PAHs from the aromatic fraction, was particularly enhanced due to the presence of biosurfactants. Cubitto et al. (2004) showed that surfactin from *Bacillus subtilis* O9 stimulated the growth of microbial population which biodegraded crude oil mixture in soil. The presence of surfactin accelerated the biodegradation of aliphatic hydrocarbons, but the degradation of aromatic hydrocarbons was not stimulated.

9.3.3.2 Gasoline

Small but continuous leakage of oil from gasoline stations has potential to contaminate soil and groundwater; therefore, effective strategies need to be developed to treat gasoline-contaminated soils. Falatko and Novak (1992) showed that biosurfactants increase the solubility of gasoline hydrocarbons. Using gasoline in a sand-filled column, they monitored the effect of biosurfactants on the solubility and biodegradation of the selected gasoline compounds (toluene, xylene, 1,2,4-trimethyl benzene and naphthalene) and found that an increase in solubility was greatest for the least soluble compound and least for the most soluble compound. Rahman et al. (2002) treated gasoline-spilled soil with a mixture of bacterial consortium, organic nutrients amendments and biosurfactants. Nutrients helped in the growth of microorganisms which biodegraded hydrocarbons, and the presence of biosurfactants solubilised hydrocarbons prior to microbial degradation.

9.3.3.3 Oil Sludge

Biodegradation of oil sludge by microbial consortium is enhanced by biosurfactants and nutrient supplementation. Rahman et al. (2003) showed that treatment of an oil sludge contaminated soil with a bacterial consortium supplemented with rhamnolipid and a nutrient solution containing nitrogen, phosphorus and potassium resulted in about 70–100% biodegradation of various hydrocarbons. Rhamnolipid enhanced the bioavailability of hydrocarbons to the microbial consortium leading to enhanced biodegradation. Similarly, Cameotra and Singh (2008) showed that a microbial consortium made up of two isolates of *P. aeruginosa* and one of *Rhodococcus erythropolis* could remove more than 98% of hydrocarbon from the soil contaminated with oily sludge when the consortium was supplemented with a nutrient mixture and a crude biosurfactant preparation (rhamnolipids) from one of the members of the consortium.

9.3.3.4 Diesel

Application of biosurfactants to enhance biodegradation of diesel in contaminated soils has been studied using rhamnolipid and surfactin (Whang et al. 2008). Both biosurfactants reduced surface tension and increased solubility of diesel. In a diesel-contaminated sandy loam soil, addition of rhamnolipid and surfactin resulted in total petroleum hydrocarbon–diesel biodegradation efficiency of 97 and 76%, respectively. The biosurfactants also stimulated growth of indigenous microorganisms which enhanced bioremediation of diesel-contaminated soil.

9.3.4 Biodegrading Bacteria as Biosurfactant Producers

Biosurfactants are generally added exogenously to enhance the bioremediation of hydrocarbon-polluted soils by indigenous microbes (Abalos et al. 2004; Cubitto et al. 2004). In some cases, the biodegrading bacteria themselves may be producing biosurfactants. This offers the advantage of continuous supply of biosurfactants making the process more economical. Das and Mukherjee (2007) showed that *P. aeruginosa* and *B. subtilis* enhanced the solubility and biodegradation of petroleum hydrocarbons in the soil contaminated with crude petroleum oil hydrocarbons. These bacteria produced biosurfactants which solubilised the hydrophobic oil hydrocarbons prior to biodegradation. A *Pseudomonas* sp. which is capable of degrading naphthalene has also been shown to produce a biosurfactant that enhances the solubility of naphthalene by more than 30 times than its aqueous solubility (Vipulanandan and Ren 2000). Another bacterium, *Pseudomonas fluorescens*, utilises various petroleum hydrocarbons including both aliphatics and aromatics by its ability to produce biosurfactants (Barathi and Vasudevan 2001). These substrates have low water solubility and may not be available for biodegradation but emulsification by biosurfactants help in their utilisation by the bacteria. Similarly, Lu et al. (2006) reported that *Pseudomonas* sp., *Flavobacterium* sp. and *Rhodococcus* sp. isolated from the contaminated soil near a gas station produced biosurfactants and were also capable of degrading gasoline and diesel oil. They also found that the consortium of biosurfactant-producing bacteria was more effective than individual isolates and could be used for the remediation of soils contaminated by gas station leakage.

9.3.5 *Importance of the Structure of Biosurfactants*

Structure of a biosurfactant may act as a determinant factor for the bioremediation of hydrocarbons in soil. The lactonic form of sophorolipids has been found to inhibit hexadecane biodegradation while the acidic form stimulates it (Ito and Inoue 1982; Ito et al. 1980). The methyl ester form of rhamnolipids has been found to be more effective bio-degrader of hexadecane and octadecane than the acid form (Zhang and Miller 1995). Structural differences may impart different physicochemical properties to biosurfactants that may guide the efficiency of biodegradation and bioremediation. The methyl ester form of rhamnolipid lowers interfacial tension to <0.1 dyne/cm while the acid form lowers it only up to 5 dyne/cm (Zhang and Miller 1995). The carboxyl group of the acid form of rhamnolipid confers it a negative charge which interacts better with water than with alkanes, thus it is less effective in reducing the interfacial tension. On the other hand, the methyl ester form of rhamnolipid lowers the interfacial tension more and acts as a better dispersant of alkanes. However, because of its low water solubility, the methyl ester form of rhamnolipids may not be suitable for environmental applications. Therefore, it was proposed that a mixture of methyl ester and acid forms of rhamnolipids in 1:1 ratio would be more effective for alkane biodegradation (Zhang and Miller 1995). Another study on the biodegradation of phenanthrene using monorhamnolipid and dirhamnolipid produced by *Pseudomonas* sp. also emphasised on the importance of structure of the biosurfactants in determining the solubility and bioavailability of hydrocarbons for biodegradation (Zhang et al. 1997). Although the stimulation of biodegradation was similar for both types of rhamnolipids, monorhamnolipid was found to be more effective for the solubilisation of phenanthrene while the bioavailability of phenanthrene was more within the micelles of dirhamnolipid.

9.4 Use of Biosurfactants in Removal of Heavy Metals from Soil

The most hazardous heavy metals in the EPA list of priority pollutants include cadmium, copper, lead, mercury, nickel and zinc. These metals, unlike hazardous organic contaminants, cannot be degraded or detoxified. They not only impact microbial flora in soil but also contaminate groundwater resulting in potential toxicological impact on human health with mutagenic and carcinogenic effects. Due to their anionic nature biosurfactants can be used for bioremediation of heavy metals from soil (Table 9.2). The mechanism of removal essentially involves the formation of ionic bonds between anionic biosurfactants and cationic toxic metals. If these bonds are stronger than that of metal with soil, the biosurfactant–metal complex can be easily flushed out by pumping water through soil to remove metal contaminants. Being smaller in size, biosurfactants are more effective than microbial whole cells or exopolymers used for bioremediation (Herman et al. 1995; Tan et al. 1994). Molecular weight of biosurfactants is generally <2,000 while that of exopolymers is about 10^6. The vesicles and micelles of the most commonly used biosurfactant, rhamnolipids, are <50 nm and <5 nm in diameter, respectively (Champion et al. 1995). The size of these aggregates is dependent on pH but in both acidic and basic conditions it is small enough for easy flow and not to get filtered by soil pores of size ~200 nm (Dahrazma et al. 2008). Small-sized metal–biosurfactant complex is easily removed during soil washing, whereas the bigger cells or exopolymers complexing with metals get filtered by small pores of soil. The efficiency of removal of metal contaminants from soil is dependent on various factors such as type of biosurfactant, type of soil, pH of soil, particle size and type of metal contaminant.

9.4.1 Removal of Cadmium

Cadmium is an acutely toxic heavy metal. Cadmium along with lead and mercury are the three most hazardous heavy metals for the environment and human beings. Cadmium is used in electroplating, paint pigments, and nickel–cadmium batteries. It gets accumulated in soil due to fertilisers, industrial waste disposal and sewage disposal. Biosurfactants, particularly rhamnolipids, have been shown to remove cadmium from soil (Herman et al. 1995; Tan et al. 1994; Torrens et al. 1998) (Table 9.2). Asci et al. (2008a) demonstrated that rhamnolipid from *Pseudomonas* removed 96% of cadmium from the soil components, K-feldspar and sepiolite. They suggested that their technique could be modified to remove cadmium from waste waters. In another study, they reported that rhamnolipids at pH 6.8 removed 71.9% of cadmium from another soil component, kaolin (Asci et al. 2007).

Removal of cadmium from soil is also important for effective biodegradation of organic contaminants in soil because the presence of heavy metals such as cadmium often hampers the biodegradation of organic material (Said and Lewis 1991). Sandrin et al. (2000) showed that metal-complexation by rhamnolipid reduced the cadmium toxicity and allowed enhanced biodegradation of an organic contaminant of soil, naphthalene. They demonstrated that rhamnolipid not only complexed with cadmium but it also induced the release of lipopolysaccharide that altered the surface of the cell. They proposed that metal complexation together with cell surface alteration reduced the uptake and toxicity of cadmium to biodegrading bacteria, resulting in enhanced bioremediation.

9.4.2 Removal of Lead

Lead contamination of soil is an important environmental problem. It is used in smelters, solder, paint and batteries. It gets into soil from fossil fuel combustion, landfills and battery disposal. Lead can be effectively removed from soil using biosurfactants (Herman et al. 1995; Juwarkar et al. 2007) (Table 9.2). Juwarkar et al. (2007) showed that rhamnolipid can be used to remove lead along with cadmium from soil. In their experiments, rhamnolipids removed 88% of lead and 92% of cadmium from soil within 36 h. Low concentrations of biosurfactant (0.1%) used for this purpose had no toxic effect on the microbial population of soil suggesting that biosurfactants can be used effectively for heavy metal bioremediation without destroying the soil structure. Kim and Vipulanandan (2006) showed removal of lead from water and contaminated soil (kaolinite clay) using a biosurfactant isolated from *Flavobacterium*. Their FTIR study showed that carboxyl group of the biosurfactant was effective in lead removal.

9.4.3 Removal of Arsenic

Arsenic is a toxic heavy metal which gets added to soil because of various human activities such as mining, fossil fuel combustion, pesticides and industrial waste (Wang and Mulligan 2006). Wang and Mulligan (2009a, b) explored the use of rhamnolipids in bioremediation of arsenic from mine tailings (Table 9.2). They found that rhamnolipid could be used to enhance mobilisation of arsenic in alkaline conditions. Rhamnolipid mobilised not only arsenic but also other heavy metals such as copper, lead and zinc. Arsenic mobilisation was found to be positively correlated with the mobilisation of iron and other heavy metals suggesting that mobilisation of co-existing metals may enhance arsenic mobilisation in the presence of rhamnolipids by incorporating it into aqueous organic complexes or micelles through metal-bridging mechanisms. Biosurfactant foam technology can also be used for bioremediation of arsenic contaminated soil (Wang and Mulligan 2004a, b). Nutrients or microbial cells can be delivered to the subsurface increasing availability of iron and arsenic to microorganisms.

9.4.4 Removal of Zinc and Copper

Zinc and copper are not as toxic heavy metals as cadmium or lead, but high levels of these metals may get accumulated in soil causing toxicity. Zinc enters soil from galvanizing plant effluent, burning of coal and waste and municipal waste. Increased levels of copper in soil occur due to fertilisers, pesticides, agricultural and municipal waste. Rhamnolipid has been found to remove zinc from soil at near neutral pH (Table 9.2). Asci et al. (2008b) showed removal of about 98.9% of zinc from a soil component, Na-feldspar, at pH 6.8 using rhamnolipids. They suggested that it was due to formation of small vesicles and micelles at pH > 6.0. The low interfacial tension generally prevalent in this range helped in complexation of zinc with the biosurfactant. Rhamnolipid has also been shown to remove copper from soil (Table 9.2). Mulligan et al. (2007) showed that using 2% rhamnolipids at pH 6.5, 46% of copper could be removed from a construction site soil in Canada, and 84% of copper could be removed from lake sediment from Japan.

9.4.5 Removal of Multiple Heavy Metals Using Rhamnolipids

Rhamnolipids are the most commonly used biosurfactants for washing soils polluted with heavy metals (Table 9.2). While rhamnolipids have been shown to remove individual heavy metals from soil, there are also many reports indicating its application to remove a combination of many heavy metals together from soil. For example, Herman et al. (1995) showed that cadmium, lead and zinc were removed from a sandy loam soil using rhamnolipid. Juwarkar et al. (2007) showed that 92% of cadmium and 88% of lead were removed from an artificially contaminated soil using 0.1% rhamnolipid. Dahrazma and Mulligan (2007) showed that about 37% of copper, 13% of zinc and 27% of nickel were removed from soil sediments using 0.5% of rhamnolipid. The metal removal increased with increasing concentration of the biosurfactant, and alkaline conditions (addition of 1% NaOH) also enhanced the bioremediation process up to four times.

Even higher metal removal efficiencies can be achieved if foam technology is applied to aqueous rhamnolipid solution. Formation of foam increases the flooding efficiency of surfactant flushing resulting in enhanced removal efficiency (Jeong et al. 2000). Mulligan and Wang (2006) found that transfer from aqueous solution to foam increased the efficiency of removal of heavy metals, although the concentration of the biosurfactant rhamnolipid was same (0.5%) in both aqueous solution and foam. They showed that the aqueous solution of rhamnolipid removed 61.7% of cadmium and 51% for nickel, while the rhamnolipid foam removed 73.2% of cadmium and 68.1% of nickel from an artificially contaminated soil.

9.4.6 Removal of Heavy Metals Using Other Biosurfactants

Although rhamnolipids are the mostly reported biosurfactants for removing heavy metals from soil, there are some reports available that show the capability of other biosurfactants for heavy metal remediation of soil (Table 9.2). Mulligan et al. (1999a) showed that surfactin and sophorolipid could be used to remove copper and zinc from hydrocarbon-contaminated sandy soil. Surfactin removed about 70% of copper and 50% of hydrocarbon, while sophorolipid removed 100% of zinc from the contaminated soil.

Mulligan et al. (1999b) further showed that 0.25% of surfactin with 1% NaOH removed 25% of copper and 6% of zinc from a soil highly contaminated with metals and hydrocarbons. Sequential extraction of soil with surfactin was able to remove 70% of copper and 22% of zinc. It was postulated that metal removal occurred by sorption of surfactin to soil interface, metal complexation, desorption of the complex through lowering surface tension and micellar complexation. Mulligan et al. (2001) also evaluated the capacity of three biosurfactants (surfactin, rhamnolipid and sophorolipid) to remove copper and zinc from soil sediments. In a single washing step, 0.5% of rhamnolipid removed 65% of copper and 18% of zinc, and 4% sophorolipid removed 25% of copper and 60% of zinc. Surfactin was found to be less effective and removed 15% of copper and 6% of zinc. They suggested that biosurfactant form complexes with metals, resulting in detachment of metal from soil in to the soil solution in which metal gets associated with micelles of biosurfactant. In sequential extraction, they found that rhamnolipid and sophorolipid removed organically bound copper while sophorolipid removed carbonate and oxide-bound zinc. Franzetti et al. (2009) evaluated capability of biosurfactants, BS29 bioemulsan produced by *Gordonia* sp., to remove heavy metals from an artificially contaminated soil. Using soil-washing experiments they showed that although rhamnolipids were more effective, bioemulsans were also able to remove 17% of copper, 19% of cadmium, 47% of lead, 31% of zinc and 25% of nickel from soil.

9.4.7 *Removal of Transition Metals*

The biosurfactant rhamnolipid has been implied to remove transition metal, chromium, from contaminated soil. Chromium is a hazardous contaminant of soil. Its hexavalent form, Cr(VI), is highly soluble and mobile, and is considered to be very toxic having carcinogenic and mutagenic effects. However, its trivalent form, Cr(III) is considered to be stable and immobile due to its low solubility. Cr(VI) is treated by reducing it to Cr(III) but re-oxidisation of Cr(III) to Cr(VI) can cause soil pollution. Massara et al. (2007) used rhamnolipids to remove chromium from chromium-contaminated kaolinite. They showed that rhamnolipids had the capability to remove 25% of Cr(III) and enhance removal of Cr(VI) from soil. Rhamnolipids reduced almost 100% of extracted Cr(VI) to Cr(III) in a period of 24 days indicating that biosurfactants can be used for removal and conversion of Cr(VI) to Cr(III).

9.5 Conclusion

Hydrocarbons and metal contaminants in soil are of great concern. A range of aliphatic and aromatic hydrocarbons, including highly toxic PAHs, find their way into soil by industrial disposal, and coal and crude oil processing. Heavy metals get accumulated into soil by burning of fossil fuel and disposal of sewage, domestic and industrial waste. These pollutants get sorbed to the soil particles which decreases their solubility and bioavailability. Soil remediation is dependent on increasing their solubility and bioavailability into the aqueous phase of soil to facilitate their removal or biodegradation. The synthetic surfactants used for enhancing solubility may be toxic or resist biodegradation. Research indicates that biosurfactants are eco-friendly alternative to the synthetic surfactants due to their low toxicity and high biodegradability. Soil bioremediation is carried out by the addition of biosurfactants, biosurfactant-producing microorganisms or nutrients which encourage the growth of biosurfactant-producing microorganisms. Many lab scale studies performed on the artificially contaminated soil add to the knowledge of capabilities of different types of biosurfactants to remove various types of contaminants. More lab and field investigations on the soil collected from the contaminated sites will be beneficial for application of biosurfactants in large-scale environmental applications.

High production cost of biosurfactants is a limiting factor for their wide applicability. There are many considerations that can help. Crude biosurfactant preparations can be used because of the lesser requirement for the purity of compounds, thereby reducing the cost of purification. Indigenous microorganisms can be stimulated to produce biosurfactants by nutrient supplementation which is also a clear advantage over synthetic surfactants. Recovery and reuse of biosurfactants in soil-washing method can also reduce the economic cost considerably. Advances in research on the cost-effective production of biosurfactants using cheap and renewable substrates will make the process economical.

References

Abalos A, Vinas M, Sabate J, Manresa MA, Solanas AM (2004) Enhanced biodegradation of Casablanca crude oil by a microbial consortium in presence of a rhamnolipid produced by *Pseudomonas aeruginosa* AT10. Biodegradation 15:249–260

Arima K, Kahinuma A, Tamura G (1968) Surfactin, a crystalline peptide lipid surfactant produced by *Bacillus subtilis*: isolation, characterization and its inhibition of fibrin clot formation. Biochem Biophys Res Commun 31:488–494

Asci Y, Nurbas M, Acikel YS (2007) Sorption of Cd(II) onto kaolin as a soil component and desorption of Cd(II) from kaolin using rhamnolipid biosurfactant. J Hazard Mat B 139:50–56

Asci Y, Nurbas M, Acikel YS (2008a) A comparative study for the sorption of Cd(II) by K-feldspar and sepiolite as soil components, and the recovery of Cd(II) using rhamnolipid biosurfactant. J Environ Manage 88:383–392

Asci Y, Nurbas M, Acikel YS (2008b) Removal of zinc ions from a soil component Na-feldspar by a rhamnolipid biosurfactant. Desalination 223:361–365

Bai G, Brusseau ML, Miller RM (1997) Biosurfactant-enhanced removal of residual hydrocarbon from soil. J Contam Hydrol 25:157–170

Bai G, Brusseau ML, Miller RM (1998) Influence of cation type, ionic strength, and pH on solubilisation and mobilization of residual hydrocarbon by a biosurfactant. J Contam Hydrol 30:265–279

Banat IM, Makkar RS, Cameotra SS (2000) Potential commercial applications of microbial surfactants. Appl Microbiol Biotechnol 53:495–508

Barathi S, Vasudevan N (2001) Utilization of petroleum hydrocarbons by *Pseudomonas fluorescens* isolated from a petroleum-contaminated soil. Environ Int 26:413–416

Barkay T, Navon-Venezia S, Ron EZ, Rosenberg E (1999) Enhancement of solubilization and biodegradation of polyaromatic hydrocarbons by the bioemulsifier alasan. Appl Environ Microbiol 65:2697–2702

Cameotra SS, Singh P (2008) Bioremediation of oil sludge using crude biosurfactants. Int Biodeterior Biodegradation 62:274–280

Champion JT, Gilkey JC, Lamparski H, Retterer J, Miller RM (1995) Electron microscopy of rhamnolipid (biosurfactant) morphology: effects of pH, cadmium, and octadecane. J Colloid Interface Sci 170:569–574

Chang J-H, Qiang Z, Huang C-P, Ellis AV (2009) Phenanthrene removal in unsaturated soils treated by electrokinetics with different surfactants–Triton X-100 and rhamnolipid. Colloids Surf A 348:157–163

Christofi N, Ivshina IB (2002) Microbial surfactants and their use in field studies of soil remediation. J Appl Microbiol 93:915–929

Cubitto MA, Moran AC, Commendatore M, Chiarello MN, Baldini MD, Sineriz F (2004) Effects of *Bacillus subtilis* O9 biosurfactant on the bioremediation of crude oil-polluted soils. Biodegradation 15:281–287

Dahrazma B, Mulligan CN (2007) Investigation of the removal of heavy metals from sediments using rhamnolipid in a continuous flow configuration. Chemosphere 69:705–711

Dahrazma B, Mulligan CN, Nieh M-P (2008) Effects of additives on the structure of rhamnolipid (biosurfactant): a small-angle neutron scattering (SANS) study. J Colloid Interface Sci 319:590–593

Das K, Mukherjee AK (2007) Crude petroleum-oil biodegradation efficiency of *Bacillus subtilis* and *Pseudomonas aeruginosa* strains isolated from a petroleum-oil contaminated soil from North-East India. Bioresour Technol 98:1339–1345

Desai JD, Banat IM (1997) Microbial production of surfactants and their commercial potential. Microbiol Mol Biol Rev 61:47–64

Falatko DM, Novak JT (1992) Effects of biologically produced surfactants on the mobility and biodegradation of petroleum hydrocarbons. Water Environ Res 64:163–169

Franzetti A, Caredda P, Ruggeri C, La Colla P, Tamburini E, Papacchini M, Bestetti G (2009) Potential applications of surface active compounds by *Gordonia* sp. strain BS29 in soil remediation technologies. Chemoshpere 75:801–807

Gu MB, Chang ST (2001) Soil biosensor for the detection of PAH toxicity using an immobilized recombinant bacterium and a biosurfactant. Biosens Bioelectron 16:667–674

Guerra-Santos L, Kappeli O, Fiechter A (1984) *Pseudomonas aeruginosa* biosurfactant production in continuous culture with glucose as carbon source. Appl Environ Microbiol 48:301–305

Herman DC, Artiola JF, Miller RM (1995) Removal of cadmium, lead, and zinc from soil by a rhamnolipid biosurfactant. Environ Sci Technol 29:2280–2285

Ishigami Y, Gama H, Nagahora H (1987) The pH-sensitive conversion of molecular aggregated of rhamnolipid biosurfactant. Chem Lett 5:763–766

Ito S, Inoue S (1982) Sophorolipids from *Torulopsis bombicola*: possible relation to alkane uptake. Appl Environ Microbiol 43:1278–1283

Ito S, Kinta M, Inoue S (1980) Growth of yeasts on n-alkanes: inhibition by a lactonic sophorolipid produced by *Torulopsis bombicola*. Agric Biol Chem 44:2221–2223

Jain DK, Lee H, Trevors JT (1992) Effect of addition of *Pseudomonas aeruginosa* UG2 inocula or biosurfactants on biodegradation of selected hydrocarbons in soil. J Ind Microbiol 10:87–93

Jarvis FG, Johnson MJ (1949) A glyco-lipid produced by *Pseudomonas aeruginosa*. J Am Chem Soc 71:4124–4126

Jeong SW, Corapcioglu MY, Roosevelt SE (2000) Micromodel study of surfactant foam remediation of residual trichloroethylene. Environ Sci Technol 34:3456–3461

Juwarkar AA, Nair A, Dubey KV, Singh SK, Devotta S (2007) Biosurfactant technology for remediation of cadmium and lead contaminated soils. Chemosphere 68:1996–2002

Kim J, Vipulanandan C (2006) Removal of lead from contaminated water and clay soil using a biosurfactant. J Environ Eng 132:777–786

Kuyukina MS, Ivshina IB, Makarov SO, Litvinenko LV, Cunningham CJ, Philp JC (2005) Effect of biosurfactants on crude oil desorption and mobilization in a soil system. Environ Int 31:155–161

Lai C-C, Huang Y-C, Wei Y-H, Chang J-S (2009) Biosurfactant-enhanced removal of total petroleum hydrocarbons from contaminated soil. J Hazard Mater 167:609–614

Lang S, Wagner F (1987) Structure and properties of biosurfactants. In: Kosaric N, Cairns WL, Gray NCC (eds) Biosurfactants and biotechnology. Dekker, NewYork, pp 21–45

Lang S, Wullbrandt D (1999) Rhamnose lipids-biosynthesis, microbial production and application potential. Appl Microbiol Biotechnol 51:22–32

Lu S-J, Wang H-Q, Yao Z-H (2006) Isolation and characterization of gasoline-degrading bacteria from gas station leaking-contaminated soils. J Environ Sci 18:969–972

Massara H, Mulligan CN, Hadjinicolaou J (2007) Effect of rhamnolipids on chromium-contaminated kaolinite. Soil Sediment Contam 16:1–14

Mulligan CN (2005) Environmental applications for biosurfactants. Environ Pollut 133: 183–198

Mulligan CN (2009) Recent advances in the environmental applications of biosurfactants. Curr Opin Colloid Interface Sci 14:372–378

Mulligan CN, Wang S (2006) Remediation of heavy metal-contaminated soil by a rhamnolipid foam. Eng Geol 85:75–81

Mulligan CN, Yong RN, Gibbs BF (1999a) On the use of biosurfactants for the remocal of heavy metals from oil-contaminated soil. Environ Prog 18:50–54

Mulligan CN, Yong RN, Gibbs BF, James S, Bennett HPJ (1999b) Metal removal from contaminated soil and sediments by the biosurfactant surfactin. Environ Sci Technol 33:3812–3820

Mulligan CN, Yong RN, Gibbs BF (2001) Heavy metal removal from sediments by biosurfactants. J Hazard Mat 85:111–125

Mulligan CN, Oghenekevwe C, Fukue M, Shimizu Y (2007) Biosurfactant enhanced remediation of a mixed contaminated soil and metal contaminated sediment. In: Seventh geoenvironmental engineering seminar, Grenoble, France, May 19–24

Oberbremer A, Muhller-Hurtig R, Wagner F (1990) Effect of addition of microbial surfactant on hydrocarbon degradation in soil population in a stirred reactor. Appl Microbiol Biotechnol 32:485–489

Ochoa-Loza FJ, Noordman WH, Jannsen DB, Brusseau ML, Maier RM (2007) Effect of clays, metal oxides, and organic matter on rhamnolipid biosurfactant sorption by soil. Chemosphere 66:1634–1642

Parra JL, Guinea J, Manresa MA, Robert M, Mercade ME, Comelles F, Bosch MP (1989) Chemical characterization and physicochemical behaviour of biosurfactants. J Am Oil Chem Soc 66:141–145

Pei X, Zhan X, Zhou L (2009) Effect of biosurfactant on the sorption of phenanthrene onto original and H_2O_2-treated soils. J Environ Sci 21:1378–1385

Perfumo A, Banat IM, Marchant R, Vezzulli L (2007) Thermally enhanced approaches for bioremediation of hydrocarbon-contaminated soils. Chemosphere 66:179–184

Rahman KSM, Banat IM, Thahira J, Thayumanavan T, Lakshmanaperumalsamy P (2002) Bioremediation of gasoline contaminated soil by a bacterial consortium amednded with poultry litter, coir pith and rhamnolipid biosurfactant. Bioresour Technol 81:25–32

Rahman KSM, Rahman TJ, Kourkoutas Y, Petsas I (2003) Enhanced bioremediation of n-alkane in petroleum sludge using bacterial consortium amended with rhamnolipid and micronutrients. Bioresour Technol 90:159–168

Robert M, Mercade ME, Bosch MP, Parra JL, Espuny MJ, Manresa MA, Guinea J (1989) Effect of the carbon source on biosurfactant production by *Pseudomonas aeruginosa* 44T1. Biotechnol Lett 11:871–874

Rodrigues L, Banat IM, Teixeira J, Oliveira R (2006) Biosurfactants: potential applications in medicine. J Antimicrob Chemother 57:609–618

Said WA, Lewis DL (1991) Quantitative assessment of the effects of metals on microbial degradation of organic chemicals. Appl Environ Microbiol 57:1498–1503

Sandrin TR, Chech AM, Maier RM (2000) A rhamnolipid biosurfactant reduces cadmium toxicity during naphthalene biodegradation. Appl Environ Microbiol 66:4585–4588

Scheibenbogen K, Zytner RG, Lee H, Trevors JT (1994) Enhanced removal of selected hydrocarbons from soil by *Pseudomonas aeruginosa* UG2 biosurfactants and some chemical surfactants. J Chem Technol Biotechnol 59:53–59

Schippers C, Gessner K, Muller T, Scheper T (2000) Microbial degradation of phenanthrene by addition of a sophorolipid mixture. J Biotechnol 83:189–198

Shin K-H, Kim K-W, Ahn Y (2006) Use of biosurfactant to remediate phenanthrene-contaminated soil by the combined solubilisation-biodegradation process. J Hazard Mater 137:1831–1837

Singh A, Van Hamme JD, Ward OP (2007) Surfactants in microbiology and biotechnology: Part 2. Application aspects. Biotechnol Adv 25:99–121

Tan H, Champion JT, Artiola JF, Brusseau ML, Miller RM (1994) Complexation of cadmium by a rhamnolipid biosurfactant. Environ Sci Technol 28:2402–2406

Torrens JL, Herman DC, Miller-Maier RM (1998) Biosurfactant (rhamnolipid) sorption and the impact on rhamnolipid-facilitated removal of cadmium from various soils under saturated flow conditions. Environ Sci Technol 32:776–781

Urum K, Pekdemir T (2004) Evaluation of biosurfactants for crude oil contaminated soil washing. Chemosphere 57:1139–1150

Urum K, Pekdemir T, Gopur M (2003) Optimum conditions for washing of crude oil-contaminated soil with biosurfactant solutions. Process Saf Environ Prot 81:203–209

Vipulanandan C, Ren X (2000) Enhanced solubility and biodegradation of naphthalene with biosurfactant. J Environ Eng 126:629–634

Wang S, Mulligan CN (2004a) Rhamnolipid foam enhanced remediation of cadmium and nickel contaminated soil. Water Air Soil Pollut 157:315–330

Wang S, Mulligan CN (2004b) An evaluation of surfactant foam technology in remediation of contaminated soil. Chemosphere 57:1079–1089

Wang S, Mulligan CN (2006) Occurrence of arsenic contamination in Canada: sources, behavior, and distribution. Sci Total Environ 366:701–721

Wang S, Mulligan CN (2009a) Rhamnolipid biosurfactant-enhanced soil flushing for the removal of arsenic and heavy metals from mine tailings. Process Biochem 44:296–301

Wang S, Mulligan CN (2009b) Arsenic mobilization from mine tailings in the presence of a biosurfactant. Appl Geochem 24:928–935

Whang L-M, Liu P-W, Ma C-C, Cheng S-S (2008) Application of biosurfactants, rhamnolipid, and surfactin, for enhanced biodegradation of diesel-contaminated water and soil. J Hazard Mat 151:155–163

Zhang YM, Miller RM (1992) Enhanced octadecane dispersion and biodegradation by a *Pseudomonas* rhamnolipid surfactant (biosurfactant). Appl Environ Microbiol 58:3276–3282

Zhang YM, Miller RM (1995) Effect of rhamnolipid (biosurfactant) structure on solubilisation and biodegradation of n-alkanes. Appl Environ Microbiol 61:2247–2251

Zhang YM, Maier WJ, Miller RM (1997) Effect of rhamnolipids on the dissolution, bioavailability and biodegradation of phenanthrene. Environ Sci Technol 31:2211–2217

Part II
Biocontrol

Chapter 10
Biological Control of Pests

Anu Kalia* and Rajinder K. Mudhar

10.1 Introduction

Plants are multicellular autotrophic eukaryotes that exhibit the characteristics such as getting sick, exhibiting symptoms of disease and senescence, and finally death occurs as it happens to animals and humans. The feature of getting sick reflects onset and progression of a plant disease for which there are a series of causes both biotic and abiotic factors such as environmental stresses, genetic or physiological disorders, and infectious agents including bacteria, fungi, viruses, and viroids (Montesinos 2000). However, among these varied factors microbes offer a dual role; i.e., these could be the causative agents of a disease or could also act as biocontrol agents by virtue of several unique mechanisms, pathways, secretions, or products that emerge during plant–microbe interactions.

There are four types of biological control strategies, namely, conservation (application of natural enemies occurring at a particular site), classical (introduction of exotic natural enemies to a new locale where they did not originate or do not occur naturally), augmentative (supplemental release of natural enemies at critical time in season, i.e., inoculative release or simply huge release at one time, i.e., inundative release), and importation biological control (cost-effective alternative to chemical control for basic food crops of resource-poor farmers) (Bentely and O'Neil 1997).

The research and development activities on production, formulation, and use of microbes as biocontrol agents have gained impetus in recent years for the sake of sustainable agriculture. In modern agriculture, the increasing use of chemical pesticide inputs has several drastic negative aftermaths, i.e., development of pest/pathogen resistance to applied agents, their growing cost, and nontarget environmental impacts

A. Kalia* (✉) and R.K. Mudhar
Department of Microbiology, PAU, Ludhiana, Punjab-141004
and
*Present affiliation
Electron Microscopy and Nanoscience Laboratory, PAU, Ludhiana, Punjab-141004
e-mail: kaliaanu@gmail.com

A. Singh et al. (eds.), *Bioaugmentation, Biostimulation and Biocontrol*, Soil Biology 28,
DOI 10.1007/978-3-642-19769-7_10,

(Gerhardson 2002) besides consumer demand of pesticide-free food. That is why biocontrol agents have emerged as an alternative or as supplemental forms for the reduction of pesticide input or in certain instances biological substitutes to these chemical poisons (Postma et al. 2003). The biocontrol measures, more appropriately microbial control agents, include bacterial, actinomycetal, viral, and fungal forms functioning in diverse ways to combat several pests, pathogens, weeds, and plant parasitic nematodes that affect the productivity of crop, horticultural, and many other plants. In simple terms, biological control could be defined as using biota to reduce biota in cost-effective and environmentally safe manner. Natural biological control is the reduction of pest organisms that occurs 'for free' since the evolution of the first ecosystem some 500 million years ago, can be found in all ecosystems, and takes place without human interventions (van Lenteren 2006).

10.2 Biological Control Agents

Biological control is defined as the reduction or protection of pest populations by natural enemies such as use of predators (such as lady beetles and lacewings), parasitoids (wasps species and some flies), and pathogens (bacteria, fungi, and viruses) to curb insect pests, use of antagonists to control plant disease pathogens, and application of herbivores and plant pathogens for curbing weeds (Van Driesche et al. 2008). The buildup of the natural enemy populations could be done by following management techniques such as plantation of cover crops, providing nectar-producing plants and sources of alternate hosts in and around fields, and interplanting different crops to provide habitat diversity, which results in enhanced biological control of pests. Biological control thus involves the active manipulation of natural phenomena making it most harmless, nonpolluting, and self-perpetuating control method to control pests, pathogens, and weeds. In this chapter, more emphasis would be on the types and application of microbial biocontrol agents.

Plant growth promoting rhizomicrobes (PGPR) and endophytes are natural biocontrol agents. Based on the mode of action, Bashan and Holguin (1998) divided PGPRs to ones that promote plant growth and the others with bioprotectant action. Though the PGPRs are naturally present in the bulk soil, the number and activity is meager in comparison to the rhizosphere, which is relatively rich in nutrients as 40% of plant photosynthates are lost from roots in this region only (Bloemberg and Lugtenberg 2001).

Endophytes are bacterial and fungal forms which are present entirely within living host plant tissues asymptomatically and do possess potential as probable biocontrol agents (Wilson 1995). The common bacterial forms include the broader range of the PGPRs that enter root interior by crack entry or entry at the site of injury and establish intercellular endophytic populations and may involve the interplay of chemical signals between the endophyte and host as in the case of root nodulating bacteria/mycorrhiza (Gray and Smith 2005). The entry of the endophytic fungi and bacteria could also occur through other open gateways such as stomata, lenticels, wounded aerial parts, and floral parts, while the extent of infection to the internal most tissues reflects the fine-tuned selective adaptations of the endophyte(s) to inhabit these specific niches. These endophytes may be seed-borne or rhizosphere-derived and could exhibit plant or host specificity or could also exhibit nonhost specificity particularly for the opportunistic endophytes. The common root tissues inhabited by the endophytic bacteria after entry are the interspaces between the epidermal cells, below collapsed epidermal cells, within epidermal cells, and inside intercellular spaces in the root cortex.

Sessitsch et al. (2004) isolated seven endophytes from potato that were antagonistic to the fungal and the bacterial pathogens and were considered as promising biocontrol agents. Endophytic fungi have been identified in woody plants, trees, shrubs, ferns, and grasses (Saikkonen et al. 1998), which range from ecto- to endomycoorhizal as well as other fungal forms that inhabit the host tissues to either enhance the systemic resistance or block the receptor sites for attachment or adsorption during entry of the pathogen. The best-studied plant–fungal association is the mycorrhizal forms, particularly arbuscular mycorrhizal (AM) association to be the most common (Harrier 2001).

10.3 Mechanism of Biocontrol

Biological control using microbial agents can occur through various modes; however, the chief one includes competition for an ecological niche or a substrate, production of inhibitory allelochemicals, and development of induced systemic resistance, i.e., ISR (Compant et al. 2005).

10.3.1 Competition for an Ecological Niche or a Substrate

Competition for nutrients supplied by root exudates occurs between beneficial rhizobacteria and pathogens on the root. Similarly, the introduced microbial biocontrol agents establish large populations on the surface of the planting material and roots, thereby acting as partial sink for nutrients in the rhizosphere. This may result in reduction in the amount of carbon and nitrogen available to stimulate germination of spores of fungal pathogens or for subsequent colonization of the roots. Fluorescent pseudomonads are especially suited for rapid uptake or scavenging of nutrients, since they are nutritionally versatile and grow rapidly in the rhizosphere. The rapid colonization of the biocontrol agent results in niche hijacking, which not only helps in adsorption of the inoculated biocontrol agent onto clay particles, root surface, and internal tissues but also helps in evading the soil-borne pathogens by clogging of the surface binding ligands/receptors on the root surface and restricts the attachment and entry of the pathogen. The colonization of the inoculated biocontrol agent could be selective or preferential due to production of intricate host–microbe interaction signals.

10.3.2 Iron Competition and Role of Siderophores

Among various mechanisms of biocontrol, the pathogen could be suppressed by depriving it of nutrients and iron competition is one such mechanism. Iron, though abundant in earth's crust, exists in highly insoluble form (ferric hydroxide) and is sparingly available ($<10^{-8}$ M in soil solutions at neutral pH) to organisms. This helped in the evolution of high-affinity iron uptake system in bacteria, i.e., siderophores (low-molecular weight Fe-binding ligand) to shuttle iron into the cell, e.g., production of pseudobactin siderophore by fluorescent pseudomonads. Siderophore synthesis is affected by a myriad of environmental factors such as pH, level, and form of iron/iron ions, the presence of other trace elements, and an adequate supply of essential nutrients such as C, N, and P (Timms-Wilson et al. 2000). Bacterial siderophores contribute to the suppression of certain fungal and oomycete diseases (Buysens et al. 1996) by depriving the pathogenic fungus of the essential iron element as fungal siderophores have lower affinity to sequester iron (Loper and Henkels 1999). The siderophores could provide iron not only to the microbial cell producing it but also to the other neighboring microbes in the ecological niche and even plants, i.e., the phenomenon of heterologous siderophores.

10.3.3 Production of Allelochemicals for Suppression of Pathogen

The biocontrol action of majority of microbes is widely due to the phenomena of antibiosis or production of secondary metabolites or other compounds/products that either kill the pathogen or suppress one/more stages of life cycle/growth. Usually, colonization or even the initial population size of the biocontrol agent suppresses/modulates the pathogen from farther distances by production of quenchers or allelochemicals.

Antibiosis by production of antibiotics and bacteriocins is the most common mechanism of biocontrol, which may be because it is effective for suppressing pathogens in the rhizosphere and is often attributed to the production of antibiotics or other secondary metabolites. The fluorescent pseudomonads represent a unique example of antibiosis against root pathogens as these rhizobacteria produce a variety of antibiotics on roots grown in soil (natural niches), particularly the phenazine derivatives active against the take-all disease in wheat. The most frequently detected classes of antibiotics or antifungal metabolites (AFMs) produced by *Pseudomonas* biocontrol strains include phenazines, pyrrolnitrin, 2,4-diacetylphloroglucinol, and pyoluteorin, while new AFMs belonging to the class of cyclic lipopeptides, such as viscosinamide (Nielsen et al. 1999) and tensin (Nielsen et al. 2000), have also been discovered. The *Pseudomonas fluorescens* strain CHAO produces 2,4-diacetylphloroglucinol and pyoluteorin, which directly interfere with growth of various pathogens and contribute to disease suppression (Maurhofer et al. 1994) and also there exists a quantitative relationship between antibiotic production and disease suppression. Bacilli, the endospore forming bacteria, are the other common bacterial forms that have great potential as biocontrol agents. *Bacillus cereus* suppresses diseases caused by the oomycetes fungi by production of two antibiotics: zwittermicin A (aminopolyol) and kanosamine (aminoglycoside) as reported by Milner et al. (1996). The antimicrobial compounds produced by the inoculated biocontrol agent may induce fungistasis by inhibiting the spore germination or may exert fungicidal effects by causing lysis of fungal mycelia.

Another mechanism of antibiosis is by production of bacteriocins, which are proteinaceous antagnostic substances belonging to important class of antibiotics produced by bacteria that are lethal to other bacteria. These are usually peptides or proteins that selectively kill related bacteria (of the same species or genus) but do not affect other organisms. Bacteriocin-mediated antagonism is believed to occur in virtually any niche colonized by bacteria. Bacteriocin production is detectable in most strains of the human opportunistic pathogen *Pseudomonas aeruginosa* and has been used to differentiate clinical isolates. Bacteriocins producing strains have been identified among natural isolates of *Rhizobium trifolii*, *R. leguminosarum*, *R. japonicum*, and cowpea rhizobia.

A number of alternate mechanisms have also been shown for the increased plant growth and resistance to disease in plants inoculated with biocontrol bacteria. The production of potent extracellular lytic enzymes such as chitinases and laminarinase, capable of destroying fungal cell walls of *Fusarium* and other fungal pathogens, by some isolates of rhizospheric bacteria such as *Pseudomonas stutzeri* and *Serratia marscens*, (cultures or cell-free extracts) is one of the several possible mechanisms of disease control. Some of the specialized compounds and enzymes are produced by biocontrol agents that are discussed here.

Biosurfactants are organic compounds produced by a variety of microorganisms, including bacteria, fungi, and yeasts that alter the conditions prevailing at a surface or interface and are also synonymously used with adjuvants or bioemulsifiers. Biosurfactants include both the low-molecular mass compounds such as glycolipids and lipopeptides (rhamnolipids, surfactin) and the high-molecular mass compounds such as proteins and lipoproteins (de Souza et al. 2003). Nielsen et al. (2002) have reported the role of rhamnolipids in the control of plant pathogenic fungi *Pythium aphanidermatum*, *Plasmopara lactucae-radicis*, and *Phytophthora capsici*. The actual mode of action of rhamnolipids is via cessation of motility and the lysis of entire zoospore populations within <1 min as well as inhibitory activity against the spore germination and hyphal growth of several fungal pathogens (Kim et al. 2000).

Common lytic enzymes excreted by certain soil bacteria (*Streptomyces*, *Serratia*, *Pseudomonas*, *Bacillus*, and *Paenibacillus*) include hydrolases, chitinases, laminarinase, β-1,3-glucanase, and proteases that help in suppression of the intruding fungal pathogen by lysis of cell wall and inhibition of spore germination/germ tube elongation and hyphal growth. The inoculation of these bacteria may help in the effective control of some of the notorious pathogens such as *Botrytis cinerea*,

Sclerotium, *Fusarium oxysporum*, *F. solani*, *Rhizoctonia solani*, and *Pythium ultimum* (Compant et al. 2005).

Virulence factors are the variety of compounds secreted by the intruding pathogen in the surroundings for hydrolysis of the host tissues, further encroachment to deeper tissues, and easier absorption of the nutrients from the site of attack. The biocontrol agents may help in curbing disease progression to deeper tissues as well as pathogens by detoxification of virulence factors due to production of virulence factor degrading/detoxifying compounds such as toxin-binding proteins (esterase) and autoinducer signal degraders that quench pathogen quorum-sensing capacity, thereby blocking the expression of numerous virulence genes (Compant et al. 2005).

Biopriming by PGPRs or preinoculation of plant roots or seeds with avirulent or weakly aggressive strains of the disease-causing fungi can induce systemic resistance (ISR) against pathogens/causative agents of disease(s). Van Peer et al. (1991) have first observed the PGPR-elicited ISR in carnation (*Dianthus caryophillus*) to *Fusarium* sp. induced wilt. Manifestation of ISR is dependent on the combination of host plant and bacterial strain (Kilic-Ekici and Yuen 2004). This induced resistance occurs in whole plant body via a salicylic acid-independent pathway involving jasmonate and ethylene signals. The ISR is induced in response to diverse compounds/macromolecules such as proteins (flagella, pilli, extrinsic or outer membrane proteins, and low-molecular weight siderophores), lipopolysaccharides, and volatile organic compounds of the bacterial cell (Van Loon et al. 1998).

The biopriming by PGPR that triggers ISR also alters host physiology and metabolic responses in terms of increased accumulation of peroxidase, phenylalanine ammonia lyase, polyphenol oxidase, and/or chalcone synthase. Simultaneously, it increases the strength of plant cell wall by outer tangential thickening of the exodermis, layer(s) of cortical cells, and deposition of callose, further leading to an enhanced synthesis of plant defense chemicals (array of phenolics) upon challenge by pathogens and/or abiotic stress factors (Nowak and Shulaev 2003). The plant defense compounds include the phytoalexins, which are low-molecular weight antimicrobial compounds produced in the infected tissue by the plant itself under stress conditions or in response to pathogenic invasion. The phytoalexin production is considered as an additional mechanism of disease suppression in plant–pathogen relationships.

10.4 Biopesticides

A large number of microbial species as well as large organisms belonging to various groups have potential as biological control agents for array of pathogens such as insects, fungi, and bacteria. Thus, these could act as bioinsecticides (due to entomopathogenic action), biofungicides (for evading fungal pathogens), and bioherbicides (for destruction of weed plants).

10.4.1 Bioinsecticides

These refer to a range of genera of entomopathogenic organisms including bacteria, viruses, protozoans, and nematodes that could be used as a tool of integrated pest management (IPM) (Gnanasambandan et al. 2000). Like other natural enemies, these pathogens can exert considerable control on the target pest population and help decrease it to safe levels (Verma and Dubey 1999).

10.4.1.1 Entomopathogenic Bacteria

Bacteria which are used for the control of insect pests are *Bacillus thuringiensis*, *Bacillus papilliae*, *Bacillus sphaericus*, *Pseudomonas fluorescens*, and *Serratia entomophila*. Of all these, that have been evaluated, the most attractive by far has been the bacterial insect pathogen, *Bacillus thuringiensis*. It is gram-positive, aerobic, sporulating rod-shaped bacterium that produces proteinaceous parasporal inclusions during sporulation including delta endotoxin, beta, alpha, and gamma exotoxins, which are highly insecticidal at very low concentrations. The specific toxicity of these crystalline inclusions against insect pests makes this organism a potential agent for biological control. There exists 34 recognized species of *B. thuringiensis* including most commonly used subspecies *kurstaki*, *israelensis*, and *tenebrionis* active against lepidoptera, diptera (mosquitoes and blackflies), and *Leptinotarsa decemlineata* (Whalon and McGaughey 1998). The crystal proteins are biodegradable and are specific to each subspecies/strain and are coded by a single plasmid-borne gene. These crystalline proteins have the basic target being the insect midgut columnar epithelial cells and result in decreased absorption of minerals and nutrition from midgut and finally death of the columnar cells. To date, Bt-based formulations (67 registered products with more than 450 formulations) occupy key position accounting for nearly 90% of total biopesticide sales (Neale 1997). Though it is observed to date, except for a few reports that *B. thuringenisis*-based products are cost-effective, very specific to target pest species, and safe to people and the environment, but the survival of the Bt products or formulations in the market is purely governed by the cost-effectiveness and field efficacy criteria followed by safety issues as development of resistance toward one or other type of the Bt subspecies.

10.4.1.2 Entomopathogenic Viruses

The entomopathogenic viruses, particularly baculoviruses, offer a promising option of biocontrol second only to bacteria in terms of population, adoption, and success. These are naturally occurring pathogens that are specific for a single or a few related insect species making them environmentally safe (produce no toxic residues, are harmless to nontarget organisms such as beneficial insects and vertebrates, and pests do not exhibit major resistance). Insect viruses are attractive as biological control agents and could be a feasible alternative to chemical insecticides in the management of insect infestations (Sun and Peng 2007).

Baculoviruses belong to virus family baculoviridae (large enveloped viruses) having two prominent members, namely, Nuclear Polyhedrosis virus (NPV) and Granulovirus (GV), which display the greatest microbial biocontrol potential (Moscardi 1999) due to their exclusive pathogenicity to arthropod insects spanning over 400 insect species of Lepidoptera (butterflies and moths) and Hymnoptera (sawflies) classes besides members of Coleoptera (beetles), Diptera, and decapoda crustaceans (Gupta et al. 2007). These viruses consist of a large, double-stranded covalently closed circular DNA (varying from 80 to 180 kbs) as genetic material (DNA viruses), which is enclosed within an occlusion body composed of polyhedron (Vail et al. 1999). The virus action initiates by afflicting the insect larvae causing their paralysis and subsequent liquefaction, which allows the release of progeny virus, produced from the insect biomass. The specificity and production of secondary inoculum make baculoviruses an attractive alternative to broad-spectrum traditional insecticides and ideal components of the IPM system (Sankarama 1999).

The problem related to long duration (more than a week) for causing infection and further killing of the insect larvae by the inoculated baculoviruses has to be amended by manipulation of the baculovirus genomes that would enhance the speed of kill by combined action of baculoviral pathogenicity with the insecticidal action of toxin, hormone, or enzyme, which is active on insects. Recombinant baculovirus technology has been used to improve the insecticidal qualities of baculoviruses and employed for the expression of foreign proteins. In general, the foreign genes are inserted into the baculoviral genome into the polyhedron gene (*polh*) locus that encodes the occlusion body protein; however, other gene insertion loci (*p10*) or the nonessential regions of the baculoviral genome have been identified as field stability of the recombinant virus requires intact *polh* loci. Genetic engineering could be performed using genes of various origins such as Bt gene, scorpion toxin (BelT/AaIT) gene, straw itch mite toxin (TxP1) gene, and insect hormone genes such as diuretic hormone from *Manduca sexta* that alters the larval fluid metabolism, eclosion hormone associated with ecdysis from *M. sexta* that causes initiation of the eclosion behavior in the inoculated larvae, and ecdysteroid UDP-glucosyltransferase (egt) juvenile hormone esterase (JHE) expression in baculovirus (Lasa et al. 2009) that results in stopping of larval feeding and commencing of molting stage (Bonning and Hammock 1996).

Baculoviruses primarily infect the larval stage of the insect with infection occurring accidentally during feeding of the larvae on the plant foliage. The viral polyhedra taken up during the feeding process are solubilized in the alkaline environment of the midgut to release the infective virions. The virions now replicate within nuclei of the epithelial cells lining the midgut to release more infecting virion units in budded form within 10–12 h of infection or the virions get occluded within the polyhedra late in the infection process. The death of the larvae commences as the larvae no longer feed on the plant tissues. Tissue liquefaction and then rupture of these cells upon death of the infected larvae liberate masses of these polyhedra in the soil environment where they lie highly stable persisting for many years (Cory and Bishop 1995).

Commercially, the baculovirus products are mostly produced in the form of concentrated wettable powders apart from liquid or granular forms. As the ultraviolet (UV) radiations ($\lambda = 290–320$ nm) completely inactivate the virus, UV protectants such as metallic oxides are used besides the addition of anti-evaporants and spreaders/wetting agents to the virus formulations. On field application, these viruses generally fail if not applied at right place and right time; thus, information regarding the insect behavior on the crop after hatching, its distribution within the crop in each instar, and the area of foliage ingested per instar all determine effective use of the virus (Simon et al. 2008). Moreover, the baculoviruses also exhibit temperature and contaminant sensitivity (Lasa et al. 2008). Since the baculoviral mass production depends on the availability of the host insect larvae, new methods of rearing and multiplication are being sort for like culturing of viruses in insect-derived cell lines (Hardin 2002; Rodas et al. 2005).

10.4.1.3 Entomopathogenic Fungi

In spite of the recent advances in insect pathology, the study of mycoses caused by entomogenous fungi has held a relatively modest position. Association of fungi with insects is well known and over 750 species of nearly 100 genera are found in the division Eumycota and in the following subdivisions: Mastigomycotina, Deuteromycotina, Zygomycotina, Ascomycotina, and Basidiomycotina (Tanada and Kaya 1993). Among these fungal classes, many entomopathogenic fungi, especially those in the order Entomophthorales, are responsible for epizootics that often successfully regulate pest insect populations and the most common method of employing fungi for insect control is through inundative means (Marrone 2002). Some of the genera that have been most intensively investigated for mycoinsecticides are *Beauveria*, *Metarhizium*, *Verticillium*, *Poecoelomyces*, *Nomuraea*, *Erynia*, *Entomophthora*, *Zoophthora*, etc. The first two genera, i.e., *Beauveria* and *Metarhizium* have been identified from 700 and 300 species of insects respectively and have been used on a large scale over a number of years in different countries (Narayanan 2002).

Differing from bacteria and viruses, fungi can infect insects not only through the gut but also through spiracles and particularly through the surface of the integument (Shah and Pell 2003). Omission/failure of these fungi at attachment site followed by spore germination and penetration of the insect cuticle will produce low virulent fungi, though they might have high toxin biosynthetic capability (Arora and Dhaliwal 2001). *Beauveria bassiana* is a common soil-borne fungus that occurs worldwide and attacks a wide range of both immature and adult insects (Khachatourians et al. 2002). It causes white muscardine disease and is useful in the control of Japanese corn borer, white flies, aphids, grasshoppers, Colorado potato beetle, Japanese beetle, boll weevil, and codling moth. Sheeba et al. (2001) have reported the efficacy of *Beauveria bassiana* for the control of rice weevil. Similarly, another epizootic fungi used as bioinsecticide is *Metarhizium anisopliae*, which is a soil-borne fungus that causes green muscardine disease and could be used to control locusts, grasshoppers, termites, curculionids, and scarbeids. Fungal activity is positively associated with high humidity and rainfall with infection levels that remain higher under well watered conditions for the introduced resting spores (Kaaya and Hassan 2000). Lacey et al. (2001) have reported the inoculation effects of mycoinsecticide *Entomophaga maimaiga* on all larval instars of gypsy moth that were susceptible to infection by *E. maimaiga*.

Technical efficacy, and to a lesser extent practical efficacy, is essential for success, and major advances have been made in the production, formulation, and application of hyphomycete fungi as mycoinsecticides. In this respect, the development of entomophthoralean fungi as mycoinsecticides has been beset by technical difficulties. Obstacles mainly relate to mass production and the size and stability of propagules for storage and formulation (Pell et al. 2001; Shah and Pell 2003).

10.4.1.4 Other Entomopathogenic Organisms

Apart from the most popular Bt, baculoviral, and fungal bioinsecticides, other types of bioinsecticides, called biochemical insecticides, are also used for certain special types of insect pests. The general classes included are the protozoans, nematodes, and even insects parasitizing on the pest insects as well as some insect chemicals against the biochemical processes involved in insect signaling/communication and reproductivity.

Insects get infested and get killed by many protozoan diseases caused by chronic infection of host-specific protozoan genera and thus these eukaryotic microbes comprise an important regulatory role in insect population. The general mode of action of these entomopathogenic protozoans is through debilitation of the overall fitness and reproduction, thereby decreasing host vigor and longevity by acting as chronic debilitating agents. *Nosemia fumiferarae* is one such typical protozoan being exploited for the management of spruce budworm. "Noloc" is the formulation based on *N. locustae* produced by a private company M/s Sandoz Inc., which is used against grasshoppers. *Varimorpha necatrix* is another protozoan that infects 36 lepidopteran pests and has peculiar abilities of high virulence and wide host range, making it an attractive candidate for functioning as better microbial control agent.

Entomopathogenic nematodes parasitizing on insects either alone or sometimes in combination with bacteria may be used as bioinsecticides (Smart 1995). A plethora of nematodes belonging to more than 30 families are observed to be associated with insects and other vertebrates among which members belonging to seven nematodal families have been most studied and are observed to have active bioinsecticidal activities. The nematodes share a symbiotic relationship with the prokaryotic bacterial genera, which equip these bacteria harboring nematodes to be highly virulent in behavior so that it can kill its host within 48 h through the action of the mutualistic bacteria. Bacteria involved in this interaction belong generally to genus *Xenorhabdus*, with *X. luminescens* and *X. nematophilus*. Bacteria alone cannot do any damage to the insect host, while nematodes in the absence of bacteria reproduce very poorly, and therefore, fail to cause any pathogenicity. It is the mutualistic nematode bacteria complex that enhances nematode reproduction and ultimately kills the insects (Kaya and Gaugler 1993). *Neoplectana carpocapsae* vectors a bacterium that builds up in the body cavity of the insect and can cause septicemia followed by insect death. The general insect pests that are targeted by nematodal biopesticides include the soil insects and those that live under cryptic habitat conditions such as citrus root weevil and black cutworm. Ishibashi and Choi (1991) have reported the efficacy *Aphelenchus avenae* to suppress infection of the turnip moth *Agrotis segetum*.

These entomopathogenic nematodes could be reared and mass produced on artificial chemically known solid/liquid monoxenic media or by in vivo cultivation using suitable host insect. However, the generally field applied/commercial formulations that demand longer shelf life of the product at room temperature (at least 6 months) and ease of application over several acres of land exist in the form of wettable dispersible granules containing infective juveniles. The five species of nematodes that are being sold commercially as bioinsecticides are *Steinernema carpocapsae*, *S. riobravis*, *S. feltiae*, *Heterorhabditis bacteriophora*, and *H. megidis*. The nematodes being sensitive to several biotic and abiotic factors could be genetically manipulated for enhanced killing by virtue of better resistance to altered environmental conditions particularly high temperature stress, increased virulence, and search capacity.

More than 150 species of natural enemies (parasitoids, predators, and pathogens) are currently commercially available for augmentative forms of biological control (van Lenteren 2005). The natural insect parasites include the organisms such as lacewings, mites, and lady beetles, which feed on a number of their natural prey insects. They often attack different life stages of the pest and even different pest species. *Trichogramma* wasps are known to act as parasitoids for many caterpillar pests and are highly effective in curbing caterpillar infestation. It parasitizes the eggs of the pest species, killing the pest before any feeding injury (Fouche et al. 2000). The advantage of using parasites is that they are usually extremely well adapted to their natural host and are very good at finding their host pests even when numbers of the pest are relatively low. Moreover, it becomes unlikely that resistance will develop to a control agent, and in many cases, the control can be self-perpetuating over long periods of time (Bale et al. 2008).

Spiders are voracious predators of insects. They are well adapted to certain habitats because of their ability to withstand periods of low food availability and also to take advantage of periods of prey abundance. Spiders are important predators of pests of cotton, rice, apple, banana, and various other crops and plantations. In India and other tropical countries, the giant crab spider, *Heteropoda venatoria*, is a commonly found predator of cockroaches in crevices and cracks (http://tcdc.undp.org/Sie/experiences/vol4/Rearing%20spiders.pdf). Suppression of insect pests such as plant hoppers and leafhoppers in the rice fields can also be performed by spiders. Moreover, the pest resurgence after insecticide spraying could be linked to the negative impact of insecticides on spiders and other natural enemies (Sigsgaard 2000).

These are naturally occurring substances that control pests by nontoxic mechanisms and in contrast to conventional pesticides that directly kill or inactivate the pest. Biochemical pesticides include substances, such as insect sex pheromones, that interfere with mating (mating disruption), as well as various scented plant extracts that attract insect pests to traps. If an area is saturated with the pheromone of the pest insect, the message given out by individual insects is swamped, which vigorously affects the mating process that now entirely depends on chance juxtaposition of a male and a female type of the target insect. Thus, either no progeny, i.e., eggs are laid by the unmated insects, or even if the eggs are laid, they are infertile.

10.4.2 Bioherbicides

Bioherbicides are the biocontrol agents applied for the control and eradication of weed plants. Thus, the biological weed control involves inoculating/using living organisms, such as insects, nematodes, bacteria, or fungi, to reduce weed populations. The basic purpose as well as mechanism of this control is similar to the conventional chemical herbicides (Bishop 2000).

Since a variety of biocontrol agents could be used for curbing the growth of weeds, there is a substantial variation in the mechanism of action followed by a particular type of organism. Few biological control agents attach to roots and thereby stunt plant growth particularly the root surface inhabiting bacteria (deleterious rhizobacteria) that release toxins, which causes stunting of the root. However, the fungal bioherbicides exhibit first the infection of the roots followed by disruption of the water transport system of the root, which reduces leaf growth. Biocontrol of herbicides also involves the use of macroorganisms such as beneficial insects and nematodes that feed directly on the weed roots, thereby causing injury which allows bacteria and fungi to penetrate or may voraciously devour on the aerial stem and leaves of the inoculated plant, thus reducing the leaf surface available for energy capture.

Two types of bioherbicides that have been proposed and utilized commercially for control of weeds include natural plant products, i.e., allelochemicals and mycoherbicides (active fungal cultures for target-specific control). Mycoherbicides have gained impetus regarding their application and production due to the target host specificity of the plant pathogenic fungus as well as the convenient culturability of these microbes under standard lab conditions because most of them are not fastidious in nutritional requirements for their growth and mass production.

Green (2003) reviewed the possible use of fungal herbicides for biocontrol of forest weeds that compete with the young tree saplings in woodlands and decrease the growth in commercial tree plantations in the UK. He reported the greatest potential for the application of a wood-rotting fungus as a bioherbicide stump treatment for *Rhododendron*. The basic mechanism of mycoherbicidal control is invasion of vascular tissue once host tissues are encountered resulting in stunting or killing of the invaded plant; however, the mortality is less (around 25%) under field conditions. Thus, to increase the mortality rate of the weed plants, it is prerequisite to increase the virulence of the fungus. Tiourebaev et al. (2000) have reported improvement in virulence of mycoherbicides by method of amino acid excretion and suggested incremental increase in virulence by selection for additional excretion of the same or different amino acid. Ortiz Ribbing and Williams (2006) have reported the use of two fungal pathogens for biological weed management of several *Amaranthus* species, some of the biotypes of which have developed resistance to multiple herbicide families. They observed 80–100% seedling mortality for *A. albus* and *A. blitoides*, 14–15 days after transplantation for the mixture or *Microsphaeropsis amaranthi* alone, in greenhouse and field trials.

10.4.3 Biofungicides

Biofungicides are the type of biopesticides used to curb and control fungal plant pathogens by introduction/inoculation of microbial live cells in or onto the plant. It also includes engineering of the genes into the plant genome, resulting in the production of compounds that lead to fungal pathogen control or the generation of a hypersensitive response through resistance genes/manipulation of genes involved in systemic acquired resistance. The general inoculant used and supplied commercially as biofungicide is *Trichoderma* sp.; however, newer candidates for being better biofungicides are now searched, reviewed, and analyzed to check efficacy upon field release. Lahdenpera (2003) reported the use of *Streptomyces griseovirdis* K61, a powdered formulation of dried spores and mycelia of the soil actinomycete (trade name Mycostop), as a potent biofungicide having the ability to curb damping off of cauliflower caused by *Alternaria*.

The agroecological management strategies would not only help in optimal recycling of nutrients and organic matter turnover, closed energy flows, and water and soil conservation but also would help in balancing the pest-natural enemy populations (Vandermeer 1995). Agricultural advantages of mixed cropping could be obtained by the biological effects such as light competition, by offering weed-suppressing capacities, or by diversification of plant covers to break development cycles of pests.

10.5 Future Perspectives

The future perspectives in biopesticides are leaning toward development and release of transgenic crop plants that have the acquired ability to annihilate the pest attack. Several molecular tools/techniques and protocols are helping in pinpointing a relationship between the introduced foreign gene (of biopesticidal action) and the candidate host line/variety that exhibits maximum expression of the introduced gene and thus enhanced mortality of the target pest. The correlation studies between the native resistance genes in plant and the introduced transgene say Bt gene have showed that resistant variety with transgenic insertion exhibits higher mortality than susceptible line having trangene (Walker et al. 2004).

Thus, the strategy of gene pyramiding using multiple Bt genes, Bt and unrelated transgenes, or Bt and native plant genes could help rescue the development of resistance to known arsenal of biopesticides. Walker et al (2004) have used a multiple resistance gene pyramiding strategy to obtain the soybean variety possessing two quantitative trait loci (QTLs) exhibiting positive correlation with the Bt transgene.

Being transgenic, these new generation plants carry genes and produce compounds foreign to their environment that holds concern about their environmental use and potential ecological effects. The issue of direct and indirect effects on nontarget organisms and ecosystems is particularly important because many transgenic plants are being developed that have new or enhanced antimicrobial properties for protection from phytopathogens. Microcosm and field studies showed that exposure to transgenic plants produced changes in the population levels and composition of some soil and plant microorganisms (Dunfield and Germida 2004). GM potatoes consistently altered the physiological profile of the rhizosphere microbial community at harvest, but the effect did not persist from one season to the next (O'Callaghan and Glare 2001). The transgenic crops alter both the population levels and species composition of bacteria and fungi. Moreover, transgenic plants cause transient but significant increase in levels of culturable, aerobic bacteria and fungi (Donegan et al. 1995).

10.6 Conclusions

Biocontrol agents and biopesticides are the integral components of IPM strategies and are efficient agencies to get rid of notorious crop pests resulting in decreasing the net profits due to damage at the pre- as well postharvest stages of the crop production. Crop pests include a wide range of insect, fungal, bacterial, viral agents, rodents, birds, and other animals that tremendously damage the crop at various stages of maturation and storage among which insect pests are most noteworthy. In the currently still pesticide-dominated agriculture, biological control has found its place in the form of augmentative releases. Augmentative releases can be used for management of pests that are hard to control with pesticides. The use of biopesticides will not only help in preventing the dumping of thousands of tons of agrochemical on the earth but also will provide the residue free food and a safe environment to live. Research and development of protocols and processes in the formulation of an effective biocontrol agent specifically a biopesticide will pave toward sharpening the action as well as host spectrum of the biocontrol agent probably by using the process and protocols of genetic engineering that help combine two or more effective lethal processes to finally tailor them into one agent/organism. This would decrease the probability of development of resistant strains due to effective mortality and thus little morbidity. Several government and public agencies are manufacturing biopesticides at the commercial scale and are providing farmers these weapons for fighting against crop pathogens and pests. Genetically modified microbes and transgenic crops are the recent episodes in the chapter of IPM strategies and are gaining popularity regarding their efficacy in eradicating pests, but stern concerns are evident regarding the potential environmental hazards (vertical/horizontal transfer of prokaryotic genes) of their long-term use.

Undoubtedly, all the efforts directed toward the biocontrol agents, may be search or research, introduction or production, development or improvement, on-farm failure or success, low or high cost, and finally adoption or rejection by the end user, fall into the domain of modern agriculture biotechnology. Unless, it does not meet the needs of the poor farmer, the objective remains half met. Thus, it is imperative for a biocontrol scientist as well as its production technologist to follow a midway approach so that if biological control agents help to preserve the environment safety on the one hand, it must also fulfill the aspirations of the farming community in terms of production and profitability from agriculture.

References

Arora R, Dhaliwal GS (2001) Integrated pest management. Concept and approaches. Microbial control. Kalyani, New Delhi, pp 197–233

Bale JS, van Lenteren JC, Bigler F (2008) Biological control and sustainable food production. Philos Trans R Soc B 363:761–776

Bashan Y, Holguin G (1998) Proposal for the division of plant growth-promoting rhizobacteria into two classifications: biocontrol-PGPB (plant growth promoting bacteria) and PGPB. Soil Biol Biochem 30:1225–1228

Bentely JW, O'Neil RJ (1997) On the ethics of biological control of insect pests. Agric Human Values 14:283–289

Bishop A (2000) Bioherbicides: new products and new modes of action. Tas Reg 6:41–42

Bloemberg GV, Lugtenberg BJJ (2001) Molecular basis of plant growth promotion and biocontrol by rhizobacteria. Curr Opin Plant Biol 4:343–350

Bonning BC, Hammock BD (1996) Development of recombinant baculoviruses for insect pest control. Annu Rev Entomol 41:191–210

Buysens S, Heungens K, Poppe J, Hofte M (1996) Involvement of pyochelin and pyoverdin in suppression of *Pythium*-induced damping-off of tomato by *Pseudomonas aeruginosa* 7NSK2. Appl Environ Microbiol 62:865–871

Compant C, Duffy B, Nowak J, Clément C, Barka A (2005) Use of plant growth-promoting bacteria for biocontrol of plant diseases: Principles, mechanisms of action, and future prospects. Appl Environ Microbiol 71:4951–4959

Cory JS, Bishop DHL (1995) Methods in molecular biology (Richardson CD eds), Humana Press, Totowa, NJ. 39: 277–294 de Souza JT, de Boer M, de Waard P, van Beek TA, Raaijmakers JM (2003) Biochemical, genetic, and zoosporicidal properties of cyclic lipopeptide surfactants produced by *Pseudomonas fluorescens*. Appl Environ Microbiol 69:7161–7172

Donegan KK, Palm CJ, Fieland VJ, Porteous LA, Ganio LM, Schaller DL, Bucao LQ, Seidler RJ (1995) Changes in levels, species, and DNA fingerprints of soil microorganisms associated with cotton expressing the *Bacillus thuringiensis* var. *kurstaki* endotoxin. Appl Soil Ecol 2:111–124

Dunfield KE, Germida JJ (2004) Impact of genetically modified crops on soil- and plant-associated microbial communities. J Environ Qual 33:806–815

Fouche C, Gaskell M, Koike ST, Mitchell J, Smith R (2000) Insect pest management for organic crops. Publication 7251, Division of Agriculture and Natural Resources, University of California, pp 1–5

Gerhardson B (2002) Biological substitutes for pesticides. Trends Biotechnol 26:338–343

Gnanasambandan S, Balakrishnamurthy P, Pillai KS (2000) Integrated pest management in 21st century. Pestol XX1V:9–11

Gray EJ, Smith DL (2005) Intracellular and extracellular PGPR: commonalities and distinctions in the plant-bacterium signaling processes. Soil Biol Biochem 37:395–412

Green S (2003) A review of the potential for the use of bioherbicides to control forest weeds in the UK. Forestry 76:285–298

Gupta RP, Kalia A, Kapoor S (2007) Biopesticides. In: Bioinoculants: a step towards sustainable agriculture. New India Publishing Agency, New Delhi, pp 223–256

Hardin B (2002) Giving baculoviruses a better edge. Agricul Res (January), pp 14–15

Harrier LA (2001) The arbuscular mycorrhizal symbiosis: a molecular review of the fungal dimension. J Exp Bot 52:469–478

Ishibashi N, Choi DR (1991) Biological control of soil pests by mixed application of entomopathogenic and fungivorous nematodes. J Nematol 23:175–181

Kaaya GP, Hassan S (2000) Entomogenous fungi as promising biopesticides for tick control. Exp Appl Acarol 24:913–926

Kaya HK, Gaugler R (1993) Entompathogenic nematodes. Annu Rev Entomol 38:181–206

Khachatourians GG, Velencia EP, Miranpuri GS (2002) *Beauveria bas*siana and other entomopathogenic fungi in management of insect pests. In: Opender Kaul, Dhaliwal GS (eds) Microbial Biopesticides. Taylor and Francis, New York/London, pp 239–275

Kilic-Ekici O, Yuen GY (2004) Comparison of strains of *Lysobacter enzymogenes* and PGPR for induction of resistance against *Bipolaris sorokiniana* in tall fescue. Biol Control 30:446–455

Kim BS, Lee JY, Hwang BK (2000) *In vivo* control and in vitro antifungal activity of rhamnolipid B, a glycilipid antibiotic, against *Phytophthora capsici* and *Colletotrichum orbiculare*. Pest Manage Sci 56:1029–1035

Lacey LA, Frutos R, Kaya HK, Vails P (2001) Insect pathogens as biological control agents: do they have a future? Biol Control 21:230–248

Lahdenpera ML (2003) *Streptomyces* biofungicides in seed application. Veradera Infoletter 12:1–3

Lasa R, Williams T, Caballero P (2008) Insecticidal properties and microbial contaminants in a *Spodoptera exigua* multiple nucleopolyhedrovirus (SeMNPV, Baculoviridae) formulation stored at different temperatures. J Econ Entomol 101:42–49

Lasa R, Moreno I, Caballero P, Williams T (2009) Application of juvenile hormone analogue and optical brightener technologies to the production of *Spodoptera frugiperda* multiple nucleopolyhedrovirus. IOBC Wprs Bull 45:153–156

Looper JE, Henkels MD (1999) Utilization of heterologous siderophores enhances levels of iron available to *Pseudomonas putida* in the rhizosphere. Appl Environ Microbiol 65:5357–5363

Marrone PG (2002) An effective biofungicide with novel modes of action. Pesticide Outlook (October):193–194

Maurhofer M, Hase C, Meuwly P, Metraux JP, Defago G (1994) Induction of systemic resistance of tobacco to tobacco necrosis virus by the root-colonizing *Pseudomonas fluorescens* strain CHA0: influence of the gacA gene and of pyoverdine production. Phytopathology 84:139–146

Milner JL, Stohl EA, Handelsman J (1996) Zwittermicin A resistance gene from *Bacillus cereus*. J Bacteriol 178:4266–4272

Montesinos E (2000) Pathogenic plant-microbe interactions. What we know and how we benefit. Int Microbiol 3:69–70

Moscardi F (1999) Assessment of the application of baculoviruses for control of Lepidoptera. Annu Rev Entomol 44:257–289

Narayanan K (2002) Microbial control of insect pests. Role of genetic engineering and tissue culture. In: Dhaliwal GS, Koul O (eds) Microbial biopesticides. Taylor and Francis, New York/ London, pp 117–180

Neale MR (1997) Biopesticides-harmonisation of regulation requirements within EU Directive 91/ 414: an industry view. Bull OEPP 27:89–93

Nielsen TH, Christopheresen C, Anthoni U, Sorensen J (1999) Viscosinamide, a new cyclic depsipeptide with surfactant and antifungal properties produced by *Pseudomonas fluorescens* DR54. J Appl Microbiol 87:80–90

Nielsen TH, Thrane C, Christophersen C, Anthoni U, Sorensen J (2000) Structure, production characteristics and fungal antagonism of tensin-a new antifungal cyclic lipopeptide from *Pseudomonas fluorescens* strain 96.578. J Appl Microbiol 89:992–1001

Nielsen TH, Sorensen D, Tobiasen C, Andersen JB, Christophersen C, Givskov M, Sorensen J (2002) Antibiotic and biosurfactant properties of cyclic lipopeptides produced by fluorescent *Pseudomonas* spp. from the sugar beet rhizosphere. Appl Environ Microbiol 68:3416–3423

Nowak J, Shulaev V (2003) Priming for transplant stress resistance in in vitro propagation. In Vitro Cell Dev Biol Plant 39:107–124

O'Callaghan M, Glare TR (2001) Impacts of transgenic plants and microorganisms on soil biota. N Z Plant Protec 54:105–110, cited online www.hortnet.co.nz/publications/nzpps

Ortiz Ribbing LM, Williams M (2006) Potential of *Phomopsis amaranthicola* and *Microsphaeropsis amaranthi*, as bioherbicides for several weedy *Amaranthus* species. Crop Prot 25:39–46

Pell JK, Eilenberg J, Hajek AE, Steinkraus DC (2001) Biology, ecology and pest management potential of entomophthorales. In: Butt TM, Jackson C, Magan N (eds) Fungi as biocontrol agents: progress, problems and potential. CAB International, Wallingford, pp 71–153

Postma J, Montanari M, van den Boogert PHJF (2003) Microbial enrichment to enhance the disease suppressive activity of compost. Eur J Soil Biol 39:157–163

Rodas VL, Marques FH, Honda MT, Soares DM, Jorge SAC, Antoniazzi MM, Medugno C, Castro MEB, Ribeiro BM, Souza ML, Tonso A, Pereira CA (2005) Cell culture derived AgMNPV bioinsecticide: biological constraints and bioprocess issues. Cytotechnology 48:27–39

Saikkonen K, Faeth SH, Helander M, Sullivan TJ Fungal endophytes: a continuum of interactions with host plants. Annu Rev Ecol Syst 29:319–343

Sankarama A (1999) Integrated pest management: looking back and forward. Curr Sci 77

Sessitsch A, Reiter B, Berg G (2004) Endophytic bacterial communities of field-grown potato plants and their plant growth-promoting and antagonistic abilities. Can J Microbiol 50:239–249

Shah PA, Pell JK (2003) Entomopathogenic fungi as biological control agents. Appl Microbiol Biotechnol 61:413–423

Sheeba G, Seshadri S, Raja N, Janarthanan S, Ignacimuthm S (2001) Efficacy of *Beauveria bassiana* for the control of rice weevil *Sitophilus oryzae* (L) (Coleoptera:Cuculionidae). Appl Ent Zool 36:117–120

Sigsgaard L (2002) Early season natural biological control of insect pests in rice by spiders – and some factors in the management of the cropping system that may affect this control. In: European Arachnology 2000 (eds. S. Toft and N. Scharff), Aarhus University Press, Aarhus, pp 57–64

Simon O, Williams T, Lopez-Ferber M, Taulemesse JM, Caballero P (2008) Population genetic structure determines the speed of kill and occlusion body production in *Spodoptera frugiperda* multiple nucleopolyhedrovirus. Biol Control 44:321–330

Smart GC Jr (1995) Entomopathogenic nematodes for the biological control of insects. J Nematol 27:529–534

Sun XL, Peng HY (2007) Recent advances in biological control of pest insects by using viruses in China. Virol Sin 22:158–162

Tanada Y, Kaya HK (1993) Insect pathology. Academic/Javanovich, Narcourt Brace/San Diego

Timms-Wilson TM, Ellis RJ, Renwick A, Rhodes DJ, Mavrodi D, Weller DM, Thomashow LS, Bailey MJ (2000) Chromosomal insertion of phenazine-1-carboxylic-acid biosynthetic pathway enhances efficacy of damping-off disease control by *Pseudomonas fluorescens*. Mol Plant Microbe Interact 13:1293–1300

Tiourebaev KS, Nelson S, Zidack NK, Kaleyva T, Pilge AL, Anderson TW, Sands DC (2000) Amino acid excretion enhances virulence of bioherbicides. In: Neal R. Spencer (ed) Proceedings of the 10th international symposium on biological control of weeds, 4–14 July 1999, Montana State University, Bozeman, Montana, USA, pp 295–299

Vail PV, Hostetter DL, Hoffmann DF (1999) Development of the multi-nucleocapsid nucleopolyhedro viruses (MNPVs) infectious to loopers (Lepidopetra, Noctuidae: Plusiinae) as microbial controls agents. Int Pest Manage Revs 4:231–257

Van Driesche R, Hoddle M, Center TD (2008) Control of pests and weeds by natural enemies: an introduction to biological control. Blackwell, Oxford, p 473

van Lenteren JC (2005) Early entomology and the discovery of insect parasitoids. Biol Control 32:2–7

van Lenteren JC (2006) Ecosystem services to biological control of pests: why are they ignored? Proc Neth Entomol Soc Meet 17:103–111

Van Loon LC, Bakker PAHM, Pieterse CMJ (1998) Systemic resistance induced by rhizosphere bacteria. Ann Rev Phytopathol 36:453–483

Van Peer R, Niemann GJ, Schippers B (1991) Induced resistance and phytoalexin accumulation in biological control of *Fusarium* wilt of carnation by *Pseudomonas* sp. strain WCS417r. Phytopathology 81:728–734

Vandermeer J (1995) The ecological basis of alternative agriculture. Ann Rev Ecol Syst 26:201–224

Verma J, Dubey NK (1999) Prospective of botanical and microbial products as pesticides of tomorrow. Curr Sci 76:172–178

Walker DR, Narvel JM, Boerma HR, All JN, Parrott WA (2004) A QLT that enhances and broadens Bt insect resistance in soybean. Theor Appl Genet 109:1051–1057

Whalon ME, McGaughey WH (1998) *Bacillus thuringiensis*: use and resistance management. In: Ishaaya I, Degheele D (eds) Insecticides with novel modes of action: mechanism and application. Springer, Berlin, pp 106–137

Wilson D (1995) Fungal endophytes which invade insect galls: insect pathogens, benign saprophytes, or fungal inquilines? Oecologia 103:255–260

Chapter 11
Induced Systemic Resistance in Biocontrol of Plant Diseases

Sudhamoy Mandal and Ramesh C. Ray

11.1 Introduction

Induced resistance (IR) is the general term for all types of elicited responses that lead to enhanced protection against disease including both locally and systemically induced resistance (Hammerschmidt et al. 2001). One of the classic forms of induced resistance is systemic acquired resistance (SAR) controlled by a signaling pathway that depends on endogenous accumulation of salicylic acid (SA) (Durrant and Dong 2004). SAR is defined as an induced system of resistance triggered by pathogens or elicitors, which give long-lasting protection against a broad spectrum of pathogens (Chester 1933; Durrant and Dong 2004). It has been demonstrated that SAR can be induced by the plant hormone SA in *Arabidopsis*, tobacco, cucumber, rice, beans, and tomato (Malamy et al. 1990; Sticher et al. 1997; Edgar et al. 2006; Mandal et al. 2009). SAR is characterized by the activation of SAR genes, including genes that encode pathogenesis-related (PR) proteins (Linthorst 1991), which are often used as markers for the state of induced resistance. Plant resistance can be induced by the application of synthetic compounds such as 2,6-dichloroisonicotinic acid (INA) (Métraux et al. 1991) and benzothiadiazole-7-carbothioic acid (BTH) (Lawton et al. 1996). Induction of systemic resistance to pathogens is a promising approach for controlling plant diseases. In addition to SAR, an alternative approach to inducing systemic resistance was reported for the first time in common bean against *Pseudomonas syringae* pv. *phaseolicola* (Alström 1991), in carnation against *Fusarium oxysporum* f. sp. *dianthi* (Van Peer et al. 1991), and in cucumber against *Colletotrichum orbiculare* (Wei et al. 1991). These reports unequivocally established that some strains of plant growth-promoting rhizobacteria (PGPR) can induce plant resistance against different pathogens. This phenomenon is termed

S. Mandal (✉)
Plant Pathology Laboratory, Central Horticultural Experiment Station, Aiginia, Bhubaneswar 751019, India
e-mail: sudhamoy1331971@rediffmail.com

R.C. Ray
Regional Centre, Central Tuber Crops Research Institute, Dumduma, Bhubaneswar 751019, India

A. Singh et al. (eds.), *Bioaugmentation, Biostimulation and Biocontrol*, Soil Biology 28,
DOI 10.1007/978-3-642-19769-7_11, © Springer-Verlag Berlin Heidelberg 2011

induced systemic resistance (ISR). The bacteria that are derived from and exert stimulatory effect on the plant root are generally designated as PGPR. Mechanisms of biological control by which rhizobacteria can promote plant growth indirectly by reducing the level of disease include antibiosis, ISR, and competition for nutrients and niches (Lugtenberg and Kamilova 2009). In contrast to SAR, ISR develops as a result of the colonization of plant roots by PGPR and is mediated by a jasmonate- or ethylene-sensitive pathway (Pieterse et al. 1998). SAR and ISR are two forms of induced resistance where plant defenses are preconditioned by prior infection or treatment that results in resistance against subsequent challenge by a pathogen. ISR-expressing plants are primed for enhanced expression of predominantly JA- and ET-regulated genes upon pathogen infection (Verhagen et al. 2004; Cartieaux et al. 2008).

Before discovery of the ISR phenomenon, the PGPR (mainly fluorescent *Pseudomonas* spp.) had been used for their ability to control soil-borne pathogens through such mechanisms as competition for nutrients, siderophore-mediated competition for iron, or antibiosis (Bakker et al. 1991; Schippers 1992; Thomashaw and Weller 1995). So far PGPR-mediated ISR has been demonstrated to occur in several plant species and it has been shown to be effective against different types of pathogens including fungi, bacteria, and viruses. ISR confers on the plant an enhanced defensive capacity against pathogens (Van Loon and Bakker 2005). This elevated defensive capacity of the plant is manifested as a reduction in the rate of disease development upon infection with a pathogen, resulting in fewer diseased plants or in lesser disease severity. Unlike SAR which is dependent on SA, ISR is dependent on jasmonic acid (JA) and ethylene (ET) signaling in the plant (Van Loon 2007). In inducing ISR, contrary to biocontrol mechanisms, extensive colonization of the root system by the organism is not required (Dekkers et al. 2000; Kamilova et al. 2005). In contrast to R-gene-mediated resistance, it is not specific but active against all types of pathogens, as well as against several nematodes and insects. Once induced, plants may remain protected for a considerable part of their lifetime, indicating that when the state of ISR has been triggered in the plant, it is rather stable (Van Loon et al. 1998).

11.2 Induction of Systemic Resistance by *Pseudomonas* spp.

Several root-colonizing microorganisms are known to suppress diseases by ISR in plants. Among these microorganisms, particularly important is systemic resistance induced by nonpathogenic PGPR belonging to the genus *Pseudomonas*. *Pseudomonas fluorescens* strain WCS417 was applied to the roots of carnation, and plants were challenged 1 week later by stem inoculation with *F. oxysporum* f. sp. *dianthi*. As a result, both the number of diseased plants and disease severity were significantly reduced compared to plants not treated with the bacteria (Van Peer et al. 1991). Similar observations have been made in cucumber (Wei et al. 1991), tobacco (Maurhofer et al. 1994), radish (Leeman et al. 1995), *Arabidopsis* (Pieterse et al. 1996; Van Wees et al. 1997), tomato (Duijff et al. 1997), and bean (Bigirimana and Höfte 2002) against various plant pathogens. Like WCS417, *P. fluorescens* FPT9601-T5 was found to trigger ISR in *Arabidopsis* against *P. syringae* pv. *tomato.* Using an Affymetrix GeneChip probe array containing approximately 22,800 genes, Wang et al. (2005) detected 95 and 105 genes that were up- and downregulated, respectively, in leaves of soil-grown plants that had been root-dipped in a suspension of the bacteria.

Another PGPR with the ability to induce systemic resistance is *Pseudomonas aeruginosa* 7NSK2. This strain produces pyoverdin, pyochelin, and SA under iron-limiting conditions (Buysens et al. 1996). The production of SA by 7NSK2 seems to be required for the induction of systemic resistance in tobacco against tobacco mosaic virus (TMV) (De Meyer et al. 1999) and has also been implicated in the systemic resistance induced by this strain against *Botrytis cinerea* in bean (De Meyer and Höfte 1997). However, Audenaert et al. (2002) described that production of the phenazine compound pyocyanin, together with the SA-containing siderophore pyochelin, is required for ISR by 7NSK2 against *B. cinerea* in tomato. *P. aeruginosa* strain 7NSK2 is a producer of SA, and its induction of resistance against *B. cinerea* in bean was reported to be reduced when it had lost the ability to produce SA (De Meyer and Höfte 1997). Recently, it has been reported that *P. aeruginosa* 7NSK2 induces ISR in rice against *Pyricularia grisea* through production of pyocyanin (De Vleesschauwer et al. 2006).

P. fluorescens WCS374 induces ISR against *Ralstonia solanacearum* in *Eucalyptus*, and the ISR by WCS374 in *Eucalyptus urophylla* is triggered by its pseudobactin siderophore (Ran et al. 2005a). This strain was earlier reported to trigger ISR against *Fusarium* wilt in radish (Leeman et al. 1995). *Pseudomonas putida* WCS358 was originally isolated from potato tuber surface. The strain WCS358 induces ISR against *R. solanacearum* in *Eucalyptus*, and the ISR by WCS374 in *E. urophylla* is triggered by its pseudobactin siderophore (Ran et al. 2005a). This strain was found to induce ISR in *Arabidopsis* against *P. syringae* pv. *tomato* (Bakker et al. 2003; Meziane et al. 2005), tomato against *B. cinerea*, and bean against *B. cinerea* and *Colletotrichum lindemuthianum* (Meziane et al. 2005).

The biocontrol bacterium, *P. fluorescens* WCS365, induces ISR in tomato (Kamilova et al. 2005). De Weert et al. (2002) reported that this biocontrol bacterium shows strong chemotaxis toward the major tomato root exudate component, citric acid. In roots of the legume species *Medicago trunculata*, Sanchez et al. (2005) found 58 genes to be upregulated in response to colonization by the growth-promoting strain *P. fluorescens* C7R12, a number in line with that found by Verhagen et al. (2004) in *Arabidopsis* roots colonized by strain *P. fluorescens* WCS417. Using cDNA microarrays representing approximately 14,300 genes, Cartieaux et al. (2003) monitored gene expression in both leaves and roots of axenic *Arabidopsis* plants infected by resistance-inducing *Pseudomonas thivervalensis* strain MLG45. Plants colonized by this rhizobacterium showed decreased photosynthetic rates and reduced growth, indicating that *P. thivervalensis* acted as a minor pathogen rather than a PGPR. This conclusion was supported by the changes in gene expression observed. Kim et al. (2004), using subtractive hybridization, did not detect any changes in leaves of cucumber plants grown in sterilized soilless growing medium from seeds coated with *Pseudomonas chlororaphis* O6, a strain that was effective in inducing systemic resistance against target leaf spot caused by *Corynespora cassiicola.*

Induction of resistance by *P. fluorescens* strain CHA0 in tobacco against Tobacco Necrosis Virus (TNV) was reported to be associated with PR protein accumulation, suggesting that nonpathogen-induced ISR and pathogen-induced SAR share similar mechanisms (Maurhofer et al. 1994). However, the induced plants were slightly stunted. Strain CHA0 is a producer of the antibiotics diacetylphloroglucinol and pyrrolnitrin, as well as of HCN, substances with toxicity to plants. However, PR proteins did not accumulate in radish plants expressing ISR elicited by *P. fluorescens* strain WCS417r (Hoffland et al. 1995; 1996). Moreover, Pieterse et al. (1996) demonstrated that in *Arabidopsis*, ISR induced by WCS417r was not associated with PR gene activation and was elicited in transgenic *Arabidopsis* plants unable to accumulate SA. This indicates that in contrast to pathogen-induced SAR, WCS417r-mediated ISR is controlled by an SA-independent signaling pathway. In *Arabidopsis-Peronospora parasitica*, this strain induces ISR through the antifungal determinant 2,4-diacetyl phloroglucinol (Iavicoli et al. 2003).

P. fluorescens strains WCS374 and WCS417 are likewise able to produce SA under iron-limiting conditions. Mutants that have lost the O-antigenic side chain of the lipopolysaccharide (LPS) no longer induced resistance in radish under iron-sufficient conditions, but did so in the presence of an iron-chelating compound, indicating an additional bacterial determinant to be active under low-iron conditions (Leeman et al. 1996). ISR elicited by almost all strains was found to be SA-independent, also by strains such as *P. fluorescens* CHA0 and *Serratia marcescens* 90-166, which can themselves produce SA as an additional siderophore (Van Loon and Bakker 2005). Only the systemic resistance induced by *P. aeruginosa* strain 7NSK2 was SA-dependent, since 7NSK2 no longer induced resistance in *NahG* tobacco (De Meyer et al. 1999) and *NahG* tomato plants (Audenaert et al. 2002). However, it is not associated with pathogenesis-related protein expression (De Meyer et al. 1999). Whereas generally rhizobacteria are not dainty in colonizing roots of different plant species, the perception by the plant of bacterial determinants that trigger ISR appears to be quite specific (Bakker et al. 2003; Meziane et al. 2005; Van Loon and Bakker 2005). Apparently, one or more bacterial components need to be recognized by specific plant receptors. Of the three strains, *P. putida* WCS358, *P. fluorescens* WCS374, and *P. fluorescens* WCS417, none is active in eliciting ISR in all out of six plant species, even though levels of root colonization are similar. Remarkably, in *Arabidopsis* strain WCS374 was differentially active in eliciting ISR against different pathogens depending on bioassay conditions, suggesting that the type and effectiveness of the systemic resistance that is induced by this rhizobacterium is variable.

11.3 Induction of Systemic Resistance by *Bacillus* spp.

Another important nonpathogenic PGPR-eliciting ISR in plants belongs to the genus *Bacillus*. Although commercialization of PGPR is mainly proceeding with *Bacillus* spp. rather than *Pseudomonas* spp., the preponderance of research on

PGPR as elicitors of growth promotion or ISR employs PGPR strains that are fluorescent pseudomonads (Kloepper et al. 2004). Several strains of the species *Bacillus amyloliquefaciens*, *B. subtilis*, *B. pasteurii*, *B. cereus*, *B. pumilus*, *B. mycoides*, and *B. sphaericus* elicit ISR against various diseases on a variety of hosts. Protection of plants from diseases by *Bacillus* spp. has been reported against fungal, bacterial, and viral pathogens, as well as against root-knot nematodes (Choudhary and Johri 2009). Here, few examples of work on ISR elicited by *Bacillus* spp. in different crops against pathogens are cited.

Systemic resistance was reported to be induced by *B. subtilis* AF1 against *Aspergillus niger* on peanut (*Arachis hypogaea*). Inoculation of peanut seeds with *B. subtilis* AF1 in soil containing *A. niger* resulted in a significant reduction in the incidence of crown rot of seedlings. This biological control was associated with the induction of lipoxygenase activity in seedlings, suggesting that AF1 elicited ISR in peanut (Sailaja et al. 1997). *B. subtilis* has been reported to induce ISR against some vectors of plant viruses, which may augur well for management of vector-borne virus diseases. Inoculation of tomato roots with a *B. subtilis* strain BEB-DN isolated from the rhizosphere of potato plants was found to generate an ISR response against virus-free *Bemisia tabaci*. However, the observed resistance appeared to represent a combination of JA-dependent and JA-independent responses, since the retardation effect by this strain on *B. tabaci* development was still effective in the highly susceptible *spr2* tomato mutants with an impaired capacity for JA biosynthesis (Valenzuela-Soto et al. 2010).

Two strains of *B. pumilus* (strains 203-6 and 203-7) and one of *B. mycoides* (strain Bac J) reduced the severity of *Cercospora* leaf spot of sugar beet through the induction of systemic resistance (Bargabus et al. 2002; 2004). *B. pumilus* strain SE34 could, when incorporated into the potting medium, provide tomato plants systemic protection against late blight caused by *Phytophthora infestans* (Yan et al. 2003). Selected *Bacillus* PGPR strains emit volatile compounds that can trigger ISR in plants. In *Arabidopsis* seedlings exposed to bacterial volatile blends from *B. subtilis* GB03 and *B. amyloliquefaciens* IN937a, disease severity by the bacterial pathogen *Erwinia carotovora* subsp. *carotovora* was significantly reduced compared with seedlings not exposed to bacterial volatiles before pathogen inoculation (Ryu et al. 2004).

Bacillus spp. could reduce the severity of blue mold of tobacco caused by *Peronospora tabacina*. ISR was elicited by *B. pasteurii* C-9 and *B. pumilus* SE34 and T4 strains when the time interval between the last application of bacteria and challenge inoculation with *P. tabacina* was 6 weeks. The results indicated an association between the capacity of the tested strains of *Bacillus* spp. to promote growth and elicit ISR (Zhang et al. 2004). Another *Bacillus* sp., *B. vallismortis* strain EXTN-1, has been proved in eliciting ISR in several crops against many pathogens, including viruses. *B. vallismortis* strain EXTN-1 induced systemic resistance in potato plants against Potato Virus Y and X, resulting in significant disease suppression in the field (Park et al. 2006).

Recently, it has been reported that *B. cereus* strain BS 03 was able to induce systemic resistance in pigeon pea against fusarial wilt. However, induction of ISR by BS 03 was better in combination with a rhizobial strain RH 2 (Dutta et al. 2008).

11.4 Induction of Systemic Resistance by Other Microorganisms

Besides *Pseudomonas* and *Bacillus*, there have been an increasing number of microorganisms (fungi and bacteria) inducing resistance in plants against a number of pathogens.

11.4.1 Trichoderma *spp.*

Species of *Trichoderma* have the ability to control numerous foliar, root, and fruit pathogens and even invertebrates such as nematodes. They also have many other capabilities such as ameliorating abiotic stresses, alleviating physiological stresses, enhancing nutrient uptake in plants, increasing nitrogen-use efficiency in crops, and improving photosynthetic efficiency. All of these capabilities are a consequence of their abilities to reprogram plant gene expression, probably through activation of a limited number of general plant pathways (Shoresh et al. 2010). Some rhizosphere-competent *Trichoderma* strains colonize entire root surfaces of plants with morphological features reminiscent of those seen during mycoparasitism. *Trichoderma* strains capable of establishing such interaction induce metabolic changes in plants that increase resistance to a wide range of plant-pathogenic microorganisms and viruses (Harman et al. 2004). What was probably the first clear demonstration of induced resistance by *Trichoderma* was published in 1997 by Bigirimana et al. (1997). They showed that treating soil with *Trichoderma harzianum* strain T-39 made leaves of bean plants resistant to diseases that are caused by the fungal pathogens *B. cinerea* and *C. lindemuthianum*, even though T-39 was present only on the roots and not on the foliage. The same group extended their findings from *B. cinerea* to other dicotyledonous plants (De Meyer et al. 1998). Similar studies have now been carried out with a wide range of plants, including both monocotyledons and dicotyledons, and with different *Trichoderma* species and strains. In cucumber, root colonization by *Trichoderma asperellum* strain T-203 causes an increase in phenolic glucoside levels in leaves, which are strongly inhibitory to a range of bacteria and fungi spp. (Yedidia et al. 2003). The systemic response in plants occurs through the JA/ET signaling pathway in a manner similar to the rhizobacteria-induced systemic resistance (Shoresh et al. 2005). The protection afforded by the biocontrol agent is associated with the accumulation of mRNA of two defense genes: the phenylpropanoid pathway gene phenylalanine ammonia lyase (PAL) and the lipoxygenase pathway gene hydroxyperoxide lyase (HPL) (Yedidia et al. 2003). In *Trichoderma*-treated cucumber seedlings upon pathogen challenge, increased levels of other defense-related plant enzymes, such as peroxidases, chitinases, and β-1,3-glucanases, have been recorded (Shoresh et al. 2005). This potentiation in the gene expression enables *Trichoderma*-treated plants to be more resistant to subsequent pathogen infection. The MAPK signal transduction pathways, of both the plant and *Trichoderma*, are important for the induction of systemic resistance (Viterbo et al. 2005).

11.4.2 Serratia *spp.*

S. marcescens 90-166 is another microorganism that can induce resistance to fungal, viral, and bacterial pathogens in cucumber such as *C. orbiculare*, *F. oxysporum* f. sp. *cucumerinum*, Cucumber Mosaic Virus, *P. syringae* pv. *lachrymans*, and *Erwinia tracheiphila* (Höfte and Bakker 2007). High levels of disease control were achieved in cucumber with *S. marcescens* 90-166 providing 89% control of *E. tracheiphila* (Zehnder et al. 2001). *S. marcescens* 90-166 is known to produce SA, but mutants deficient in SA production retained ISR activity in cucumber against *C. orbiculare* (Press et al. 1997). Analysis of the reaction of tomato to the ISR-eliciting strain *Serratia liquefaciens* MG1, using a macroarray containing cDNA probes of 70 defense-related and signaling genes, revealed enhanced expression of 12 genes. Seven of those coded for PRs, whereas the others were involved in oxidative stress, ethylene signaling, or metabolism (Shuhegger et al. 2006).

11.4.3 Nonpathogenic Strains of F. oxysporum

Several reports have documented the induction of resistance to *Fusarium* wilt by using either nonpathogenic strains of *F. oxysporum* (npFo), as in the case of cucumber (Mandeel and Baker 1991; Benhamou et al. 2002), chickpea (Hervás et al. 1995; Kaur and Singh 2007), tomato (Fuchs et al. 1999), and banana (Nel et al. 2006), or formae speciales of *F. oxysporum*, such as f. sp. *melonis* in cucumber (Gessler and Kuć 1982; Freeman et al. 2002) and f. sp. *dianthi* in tomato (Kroon et al. 1991). However, in contrast to published results, a nonpathogenic mutant of a pathogenic strain of *F. oxysporum* f. sp. *melonis* (rev127) did not protect muskmelon when coinoculated with the parental strain. Even another nonpathogenic mutant of the same pathogen (Rev127) was also unable to protect the host plant (L'Haridon et al. 2007). The npFo strain Fo47 was shown to induce systemic resistance to *Fusarium* wilt in tomato (Fuchs et al. 1997). In another study, npFo isolates were shown to be effective inducers of systemic acquired resistance and pathogen defense responses in *Asparagus officinalis*. This induced resistance by np*Fo* was associated with the activation of defense-related enzymes such as peroxidase and phenylalanine ammonia lyase and accumulation of lignin (He et al. 2002). Like *Trichoderma*, npFo has several different mechanisms for the direct antagonism of plant pathogens and the induction of plant resistance, and all of these mechanisms are probably important in biocontrol (Fravel et al. 2003). Soil-borne protective strains of *F. oxysporum* are usually effective through an association of modes of action including competition for nutrients in the rhizosphere, competition for root colonization, and induced resistance (Alabouvette et al. 2007).

11.4.4 Pythium oligandrum

Another mycoparasite receiving considerable attention as a potential biocontrol agent of a number of soil-borne plant pathogens is *Pythium oligandrum*. Tomato plants previously inoculated with *P. oligandrum* afford increased resistance to *F. oxysporum* f. sp. *radicis-lycopersici* attack. This resistance is mainly associated with a strong antagonistic activity in the rhizosphere and in planta as well as with the induction of structural and biochemical barriers that adversely affect pathogen growth and development. These observations provide the first convincing evidence that *P. oligandrum* has the potential to induce plant defense reactions in addition to acting as a mycoparasite (Benhamou et al. 1997). Pretreatment of tomato plants with oligandrin, the elicitin-like protein produced by the mycoparasite *P. oligandrum*, reduces disease incidence caused by *F. oxysporum* f. sp. *radicis-lycopersici* by sensitizing the plants to elaborate an efficient defense strategy (Benhamou et al. 2001). Evidence suggests that the cell wall protein fractions isolated from *P. oligandrum* display the ability to induce resistance in sugar beet against *Rhizoctonia solani*, which is one of the pathogens of root rot and damping-off, and in wheat against *Fusarium graminearum*, which is one of the pathogens of head blight (Takenaka et al. 2003).

11.4.5 Penicillium oxalicum

It was reported that treatment of tomato plants with conidia of *Penicillium oxalicum* induced resistance against tomato wilt. *P. oxalicum* and *F. oxysporum* f. sp. *lycopersici* were placed at separate sites on tomato plants or in soil, avoiding a direct interaction between the fungi (De Cal et al. 1997). An aqueous extract of the mycelium of *Penicillium chrysogenum* induced early defense-related responses such as an extracellular alkalinization in cell cultures and ethylene production in leaf slices of numerous mono- and dicotyledon plant species, including *Arabidopsis thaliana*, tomato, tobacco, and rice. The authors concluded that *P. chrysogenum* contains at least one unidentified elicitor, most likely a protein or a glycoprotein, inducing resistance via signal transduction pathways different from classical SA/NPR1- or JA/ET-dependent pathways (Thuerig et al. 2006).

11.4.6 Rhizoctonia *sp.*

Similarly, the fungal genus *Rhizoctonia* contains both plant-pathogenic and nonpathogenic species and strains, with those that are nonpathogenic frequently acting as biocontrol agents; again, these organisms induce plant resistance (Hwang and Benson 2003). Induction of systemic resistance was demonstrated by binucleate *Rhizoctonia* to a foliar pathogen *B. cinerea* of geranium in an integrated management program (Olson and Benson 2007). The obligate plant-symbiotic mycorrhizal fungi might initially suppress plant resistance during the infection process (Guenoune et al. 2001), but enhanced systemic resistance sometimes occurs once mycorrhizal fungi are established in plant roots (Pozo et al. 2002).

11.4.7 Colletotrichum magna

A nonpathogenic mutant of *Colletotrichum magna* (path-1) was shown to protect watermelon and cucumber seedlings from anthracnose disease elicited by wild-type *C. magna* (Freeman and Rodriguez 1993). The mechanism(s) that allows path-1 to protect plants against disease appears to involve an interaction between the mutant and the plant defense system. The path-1 mutant may result in the development of a novel, long-term biocontrol strategy for plant protection (Redman et al. 1999).

11.4.8 Others

Induced resistance was found to be a mechanism for biological control of leaf spot, caused by *Bipolaris sorokiniana*, in tall fescue (*Festuca arundinacea*) using the bacterium *Lysobacter enzymogenes* strain C3 (Kilic-Ekici and Yuen 2003). Application of live or heat-killed cells to tall fescue leaves resulted only in localized resistance confined to the treated leaf, whereas treatment of roots resulted in systemic resistance expressed in the foliage. Induced resistance by C3 was not host or pathogen specific; foliar application of heat-killed C3 cells controlled *B. sorokiniana* on wheat and also was effective in reducing the severity of brown patch, caused by *R. solani*, on tall fescue. *Piriformospora indica*, a model organism for species of the recently described order Sebacinales, increases biomass and grain yield of crop plants. In barley, the endophyte induces root resistance against *Fusarium culmorum*, one of the fungal species causing head blight, and systemic resistance to barley powdery mildew *Blumeria graminis* f. sp. *hordei* via an unknown mechanism probably independent of salicylate or jasmonate accumulation. (Waller et al. 2005). In contrast to AM fungi, *P. indica* colonizes *Arabidopsis*, and recent results provide evidence that the fungus induces systemic resistance in this model plant similar to the resistance provided to the powdery mildew fungus in barley (Deshmukh et al. 2006; Stein et al. 2008; Molitor and Kogel 2009).

11.5 Mechanisms of Induced Systemic Resistance

PGPR may activate inducible defense mechanisms in the plant in a way similar to pathogenic microorganisms. Such mechanisms can include reinforcement of plant cell walls, production of antimicrobial phytoalexins, synthesis of PRs

(Hammond-Kosack and Jones 1996), as well as an enhanced capacity to express these defense responses upon challenge inoculation with a pathogen, a mechanism known as "sensitization," "priming," or "potentiation" (Conrath et al. 2006). Activation of defense reactions suggests that even a beneficial rhizobacterium may be perceived by the plant as a potential threat and that such perception involves production of resistance-eliciting compounds that act mechanistically similar to elicitors produced by plant pathogenic fungi and bacteria. Plants possess sensitive mechanisms to perceive both fungi and bacteria through conserved components that are specific to their kingdoms and act as general elicitors. These are commonly referred to as pathogen-associated molecular patterns (PAMPs) (Nürnberger and Lipka 2005).

Van Peer and Schippers (1992) and Leeman et al. (1995) showed that the O-antigenic side chain of the outer membrane LPS of strain WCS417r is the main determinant for the induction of ISR against *Fusarium* wilt disease in both carnation and radish. A bacterial mutant lacking the O-antigenic side chain did not induce resistance, whereas LPS-containing cell walls and purified LPS of WCS417r induced ISR to the same extent as living bacteria. Other bacterial determinants suggested to contribute to ISR are siderophores and SA (Leeman et al. 1996; Maurhofer et al. 1994). It has been demonstrated that ISR-inducing fluorescent *Pseudomonas* spp. are differentially active in eliciting ISR in *Arabidopsis*. Furthermore, it was found that in contrast to what was observed in carnation and radish, the LPS of WCS417r played only a minor role in the elicitation of ISR in *Arabidopsis*, indicating that WCS417r possessed more than a single ISR-inducing determinant (Van Wees et al. 1997). Low inoculum densities of WCS417r or inoculum cultivated at elevated temperatures induced resistance against a broad spectrum of pathogens with different parasitic habits (Ran et al. 2005b). This wide range of effectiveness of WCS374r-elicited ISR strongly suggests that multiple resistance responses are involved. Recent studies by De Vleesschauwer et al. (2008) demonstrated conclusively that *P. fluorescens* WCS374r-elicited ISR in rice against *Magnaporthe oryzae* is based on Pseudobactin-mediated priming for a salicylic acid-repressible multifaceted defense response (see Table 11.1 for bacterial determinants).

The modes of action that play a role in disease suppression by PGPR include siderophore-mediated competition for iron, antibiosis, production of lytic enzymes, and ISR (Bakker et al. 2007). The ability to act as bioprotectants via ISR has been demonstrated for both rhizobacteria and bacterial endophytes, and considerable progress has been made in elucidating the mechanisms of plant–PGPR–pathogen interaction (Compant et al. 2005). Several bacterial components have been reported to induce ISR, such as flagella, lipopolysaccharides (LPS), salicylic acid, and siderophores (Van Loon 2007). Besides these, cyclic lipopeptides (Ongena et al. 2007), the antifungal factor 2,4-diacetyl phloroglucinol (Phl) (Iavicoli et al. 2003), the signal molecule *N*-acyl homoserine lactone (AHL) (Shuhegger et al. 2006), pyochelin and pyocyanin (Audenaert et al. 2002), volatile blends produced by *B. subtilis* GB03, and also the individual volatiles acetoin and 2,3-butanediol (Ryu et al. 2003) have been found to elicit ISR in plants against various pathogens. Now it has been demonstrated that *P. fluorescens* CHA0 and *P. aeruginosa* 7NSK2 induce ISR in grapevine against *B. cinerea* and trigger an oxidative burst and phytoalexin (i.e., resveratrol

Table 11.1 Examples of bacterial determinants involved in induced systemic resistance by plant growth-promoting rhizobacteria (PGPR) in different host–pathogen combinations

PGPR strain	Determinant(s)	Host–pathogen	References
Pseudomonas fluorescens CHA0	Pseudobactin siderophore	Tobacco–Tobacco mosaic virus	Maurhofer et al. (1994)
P. fluorescens WCS374		Radish–Fusarium	Leeman et al. (1996)
P. putida WCS358		*Arabidopsis–P. syringae* pv. *tomato*	Meziane et al. (2005)
P. putida WCS358		Bean–*Botrytis cinerea*	Meziane et al. (2005)
P. putida WCS358		Eucalyptus–*Ralstonia solanacearum*	Ran et al. (2005a)
P. putida WCS358		Tomato–*Botrytis cinerea*	Meziane et al. (2005)
P. fluorescens WCS374r		Rice–*Magnaporthe oryzae*	De Vleesschauwer et al. (2008)
P. putida WCS358	Flagella	*Arabidopsis–P. syringae* pv. *tomato*	Meziane et al. (2005)
P. fluorescens WCS374	Lipopolysaccharides	Radish–Fusarium	Leeman et al. (1996)
P. fluorescens WCS417		*Arabidosis–P. syringae* pv. *tomato*	Van Wees et al. (1997)
P. putida WCS358		Carnation–Fusarium wilt	Van Peer and Schippers (1992)
		Radish–Fusarium	Leeman et al. (1996)
		Arabidopsis–P. syringae pv. *tomato*	Meziane et al. (2005)
		Bean–*Botrytis cinerea*	Meziane et al. (2005)
P. aeruginosa 7NSK2	Salicylic acid	Bean–*Colletotrichum lindemuthianum*	Bigirimana and Höfte (2002)
P. aeruginosa 7NSK2		Tobacco	De Meyer et al. (1999)
P. fluorescens P3 *pchBA*		Tobacco	Maurhofer et al. (1998)
P. aeruginosa 7NSK2	Pyochelin, pyocyanin	Tomato–*Botrytis cinerea*	Audenaert et al. (2002)
	Pyocyanin	Rice–*Pyricularia grisea*	De Vleesschauwer et al. (2006)
P. putida BTP1	*N*-alkylated benzylamine derivative	Bean–*Botrytis cinerea*	Ongena et al. (2005b)
P. fluorescens CHA0	2,4-Diacetylphloroglucinol	*Arabidopsis–Peronospora parasitica*	Iavicoli et al. (2003)
Bacillus subtilis M4	Cyclic lipopeptides (e.g., fengycin)	Bean–*Pythium ultimum*	Ongena et al. (2007)
B. subtilis GB03	2,3-Butanediol	*Arabidopsis–Erwinia carotovora* ssp. *carotovora*	Ryu et al. (2004)
B. amyloquefaciens IN937a	2,3-Butanediol	*Arabidopsis–Erwinia carotovora* ssp. *carotovora*	Ryu et al. (2004)
B. subtilis FB17	L-Malic acid	*Arabidopsis–P. syringae* pv. *tomato*	Rudrappa et al. (2008)
P. putida IsoF	*N*-Acyl homoserine lactone	Tomato–*Alternaria alternata*	Shuhegger et al. (2006)
Serratia marcescens MG1	*N*-Acyl homoserine lactone	Tomato–*Alternaria alternata*	Shuhegger et al. (2006)
Serratia marcescens 90-166	Catechol-type siderophore	Cucumber–*Colletotrichum orbiculare*	Press et al. (2001)

Adapted from Höfte and Bakker (2007) and Bakker et al. (2007)

and viniferin) accumulation in grape cells and prime leaves for accelerated phytoalexin production upon challenge with *B. cinerea*. This report also highlights the importance of SA, pyochelin, and/or pyoverdin in priming phytoalexin responses and induced grapevine resistance by *P. aeruginosa* 7NSK2 against *B. cinerea* (Verhagen et al. 2010).

It has been suggested that effective competition for ferric iron could be the main mode of action of *P. putida* WCS358 (Bakker et al. 1993). *P. putida* WCS358 cannot trigger ISR in carnation (Duijff et al. 1993) or radish (Leeman et al. 1995), but it induces ISR in *A. thaliana* (Van Wees et al. 1997), *E. urophylla* (Ran et al. 2005a), bean, and tomato (Meziane et al. 2005). Recently, it has been revealed through microarray analysis that the R2R3-MYB-like transcription factor gene MYB72 is specifically activated in the roots upon colonization by *P. fluorescens* strains WCS417r, and MYB72 is essential to establish broad-spectrum ISR against the pathogens *P. syringae* pv. *tomato*, *Hyaloperonospora parasitica*, *Alternaria brassicicola*, and *B. cinerea* (Van der Ent et al. 2008).

Strains of *Bacillus* produce several cyclic peptides, aminopolyols, and aminogycosides, which are having an influence on the ISR development (Yu et al. 2002). It has been reported that volatile organic compounds emanated from PGPR may play a key role in ISR process (Ping and Boland. 2004; Ryu et al. 2004). For example, volatiles secreted by *B. subtilis* GB03 and *B. amyloquefaciens* IN937a were able to activate an ISR pathway in *Arabidopsis* seedlings challenged with the soft-rot pathogen *E. carotovora* ssp. *carotovora* (Ryu et al. 2004). In the study, it was found that infection of leaves of *A. thaliana* seedlings with the foliar pathogen *P. syringae* pv. *tomato* Pst DC3000 results in enhanced secretion of L-malic acid by the roots and that the enhanced level of L-malic acid selectively signals and recruits the beneficial rhizobacterium *B. subtilis* FB17, which is a biocontrol bacterium that protects the plant through ISR (Rudrappa et al. 2008). In another study, involvement of fengycin (a cyclic polypeptide) was implicated in the ISR-eliciting effect of strain M4, as these molecules may induce the synthesis of plant phenolics involved in or derived from the defense-related phenylpropanoid metabolism (Ongena et al. 2005a). In a recent study, it has been concluded that surfactin (another cyclic lipopeptide) could activate a biochemical cascade of molecular events leading to defensive responses in tobacco cell suspensions. According to the authors, this study sheds new light not only on defense-related events induced following recognition of amphiphilic lipopeptides from *Bacillus* spp. but also more globally on the way elicitors from beneficial bacteria can be perceived by host plant cells (Jourdan et al. 2009).

11.6 Conclusion

Plant disease poses threat to global food security. This stems from the fact that at least 10% of global food production is lost due to plant diseases. Catastrophic plant diseases exacerbate the current deficit of food supply in which at least 800 million people are inadequately fed (Strange and Scott 2005). According to the FAO estimates, plant diseases could cost the US alone $33 billion per year (Maor and Sirashu 2005). These facts explain why diseases affecting plants have been feared as much as human diseases and war throughout history.

During the last few decades, control of plant diseases has become increasingly difficult. In spite of the great advantages they have brought to agricultural development, the excessive use of pesticides including fungicides has taken its toll environmentally and on human health. In addition, the emergence of fungicide-resistant strains of pathogens and the rigorous regulation of fungicide use and disposal has reduced the possibility to conceive control strategies based on chemicals. Climate change is expected to impact positively on plant disease spread. Hence, there is concerted effort worldwide to explore new alternatives that control pre- and postharvest pathogenic diseases, giving priority to methods that reduce disease incidence and avoid negative and side effects on human health. In this highly charged scenario, the concept of ISR offers a broad-spectrum disease management strategy, which needs to be realized with full scientific backing in countering the threat posed by plant disease. In the recent past, the complete sequence of the 7.1-Mb size Pf-5 genome was determined. *P. fluorescens* Pf-5 is a plant commensal bacterium that inhabits the rhizosphere and produces secondary metabolites that suppress soil-borne plant pathogens (Paulsen et al. 2005). Literally, ISR research is only 20 years old and impressive progress has been made in its different dimensions. However, there remain many scientific challenges for research in the field of ISR by PGPR and other plant-beneficial microorganisms. Haas and Défago (2005) suggested that it would be important to exploit molecular techniques to study the genome expression of plant-beneficial and plant-pathogenic microorganisms *in situ* and to obtain a fuller picture of rhizosphere biodiversity. It is now generally felt that despite the best research efforts devoted to the identification and characterization of bacteria-derived elicitors of ISR, much remains to be discovered about how these determinants are perceived and ultimately give rise to ISR. In addition to unraveling the molecular regulation of induced resistance, we need to endeavor to decipher microbial signals that are most effective in eliciting pathogen resistance in plants through ISR.

References

Alabouvette C, Olivain C, L'Haridon F, Aime S, Steinberg C (2007) Using strains of *Fusarium oxysporum* to control *Fusarium* wilts: dream or reality? In: Vurro M, Gressel J (eds) Novel biotechnologies for biocontrol agent enhancement and management, NATO-ASI Series. Springer, Germany, pp 157–177

Alström S (1991) Induction of disease resistance in common bean susceptible to halo blight bacterial pathogen after seed bacterization with rhizosphere pseudomonads. J Gen Appl Microbiol 37:495–501

Audenaert K, Pattery T, Cornelis P, Höfte M (2002) Induction of systemic resistance to *Botrytis cinerea* in tomato by *Pseudomonas aeruginosa* 7NSK2: role of salicylic acid, pyochelin, and pyocyanin. Mol Plant Microbe Interact 15:1147–1156

Bakker PAHM, Van Peer R, Schippers B (1991) Suppression of soil-borne plant pathogens by fluorescent pseudomonads: mechanisms and prospects. In: Beemster ABR, Bollen GJ, Gerlagh M, Ruissen MT, Schippers B, Tempel A (eds) Developments in agricultural and managed-forest ecology, vol 23. Elsevier, Amsterdam, pp 217–230

Bakker PAHM, Raaijmakers JM, Schippers B (1993) Role of iron in the suppression of bacterial plant pathogens by fluorescent pseudomonads. In: Barton LL, Hemming BC (eds) Iron chelation in plants and soil microorganisms. Academic, San Diego, pp 269–278

Bakker PAHM, Ran LX, Pieterse CMJ, Van Loon LC (2003) Understanding the involvement of rhizobacteria-mediated induction of systemic resistance in biocontrol of plant diseases. Can J Plant Pathol 25:5–9

Bakker PAHM, Pieterse CMJ, Van Loon LC (2007) Induced systemic resistance by fluorescent *Pseudomonas* spp. Phytopathology 97:239–243

Bargabus RL, Zidack NK, Sherwood JW, Jacobsen BJ (2002) Characterization of systemic resistance in sugar beet elicited by a non-pathogenic, phyllosphere-colonizing *Bacillus mycoides*, biological control agent. Physiol Mol Plant Pathol 61:289–298

Bargabus RL, Zidack NK, Sherwood JW, Jacobsen BJ (2004) Screening for the identification of potential biological control agents that induce systemic acquired resistance in sugar beet. Biol Control 30:342–350

Benhamou N, Rey P, Chérif M, Hockenhull J, Tirilly Y (1997) Treatment with the mycoparasite *Pythium oligandrum* triggers induction of defense-related reactions in tomato roots when challenged with *Fusarium oxysporum* f. sp. *radicis-lycopersici*. Phytopathology 87:108–122

Benhamou N, Bélanger RR, Rey P, Tirilly Y (2001) Oligandrin, the elicitin-like protein produced by the mycoparasite *Pythium oligandrum*, induces systemic resistance to *Fusarium* crown and root rot in tomato plants. Plant Physiol Biochem 39:681–698

Benhamou N, Garand C, Goulet A (2002) Ability of nonpathogenic *Fusarium oxysporum* strain Fo47 to induce resistance against *Pythium ultimum* infection in cucumber. Appl Environ Microbiol 68:4044–4060

Bigirimana J, Höfte M (2002) Induction of systemic resistance to *Colletotrichum lindemuthianum* in bean by a benzothiadiazole derivative and rhizobacteria. Phytoparasitica 30:159–168

Bigirimana J, De Meyer G, Poppe J, Elad Y, Höfte M (1997) Induction of systemic resistance on bean (*Phaseolus vulgaris*) by *Trichoderma harzianum*. Med Fac Landbouww Univ Gent 62:1001–1007

Buysens S, Huengens K, Poppe J, Höfte M (1996) Involvement of pyochelin and pyoverdin in suppression of *Pythium*-induced damping-off of tomato by *Pseudomonas aeruginosa* 7NSK2. Appl Environ Microbiol 62:865–871

Cartieaux F, Thibaud MC, Zimmerli L, Lessard P, Sarrobert C, David P, Gerbaud A, Robaglia C, Somerville S, Nussaume L (2003) Transcriptome analysis of Arabidopsis colonized by a plant-growth promoting rhizobacterium reveals a general effect on disease resistance. Plant J 36:177–188

Cartieaux F, Contesto C, Gallou A, Desbrosses G, Kopka J, Taconnat L, Renou JP, Touraine B (2008) Simultaneous interaction of *Arabidopsis thaliana* with *Bradyrhizobium* sp strain ORS278 and *Pseudomonas syringae* pv. *tomato* DC3000 leads to complex transcriptome changes. Mol Plant Microbe Interact 21:244–259

Chester KS (1933) The problem of acquired physiological immunity in plants. Q Rev Biol 8:275–324

Choudhary DK, Johri BN (2009) Interactions of *Bacillus* spp. and plants: with special reference to induced systemic resistance (ISR). Microbiol Res 164:493–513

Compant S, Duffy B, Nowak J, Clément C, Barka EA (2005) Use of plant growth-promoting bacteria for biocontrol of plant diseases: principles, mechanisms of action, and future prospects. Appl Environ Microbiol 71:4951–4959

Conrath U, Beckers GJM, Flors V, García-Agustín P, Jakab G, Mauch F, Newman MA, Pieterse CMJ, Poinssot B, Pozo MJ, Pugin A, Schaffrath U, Ton J, Wendehenne D, Zimmerli L, Mauch-Mani B (2006) Priming: getting ready for battle. Mol Plant Microbe Interact 19:1062–1071

De Cal A, Pascual S, Melgarejo P (1997) Involvement of resistance induction by *Penicillium oxalicum* in the biocontrol of tomato wilt. Plant Pathol 46:72–79

De Meyer G, Höfte M (1997) Salicylic acid produced by the rhizobacterium *Pseudomonas aeruginosa* 7NSK2 induces resistance to leaf infection by *Botrytis cinerea* on bean. Phytopathology 87:588–593

De Meyer G, Bigirimana J, Elad Y, Höfte M (1998) Induced systemic resistance in *Trichoderma harzianum* T39 biocontrol of *Botrytis cinerea*. Eur J Plant Pathol 104:279–286

De Meyer G, Audenaert K, Höfte M (1999) *Pseudomonas aeruginosa* 7NSK2-induced systemic resistance in tobacco depends on in planta salicylic acid accumulation but is not associated with PR1a expression. Eur J Plant Pathol 105:513–517

De Vleesschauwer D, Cornelis P, Höfte M (2006) Redox-active pyocyanin secreted by *Pseudomonas aeruginosa* 7NSK2 triggers systemic resistance to *Magnaporthe grisea* but enhances *Rhizoctonia solani* susceptibility in rice. Mol Plant Microbe Interact 19:1406–1419

De Vleesschauwer D, Djavaheri M, Bakker PAHM, Höfte M (2008) *Pseudomonas fluorescens* WCS374r-induced systemic resistance in rice against *Magnaporthe oryzae* is based on Pseudobactin-mediated priming for a salicylic acid-repressible multifaceted defense response. Plant Physiol 148:1996–2012

De Weert S, Vermeiren H, Mulders IHM, Kuiper I, Hendrickx N et al (2002) Flagella-driven chemotaxis towards exudate components is an important trait for tomato root colonization by *Pseudomonas fluorescens*. Mol Plant Microbe Interact 15:1173–1180

Dekkers LC, Mulders CHM, Phoelich CC, Chin-A-Woeng TFC, Wijfjes AHM, Lugtenberg BJJ (2000) The *sss* colonization gene of the tomato-*Fusarium* f.sp. *radicis-lycopersici* biocontrol strain *Pseudomonas fluorescens* WCS365 can improve root colonization of other wild type *Pseudomonas* spp. bacteria. Mol Plant Microbe Interact 13:1177–1183

Deshmukh S, Hückelhoven R, Schäfer P, Imani J, Sharma M, Weiss M, Waller F, Kogel K-H (2006) The root endophytic fungus *Piriformospora indica* requires host cell death for proliferation during mutualistic symbiosis with barley. Proc Natl Acad Sci U S A 103:18450–18457

Duijff BJ, Meijer JW, Bakker PAHM, Schippers B (1993) Siderophore-mediated competition for iron and induced resistance in the suppression of fusarium wilt of carnation by fluorescent *Pseudomonas* spp. Neth J Plant Pathol 99:277–289

Duijff BJ, Gianinazzi-Pearson V, Lemanceau P (1997) Involvement of the outer membrane lipopolysaccharides in the endophytic colonization of tomato roots by biocontrol *Pseudomonas fluorescens* strain WCS417r. New Phytol 135:325–334

Durrant WE, Dong X (2004) Systemic acquired resistance. Annu Rev Phytopathol 42:185–209

Dutta S, Mishra AK, Dileep Kumar BS (2008) Induction of systemic resistance against fusarial wilt in pigeon pea through interaction of plant growth promoting rhizobacteria and rbizobia. Soil Biol Biochem 40:452–461

Edgar CI, McGrath KC, Dombrecht B, Manners JM, Maclean DC, Schenk PM, Kazan K (2006) Salicylic acid mediates resistance to the vascular wilt pathogen *Fusarium oxysporum* in the model host *Arabidopsis thaliana*. Australas Plant Pathol 35:581–591

Fravel D, Olivain C, Alabouvette C (2003) *Fusarium oxysporum* and its biocontrol. New Phytol 157:493–502

Freeman S, Rodriguez RJ (1993) Genetic conversion of a fungal plant pathogen to a nonpathogenic, endophytic mutualist. Science 260:75–78

Freeman S, Zveibil A, Vintal H, Maymon M (2002) Isolation of nonpathogenic mutants of *Fusarium oxysporum* f. sp. *melonis* for biological control of Fusarium wilt in cucurbits. Phytopathology 92:164–168

Fuchs JG, Moënne-Loccoz Y, Défago G (1997) Nonpathogenic *Fusarium oxysporum* strain Fo47 induces resistance to Fusarium wilt in tomato. Plant Dis 81:492–496

Fuchs JG, Moënne-Loccoz Y, De´fago G (1999) Ability of nonpathogenic Fusarium oxysporum Fo47 to protect tomato against Fusarium wilt. Biol Control 14:105–110

Gessler C, Kuć J (1982) Induction of resistance to Fusarium wilt in cucumber by root and foliar pathogens. Phytopathology 72:1439–1441

Guenoune D, Galili S, Phillips DA, Volpin H, Chet I, Okon Y, Kapulnik Y (2001) The defense response elicited by *Rhizoctonia solani* is suppressed by colonization of the AM fungus *Glomus intraradices*. Plant Sci 160:925–932

Haas D, Défago G (2005) Biological control of soil-borne pathogens by fluorescent pseudomonads. Nat Rev Microbiol 3:307–319

Hammerschmidt R, Métraux JP, van Loon LC (2001) Inducing resistance: a summary of papers presented at the first international symposium on induced resistance to plant diseases, Corfu, May 2000. Eur J Plant Pathol 107:1–6

Hammond-Kosack KE, Jones JDG (1996) Resistance gene-dependent plant defense responses. Plant Cell 8:1773–1791

Harman GE, Howell CR, Viterbo A, Chet I, Lorito M (2004) *Trichoderma* spp.: opportunistic avirulent plant symbionts. Nat Rev Microbiol 2:43–56

He CY, Hsiang T, Wolyn DJ (2002) Induction of systemic disease resistance and pathogen defense responses in *Asparagus officinalis* by nonpathogenic strains of *Fusarium oxysporum*. Plant Pathol 51:225–230

Hervás A, Trapero-Casas JL, Jiménez-Díz RM (1995) Induced resistance against Fusarium wilt of chickpea by nonpathogenic races of *Fusarium oxysporum* f. sp. *ciceris* and nonpathogenic isolates of *F. oxysporum*. Plant Dis 79:1110–1116

Hoffland E, Pieterse CMJ, Bik L, Van Pelt JA (1995) Induced systemic resistance in radish is not associated with accumulation of pathogenesis-related proteins. Physiol Mol Plant Pathol 46:309–320

Hoffland E, Hakulinen J, van Pelt JA (1996) Comparison of systemic resistance induced by avirulent and nonpathogenic *Pseudomonas* species. Phytopathology 86:757–762

Höfte M, Bakker PAHM (2007) Competition for iron and induced systemic resistance by siderophores of plant growth promoting rhizobacteria. In: Varma A, Chincholkar SB (eds) Soil biology: microbial siderophores, vol 12. Springer, Berlin, pp 121–133

Hwang J, Benson DM (2003) Expression of induced resistance in poinsietta cuttings against *Rhizoctonia* stem rot by treatment of stock plants with binucleate *Rhizoctonia*. Biol Control 27:73–80

Iavicoli A, Boutet E, Buchala A, Métraux J-P (2003) Induced systemic resistance in *Arabidopsis thaliana* in response to root inoculation with *Pseudomonas fluorescens* CHA0. Mol Plant Microbe Interact 16:851–858

Jourdan E, Henry G, Duby F, Dommes J, BarthélemyJP TP, Ongena M (2009) Insights into the defense-related events occurring in plant cells following perception of surfactin-type lipopeptide from *Bacillus subtilis*. Mol Plant Microbe Interact 22:456–468

Kamilova F, Validov S, Azarova T, Mulders I, Lugtenberg B (2005) Enrichment for enhanced competitive plant root tip colonizers selects for a new class of biocontrol bacteria. Environ Microbiol 7:1809–1817

Kaur R, Singh RS (2007) Study of induced systemic resistance in *Cicer arietinum* L. due to nonpathogenic *Fusarium oxysporum* using a modified split root technique. J Phytopathol 155:694–698

Kilic-Ekici O, Yuen GY (2003) Induced resistance as a mechanism of biological control by *Lysobacter enzymogenes* strain C3. Phytopathology 93:1103–1110

Kim MS, Kim YC, Cho BH (2004) Gene expression analysis in cucumber leaves primed by root colonization with Pseudomonas chlororaphis O6 upon challenge inoculation with *Corynespora cassiicola*. Plant Biol 6:105–108

Kloepper JW, Ryu CM, Zhang S (2004) Induced systemic resistance and promotion of plant growth by *Bacillus* spp. Phytopathology 94:1259–1266

Kroon BAM, Scheffer RJ, Elgersma DM (1991) Induced resistance in tomato plants against Fusarium wilt invoked by *Fusarium oxysporum* f. sp. *dianthi*. Neth J Plant Pathol 97:401–408

L'Haridon F, AiméS AC, Olivain C (2007) Lack of biocontrol capacity in a non-pathogenic mutant of *Fusarium oxysporum* f. sp. *melonis*. Eur J Plant Pathol 118:239–246

Lawton KA, Friedrich L, Hunt M, Weymann K, Delaney T, Kessmann H, Staub T, Ryals J (1996) Benzothiadiazole induces disease resistance in *Arabidopsis* by activation of the systemic acquired resistance signal transduction pathway. Plant J 10:71–82

Leeman M, van Pelt JA, den Ouden FM, Heinsbroek M, Bakker PAHM, Schippers B (1995) Induction of systemic resistance against fusarium wilt of radish by lipopolysaccharides of *Pseudomonas fluorescens*. Phytopathology 85:1021–1027

Leeman M, Den Ouden FM, Van Pelt JA, Dirkx FPM, Steijl H, Bakker PAHM, Schippers B (1996) Iron availability affects induction of systemic resistance to *Fusarium* wilt of radish by *Pseudomonas fluorescens*. Phytopathology 86:149–155

Linthorst HJM (1991) Pathogenesis-related proteins of plants. Crit Rev Plant Sci 10:123–150

Lugtenberg B, Kamilova F (2009) Plant-growth-promoting rhizobacteria. Annu Rev Microbiol 63:541–556

Malamy J, Carr JP, Klessig DF, Raskin I (1990) Salicylic acid: a likely endogenous signal in the resistance response of tobacco to viral infection. Science 250:1002–1004

Mandal S, Mallick N, Mitra A (2009) Salicylic acid-induced resistance to *Fusarium oxysporum* f. sp. *lycopersici* in tomato. Plant Physiol Biochem 47:642–649

Mandeel Q, Baker R (1991) Mechanisms involved in biological control of Fusarium wilt of cucumber with strains of nonpathogenic *Fusarium oxysporum*. Phytopathology 81:462–469

Maor R, Sirashu K (2005) The arms race continues: battle strategies between plants and fungal pathogens. Curr Opin Microbiol 43:399–404

Maurhofer M, Hase C, Meuwly P, Métraux JP, Défago G (1994) Induction of systemic resistance of tobacco to tobacco necrosis virus by the root-colonizing *Pseudomonas fluorescens* strain CHA0: influence of the *gacA* gene and of pyoverdine production. Phytopathology 84:139–146

Maurhofer M, Reimmann C, Schmidli-Sacherer P, Heeb S, Haas D, Défago G (1998) Salicylic acid biosynthetic genes expressed in *Pseudomonas fluorescens* strain P3 improve the induction of systemic resistance in tobacco against tobacco necrosis virus. Phytopathology 88:678–684

Métraux J-P, Ahl-Goy P, Staub T, Speich J, Steinemann A, Ryals J, Ward E (1991) Induced resistance in cucumber in response to 2,6-dichloroisonicotinic acid and pathogens. In: Hennecke H, Verma DPS (eds) Advances in molecular genetics of plant-microbe interactions, vol 1. Kluwer, Dordrecht, The Netherlands, pp 432–439

Meziane H, Van der Sluis I, Van Loon LC, Höfte M, Bakker PAHM (2005) Determinants of *Pseudomonas putida* WCS358 involved in inducing systemic resistance in plants. Mol Plant Pathol 6:177–185

Molitor A, Kogel K-H (2009) Induced resistance triggered by *Piriformospora indica*. Plant Signal Behav 4:215–216

Nel B, Steinberg C, Labuschagne N, Viljoen A (2006) The potential of nonpathogenic *Fusarium oxysporum* and other biological control organisms for suppressing *Fusarium* wilt of banana. Plant Pathol 55:217–223

Nürnberger T, Lipka V (2005) Non-host resistance in plants: new insights into an old phenomenon. Mol Plant Pathol 6:335–345

Olson HA, Benson DM (2007) Induced systemic resistance and the role of binucleate *Rhizoctonia* and *Trichoderma hamatum* 382 in biocontrol of Botrytis blight in geranium. Biol Control 42:233–241

Ongena M, Jacques P, Touré Y, Destain J, Jabrane A, Thonart P (2005a) Involvement of fengycin-type lipopeptides in the multifaceted biocontrol potential of *Bacillus subtilis*. Appl Microbiol Biotechnol 69:29–38

Ongena M, Jourdan E, Schäfer M, Kech C, Budzikiewicz H, Luxen A, Thonart P (2005b) Isolation of an *N*-alkylated benzylamine derivative from *Pseudomonas putida* BTP1 as elicitor of induced systemic resistance in bean. Mol Plant Microbe Interact 18:562–569

Ongena M, Jourdan E, Adam A, Paquot M, Brans A, Joris B, Arpigny JL, Thonart P (2007) Surfactin and fengycin lipopeptides of *Bacillus subtilis* as elicitors of induced systemic resistance in plants. Environ Microbiol 9:1084–1090

Park K, Paul D, Ryu KR, Kim EY, Kim YK (2006) *Bacillus vallismortis* Strain EXTN-1 mediated systemic resistance against *Potato virus Y* and *X* in the field. Plant Pathol J 22:360–363

Paulsen IT, Press CM, Ravel J, Kobayashi DY, Myers GSA, Mavrodi DV, DeBoy RT, Seshadri R, Ren Q, Madupu R, Dodson RJ, Durkin AS, Brinkac LM, Daugherty SC, Sullivan SA, Rosovitz MJ, Gwinn ML, Zhou L, Schneider DJ, Cartinhour SW, Nelson WC, Weidman J, Watkins K, Tran K, Khouri H, PiersonEA PIIILS, Thomashow LS, Loper JE (2005) Complete genome sequence of the plant commensal *Pseudomonas fluorescens* Pf-5. Nat Biotechnol 23:873–878

Pieterse CMJ, Van Wees SCM, Hoffland E, Van Pelt JA, Van Loon LC (1996) Systemic resistance in Arabidopsis induced by biocontrol bacteria is independent of salicylic acid accumulation and pathogenesis-related gene expression. Plant Cell 8:1225–1237

Pieterse CMJ, Van Wees SCM, Van Pelt JA, Knoester M, Laan R, Gerrits H, Weisbeck PJ, Van Loon LC (1998) A novel signaling pathway controlling induced systemic resistance in *Arabidopsis*. Plant Cell 10:1571–1580

Ping L, Boland W (2004) Signals from the underground: bacterial volatiles promote growth in *Arabidopsis*. Trends Plant Sci 9:263–269

Pozo MJ et al (2002) Localized versus systemic effect of arbuscular mycorrhizal fungi on defence responses of *Phytophthora* infection in tomato plants. J Exp Bot 53:525–534

Press CM, Wilson M, Tuzun S, Kloepper JW (1997) Salicylic acid produced by *Serratia marcescens* 90-166 is not the primary determinant of induced systemic resistance in cucumber or tobacco. Mol Plant Microbe Interact 10:761–768

Press CM, Loper JE, Kloepper JW (2001) Role of iron in rhizobacteria-mediated induced systemic resistance of cucumber. Phytopathology 91:593–598

Ran LX, Li ZN, Wu GJ, Van Loon LC, Bakker PAHM (2005a) Induction of systemic resistance against bacterial wilt in *Eucalyptus urophylla* by fluorescent *Pseudomonas* spp. Eur J Plant Pathol 113:59–70

Ran LX, Van Loon LC, Bakker PAHM (2005b) No role for bacterially produced salicylic acid in rhizobacterial induction of systemic resistance in Arabidopsis. Phytopathology 95: 1349–1355

Redman RS, Freeman S, Clifton DR, Morrel J, Brown G, Rodriguez RJ (1999) Biochemical analysis of plant protection afforded by a nonpathogenic endophytic mutant of *Colletotrichum magna*. Plant Physiol 119:795–804

Rudrappa T, Czymmek KJ, Paŕe PW, Bais HP (2008) Root-secreted malic acid recruits beneficial soil bacteria. Plant Physiol 148:1547–1556

Ryu C-M, Farag MA, Hu C-H, Reddy MS, Wie H-X, Paré PW, Kloepper JW (2003) Bacterial volatiles promote growth of *Arabidopsis*. Proc Natl Acad Sci U S A 100:4927–4932

Ryu CM, Farag MA, Hu CH, Reddy MS, Kloepper JW, Paré PW (2004) Bacterial volatiles induce systemic resistance in *Arabidopsis*. Plant Physiol 134:1017–1026

Sailaja PR, Podile AR, Reddanna P (1997) Biocontrol strain *Bacillus subtilis* AF1 rapidly induces lipoxygenase in groundnut (*Arachis hypogaea* L.) compared to crown rot pathogen *Aspergillus niger*. Eur J Plant Pathol 104:125–132

Sanchez L, Weidmann S, Arnould C, Bernard AR, Gianinazzi S, Gianinazzi-Pearson V (2005) *Pseudomonas fluorescens* and *Glomus mosseae* trigger DMI3-dependent activation of genes related to a signal transduction pathway in roots of *Medicago trunculata*. Plant Physiol 139:1065–1077

Schippers B (1992) Prospects for management of natural suppressiveness to control soilborne pathogens. In: Tjamos EC, Papavizas GC, Cook RJ (eds) Biological control of plant diseases, progress and challenges for the future. Plenum, New York, USA, pp 21–34

Shoresh M, Yedidia I, Chet I (2005) Involvement of jasmonic acid/ethylene signaling pathway in the systemic resistance induced in cucumber by *Trichoderma asperellum* T203. Phytopathology 95:76–84

Shoresh M, Harman GE, Mastouri F (2010) Induced systemic resistance and plant responses to fungal biocontrol agents. Annu Rev Phytopathol 48 http://www.annualreviews.org/catalog/pubdates.aspx

Shuhegger R, Ihring A, Gantner S, Bahnweg G, Knappe C, Vogg G, Hutzler P, Schmid M, Van Breusegem F, Eberl L, Hartmann A, Langebartels C (2006) Induction of systemic resistance in tomato by *N*-acyl-L-homoserine lactone-producing rhizosphere bacteria. Plant Cell Environ 29:909–918

Stein E, Molitor A, Kogel K-H, Waller F (2008) Systemic resistance in *Arabidopsis* conferred by the mycorrhizal fungus *Piriformospora indica* requires jasmonic acid signaling and the cytoplasmic function of NPR1. Plant Cell Physiol 49:1747–1751

Sticher L, Mauch-Mani B, Métraux JP (1997) Systemic acquired resistance. Annu Rev Phytopathol 35:235–270

Strange RN, Scott PR (2005) Plant disease: a threat to global food security. Annu Rev Phytopathol 43:83–116

Takenaka S, Nishio Z, Nakamura Y (2003) Induction of defense reactions in sugar beet and wheat by treatment with cell wall protein fractions from the mycoparasite *Pythium oligandrum*. Phytopathology 93:1228–1232

Thomashaw LS, Weller DM (1995) Current concepts in the use of introduced bacteria for biological control: mechanisms and antifungal metabolites. In: Stacey G, Keen N (eds) Plant-microbe interactions, vol 1. Chapman and Hall, New York, USA, pp 187–235

Thuerig B, Felix G, Binder A, Boller T, Tamm L (2006) An extract of *Penicillium chrysogenum* elicits early defense-related responses and induces resistance in *Arabidopsis thaliana* independently of known signaling pathways. Physiol Mol Plant Pathol 67:180–193

Valenzuela-Soto JH, Estrada-Hernández MG, Ibarra-Laclette E, Délano-Frier JP (2010) Inoculation of tomato plants (*Solanum lycopersicum*) with growth-promoting *Bacillus subtilis* retards whitefly *Bemisia tabaci* development. Planta 231:397–410

Van der Ent SV, Verhagen BWM, Doorn RV, Bakker D, Verlaan MG, Pel MJC, Joosten RG, Proveniers MCG, Van Loon LC, Ton J, Pieterse CMJ (2008) MYB72 is required in early signaling steps of rhizobacteria-induced systemic resistance in Arabidopsis. Plant Physiol 146:1293–1304

Van Loon LC (2007) Plant responses to plant growth-promoting bacteria. Eur J Plant Pathol 119:243–254

Van Loon LC, Bakker PAHM (2005) Induced systemic resistance as a mechanism of disease suppression by rhizobacteria. In: Siddiqui ZA (ed) PGPR: biocontrol and biofertilization. Springer Science, Dordrecht, The Netherlands, pp 39–66

Van Loon LC, Bakker PAHM, Pieterse CMJ (1998) Systemic resistance induced by rhizosphere bacteria. Annu Rev Phytopathol 36:453–483

Van Peer R, Schippers B (1992) Lipopolysaccharides of plant growth-promoting *Pseudomonas* spp. strain WCS417r induce resistance in carnation to Fusarium wilt. Neth J Plant Pathol 98:129–139

Van Peer R, Niemann GJ, Schippers B (1991) Induced resistance and phytoalexin accumulation in biological control of Fusarium wilt of carnation by *Pseudomonas* sp. strain WCS417r. Phytopathology 81:728–734

Van Wees SCM, Pieterse CMJ, Trijssenaar A, Vańt Westende Y, Hartog F, Van Loon LC (1997) Differential induction of systemic resistance in *Arabidopsis* by biocontrol bacteria. Mol Plant Microbe Interact 10:716–724

Verhagen BWM, Glazebrook J, Zhu T, Chang HS, Van Loon LC, Pieterse CMJ (2004) The transcriptome of rhizobacteria-induced systemic resistance in *Arabidopsis*. Mol Plant Microbe Interact 17:895–908

Verhagen BWM, Trotel-Aziz P, Couderchet M, Höfte M, Aziz A (2010) *Pseudomonas* spp.- induced systemic resistance to *Botrytis cinerea* is associated with induction and priming of defence responses in grapevine. J Exp Bot 61:249–260

Viterbo A, Harel M, Horwitz BA, Chet I, Mukherjee PK (2005) Trichoderma MAP-kinase signaling is involved in induction of plant systemic resistance. Appl Environ Microbiol 71:6241–6246

Waller F, Achatz B, Baltruschat H, Fodor J, Becker K, Fischer M, Heier T, Hückelhoven R, Neumann C, von Wettstein D, Franken P, Kogel KH (2005) The endophytic fungus *Piriformospora indica* reprograms barley to salt-stress tolerance, disease resistance, and higher yield. Proc Natl Acad Sci U S A 102:13386–13391

Wang Y, Ohara Y, Nakayashiki H, Tosa Y, Mayama S (2005) Microarray analysis of the gene expression profile induced by the endophytic plant growth-promoting rhizobacteria, *Pseudomonas fluorescens* FPT9601-T5 in *Arabidopsis*. Mol Plant Microbe Interact 18:385–396

Wei G, Kloepper JW, Tuzun S (1991) Induction of systemic resistance of cucumber to *Colletotrichum orbiculare* by select strains of plant growth-promoting rhizobacteria. Phytopathology 81:1508–1512

Yan Z, Reddy MS, Kloepper JW (2003) Survival and colonization of rhizobacteria in a tomato transplant system. Can J Microbiol 49:383–389

Yedidia I, Shoresh M, Kerem Z, Benhamou N, Kapulnik Y, Chet I (2003) Concomitant induction of systemic resistance to *Pseudomonas syringae* pv. *lachrymans* in cucumber by *Trichoderma asperellum* (T-203) and the accumulation of phytoalexins. Appl Environ Microbiol 69:7343–7353

Yu GY, Sinclair B, Hartman GL, Bertagnolli BL (2002) Production of iturin A by *Bacillus amyloliqefaciens* suppressing *Rhizoctonia solani*. Soil Biol Biochem 34:955–963

Zehnder GW, Murphy EJ, Sikora EJ, Kloepper JW (2001) Application of rhizobacteria for induced resistance. Eur J Plant Pathol 107:39–50

Zhang S, Reddy MS, Kloepper JW (2004) Tobacco growth enhancement and blue mold disease protection by rhizobacteria: relationship between plant growth promotion and systemic disease protection by PGPR strain 90-166. Plant Soil 262:277–288

Chapter 12
Biological Control of Termites by Antagonistic Soil Microorganisms

Satyavir S. Sindhu, Y.S. Rakshiya, and M.K. Verma

12.1 Introduction

Termites belong to the insect order Isoptera and are characterized by their colonial behavior. The word Isoptera originated from the Greek, in which "isos" means equal and "pteron" means wing, and refers to the two pairs of identical wings in the adult (Thorne and Carpenter 1992). Termites are medium-sized, soft-bodied, light-colored, polymorphic, cellulose-eating social insects living in large communities of several hundred to several million individuals. The fossil record indicates that termites evolved about 220 million years ago (Collins 1988; Thorne and Carpenter 1992). They are said to be derived from a primitive group of wood-dwelling cockroaches, clearly seen in the obligate dependence on mutualistic intestinal protists and, in some higher forms, externally cultivated basidiomycete fungi (Bignell and Eggleton 1998). They are widely dispersed throughout the tropics as well as some temperate regions and achieve their highest diversities and abundance in the rain forests of Africa, South America, and South-east Asia (Collins 1988; Bignell and Eggleton 1998). There are about 2,650 species of termites in 280 genera and seven families worldwide (Kambhampati and Eggleton 2000). Of these, about 323 species from 52 genera have been recorded in the Indo-Malayan (oriental) region (Tho 1992) and about 104 species (33 genera) have been recorded from Sabah (Thapa 1981).

Termites live in highly organized colonies. The number of individuals and ratios of each caste in a colony is very difficult to determine, varies between species, and also depends on the age as well as size of the colony (Bignell and Eggleton 1998). The individuals are differentiated morphologically into distinct forms, i.e., reproductives (king and queen), workers, and soldiers. The parent termites, the king and queen are the functional reproductives. The queen is also involved in pheromonal regulation of control for the production of each caste in a colony (Noirot and Noirot-Timothee 1970). Soldiers and workers are wingless and can be either sterile

S.S. Sindhu (✉), Y.S. Rakshiya, and M.K. Verma
Department of Microbiology, CCS Haryana Agricutural University, Hisar 125004, Haryana, India
e-mail: sindhuss@hau.ernet.in

A. Singh et al. (eds.), *Bioaugmentation, Biostimulation and Biocontrol*, Soil Biology 28,
DOI 10.1007/978-3-642-19769-7_12,

male or female. Soldiers usually represent one tenth of the population whose major role is only to defend the colony (Bignell and Eggleton 1998). The soldier has large, dark, elongated, highly sclerotized head adapted in various ways for defense. The workers are most numerous individuals in a colony and perform the work of nest building and repair, foraging, and feeding (Harris 1957). Winged reproductives or alates of both sexes are produced in large numbers in a mature colony. These alates swarm out from mature nests at particular times of the year (often during or just before rains) (Bignell and Eggleton 1998). They make short, often rather feeble, dispersal flights and then pair up on the ground after the wings have been shed (dealation). The paired termites will then select a new nesting site and once they are established, mating takes place. The first batch of eggs is produced by the female within a few days.

Termites play an important role in the tropical ecosystems by decomposing dead wood and other plant material rich in cellulose (the most abundant organic matter on the earth) (Wood and Sands 1978; Abe 1995). With their ability of utilizing dead plant, they become important in processes such as decomposition of organic matter (Peakins and Josens 1978; Wood and Johnson 1986), supplying material for many food chains, and soil engineering (translocating and altering soils physically and chemically, and maintaining soil fertility) (Lee and Wood 1971; Wood 1988). Termites also provide a possible input of nitrogen through symbiont fixation (Wood and Sands 1978; Collins 1984) and also contribute to carbon flux (Jones 1990). They also release a high amount of methane in the atmosphere worldwide. Due to the feeding function, the worker caste causes the widespread destruction resulting into major economic losses in tropical and subtropical areas by destroying agricultural crops, live trees, and wooden structures in the houses. About 300 species have been reported to cause significant damage to agricultural crops and have been recorded as pests (Edwards and Mill 1986; Su and Scheffrahn 1998). Species of *Microtermes* and *Odontotermes* have been found to damage different crops such as sugarcane, wheat, barley, maize, vegetables, garden crops, valuable ornamental crops, and even forest trees (Lai et al. 1983; Tamashiro et al. 1987).

Termites also feed on and often destroy various other structures or materials that people use, i.e., wooden portions of buildings, furniture, books, utility poles, fence posts, many fabrics, and other useful materials. The termites also attack telephone poles, boats, and other finished goods (e.g., paper and fabric), valuable manuscripts, and paintings. Sometimes, even inorganic materials such as buried electrical and telephone cables have been reportedly damaged by termites (Henderson and Dunaway 1999). In Hargobindpur (Punjab, India), people deserted an entire street as buildings started collapsing due to heavy termite attack (Roonwal 1955). The poorer areas of the tropics and subtropics probably suffer more termite damage and therefore, economic impact of termites has received the most attention (Su and Scheffrahn 1998). It has been reported that 95% of damage to wood and wood products in the USA is caused by subterranean termites (family Rhinotermitidae) and its estimated cost annually exceeds $750 million to $1 billion (Mauldin 1986; Su and Scheffrahn 1990). Moreover, subterranean termites account for 80% of the approximately $1.5 billion spent annually for termite control in the USA (Su 1993).

Termite control by chemicals involves their application to the wood or to the soil. The best method of eliminating dry wood termites is by chemical fumigation using the fumigant, usually sulfuryl fluoride, methyl bromide, or a combination of methyl bromide and carbon dioxide is pumped into the building. Insecticides such as chlorpyrifos, bifenthrin, imidacloprid, endosulfan, and lindane are currently being used for control of termites in stored wood as well as for crops (Su et al. 1999). For example, seed treatment with chlorpyrifos, lindane, or thiomethoxam resulted in lesser termite damage in cotton, maize, rice, sorghum, and sugarcane crop and was found effective against three species of termites: *Trinernitermes trinervius*, *Odontotermes smeathmani*, and *Amitermea evuncifer*, which are the principal pests of these tropical food crops (Bhanot and Singal 2007). The effective dose of thiomethoxam for these termites was 0.03 ppm, which resulted in 100% mortality within 2–8 days depending upon the species studied. Seed treatment with insecticides such as chlorpyrifos, endosulfan, formothion, and monocrotophos has been standardized in wheat (Bhanot et al. 1991a), barley (Bhanot et al. 1991b), and gram (Bhanot et al. 1995). However, the control of pest insects with chemical insecticides has generated several problems including insecticide resistance, outbreaks of secondary pests normally held in check by natural enemies, safety risks for humans and domestic animals because of their long persistence in soil and due to the entry of residual toxic chemicals in food chain, contamination of groundwater, decrease in biodiversity, and other environmental concerns. These problems and sustainability of programs based predominantly on conventional insecticides have stimulated increased interest in integrated pest management. Sustainable agriculture in the twenty-first century will rely increasingly on alternative interventions for pest management that are environment friendly and reduce the amount of human contact with chemical pesticides.

Biological control agents are effective, environment friendly, economically viable, and socially acceptable method of pest management. The moist and warm microenvironment preferred by subterranean termites supports epizootics, enhancing the potential for biological control (Verma et al. 2009). Biological control agents from areas of the termite's origin are being searched to control exotic species. Several fungi, namely, species of *Termitaria*, *Neotermus*, *Metirolella*, *Laboulbenia*, *Antennopsis*, *Metarhizium*, *Baeuveria*, *Leboulbeniopsis*, and *Coreomyceptosis* are known to be parasitic to termites (Staples and Milner 1996). Delate et al. (1995) achieved complete mortality of *Coptotermes formosanus* termites within 15 days with isolates of *Baeuveria bassiana* and *Metarhizium anisopliae* fungi. Wright et al. (2005) reported that a fungal isolate, C4-B, taxonomically identified as *M. anisopliae* (Metschnikoff) caused rapid mortality in Formosan subterranean termite alates. A nuclear polyhedrosis virus (NPV) isolated from the cotton leaf worm (*Spodoptera litoralis*) has also been reported to infect termites (Al Fazairy and Hassan 1988). Termites died 2–10 days post-infection and the authors suggested that control of *Kalotermes flavicollis* with NPV might be feasible.

Termites are also susceptible to infection by the bacterium *Serratia marcescens* (Khan et al. 1977a) and *Bacillus thuringiensis* (Khan et al. 1985). Khan et al. (1992) reported that the mortality of *Microcerotermes championi*, *Heterotermes indicola*, and *Coptotermes heimi* (Rhinotermitidae) by the pathogenic *Pseudomonas aeruginosa* (Schroeter) ranged from 25–52% at 7 days post-inoculation to 84–100% at 25 days post-inoculation in the laboratory. Connick et al. (2001) reported that *S. marcescens* isolate T8 was highly virulent to the *C. formosanus* and termite mortality was 24% by 2 days, and 99% after 19 days of the experiment. Other bacterial isolates involved in the mortality of termites included *Acinetobacter calcoacet*, *Aeromonas caviae*, *Alcaligenes latus*, *Arthrobacter* sp., *Bacillus* sp., *Chromobacterium* sp., *Corynebacterium urealyticum*, *Enterobacter gergoviae*, *Micrococcus*, *Neisseria*, and *Rhizobium radiobacter* (Osbrink et al. 2001; Kanchana Devi et al. 2007; Yuvraj Singh 2007). Nematodes belonging to the two families, Steinernematidae and Heterorhabditidae, have shown promise for use in termite control programs (Kaya and Gaugler 1993; Yu et al. 2006). These studies demonstrated the potential of microorganisms in reducing termite populations successfully in the laboratory.

Recent emphasis on biological processes able to improve agricultural productivity, while minimizing soil pollution, is essential for sustainability in agriculture system. A better understanding of ecology of termites' pathogens, optimization of conditions leading to better management or killing of termites, and appropriate formulations of microbial preparations is crucial for sustaining agricultural ecosystems. This chapter reviews the various physical, chemical, and biological methods for termite control. Recent developments and past research done on termite control emphasizing biological control agents are reviewed. The relationship between production of various metabolites by termite pathogens and their possible contribution toward termiticidal activity is also discussed. The possibilities of ensuring the survival of introduced termite's pathogens in the soil or to prepare efficacious and potent biocidal formulations are explored in this review.

12.2 Biology and Diversity of Termites

A basic understanding of termite diversity and biology is a prerequisite for adequate pest management (Krishna and Weesner 1970; Lee and Wood 1971; Grasse 1986; Donavan et al. 2007). The diversity of termites could be explained on the basis of their morphological differentiation, capacity to form different kinds of nest/mound, and their consumption of a wide range of freshly dead or decaying plant material.

12.2.1 Morphological Differentiation

The individuals in the organized colonies are differentiated morphologically into distinct forms or castes that perform different biological functions. Colonies of *Macrotermes michaefsoni* species of Australia may reach three million individuals. Most termites originally have a single king and queen, which carry out colony reproduction for 10–20 years or more. However, if one or both of this royal pair die or if the colony size becomes large and dispersed, reproductive replacements may come from developing immatures or adults, particularly in tropical species. Most swarming reproductives produce only four or five dozen offsprings in the first year. However, the egg production by the queen increases in subsequent years from three to six eggs per day to about 30,000 or more eggs per day in physogastric (swollen with eggs) *Macrotermes* queen. Such an enlarged queen becomes more than 300 times larger than the workers. The workers are pale and soft-bodied, with mouth parts adapted for chewing. They perform most of the work of the colony: nest building and repair, foraging, and grooming other members of the colony. Because of the feeding function, the worker caste causes the widespread destruction for which termites are notorious. The soldiers are adapted in various ways for defense. Morphologically, the soldiers are bigger in size and have defensive adaptations such as enlarged mandibles or stopper-like heads. Besides having mandibles, and a sclerotized head, soldiers of some genera such as *Coptotermes* have a frontal gland that discharges a defensive secretion through a frontal pore (Richards and Davies 1978). This secretion can be toxic or repellent to intruders, such as ants and entangle their legs and antennae.

12.2.2 Nest/Mound Formation

All termites live in highly organized and integrated societies or colonies within the confines of excavations within wood above-ground or in subterranean and epigeal nest systems. Some subterranean termite species often build the spectacular mounds for which these insects are renowned (Fig. 12.1). A variety of nests or mounds are constructed, varying from random shapes to subterranean forms, to oval nests on trees in tropical areas, and to elaborate ground nests in semiarid regions. Some of

Fig. 12.1 A termite nest in the University campus

the nests in Africa may extend upward more than 20 ft and measure 12 ft in diameter. These are produced by cementing soil and faecal material with secretions either from specialized frontal glands in the termite's head or from the proctodeum. Several subterranean termite species such as *C. formosanus* infest and nest within living trees. It has been estimated that as many as 30% of living trees in New Orleans, Los Angeles, are infested with termites. The termites frequently construct a basal carton nest within the trunk of the tree causing substantial structural weakening of the tree. Termites nest systems can be classified as mentioned below.

12.2.2.1 Wood Nesters

Termites in this group live in or around standing trees or dead logs. Sometimes, the dead wood is gradually replaced with wood carton, a woody substance with low nutrient concentrations, high levels of lignin, and other undigested components (Collins 1989). This includes the Kalotermids (*Kalotermes* and *Glyptotermes*), some Rhinotermids (*Schedorhinotermes*, *Parrhinotermes*, *Heterotermes*, and also *Coptotermes*), and some Termitidae members such as *Microcerotermes* and *Termes*.

12.2.2.2 Hypogeal or Subterranean Nesters

Termites whose colony centers are below the ground without any indication of their presence (Wood and Johnson 1986). They use their faeces or a mixture of faeces and mineral soil in nest construction. In some Macrotermitinae, *Apicotermes* and *Homallotermes*, a little internal structure or surface holes are present together with their complex underground nests. This enables the foragers to forage on above-ground vegetation. This group also includes many species that are facultative secondary inhabitants of epigeal mounds: *Microcerotermes*, *Pericapritermes*, and soldierless Apicotermitinae (Eggleton et al. 1996).

12.2.2.3 Epigeal Mound Builders

Termites whose colony centers associated with living or dead vegetation above ground are commonly known as mound builders (Jones 1990; Eggleton et al. 1996). The mounds are usually well characterized, often with very complex structures. Materials used for construction are of three main types: subsoil with relatively low organic content added with salivary secretion (*Macrotermes* and *Cornitermes*), wood carton (a mixture of faeces and macerated wood with a high lignin content), or a mixture of faeces and organic-rich topsoil (many soil feeders). Epigeal mound structure can differ widely within genera and also between regions within widely distributed species. Macrotermitinae and *Dontotermes* are known to build huge mounds of selected clay-rich subsoil (Wood and Johnson 1986).

12.2.2.4 Arboreal Nesters

Nests are attached outwardly to trees at different heights. These nests are normally made of wood carton. In most cases, the nests are connected to the ground by covered runways. This may assist in distinguishing some arboreal termite nests from those of ants. Nonetheless, some arboreally nesting Nasutitermitinae (e.g., *Hospitalitermes*) form open foraging columns without any connecting runways between the nest and foraging sites.

12.2.3 Trophic Groups

Termites consume wide range of freshly dead or decaying plant material including dry grass, leaf litter, decaying wood, dung, and humus. Living plant tissues, including lichen and mosses, are taken by a few species. Another feeding group that may be common and important in many tropical forests is the soil-feeding termites. Termite species can be categorized into five broad trophic categories according to their food, foraging galleries or columns, color of the abdomen, and known biology (de Souza and Brown 1994; Eggleton et al. 1995, 1997; Bignell et al. 1997).

12.2.3.1 Wood Feeders

These primitive wood eating termites feed on wood and woody litter, including dead branches still attached to trees, and they may live in their feeding galleries, which in some cases become colony centers (Eggleton et al. 1996, 1997). The wood taken may include living trees (*Coptotermes*, *Schedorhinotermes*, and *Microcerotermes dubius*), sound dead wood (Kalotermitidae), and/or fungus-attacked wood (Nasutitermitinae, some Termitinae, and Macrotermitinae) (Wood 1976; Collins 1984). Most of these termites are arboreal (attached to trees), subterranean, or epigeal nesters (Bignell et al. 1997; Eggleton et al. 1997).

12.2.3.2 Soil Feeders

These termites feed on the upper mineral soil rich in organic matter, with some degree of selection of silt and clay fractions (Sleaford et al. 1996). They are normally distributed in the soil profile, in the organic litter layer (leaves and twigs), and/or in epigeal mounds (Bignell et al. 1997; Eggleton et al. 1995, 1997). Soil feeders are very common and abundant in many tropical rain forests (Wood 1976). In South-east Asian regions, soil feeders are dominated by the Termitinae with a small number of Nasutitermitinae and Apicotermitinae (Abe 1987).

12.2.3.3 Soil–Wood Interface Feeders

Termites in this group feed on highly decayed wood, the soil under logs or soil plastered to logs, or soil mixed with leaf litter in stilt-root complexes (Eggleton et al. 1996, 1997). Soil/wood interface feeders are only found in the Termitinae, Apicotermitinae, and Nasutitermitinae. Most of them nest within dead logs, build epigeal nest, or form colony centers in the soil (Eggleton et al. 1997).

12.2.3.4 Litter-Foragers

Termites in this trophic group forage for leaf litter and small woody items litter in various stages of decay. Food sources are often taken back and stored temporarily in the nest. This group includes some subterranean and other mound-building Macrotermitinae (with fungal association), as well as certain Nasutitermitinae that forage on the surface of the ground or litter layers (Bignell et al. 1997; Eggleton et al. 1997). Genera such as *Laccessititermes* and *Longipeditermes* are also known as arboreal forages.

12.2.3.5 Micro-Epiphyte Feeders

Termites of this group forage for moss, algae, lichens, and fungi on tree barks. In South-east Asia, *Hospitalitermes hospitalis* is known to feed on lichen (Jones and Gathrone-Hardy 1995; Eggleton et al. 1997). Grass feeders also take dung and may sometimes scavenge vertebrate corpses. Grass feeders are mainly of the family Hodotermitidae, found only in savanna and deserts (Krishna 1970).

The gut microbiota of termites represents all aspects of microbial relationships from pathogenic to obligate mutualism (Dillon and Dillon 2004). Breznak (1982) reported that a dense and morphologically diverse bacterial flora colonizes the hindgut of lower and higher termites. Most bacterial isolates were found to belong to the species of strict or facultative anaerobes including *Streptococcus*, *Bacteroides*, various Enterobacteriaceae members, *Staphylococcus*, and *Bacillus*. For example, spirochetes provide the carbon, nitrogen, and energy requirements of termite nutrition via acetogensis and nitrogen fixation. In fact, microbial nitrogen fixation accounts for 60% of the nitrogen in some termite colonies (Tayasu et al. 1994). In the lower termites, flagellate protozoa essentially inhabit the insect gut as symbionts to break down cellulose and convert it into soluble and digestible substrate. Several lower termites maintain an obligate exosymbiosis with the fungus *Termitomyces* spp. that they cultivate within the nest on fungus "combs" constructed from faecal material (Wood and Thomas 1989). The fungus is cellulolytic and is responsible for external digestion, while there are also bacteria and protozoa in the gut, which may or may not be symbiotic. Several fungi perform the function of decomposition of termite excreta, which is recycled in the colony. Fungal combs not only provide the decomposed residue to the termite nests but also elevate temperature due to metabolic activity in decaying organic residues outside the termite gut.

12.3 Classification of Termites

Termites can be classified taxonomically using many different features: external morphology, internal features, food and nest type, and chemical and behavioral differences. Soldiers play an important part in termite classification. The most obvious differences are the shape and size of the head and mandibles. There are major differences in distribution, biology, and pest status among the seven conventionally recognized families of termites (Table 12.1) and these families are subdivided to include 15 subfamilies, 270 genera, and over 2,650 species (Kambhampati et al. 1996). Based on the composition of the symbiont microbiota in the gut, termites are divided into two groups: "lower termites" and "higher termites". Lower termites house flagellate protozoans and bacteria, whereas higher termites house a variety of prokaryotic microbes, but no flagellates. Some Termitinae also house cellulolytic amoebae.

12.3.1 Lower Termites

These are the primitive termites and phylogenetically known to harbor cellulose- and/or xylan-digesting flagellate protozoans and bacteria in their hindgut to aid in cellulose decomposition (Breznak and Brune 1994). The protozoans are mainly cellulolytic anaerobes and have the ability to degrade cellulose and other polysaccharides (Bignell and Eggleton 1998). Lower termites are generally found

Table 12.1 Classification of termites

Family	Subfamilies	Important genera
Kalotermitidae	–	*Postelectrotermes, Neotermes, Glyptotermes, Kalotermes, Bifiditermes, Cryptotermes*
Rhinotermitidae	Psammotermitinae, Heterotermitinae, Stylotermitinae, Coptotermitinae, Termitogetoninae, Rhinotermitinae	*Psammotermes, Heterotermes, Reticulitermes, Coptotermes, Schedorhinotermes, Rhinotermes*
Mastotermitidae	–	*Mastotermes*
Termopsidae	Termopsinae, Porotermitinae, Stolotermitinae	*Archotermopsis*
Hodotermitidae	Cretatermitinae, Hodotermitinae	*Anacanthotermes*
Serritermitidae	–	*Serritermes*
Termitidae	Termitinae, Apicotermitinae, Nasutitermitinae, Macrotermitinae	*Eurytermes, Speculitermes, Dicuspiditermes, Hypotermes, Nasutitermes, Termes, Macrotermes, Hospitalitermes, Trinervitermes, Microtermes, Odontotermes*

Source: (Logan et al. 1990)

outside forests or in marginal habitats within forests and mostly all are wood feeders except for Hodotermitidae, which are grass feeders (Collins 1989).

12.3.1.1 Kalotermitidae

Species in this family are often referred to as the dry wood termites from their nesting habit in sound wood and are believed to be a sister group to Rhinotermitidae and Termitidae (Kambhampati et al. 1996). This is the largest family of lower termites, with 25 genera and 350 species (Wood and Johnson 1986). These termites occur in small numbers in rain forests, mainly confined to dead limbs and trunks in the forest canopy (Collins 1988). Many species in this family are serious pests of forest products and two common species found in Malaysian forest are *Cryptotermes cynocephalus* and *C. domesticus*.

12.3.1.2 Rhinotermitidae

Rhinotermids contain six subfamilies: Coptotermitinae, Heterotermitinae, Psammotermitinae, Termitogetoninae, Stylotermitinae, and Rhinotermitinae. This is the most important family of lower termites in Malaysian forest and is often referred to as damp wood termites. They are found in standing or fallen trunks and limbs, and can cause severe damage to timber and living trees. Some of the common genera found in Malaysian and Bornean forests are *Coptotermes*, *Heterotermes*, *Termitogeton*, *Prorhinotermes*, *Parrhinotermes*, and *Schedorhinotermes* (Thapa 1981; Collins 1988; Tho 1992). Important pest species are *Coptotermes curvignathus* Holmgren, infesting rubber trees and pine trees, and *Curvignathus borneensis* Oshima (Collins 1988; Chey 1996). *Heterotermes* is the most primitive genera and contains 23 species, of which seven are oriental. *H. indicola* and *H. malabaricus* occur in the northern and southern regions of India, respectively. *Coptotermes*, essentially a tropical genus, is distributed in Far East, South Asia, Australia, parts of South Africa, and southern regions of the USA.

12.3.2 Higher Termites

This is the largest group dominating the order with over 80% of the genera and 74% of the species (Edwards and Mill 1986; Wood and Johnson 1986). Phylogenetically, they form mutualistic relationship with other microorganisms, usually fungi and bacteria or bacteria alone, despite the presence of endogenous cellulase to digest their food (Slaytor et al. 1997). These termites exhibit a wide variety of social specializations (Breznak 1982). They have a more elaborate external and internal anatomy and social organization compared to lower termites (Breznak and Brune

1994). Higher termites predominate in tropical forest systems as litter, wood, and soil feeders.

12.3.2.1 Termitidae

The family Termitidae contains three quarters of all known species, comprising four subfamilies: Macrotermitinae, Apicotermitinae, Termitinae, and Nasutitermitinae (Wood and Johnson 1986; Collins 1988). One of the important subfamilies is the Macrotermitinae. Genera in this subfamily are known to cultivate species of the symbiotic basidiomycete fungus *Termitomyces* on faecal combs within their nests. These termites have a correspondingly greater impact on decomposition processes than other termites. Macrotermitinae are known to originate from Africa, and of the 13 recorded genera, only four genera occur in Malaysian forests: *Macrotermes, Odontotermes, Hypotermes*, and *Microtermes*. The Macrotermitinae, fungus-growers, are widely distributed throughout the world except at high altitudes and in desert areas. Most feed on dead plant material, but some also feed on living plants causing serious and widespread damage, particularly in semiarid Africa and India. They attack a wide range of crops (Sands 1977) and are major pests of exotic forestry (Cowie et al. 1989). *Microtermes* and *Odontotermes* species are confined to oriental regions including India and Pakistan, which build both subterranean and epigeal nests. Subfamilies Termitinae and Nasutitermitinae include both wood- and soil-feeding species, and they dominate most tropical forest ecosystems. The *Termes*-group belonging to the subfamily Termitinae are some of the soil feeders, possess snapping mandibles, and are widely distributed in Peninsular Malaysia and Borneo. Nasutitermitinae is a highly specialized form of higher termites. Most of these genera found in Malaysian forests are wood feeders and they are able to consume large quantities of dead wood.

The damage-causing termites most frequently belong to the genera *Macrotermes*, *Microtermes*, and to a lesser extent *Odontotermes* and *Ancistrotermes* in Africa and predominantly to *Microtermes* and *Odontotermes* in the Indian subcontinent. *Macrotermes* spp. build large epigeal nests (mounds) from which they forage outward for distances up to 50 m in galleries/runways either just below or on the soil surface (Darlington 1982). They attack plants at the base of the stem, ring-bark, or cutting them through completely (Sands 1973; Cowie and Wood 1989). *Microtermes* genus is comprised of small termites which are earth dwelling and sometimes occurs in the mounds of other termites, especially in the *Odontotermes* genus. Some species are serious pests of crops and forest nurseries. Of the 58 species known, 13 are oriental and found in India, Pakistan, Bengladesh, Sri Lanka, Thailand, Burma, and Indonesia, etc., and damage different crops such as sugarcane, wheat, barley, maize, vegetables, and garden crops. Important species of this genus are *Microtermes incertoides, M. inseparatus, M. mycophagus, M. obesi, M. pakistanikus*, and *M. unicolor*.

Odontotermes is a large genus containing 169 species. Seventy-nine termite species are oriental and found throughout the region. It is confined to the Ethiopian and oriental regions. *Odontotermes* spp. build both subterranean and epigeal nests especially in India. Damage is due to either foraging under soil sheeting on the outer surface of the plants, sometimes leading to severence of the stem (Cowie and Wood 1989) or attack on the roots (Nair and Varma 1985; Mitchell et al. 1987). *Odontotermes assumuthi* Holmgren occurs in India from eastern Punjab to the Gangetic plains of Bihar, Maharashtra, and Chennai in the south. It is a serious pest of sugarcane and also attacks timber, the bark of standing logs, etc. Swarming occurs in the afternoon in the month of June and July from holes in the ground. They attack crops such as sugarcane, cotton, wheat, barley, forest trees, coconut, eucalyptus, tea palm, and dead wood. Other important species of this genus are *Odontotermes ceylonicus*, *O. formosanus*, *O. obesus*, *O. javanicus*, and *O. bengalensis*.

12.4 Control Measures for Termites

Subterranean termites have been around for millions of years and are well adapted to the environmental stresses. Being underground, they are fairly well protected from common hazards and the possibility of direct termiticide application. Wood and moisture are two primary requisites for successful establishment of subterranean termite infestations. It has been recommended that for construction purposes, one should use pressure-treated wood (treated with creosote, pentachlorophenol, or certain inorganic salts). This is particularly important where wood comes into contact with soil. Surface treatment of wood with borates did not protect it from *C. formosanus* (Grace and Yamamoto 1994), although impregnation of wood with borate salts under pressure provides protection from attack if moisture, which can leach the water-soluble salts, can be controlled. To prevent access of subterranean termite species to plants, different strategies are used to prevent the infestation of structures (Su and Tamashiro 1987; Su and Scheffrahn 1990; Yates et al. 1997). Usually, no attempt is made to find and destroy the colony, which may be situated anywhere within a volume of soil of hundreds of millions of cubic meters (Su and Tamashiro 1987). Treatment of soil around the structure, therefore, kills only a small proportion of the colony, the majority of the population surviving to reinfest the structure, either by flying reproductives or by workers entering through an untreated or inadequately treated portion of soil (Su and Scheffrahn 1990).

Until recently, termite management practices had focused on protecting individual structures through application of physical or chemical barriers to exclude termites from the structure, while doing little to mitigate populations of the termites in the surrounding area. Present management paradigms have changed with the advent of new termite control methods to that of reducing termite populations below economic threshold while protecting house, wood products, and trees. The new management paradigms can only be achieved through a combination of existing technologies and continued development of new technologies and integrated pest management concepts.

12.4.1 Cultural Practices

Before the advent of organic synthetic insecticides, the termite control in crops was largely done by adopting suitable cultural practices (Logan et al. 1990). These practices involved crop rotation, breaking up of termite galleries, queen removal, application of wood ash and other materials, and cultivation of termite-resistant crops.

12.4.1.1 Crop Rotation

When natural vegetation is cleared and the land is cultivated, the nests of mound-building termite species are destroyed; and termite species dependent on trees and woody litter eventually disappear. Species with deep subterranean nests and with the ability to survive on particular living crops and crop residues remain and their population increase (Kooyman and Onck 1987; Black and Wood 1989). Crop rotation and fallow periods should prevent the rapid buildup of these species to high levels. Tree crops (e.g., rubber) may not be seriously attacked by termites dependent on wood for nests and food (e.g., *Coptotermes* spp.) if planted on land previously used for several years for annual nonwoody crops (Harris 1971). However, in a number of cases, rotation seems to have led to greater levels of attack. Short-term rotations led to higher attack than longer rotations in cotton in Sudan (Tarr 1960). Cotton in rotation following lubia (*Hyacinth* bean) or dura (sorghum) in Sudan suffered heavier tap root damage than when it followed fallow, but this was considered probably due to the remaining crop residues from the previous year allowing buildup of pest termites rather than due to the direct effects of the rotation (Crowther and Barlow 1943).

12.4.1.2 Breaking Up Termite Galleries

Termite damage is not generally a problem in crops planted on deep cracking soils because the frequent cracking prevents adequate building and maintenance of mounds, runways, and galleries (Lee and Wood 1971). Artificial breaking up of the soil may have a similar effect, reducing termite foraging activity. Deep ploughing is common practice in India (Kakde 1985), and a similar practice has been recommended in South Africa (Otto 1951) and the USA. Beeson (1941) suggested that repeated digging of forest nursery soil beds would reduce termite attack. Regular cultivation around plants to break up termite passages in the soil is effective against shallow-nesting species, but may only have a temporary effect (Coaton 1950).

12.4.1.3 Queen Removal

Removal of the queen and/or destruction of the nest have been advocated frequently for the control of mound-building termites, especially the fungus-growing

Macrotermitinae (Rajagopal 1982; Kakde 1985). The nest is readily identified and the royal chamber is easy to locate. However, if nymphs or alates are present at the time of de-queening, replacement reproductives may develop (Darlington 1985). Dawkins (1949) advised lighting a fire in the mound after queen removal. The foraging activity may be reduced even if de-queened colonies do not die since they remain inactive for 12–18 months until new queens become established (Coaton 1949). Control of *C. curvignathus* Holmgren by queen removal has been attempted, but reproductives of lower termites are replaced even more readily than those of higher termites (Wood 1968); and in forestry it is mature trees that are attacked, which would necessitate continuous destruction as mounds appear (Cowie et al. 1989).

12.4.1.4 Wood Ash and Other Materials

Wood ash heaped around the base of the trunk has been recorded as preventing termite infestation of coffee bushes (Kashyap et al. 1984) and is said to repel them from date palms (Popenoe 1973). It is also effective in protecting the seedlings if mixed into forestry nursery beds or applied as a layer below polythene planting tubes (Beeson 1941) and to protect stored yams, wooden posts, and stacks of hay and maize straw (Malaka 1972). Kerosene, diesel, and crude oil have been recommended to prevent attack on timber and on tree bark (Malaka 1972; Giridhar et al. 1988; Logan and El Bakri 1990). The addition of crude fuel or fish oil emulsion to irrigation water is considered to reduce damage to tree seedlings and sugarcane (Roonwal 1979). Painting round the trunks of fruit trees could prevent foraging on bark and subsequent ring barking (Giridhar et al. 1988).

12.4.1.5 Cultivation of Termite-Resistant Crops

Resistance against termites' functions either by inhibition of pest attack or by the ability of plants to produce normal yields despite attack of termites (Horn 1988). Generally, however, the possibility of using termite-resistant varieties or species of crops and trees has been ignored and the ready availability and high efficacy of organochlorine insecticides have until recently obviated the need for research on resistance. In general, crops showing resistance or tolerance to termites are indigenous, while the susceptible crops are exotic. For instance, in Africa sorghum and millet are more resistant to termites than maize (Cowie and Wood 1989). Cowpea and bambara nuts are not attacked, whereas groundnuts suffer serious damage (Johnson et al. 1981). Presumably, indigenous crops have evolved defense mechanisms against the local termite species. Some exotic plants, such as a mango, avocado, and citrus in South Africa, are resistant to termites (Fuller 1912). When there are major losses to termites, resistant crops could provide options by way to rotation or intercropping, but the overriding factors are likely to be socio-economic. Grafting susceptible fruit trees onto root stocks of resistant species has been practiced successfully (Fuller 1912).

Cultivars of a particular crop may also differ in susceptibility (Johnson et al. 1981; Singla et al. 1988). Amin et al. (1985) screened over 500 groundnut cultivars and found a wide range of resistance to pod scarification by termites (0–44% pods scarified). Mortality of groundnuts due to termites also varied among cultivars (Mercer 1978). Variation in susceptibility between cultivars has also been recorded for tea and sugarcane (Sivapalan et al. 1977; Kumarasinghe and Ranasinghe 1988; Singla et al. 1988). Local varieties of groundnuts in Nigeria whereas castor and cotton in India are selected by farmers over many years and they are more resistant to termite attack than introduced cultivars (Roonwal 1979; Johnson et al. 1981; Parihar 1985). Poor quality, hybrid date palms (*Phoenix dactylifera* L.) in northern Sudan are reported to be more resistant to termite attack than named varieties (Logan and El Bakri 1990).

12.4.2 Application of Physical and Chemical Barriers

Prevention of infestations may involve physical means (e.g., sand or steel-mesh barriers beneath foundations) or more commonly involve chemical controls (Su and Scheffrahn 1998). Conventionally, damage to plants by subterranean termites has been prevented by persistent insecticidal barriers in the soil around the roots or preventing termite access to the crop/tree. Dry wood termites, although considerably less important, have proved extremely difficult to control chemically, but a combination of cultural and chemical control may be effective against them in some instances (Sands 1977; Cowie et al. 1989). Control measures can be divided broadly into four categories, which attempt to (a) prevent termites gaining access to the plants, (b) reduce termite numbers in the vicinity of the plants by application of chemical and plant insecticides, (c) render the plants themselves less susceptible to attack by the termites, and (d) cost-effective management of termites by biological control.

12.4.2.1 Preventing Termite's Access to Plants: Physical Barriers

Anything nonchemical that would stop foraging termites from getting to a potential host is a physical barrier. Physical barriers are primarily used to protect houses and involve the use of particulate matter or metal screens. Particles that are too large for termites to move and yet pack compactly enough to prevent passage between them have been shown to present an effective barrier (Su et al. 1991). Use of stainless steel-mesh barriers is known to stop foraging individuals of several species of subterranean termites (Lenz and Runko 1994). Basaltic termite barrier or gravel barrier is used in Hawaii as a preconstruction treatment (Grace et al. 1996; Yates et al. 2000). A continuous copper barrier is also available for exclusions of termites from structures. Galvanized steel "termite caps" have long been used to cap piers in pier and beam constructions. While these barriers cannot totally prevent termite infestation of a structure, intact barriers force that termite to build shelter tubes around them to gain access to wood and thus expose themselves during structural inspection.

12.4.2.2 Reduction of Termite Numbers by Application of Chemical and Plant Insecticides

The insecticides are either applied in the soil before sowing of crop or applied after seed treatment. Some of the plant extracts have also been found to kill the termites. The conventional but now unacceptable application of broad-spectrum persistent organochlorine insecticides as barriers in the soil prevents termite access to crops, trees, or buildings and is in general equally effective against all subterranean termites.

Use of Chemical Insecticides

Inorganic insecticides, namely, arsenicals and mercuric compounds were used earlier to kill the termites that posed hazards to human beings and cattles. Systematic studies on termite control were, however, taken up only after the World War II. Of late, seed treatment with insecticides has been found not only cheaper but also more effective than above conventional methods. A number of chemicals currently approved for the control of termites have relatively low toxicity (e.g., imidacloprid and hexaflumuron). Pyrethroids are widely used as household insecticides and have a good safety record. Organophosphorus compounds such as chlorpyrifos are widely used in agriculture as effective insecticides, but need to be handled with caution because of their acute neurotoxicity in animals and humans.

Seed treatment of wheat (Bhanot et al. 1991a), barley (Bhanot et al. 1991b), and gram (Bhanot et al. 1995) with insecticides such as aldrin, chlorpyrifos, endosulfan, formothion, and/or monocrotophos has been perfected and recommended. Seed treatment with chlorpyrifos 20 EC @ 10 ml per kg seed has been recommended for termite control in cotton. Application of 6.25 L of chlorpyrifos 20 EC or lindane 20 EC mixed with 1,500–2,500 L of water and sprinkled over the cane sets resulted in lesser termite damage in sugarcane crop (Bhanot and Singal 2007). Thiomethoxam (Actara) was found effective against three species of termites: *T. trinervius*, *O. smeathmani*, and *A. evuncifer*, which are the pests of principal tropical food crops such as maize, rice, sorghum, and sugarcane. The effective dose of thiomethoxam for these termites was 0.03 ppm, which resulted in 100% mortality within 2–8 days depending upon the species studied. Pyrethroid-based soil termiticides are highly repellent and much more suited in high temperature and low moisture environments (Su and Scheffrahn 1998).

Pre- and postconstruction ground treatment with persistent chemicals such as cyclodienes (aldrin, dieldrin, etc.) was fairly effective against subterranean termites. Some of these chemicals had residual lives of 25–30 years in the soil. However, environmental and health concerns led to the withdrawal of their use in the mid-1980s. At present, six insecticides (four pyrethroids, one nicotinoid, and one organophosphate) are widely used in the USA. Besides being more expensive, these compounds are less persistent, requiring more frequent applications. The chloronicotinyls, of which imidacloprid is the most active compound, are not readily detected by the termites and cause discontinued feeding, disorientation, and death (Potter 1997). Another nonrepellent product having unique chemistry, fipronil, has been recently registered and distributed for termite control. Preconstruction chemical treatments include broadcast spraying of the soil subtending the slab with a fix rate of chemical. Construction activities prior to pouring the slab can also disrupt the chemical barrier. Building raised garden beds next to the house using untreated soil provides a bridge over and thus circumvents the barrier. Contrary to the practice of treating houses, chemical treatment of trees is a most recent phenomenon. This is particularly important in the attempts to save valuable trees in urban landscapes. Holes are drilled into a tree trunk and termiticides are injected into the hole, usually formulation into foam, to allow movement of the pressurized steam both upward and downward within the termites' galleries.

Bait treatment: Because of short life of currently used termiticides, the relative difficulty of providing a complete chemical barrier and due to health and environment reasons, monitoring/baiting technologies have been developed to detect the presence of termites and deploy insecticides only when termites are present (Su 1994; Su et al. 1995). Research efforts have focused on alternative means of termite management and recent studies have demonstrated the efficacy of baiting schemes employing insect growth regulators in eliminating subterranean termite colonies (Su and Scheffrahn 1993; Su et al. 1995; Grace et al. 1996). To effectively use baiting technology as a control, the toxicant must be nonrepellent and must be a slow-acting toxin so that termites do not die at the site of the treatment and become secondarily repellent. Further, such a slow-acting toxin allows termites to distribute the toxin more widely throughout the colony by grooming and trophallaxis. One group of such chemical includes the chitin synthesis inhibitors. Hajjar and Casida (1978) reported that derivatives of benzoylphenyl ureas inhibit chitin synthesis in insects and other arthropods, thereby disrupting ecdysis. Ecdysis disruption in *H. indicola* and *Reticulitermes flavipes* by diflubenzuron was demonstrated by Doppelreiter and Korioth (1981). Su and Scheffrahn (1993) showed that hexaflumuron, even though slow acting, was very effective against *C. formosanus* and *R. flavipes*.

Different criteria examined for assessing the efficacy of termite bait treatments include alate numbers, foraging activity prior to and after baiting, and foraging territory (Su and Scheffrahn 1996; Thorne and Forschler 2000). Termite presence is monitored using cellulose substrates examined on a regular basis. Several monitoring/baiting systems are commercially available for both in-ground and above-ground use. Soil monitoring stations are regularly spaced around the structure to be protected. Once termite activity is detected in a monitoring station, the bait is replaced with a toxin-laced substrate. Typically, above-ground stations containing toxin-laced matrix are placed directly in contrast to exposed galleries with active termite and as such no monitoring with untreated cellulose is required.

Use of Plant Insecticides

Locally available parts of plants, plant extracts, and other substances have frequently been claimed to be effective in termite control, although they have received little rigorous assessment in the field. Laboratory studies have identified a wide range of plants containing material toxic or repellent to termites having antifeedant properties and many have been considered for use as insecticides (Stoll 1986; Gerrits and van Latum 1988; Harborne 1988). Ideally, plant insecticides should come from readily available local plants or those which are easy to grow, preferably on poor quality land, so that they do not compete with food or cash crops. The active ingradient should be available with little or no preparation and the plants should not develop into weeds or act as hosts for crop pests. In addition, they should have low toxicity to nontarget organisms, especially beneficial insects and humans.

Many timbers contain chemicals or complex mixtures of chemicals that repel or kill termites or interfere with their gut fauna (Adams et al. 1988; Lin and Wang 1988), but these chemicals are difficult to extract (Carter and de Camargo 1983). Consequently, they are unlikely to be useful as insecticides in their own right, particularly as most of the toxic chemicals in timber are avoided by termites (Carter and Smythe 1974; Carter and de Camargo 1983). Some grasses are actively avoided by termites (Sands 1961) and are planted by farmers to keep termites away from farms and gardens in Nigeria (Malaka 1972). Waste sawdust or wood chips from tree containing repellent chemicals may provide some protection if incorporated into soil or used as a mulch round crops or trees, but this has yet to be tested.

Herbaceous plants or the leaves and fruits of trees are more likely to be effective; they are easily crushed and usually can be used without complex extraction procedures. Laboratory experiments have found numerous materials repellent or toxic to termites in such plant materials, but very few of these have been tested for their effectiveness in the field. The simplest method of application is as mulch. Several mulches using leaves or berries of the plant or oil cakes (the residue after oil such as neem or castor has been extracted) are reported to be effective. Whether this was due to toxic effects or to effects of the mulch per se on soil physical and chemical properties (Lal 1987), which in turn affected plant vigor and susceptibility to termite attack, is unknown. Beeson (1941) gave recipes for mixtures of plant extracts, which prevented termite foraging on trees and attack on wound or fire damage in India. "Gambir mixture" made from the leaves of *Uncaria gambir* (Naucleaceae) or the dried aqueous extract of *Acacia catechu* (Leguminosae) mixed with oil from *Canarium strictum* (Burseraceae), *Hopea* sp. (Dipterocarpaceae), or *Shorea* sp. painted on wounds or fire damage to trees prevented invasion by termites. "Gondal fluid" (castor oil cake mixed with plant extracts from *Gardenia gummifera* (or *G. lucidia*) (Rubiaceae), *Ferula jaeschukeana* (Umbelliferae), and aloes (Agavaceae) (*Agave vera*, *A. candela*, or *A. angustifolia*) and soaked for 2 weeks) painted round the base of a tree gave protection against termite foraging for about 8 months. Giridhar et al. (1988) showed that *Calotropis* (Asclepiadaceae) latex protected wooden pegs for 4 months. Singh et al. (2002) reported antitermiticidal effect of *Calotropis* (*Calotropis procera*) extract on infestation of termites (*O. obesus*) in sugarcane hybrid. Badshah et al. (2004) also observed toxic effects of ak (*C. procera*) plant extracts against termites (*H. indicola* and *C. heimi*) (Isoptera: Rhinotermitidae).

Morales-Ramos et al. (2003) reported that the mortality of termite groups feeding on fully and partially black-stained Alaskan yellow cedar (AYC) was significantly higher than that of groups feeding on pine at the end of 32 week. Wood consumption was significantly different among all treatment groups, with means of 4.07, 8.76, 19.81, and 29.77 mg/day in the unstained, partially, and fully black-stained AYC and loblolly pine, respectively. This suggests that toxic and feeding deterrence properties of AYC heartwood were significantly reduced by black-staining fungus infection, but were not totally lost. Chemical analysis of unstained and black-stained AYC wood showed approximately a 50% reduction in concentration of secondary chemicals in the black-stained wood. Carvacrol was totally absent in the black-stained wood. Concentrations of nootkatone in the black-stained wood were one fourth of those observed in unstained AYC wood. Ahmed et al. (2006) disrupted the bacterial activities in the gut of *M. obesi* via soil treated with seed and leaf extracts of *Withania somnifera*, *Croton tiglium*, and *Hygrophila auriculata* through change in tunneling behavior and termites' gut bacteria in the laboratory. Seed extracts of *W. somnifera* and *H. auriculata* were found highly toxic with LD_{50} (4.31 and 2.98%, respectively) in 6-day period. Reduction in the area of tunneling and the number of bacterial colonies was observed in the soil treated with seed and leaf extracts of three species. There was no tunneling in seed extracts of *W. somnifera* and *C. tiglium* at their 100% concentration. Area of tunneling was short in seed extracts of plants as compared with leaf extracts on numerical terms. The number of gut bacterial colonies was also reduced in termites inhabitant soil treated with seed extracts of *W. somnifera* and *C. tiglium*. It was suggested that extracts of three species were not only toxic to *M. obesi* but also acted arrestants of movements.

Most plant-based insecticides break down rapidly in the soil and so should not present long-term environmental problems. For the same reason, they do not give the prolonged protection to crops and trees required to control termite attack. Current research on other insecticides and formulations may provide a partial solution to the problem in specific cases with an aim to reduce dependency on insecticides and to provide cheap, locally available alternatives for small farmers/foresters.

12.5 Biological Control of Termites

Biological control can be defined as "the control or decrease in the damage caused by termites due to reduction in the number and destructive activity of termites by the use of one or more organisms or with the product of a natural biological process." The microorganisms isolated from the mounds/nests of termites' colonies or from the rhizosphere soil could be screened for their termiticidal activities for subsequent use as biocontrol agents. The biological control agents have various advantages, namely, (1) these organisms are considered safer than many of the chemicals now in use, (2) they do not accumulate in the food chain, (3) self-replication circumvents repeated applications, (4) target organism seldom develop resistance as happens when chemical control agents are used, (5) where less effective than a chemical control agent, the two sometimes can be combined, and (6) properly developed biocontrol agents are not considered harmful to the ecology.

Various organisms such as viruses, bacteria, protozoa, nematodes, and fungi have shown the potential for use in biological control of termites (Lund 1971; Grace 1997; Culliney and Grace 2000; Yu et al. 2006) and serve as alternative to broad-spectrum chemical insecticides. The synergistic combinations of microbial control agents with other technologies will have excellent potential for use in integrated pest management programs. Biological control of termites constitutes a more environmentally acceptable alternative to traditional chemical control measures. When successfully implemented, it can yield permanent, cost-effective management of pest populations with minimal environmental disturbance. Classical biological control programs (i.e., the discovery, release, and colonization of natural enemies: arthropod predators and parasitoids, and pathogens) have proved successful against a variety of pests and knowledge of the role of natural enemies in the ecology of these species has steadily accumulated (DeBach and Rosen 1991). However, any pest control program poses some threat to the environment. Natural enemies introduced for the biological control of specific pests have had negative impacts on populations of nontarget organisms either directly, through predation, parasitism, infection, or competition, or indirectly, in their influence on seemingly unconnected species (Secord and Karevia 1996).

12.5.1 Pathogens of Termites

Soil insects have been considered good candidates for microbial control (Burges 1981). The soil environment offers conditions highly favorable for sustaining infection and promoting epizootics. Biocontrol of subterranean termites by microbial pathogens may be facilitated by the warm, humid environment of the colony that is protected from ultraviolet radiation and crowded. Moreover, their sharing of food (trophollaxis), intimate contact with nest mates (e.g., allogrooming), and transporting of infected cadavers also favor infections with the pathogens (Grace 1994). However, termites have evolved a complex social structure, formidable immune response, and adaptive behavior toward infected individuals so that consistently effective biocontrol by means of a single pathogen is unlikely (Logan et al. 1990). However, few confirmed pathogens have been isolated from termites (Grace 1994; Zoberi 1995). Although many prospective termite pathogens have been tested in the laboratory, relatively few have been evaluated in field trials.

12.5.1.1 Bacteria

The most widely used microbial control agent for control of pest Lepidoptera, Coleoptera, and Diptera insects is *B. thuringiensis*. The insecticidal proteins of *B. thuringiensis* are highly specific insect gut toxins with a superior safety record in regard to their effects on nontarget organisms (Lacey and Goettel 1995). Workers of *M. championi* (Snyder) (Termitidae) and *H. indicola* (Wasmann) (Rhinotermitidae) suffered 100% mortality within 13 days of exposure to two local strains of *B. thuringiensis* in laboratory tests (Khan et al. 1977a). Khan et al. (1985) reported that *H. indicola, M. championi,* and *Bifiditermes beesoni* (Gardner) (Kalotermitidae) to be highly susceptible to infection of *B. thuringiensis* (Bt), a commercial preparation of Bt (Thuricide-HP concentrate), exhibiting 100% mortality within 6 days of exposure. Laboratory colonies of *M. championi*, *H. indicola*, and *B. beesoni* exposed to suspensions of the spore-forming bacterium *S. marcescens* Bizio succumbed completely 7–13 days following infection (Khan et al. 1977b). Khan et al. (1992) tested the pathogenicity of *P. aeruginosa* (Schroeter) against *M. championi*, *H. indicola*, and *C. heimi* (Wasmann) (Rhinotermitidae) in the laboratory. Termite mortality ranged from 25–52% 7 days post-inoculation to 84–100% 25 days post-inoculation. The authors concluded that *P. aeruginosa* is "fairly" pathogenic to the three termite species, although the bacterium's potential as a biological control agent is limited by its occasional status as a plant pathogen.

Connick et al. (2001) reported that *S. marcescens* isolate T8 was highly virulent at the concentration of 3.4×10^{10} cfu/ml to the *C. formosanus*. Termite mortality was 24% by 2 days and 99% after 19 days of the experiment. Osbrink et al. (2001) isolated biological control agents from dead termites and revealed the presence of 15 bacteria and one fungus in dead termites. Multiple strains of *S. marcescens* were isolated and six out of eight strains of *S. marcescens* were reported as biological control agents for *C. formosanus* Shiraki. Bacteria isolated from termite substrata included *C. urealyticum* Pitcher, *A. calcoacet/baumanni/Gen2* (Beijerinck), *S. marcescens*, and *E. gergoviae* Brenner. Kanchana Devi et al. (2007) found that three HCN-producing rhizobacterial species, i.e., *R. radiobacter*, *A. latus*, and *A. caviae* killed *O. obesus* subterranean termites under in vitro conditions. Khan (2006) reported enhancement in virulence of *B. thuringiensis* (about 1.5–1.8) and *S. marcescens* (1.3–1.6) by 1% potassium chloride or 1% sodium citrate against the workers of *M. championi* and caused mortality of termites. The LT_{50} LT_{90} and virulence enhancement ratio showed that 1% sodium citrate when mixed with *S. marcescens* caused quicker rate of mortality of termites as compared to the mixture of 1% potassium chloride amd *S. marcescens*. Boric acid (at 1% concentration) was also found more effective to enhance pathogenicity of *B. thuringiensis* against various species of termites.

The population of bacteria in termite nest soil was determined on nutrient agar medium and King's B medium (Yuvraj Singh 2007), and found to vary from 1.2×10^6 to 256.0×10^6 colony-forming units/g soil. A total of 270 bacterial isolates were screened for the potential to produce different enzymes, i.e., lipase, protease, and chitinase on specific media for their possible involvement in killing of termites. Only 83 bacterial isolates were found to express one or more of the enzyme activities and 12 bacterial isolates expressed all three enzyme activities. The killing frequency of different bacterial isolates varied from 5.7 to 100% at 5 days under laboratory conditions. Bacterial isolates NSY 19, NNY 23, and NKY 83 caused 100% killing of the termites, whereas 14 other bacterial isolates caused more than 82% killing at 5 days of the experiment. Based on the comparative analysis of various morphological and biochemical characteristics, the antagonistic bacteria were found to belong to the genera of *Arthrobacter*, *Bacillus*, *Chromobacterium*, *Enterobacter*, *Micrococcus*, *Neisseria*, *Pseudomonas*, and *Serratia*.

Husseneder and Grace (2005) used indigenous gut bacteria *Enterobacter cloacae* of the Formosan subterranean termite, *C. formosanus* Shiraki (Isoptera: Rhinotermitidae) as shuttle system to deliver, express, and spread foreign genes in termite colonies. The gut bacterium was transformed with a recombinant plasmid (pEGFP) containing genes encoding ampicillin resistance and green fluorescent protein (GFP). In laboratory experiments, termite workers and soldiers from three colonies were fed with filter paper inoculated with transformed bacteria. Transformed bacteria were detected in termite guts by growing the entire gut flora under selective conditions and checking the cultures visually for fluorescence. It was demonstrated that (a) transformed bacteria were ingested within a few hours and the GFP gene was expressed in the termite gut; (b) transformed bacteria established a persistent population in the termite gut for up to 11 weeks; (c) transformed bacteria were efficiently transferred throughout a laboratory colony, even when the donor (termites initially fed with transformed bacteria) to recipient (not fed) ratio was low; and (d) transformed *E. cloacae* were transferred into soil; however, they did not accumulate over time and the GFP plasmid was not transferred to other soil bacteria. In the future, transgenic bacteria may be used to shuttle detrimental genes into termite colonies for improved pest control.

So far, little research has focused on bacterial infections of termites. Despite favorable laboratory results, the potential of bacteria to reduce termite populations has not been demonstrated under field conditions. Moreover, *B. thuringiensis*, in particular, has limited potential for soil insect control because of its poor survival in soil (Burges 1981).

12.5.1.2 Fungi

In recent years, much research interest has focused on the use of fungal agents for pest control (Beal and Kais 1962; Ferron 1978; Yoshimura et al. 1992; Rath 2000; Lacey et al. 2001; Wright et al. 2005). Some 700 species of entomopathogenic fungi have been reported and at least 22 species of fungi are obligate ectoparasites of termites (Blackwell 1980; Lai et al. 1982; Blackwell and Rossi 1986), although their biological control potential apparently has not been evaluated. Many of entomopathogenic fungi, especially those in the Hyphomycetes, demonstrated activity against a broad range of insect pests and are the main contenders for commercial production. Two fungal pathogens, *Beauveria bassiana* (Balsamo)

Vuillemin and *Metarhizium anisopliae* (Metschnikoff) Sorokin (Hyphomycetales), have been extensively evaluated for termite control. *Conidiobolus coronatus* isolates were found to be pathogenic to *C. formosanus*, *R. flavipes*, and *Nasutitermes exitiosus* (Yoshimura et al. 1992; Wells et al. 1995). Recently, Wright et al. (2003) patented *Paecilomyces* spp. for controlling subterranean termites. These *Paecilomyces* strains are nonrepellent, transferred among termites, and caused rapid mortality.

In a series of laboratory studies, Hänel (1981, 1982a, b) investigated the pathogenicity and biological control potential of *M. anisopliae* on the mound-building species *N. exitiosus* (Hill) (Termitidae). There was a clear correlation between conidial concentration and percentage mortality, with average mortality exceeding 95% within 11 days at the highest dosages. Termites were observed to segregate cadavers from the rest of the colony, but, in so doing, came into contact with infective conidia. Subsequent field experiments did not confirm the potential of *M. anisopliae* to give permanent control of *N. exitiosus* colonies (Hänel and Watson 1983). In five out of seven mounds treated with conidial preparations, termite population densities were significantly reduced. Also, termites exposed to conidia at feeding sites were seen to spread infection back to their colonies, which persisted in some for at least 6 weeks. Again, densities in the affected colonies were significantly decreased. However, although the fungus persisted for as long as 15 weeks in termite mounds, its incidence and termite mortality declined progressively over time. Rosengaus and Traniello (1997) tested the pathogenicity of *M. anisopliae* on laboratory colonies of *Zootermopsis angusticollis* (Hagen) (Termopsidae). Mortality due to infection was dosage-dependent, with LT_{50} (median survival time) values ranging from 2.7 days at a concentration of 7×10^7 conidia (spores) ml^{-1} to more than 10 days at 7×10^1 condia ml^{-1}. Termites directly exposed to the highest conidia concentration were capable of infecting unexposed, healthy nest mates. Milner et al. (1998) tested 93 isolates of *M. anisopliae*, (Metschnikoff) obtained from two species of termites. Using direct inoculation of most effective isolate FI-610 by applying 3×10^{11} conidia into termite mounds, successful control of *Coptotermes acinaciformis* (Froggot) was achieved in Australia.

In other laboratory studies, effect of *M. anisopliae* and *B. bassiana* on subterranean termite populations has been found to vary with the termite species and strain of fungus tested, concentration of inoculum, and its mode of delivery to termite populations. Toumanoff and Rombaut (1965) reported 100% mortality in *Reticulitermes santonensis* (de Feytaud) (Rhinotermitidae) within 5–6 days of exposure to *B. bassiana* and *M. anisopliae*. Leong (1966) demonstrated the high pathogenicity of *M. anisopliae* to *C. formosanus*. With short exposures (e.g., 5–35 min), mortality exceeded 86%, and termites succumbed within 3–6 days post-treatment. Exposure over 40 min produced 100% mortality; longer exposure times caused death within 24 h. No significant difference between the pathogenicities of *B. bassiana* and *M. anisopliae* was found. The fungus sporulated only sparsely under conditions of complete darkness, which might limit its utility for biological control of subterranean termites, although pathogenicity was unchanged between dark and light conditions. In another study, lengthy exposure (4 h) to sporulating cultures of *B. bassiana* also led to the rapid death of *R. flavipes* generally in 24–36 h (Bao and Yendol 1971).

Kramm and West (1982) found that conidia of *B. bassiana*, *M. anisopliae*, and *Gliocladium virens* Miller, Giddens, and Foster remained infective after passage through the intestines of subterranean termites (*Reticulitermes* sp.), a significant finding since mutual grooming, trophallaxis, and cannibalism should provide efficient means for spreading the pathogens throughout the colony. Strains of *B. bassiana* and *M. anisopliae* were found about equally effective, producing complete mortality within 2–5 days.

Similarly, Wells et al. (1995) concluded that one isolate each of *B. bassiana* and *M. anisopliae*, of all strains tested, showed the greatest potential for control of *C. formosanus* populations, based on LD_{50} (median lethal dose; as conidia per insect), time to death, and conidial production. However, Lai et al. (1982) using other strains of these fungi found that *M. anisopliae* was more pathogenic than *B. bassiana* in conidial suspensions topically applied to *C. formosanus* colonies. A large increase in conidial concentration (at least tenfold) was required to increase significantly the rate and extent of infection in the colonies. Grace (1991) also found *M. anisopliae* to cause greater and more rapid mortality than *B. bassiana* in *C. formosanus* to low conidial concentrations. In their evaluation of both pathogens, Jones et al. (1996) found the *M. anisopliae* strains to be more virulent, in general, than those of *B. bassiana*. However, the *M. anisopliae* strains showed lower LT_{50} values. The authors concluded that one strain of *B. bassiana* possessed greater potential as a remedial control for *C. formosanus*, based, in sum, on its lower LC_{50} value (median lethal concentration), moderately low LT_{50}, its transmissibility, and survivorship qualities.

Sajap and Kaur (1990) showed *C. curvignathus* (Holmgren) (Rhinotermitidae) to be highly susceptible to infection by *M. anisopliae*. Termites died within 36–48 h of exposure to conidia. Another study of *M. anisopliae*, in which termites exposed to conidia were introduced into laboratory colonies, reported 100% mortality in the test colonies within 8 days, even when the termite vectors carried low conidial loads (Zoberi 1995). Spread of the disease through colonies was complete, even though some of the fungus-killed termites were buried by nest mates, impairing growth of the fungus and preventing dispersal of conidia. Results suggested that epizootics within natural subterranean colonies might be initiated by a sufficient number of vector termites contaminated with large doses of inoculum introduced directly into the soil. Rath and Tidbury (1996) found that *C. acinaciformis* (Froggatt) (Rhinotermitidae) and *N. exitiosus* were equally susceptible to direct conidial applications of both Australian and American strains of *M. anisopliae*. However, Milner (1992) reported that, in general, *C. acinaciformis* and *Coptotermes frenchi* (Hill) (Rhinotermitidae) were most susceptible to *M. anisopliae* than *N. exitiosus*. Subsequently, an isolate of fungus, which was highly pathogenic to all three termite species in the laboratory, reportedly showed some potential to control termite population in field trials (Milner et al. 1996). In experiments in which conidia of two fungi were introduced directly into galleries and onto termites subsequently returned to the nest. Lai (1977) failed to initiate an epizootic in colonies of *C. formosanus*, even though the concentration used was sufficient to kill termites within 3–7 days in the laboratory. The negative results were attributed to poor germination of conidia in the nest and elimination of conidia from treated termites through grooming or secretion of fungistatic agent. The field efficacy of the pathogens could not be demonstrated despite successful laboratory trials (Suzuki 1991, 1996; Gitonga 1996). To date, injection of large quantities of conidia directly into the termite nest has had the greatest success in the field studies (Fernandes 1991; Milner et al. 1996).

A strain of *B. bassiana* isolated from *R. flavipes* and pathogenic to that species (Zoberi and Grace 1990) and to *C. formosanus* (Grace 1991) showed that, although *R. flavipes* workers isolated fungus-killed individuals, there was no avoidance of mycelia or conidia. In a later study, Grace and Zoberi (1992) found that living *R. flavipes* workers exposed to sporulating *B. bassiana* cultures effectively spread infection to unexposed nest mates, whereas introduction of fungus-killed workers did not result in sufficient spore transfer or mycelial growth to cause significant mortality. However, the level of mortality achieved in the laboratory was not considered sufficient to control a termite infestation in the field. Similarly, Kramm et al. (1982) showed that active *Reticulitermes* sp. workers exposed to whole cultures of *M. anisopliae* transferred the pathogen to previously healthy termites through grooming. However, termites killed by the fungus were avoided by healthy individuals and were less effective in spreading the disease to nest mates.

Fungi, *B. bassiana* and *M. anisopliae*, may yet find their most effective use in the form of baits, whereby the termites themselves pick up and introduce infective agents into the colony, and research in this area is progressing (Delate et al. 1995; Jones et al. 1996). Effective baiting schemes for termite control require delivery of sufficient spore inoculum to active termites and subsequent transfer from them to nest mates without stimulating colony-defensive behaviors. For example, Preston et al. (1982) found that baits composed of *M. anisopliae* conidia and culture medium (containing a mixture of conidia, mycelia, and metabolites), incorporated in different combinations into paraffin-woodmeal composite blocks, exhibited uniform repellency to termites (*Reticulitermes* sp.), but induced differential mortality. Blocks containing conidia alone produced no detectable mortality, whereas the highest mortalities were associated with blocks containing medium, which remained effective as long as 1 month. However, the fungus could not be isolated from cadavers. Other studies have reported the conidia (Rath and Tidbury 1996) and toxins (Wahlman and Davidson 1993; Grace 1995) produced by *M. anisopliae* to be repellent to termites. Delate et al. (1995) achieved complete mortality of termites within 15 days with isolates of both fungi. However, in this case, neither avoidance of fungi by termites nor isolation of infected individuals from the rest of the colony was detected. Results suggested that fungal bait stations might be useful in termite control by providing a continuous and nonrepellent source of sporulating cultures for foraging termites to contact, although a self-perpetuating infection within the experimental colonies was not demonstrated.

Other laboratory studies of bait have assessed the infectivity of sporulating strains of *B. bassiana* and *M. anisopliae* to *C. formosanus*. In preliminary work, Grace (1993) reported that termites exposed to baits accumulated spore loads

(6–8 million conidia per individual) that greatly exceeded previously established lethal concentrations (LC_{95}; 490–20,000 conidia per individual) and were capable of transmitting the pathogens to other colony members. Sun et al. (2002) quantified the sporulation of 22 total isolates of *M. anisopliae* and *B. bassiana* on cadavers of the Formosan subterranean termite, *C. formosanus*. Conidial production increased significantly over 11 days post-death. Effects of isolates of *M. anisopliae* and *B. bassiana* on in vivo sporulation were significant. In vitro and in vivo sporulation differed by as much as 89× and 232× among the selected isolates of *M. anisopliae* and *B. bassiana*, respectively. Wright et al. (2005) observed that a single fungal isolate, C4-B, taxonomically identified as *M. anisopliae* (Metschnikoff) was found to cause rapid mortality on Formosan subterranean termite alates. In initial experiments, C4-B was more lethal to both alates and workers compared with *M. anisopliae* strains ESC 1, previously marketed as the termite biocontrol agent, BioBlast. Dose–response assays in which Formosan subterranean termite alates were exposed to a known concentration of C4-B spores revealed that 10^6 spores/μl killed 100% of the alates in 3 days, both 10^5 and 10^4 spores/μl in 6 days, 10^3 spores/μl in 9 days, and 10^0 spores/μl in 12 days. Assays with workers demonstrated that 10^6 and 10^5 spores/μl killed 100% of the workers in 6 days. In an experiment to test the transfer of inoculum from infected workers to uninfected nestmates, 62.8% of the workers died in 21 days when only 20% of the workers had been inoculated. Mortality of alates caused by C4-B was tested at two field sites by dispersing fungal spores on grassy lawns and collecting alates from the treated areas. Alates thus infected showed 100% mortality by day 5, whereas only 64.8% of untreated control alates from the same collection area were dead on that day.

Thus, fungi seem to offer the potential for at least some measure of termite control of all pathogens tested (Milner and Staples 1996). However, the use of fungi in control programs is compromised by inherent biological limitations and logistical problems. Fungi have a slow mode of action and require high levels of relative humidity; there is also the need for large quantities of infective conidia to contact the target pest population in order to yield an acceptable level of control (McCoy 1990). It has also been suggested that, because out of many fungal strains evaluated for termite control, few strains were effective even under favorable conditions and termites may have evolved a degree of resistance to fungal pathogens (Burges 1981). Because of the many difficulties involved in their use, there are, at present, few commercial fungal preparations available for pest control. Further development for biological pest control of *B. bassiana* and *M. anisopliae* continues to be hampered by a lack of cost-effective methods for mass production (Federici 1990).

12.5.1.3 Viruses

A large number of viruses offer potential as microbial control agents of insects (Payne 1982). Those with the greatest microbial control potential are in the Baculoviridae (nuclear polyhedrosis viruses and granuloviruses). More than 400 insect species, mostly in the Lepidoptera and Hymenoptera, have been reported as hosts for baculoviruses. However, viral infection of termites has been little reported. Gibbs et al. (1970) isolated a virus infecting *Coptotermes lacteus* (Froggatt) (Rhinotermitidae), which was similar to acute paralysis virus of the honey bee *Apis mellifera* Linnaeus (Hymenoptera: Apidae). A nuclear polyhedrosis virus, obtained from caterpillars of *Spodoptera littoralis* Boisduval (Lepidoptera: Noctuidae), was infective to a laboratory colony of *K. flavicollis* (Fabricius) (Kalotermitidae) (Al Fazairy and Hassan 1988). Termites died 2–10 days post-infection under laboratory conditions and the authors suggested that control of *K. flavicollis* with NPV might be feasible. However, the potential of viruses for termite control has yet to be evaluated in the field (Al Fazairy and Hassan 1993).

Accessibility of the pest to be controlled is the prime factor affecting the efficacy of viral pathogens. Insects that feed openly on the foliage of host plants are most easily treated and the most promising results have been obtained against pest of this type (e.g., caterpillars sawfly larvae) (Smith 1967). Insects living in concealed habitats, such as the soil, are most difficult to infect. The efficacy, specificity, and production of secondary inoculum make baculoviruses attractive alternative to broad-spectrum insecticides and ideal components of integrated pest management (IPM) systems due to their lack of untoward effects on beneficial insects including other biological control organisms (Cunningham 1995). Unfortunately, there are other drawbacks to the use of viruses to suppress pest populations: viruses kill their hosts slowly compared to other pathogens; environmental factors such as rainfall and solar radiation may reduce viral persistence in soil; mass production of viruses is hampered by the need for living hosts or tissue culture; lastly, viral formulations have had difficulty in competing successfully, on the basis of performance and cost, with other pest control products such as chemical insecticides or even other microbial agents (Klein 1988; Fuxa 1990).

12.5.1.4 Protozoa

Protozoan diseases of insects are ubiquitous and comprise an important regulatory role in insect populations (Brooks 1988). They are generally host specific and slow acting, most often producing chronic infections. Entomopathogenic protozoa develop only in living hosts and many species require an intermediate host. Species in Microsporida are among the most commonly observed. Their main advantages are persistence and recycling in host populations and their debilitating effect on reproduction and overall fitness of target insects.

Of the four groups of protozoa containing species parasitic to insects, the Phylum Microspora includes species that are potentially most useful in applied insect control (Henry 1990). However, the amount of research pertaining to the biological control of termites using these agents is negligible. Desportes (1963) described a gregarine (Phylum Apicomplexa) from the hemocoel of the damp wood termite *Zootermopsis nevadensis* (Hagen) (Termopsidae). No assessmemt of the microbes' potential for termite control was given. Microsporidians were found in the body cavity and proventriculus of *M. championi* collected from the roots of *Saccharum munja* Roxburgh (Poaceae) (Jafri et al. 1976). The organisms attacked fat body tissues in the midgut after ingestion with food and caused death. Although protozoa may be important agents in the natural control of many insects, they have not been used as soil-applied microbial insecticides because they tend to be slow acting and cause low levels of immediate mortality. Moreover, they are vulnerable to changes in environmental conditions (Klein 1988; Henry 1990).

12.5.1.5 Nematodes

Nematodes are ubiquitous roundworm of the Phylum Nematoda found in nearly all environments throughout the world. A plethora of nematode species in more than 30 families are associated with insects and other invertebrates (Kaya and Gaugler 1993). The roundworms of the Phylum Nematoda play an important role in the natural control of many insect populations. They have been found parasitizing species in the orders Hemiptera, Diptera, Hymenoptera, Lepidoptera, Orthoptera, Coleoptera, Thysanoptera, Siphonaptera, as well as Isoptera (Nickle and Welch 1984). Four families of nematodes – the Mermithidae, Allantonematidae, Steinernematidae, and Heterorhabditidae – have shown promise for use in insect control programs (Popiel and Hominick 1992), with most research focusing on species in the latter two (Kaya and Gaugler 1993). After infection of the host, symbiotic bacteria are released into the insect hemocoel, causing septicemia, and death (Kaya and Gaugler 1993). These entomopathogenic nematodes have a number of characteristics that make them especially suitable for biological control and for commercial production as microbial insecticides: a broad host range, especially among soil-dwelling pests; ease of production, storing, and application; a high degree of safety to vertebrates, plants, and other nontarget organisms and amenability to genetic selection (Kaya and Gaugler 1993; Kaya et al. 1993).

Few studies have addressed the potential for nematodes to control termites. In an early report, Merrill and Ford (1916) found 77% of colonies and up to 100% of individuals, of *Reticulitermes lucifugus* (Rossi) (Rhinotermitidae) to be parasitized by *Mikoletzkya aerivora* (Cobb) (Diplogastridae) in field samples. Laboratory exposure of termites to nematodes resulted in successful parasitization (47% at 4 days) and 100% mortality of infected individuals at 12 days. Fujii (1975) obtained 96% mortality in *C. formosanus* within 7 days of exposure to infective-stage *Steinernema carpocapsae* (Weiser) (Steinernematidae) in laboratory experiments. Mortality exceeding 95% was recorded by Georgis et al. (1982) for both *Zootermopsis* sp. and *Reticulitermes* sp. within 3 days after laboratory exposure to *S. carpocapsae*; termites also were found to carry infection back to their colonies. High rates of infection of *Nasutitermes costalis* and *R. flavipes* with *S. carpocapsae* were also reported under laboratory conditions (Laumond et al. 1979; Trudeau 1989).

Danthanarayana and Vitarana (1987) showed that a single application of nematodes (*Heterorhabditis* sp.; 4,000–8,000 ml^{-1} of suspension) to galleries in tea plants eliminated termite colonies within 2–3 months. Evidence suggested that nematode populations were self-perpetuating in the field, even under extreme environmental conditions. Epsky and Capinera (1988) reported an ability of the subterranean termite *Reticulitermes tibialis* (Banks) (Rhinotermitidae) to avoid contact with nematodes *Steinernema feltiae* and exploited gaps in coverage in nematode-inoculated areas to attack the bait. The field results contrasted with those from their laboratory experiments, in which no avoidance of nematode-infested areas was found. In a similar study using *Heterorhabditis heliothidis* (Heterorhabditidae) as well as *S. carpocapsae* against *Reticulitermes* spp., nematodes applied to plots were not found to reduce feeding damage to wooden bait blocks and showed no effectiveness in controlling or eliminating the termites (Mauldin and Beal 1989). Perhaps the only example of apparently successful biological control of a termite with nematodes involved not a subterranean but a drywood species, *Glyptotermes dilataus* (Bugnion and Popoff) (Kalotermitidae), a pest of tea, *Camellia sinesis* (L.) Kuntze (Theaceae).

Yu et al. (2006) reported that entomopathogenic nematodes *Steinernema riobrave* Cabanillas, Poinar, and Raulston (355 strain), *S. carpocapsae* (Weiser) (Mexican 33 strain), *S. feltiae* (Filipjev) (UK76 strain), and *Heterohabditis bacteriophora* Poinar (HP88 strain) were all capable of infecting and killing three termite species, *Heterotermes aureus* (Synder), *Gnathamitermes perplexus* (Banks), and *R. flavipes* (Kollar) in laboratory sand assays. *S. riobrave* and *S. feltiae* caused low levels of *Reticulitermes virginicus* (Banks) mortality under the same conditions. Nematode concentration and incubation time had significant effects on the mortality of worker *H. aureus*. *S. riobrave* consistently generated the highest infection levels and mortality of *H. aureus* in sand assays.

Although field infections often result from nematode treatments, the impact on target pest populations most often has been modest or negligible, even when billions of nematodes ha^{-1} were applied (Kaya and Gaugler 1993). Thus, the potential for use of nematodes in biological control may be limited to cases in which chemical insecticides are not practical or appropriate (Nickle and Welch 1984). It may be feasible in limited cases to control crop-infesting termites with nematodes; in general, the utility of nematodes in biological control programs is compromised by a number of factors that influence their effectiveness. These include soil physical and chemical properties (e.g., moisture, temperature, pore size, compaction, oxygen and carbon dioxide levels, pH, salinity, and presence of synthetic chemicals) and biotic factors such as competitive or predatory interactions with other soil organisms, limited motility, and termite behaviors (e.g., isolation of infected individuals from the rest of the colony that limit the spread of infection) (Gaugler 1988; Kaya 1990; Popiel and Hominick 1992). More in-depth information on their ecology, biology, and genetics along with their genetic engineering and combinations with other control agents offers promise in insect suppression.

12.5.2 Predators

Termites have a wide variety of predators, both vertebrate and invertebrate. Attack occurs mainly upon alate reproductives or foraging workers outside the nest (Weiser and Hardy 1962; Logan et al. 1990), but a few predators attack termites within the nest (Sheppe 1970). Among the vertebrates, fish, anurans, lizards, snakes, birds, and mammals (including humans) are known to take termite prey when available (Nutting 1969). Insect and other arthropod groups reportedly preying opportunistically on termites include Scorpiones, Solifugae, Acari, Opiliones, Araneae, Chilopoda, Thysanura, Anisoptera, Blattaria, Mantodea, Phasmida, Gryllidae, Reduviidae, Neuroptera, Carabidae, Elateridae, Staphylinidae, Diptera, and aculeate Hymenoptera (Phillipsen and Coppel 1977; Johnson and Hagen 1981; Pearce 1997), although all orders of entomophagous insects probably contain at least some species that feed on termites (Deligne et al. 1981).

Ants are the greatest enemies of termites in all regions of the world (Holldobler and Wilson 1990; Culliney and Grace 2000). A large percentage of ant species, including those from the two largest genera, *Pheidole* Westwood and *Camponotus* Mayr, are opportunistic predators of termites (Holldobler and Wilson 1990). Veeresh and Gubbaiah (1984) observed the long-legged ant *Anoplolepis longipes* (Jerdon) (Formicidae), a major agricultural pest in India, to prey facultatively upon termites of an unidentified species. Wilson and Oliver (1969) found that 16.3% of prey taken by fire ants *Solenopsis richteri* Forel (Formicidae) in cutover pine forest consisted of termites (*Reticulitermes* sp.). Laboratory experiments suggested that *C. formosanus* might be relatively more resistant to fire ant predation than *Reticulitermes* spp. owing to a higher proportion of soldiers in colonies of the former and their more aggressive response to disturbance (Cornelius and Grace 1996, 1997).

Specialized arthropod predators of termites apparently are confined to a few ant species in the subfamilies Ponerinae and Myrmicinae (Ohiagu and Wood 1976; Longhurst and Howse 1979; Lepage and Darlington 1984), some of which may serve as important regulators of termite populations in natural ecosystems. For example, *Decamorium uelense* (Sanstchi) (Formicidae) consumed up to 632 *Microtermes* spp. m^{-2} y^{-1}, more than 74% of the *Microtermes* standing crop population and approximately one eighth of annual production (Longhurst et al. 1979). Recent studies have examined the potential for use of ant semiochemicals as termite repellents and toxicants. In laboratory tests, *C. formosanus* workers avoided contact with filter paper disks treated with extracts of the ant *Ochetellus glaber* (Mayr) (Formicidae); no evidence of termite habituation to the extracts was detected (Cornelius and Grace 1994). Extract-treated sand barriers deterred tunneling completely for 2–4 days at the higher extract concentrations tested, although partial penetration was seen in succeeding days. Further work showed sand treated with *O. glaber* extracts to be toxic to *C. formosanus* workers, causing 100% mortality after 24 h at the higher concentrations used; median concentrations caused lower mortality but left survivors immobilized (Cornelius et al. 1995).

12.5.3 Parasitoids

The main advantage to employ parasitoids for pest control derives from their high degree of host specificity and parasitoids have been used successfully in biological control programs worldwide. However, little information is available on parasitoids of termites and none of it seems to have relevance to biological control. Certain species of the family Phoridae (Diptera) are parasitic on termites (Disney and Kistner 1989, 1990) and might play a role in regulating termite densities (Disney 1986). Moreover, not all termite species are equally vulnerable to predation. The dispersed and modular nature of the *C. formosanus* colony, as well as its protected underground location, may limit the impact of predation on this termite species. Similarly, evidence suggests that parasitoids may be of limited value in controlling populations of concealed (including subterranean) pests in general (Gross 1991). Thus, there would seem to be little potential for use of predators or parasitoids to control *C. formosanus*.

12.6 Mechanisms Involved in Biocontrol of Termites

12.6.1 Toxin Production

Most of the insecticidal activity of *B. thuringiensis* is associated with the proteinaceous toxins located in the parasporal inclusion bodies, also known as parasporal crystals. Collectively, the toxins found in the parasporal crystals are referred to as δ-endotoxins. The Cry1 proteins (protoxins) which are found in the crystals are biologically inactive. Following ingestion and solubilization in the alkaline midgut, cleavage by gut proteases produces a 60- to 65-kDa activated protein that recognizes specific binding sites at the brush border membrane surface of epithelial columnar cells lining the gut lumen (van Rie et al. 1989; Honee et al. 1991). The next steps are pore formation, membrane transport disruption, and cell lysis leading ultimately to insect death (Schnepf et al. 1998). Exposure of laboratory colonies of the subterranean species, *R. flavipes* (Kollar) and *Reticulitermes hesperus* Banks (Rhinotermitidae), to a mixture of soluble endotoxin, spores, and inclusion bodies of *B. thuringiensis* (Berliner) resulted in greater than 95% mortality after 6 days (Smythe and Coppel 1965). By employing a commercial preparation of Bt (Thuricide-HP concentrate), Khan et al. (1978, 1985) found *H. indicola*, *M. championi*, and *B. beesoni* (Gardner) (Kalotermitidae) to be highly susceptible to infection, exhibiting 100% mortality within 6 days of exposure.

Grace and Ewart (1996) constructed recombinant cells of the bacterium *Pseudomonas fluorescens* that are induced to express the δ-endotoxin of *B. thuringiensis* (Bt). Two commercial agricultural formulations prepared by the CellCap process were evaluated for palatability to the termite *C. formosanus*. The MVP formulation, active against Lepidoptera, contained the *P. fluorescens* encapsulated δ-endotoxin of Bt var. *kurstaki*. The M-Trak™ formulation, active against Coleoptera, contained the δ-endotoxin of Bt var. *san diego*. Papers treated with either formulation at concentrations as great as 1 g cm^{-3} were readily fed upon by *C. formosanus*. As expected, the two formulations tested were not significantly toxic to the termites, both having optimal activity at a pH range outside that of the termite gut. The palatability of the CellCap formulations indicated that the host bacterium, *P. fluorescens*, is a suitable delivery system for genetically engineered termiticides. Thus, recent work incorporating recombinant DNA and microencapsulation technologies shows promise for application to termite control strategies employing Bt strains that express termite-specific endotoxin (Grace and Ewart 1996).

Gunner et al. (1994) suggested that the spores of entomopathogenic fungi may contain toxins, which would kill the termite host when ingested. Insecticidal cyclic depsipeptides were found to be produced by entomopathogenic fungi including the destruxins from *M. anisopliae* var. *major* (Kaijiang and Roberts 1986), *Aschersonia* spp. (Krasnoff and Gibson 1996), and the beauvericins from *B. bassiana* (Jegorov et al. 1989). It has been suggested that depsipeptides are localized on the surface of spores of *Beauveria* spp. (Jegorov et al. 1989), whereas *Metarhizium* destruxins are generally associated with in vivo or in vitro mycelial growth (Chen et al. 1999).

12.6.2 Siderophore Production

Extracellular siderophores of the brown-rot wood decay fungus *Gloeophyllum arabeum* (Persoon: Fries) Murnill (Polyporaceae) were found to inhibit feeding by *C. formosanus* (Grace et al. 1992). Siderophore-treated filter paper disks showed negligible feeding, whereas untreated disks were almost completely consumed over a 3-day test period. These results suggested that natural products, such as ant semiochemicals and fungal metabolites, or their synthetic analogues, might be of value in termite control programs as repellents or insecticides in wood treatments or soil applications (Amburgey and Beal 1977). However, the development of more stable formulations, such as microencapsulation, would be necessary to ensure their long-term, residual action.

12.6.3 Protease Production

A large number of microorganisms including bacteria, fungi, and actinomycetes are capable of producing proteases from various types of natural resources (Horikoshi 1971; Manachini et al. 1988; Choi et al. 1997). Lysenko and Kucera (1971) showed that *S. marcescens* produced extracellular proteases that could be a mode of pathogenicity of these bacteria in termites. Osbrink et al. (2001) examined 15 bacteria and one fungus associated with dead termites as possible biological control agents against Formosan subterranean termites, *C. formosanus*

Shiraki. Bacterial isolates from dead termites were primarily *S. marcescens* Bizio that caused septicemia in *C. formosanus* and found to contain proteolytic enzymes. Six of the eight strains of *S. marcescens* were red and could be the candidates as biological control agents for *C. formosanus*.

12.6.4 Lipase Production

Lipases (triacylglycerol acyl hydrolases) are one of the most important class of hydrolytic enzymes that catalyze both the hydrolysis and the synthesis of ester formed from glycerol and long chain fatty acids. Lipases are ubiquitous enzymes produced by all biological systems, namely, animals, plants, and microorganisms. Jafri et al. (1976) found microsporidians in the body cavity and proventiculus of *M. championi* collected from the roots of *S. munja*. The organisms passed into the midgut after ingestion with the food, attacked fat body tissues, and caused death of termites, indicating the role of lipolytic activity in termite killing.

Yuvraj Singh (2007) reported that killing frequency of different bacterial isolates obtained from the nest of termites varied from 5.7 to 100% at 5 days of observation. Bacterial isolates NNY 23, NSY 19, and NKY 83 caused 100% killing of the termites, whereas 14 other isolates caused more than 82% killing at 5 days. The most effective bacterial isolate NNY 23 was found to possess all the three enzyme activities, i.e., lipase, protease, and chitinolytic activity. Lack of correlation between enzyme activities and termite killing indicated that besides the production of three enzymes, some other metabolites (toxin or siderophore) could also be contributing to the killing of termites.

12.6.5 Chitinase Production

Lysis by hydrolytic enzymes excreted by microorganisms is a well-known feature of mycoparasitism. Chet et al. (1990) cloned the gene encoding chitinase enzyme from *S. marcescens* and transferred into *Escherichia coli*. The partially purified chitinase caused extensive bursting of the hyphal tips of fungi. Vaidya et al. (2003) isolated hyperchitinase producing mutants of *Alcaligenes xylosoxydans* and developed a rapid technique for screening of chitinolytic bacteria using chitin-binding dye calcofluor white M2R in chitin agar. Microorganisms possessing high chitinolytic activity gave clear zone under ultraviolet light after 24–48 h of incubation. The mutant *A. xylosoxydans* EMS 33 was found to produce three to four times more chitinase then wild type. Chitinase production has also been reported in *S. marcescens*, *Pseudomonas* sp., *Bacillus* strains, *P. stutzeri*, *Paenibacillus* sp., and *Pseudomonas maltophila* (Chet et al. 1990; Lim et al. 1991; Sindhu and Dadarwal 2001). Yuvraj Singh (2007) analyzed termite killing bacteria for the chitinolytic activity and reported that chitinolytic activity was observed in some of the bacteria that killed the termites. Thus, there is immense possibility that production of chitinase enzyme could contribute toward mortality of termites.

12.6.6 Production of Other Secondary Metabolites

Hydrogen cyanide (HCN) is produced by many rhizosphere bacteria and has been demonstrated to play a role in the biological control of the plant diseases (Voisard et al. 1989). HCN-producing *P. aeruginosa* has been shown to have lethal effects on nematodes (Darby et al. 1999; Gallagher and Manoil 2001). HCN-producing bacteria could be selectively introduced into termite mounds, thereby localizing cyanide production and minimizing potential deleterious effects on other soil fauna. Kanchana Devi et al. (2007) tested three different species of HCN-producing rhizobacteria for their potential to kill subterranean termite *O. obesus*, which is an important pest of the Indian subcontinent that causes extensive damage to major agricultural crops and forest plantation trees. The three bacterial species, *R. radiobacter*, *A. latus*, and *A. caviae*, were found effective in killing the termites under in vitro conditions. *R. radiobacter* and *A. latus* caused 100% mortality of the termites following 1 h incubation. *A. caviae*, which produced significantly lower amounts of HCN, caused only 70% mortality. Termites exposed to exogenous KCN showed 80% mortality at cyanide concentrations up to 2 μg/ml. The observed HCN toxicity in termites could be by inhibition of the respiratory enzymes.

12.6.7 Suppression/Activation of Immune System

Various nonpathogenic rhizosphere bacteria have the ability to induce a state of systemic resistance in plants, which provides protection against a broad spectrum of phytopathogenic organisms including fungi, bacteria, and viruses (Kempe and Sequeira 1983). Connick et al. (2001) reported that the biological control of termites may be facilitated if their highly evolved immune systems can be suppressed. Eicosanoids (C20 polyunsaturated acids) have been found to play an important role in protecting insects from bacterial infections (Miller et al. 1994). In laboratory experiments, the eicosanoid biosynthesis inhibitors dexamethasone, ibuprofen, and ibuprofen-sodium salt were each provided along with a red-pigmented isolate of *S. marcescens* Bizio to the Formosan subterranean termite, *C. formosanus* Shiraki, by means of treated filter paper (Stanley-Samuelson et al. 1991). The increased mortality that resulted with dexamethasone and ibuprofen supported the hypothesis that the termites' immune systems were suppressed by these compounds, making the insects more vulnerable to infection by *S. marcescens* (Connick et al. 2001). This effect on mortality was noted only at 3.4×10^{10} colony-forming units ml^{-1} treatment level. A significant amount of the infection and subsequent mortality may have resulted from direct contact with the bacterium and the remainder from its ingestion.

12.7 Approaches to Increase the Efficiency of Biocontrol Agent

Recently, microbial control has been a component of integrated pest management strategies in developing countries, enjoying particular success in Asia and South America (Fuxa 1989). In the developed countries, despite considerable popularity and optimistic predictions for the commercialization of biological control agents, the anticipated demand for these products has not materialized. The market for microbial pesticides is growing but still only represents less than 1% of the total crop protection market and most of this is accounted for by products based on *B. thuringiensis* (Lisansky 1997).

Biological control of termites has received little attention. Although laboratory studies have hinted at its potential, classical biological control of termites has not yet been convincingly demonstrated. The antagonistic bacteria reported for killing of termites belong to different genera, i.e., *Acinetobacter*, *Aeromonas*, *Alcaligenes*, *Arthrobacter*, *Bacillus*, *Chromobacterium*, *Corynebacterium*, *Enterobacter*, *Micrococcus*, *Neisseria*, *Pseudomonas*, *Rhizobium*, and *Serratia* (Osbrink et al. 2001; Kanchana Devi et al. 2007; Yuvraj Singh 2007). Besides bacteria, fungi *B. bassiana* and *M. anisopliae* and nematodes have shown promise to be used as effective termiticides. There are international patents (or patent applications) which involve the use of *M. anisopliae* in control of termites (Gunner et al. 1994), and Bio-Blast™. Biological termiticide, which is based on *M. anisopliae* ESF-1 (EcoScience Corporation, NJ, USA), is on sale in the USA following US-EPA registration in 1994 (EcoScience 1997). However, the limited numbers of field trials that have been attempted largely have failed to establish the permanent, self-sustaining populations of control agents necessary to reduce termite densities below economically damaging levels. This lack of success has resulted from a combination of factors including the complex interactions between the biocontrol agent, the termites, and the environment. Intrinsic limitations of natural enemies, narrow environmental tolerances, antagonistic interactions with other organisms, limited mobility, or a low level of virulence diminish their effectiveness as agents of damage control under field conditions. Also, the effectiveness of a given biocontrol agent may be restricted to a specific location, due to the effects of soil and climate. Another factor that can contribute to inconsistent performance of biocontrol agent is variable production or inactivation in situ of bacterial/fungal/nematode metabolites responsible for killing of termites. Introduced subterranean termite enemies conceivably might adversely affect nontarget termite species or have secondary effects on other organisms coming into contact with dead or dying termites (Secord and Karevia 1996). Thus, inconsistency in performance is a major constraint to their widespread use as biocontrol agent in commercial agriculture.

Various metabolites such as production of toxins, siderophores, or HCN are directly or indirectly involved in killing of termites. A good antagonistic bacterial strain should produce the secondary metabolites under variable growth conditions. Therefore, such antagonistic strains should be selected, which show a constant and medium-independent production of secondary metabolites. In addition, genetic manipulation of biocontrol agent has the potential to construct significantly better strains with improved biocontrol efficacy. Future strategies are required to clone genes involved in the production of toxins, siderophores, and other metabolites so that these cloned genes could be transferred into the microbial strains having good colonization potential (Grace and Ewart 1996). Furthermore, indigenous gut bacteria such as *E. cloacae* could be used to deliver, express, and spread foreign cloned genes in termite colonies of the Formosan subterranean termite *C. formosanus* as shuttle systems (Husseneder and Grace 2005). In the future, transgenic bacteria may be used to shuttle detrimental genes into termite colonies for improved pest control. Further, the efficacy of biocontrol agent can be improved by developing the better cultural practices and delivery systems that favor their establishment in the nest soil/colony. Thus, the biotechnological approaches used in the manipulation of microbial traits could lead to improved biocontrol activity of termite pathogenic microorganisms. The development of more stable formulations, such as microencapsulation (Grace and Ewart 1996), would be necessary to ensure their long-term, residual action. Future research could establish the role of biological materials, particularly the use of biocontrol agents, in effective termite management strategies.

12.8 Conclusion

Effective microbial control agents that can fill the void of phased out chemicals exit in nature but will require improvements in the pathogens selection, their production, formulation, and implementation (Jones et al. 1997; Culliney and Grace 2000); better understanding of how they will fit into integrated systems, their interaction with the environment and other IPM components, greater emphasis on their efficacy, safety, selectivity, etc., and their comparison with chemical pesticides; and finally their acceptance by growers and the general public. Straus and Knight (1997) presented the potential markets and methods for encouraging the use of microbial control agents and their benefits as well as limitations. Technical issues pertaining to improvement in biopesticide production, formulations, and application have also been addressed (Chapple and Bateman 1997; Jones and Burges 1997).

The role of microbial pesticides in the integrated management of insect pests has been recently reviewed for agriculture (Lacey and Goettel 1995; Tatchell 1997), forestry (Cowie et al. 1989; van Frankenhuyzen et al. 2000), and public health (Skovmand et al. 2000). In most cases, no single microbial control agent will provide sustainable control of an insect pest or complex of pests. As components of an integerated approach in all agricultural practices, entomopathogens could provide significant and selective insect control without interfering with the effectiveness of other practices (Edwards 1990). In near future, we expect to see synergistic combination of microbial control agents with other technologies (in combination with semiochemicals, soft chemical pesticides, other natural enemies, resistant plants, remote sensing, etc.) that will enhance the effectiveness and sustainability of integrated control strategies.

The use of pathogens to suppress populations of pests over large areas containing multiple agricultural and wild host plants has not been adequately explored (Bell and Hardee 1994). Such an areawide concept could take advantage, for example, of controlling populations of pests before they became economically important in crop plants. Till now, the market for microbial insecticides hardly represents only 1% of the total crop protection market (Gaugler 1997; Lisansky 1997). In the near future, microbials will face even stiffer competition from new pesticide chemistries and transgenic plants (Georgis 1997). Improvements in microbial products, grower's awareness of the benefits that microbial control offers, and the need to develop alternatives to conventional chemical insecticides should overcome many of the obstacles that microbial control is now facing. However, there could be considerable delays in the implementation of several microbial control agents that have good potential for use in IPM programs (Lacey and Goettel 1995).

References

Abe T (1987) Evolution of life types in termites. In: Kawano S, Connell JH, Hidda T (eds) Evolution and coadaption in biotic communities. University of Tokyo Press, Tokyo, pp 125–148

Abe T (1995) The termite-symbionts system: how does it work and has it evolved as a super-efficient decomposer in tropical terrestrial ecosystems? Center for Ecological Research, Kyoto University, Kyoto

Adams RP, McDaniel CA, Carter FL (1988) Termiticidal activities in the heartwood, bark/sapwood and leaves of *Juniperus* species from the United States. Biochem Syst Ecol 16:453–456

Ahmed S, Riaz MA, Shahid M (2006) Response of *Microtermes obesi* (Isoptera: Termitidae) and its gut bacteria towards some plant extracts. J Food Agric Environ 4:317–320

Al Fazairy AA, Hassan FA (1988) Infection of termites by *Spodoptera littoralis* nuclear polyhedrosis virus. Insect Sci Appl 9:37–39

Al Fazairy AA, Hassan FA (1993) Histopathology of termite *Kalotermes flavicollis* Fabr infected with a nuclear polyhedrosis virus. Insect Sci Appl 14:127–134

Amburgey TL, Beal RH (1977) White rot fungi inhibits termite attack. Sociobiology 3:35–38

Amin PW, Singh KN, Dwivedi SL, Rao VR (1985) Sources of resistance to the jassid (*Empoasea-kerri* Pruthi), thrips [*Frankliniella schultzei* (Trybom)] and termites (*Odontotermes* sp.) in groundnut (*Arachis hypogaea* L.). Peanut Sci 12:58–60

Badshah H, Farmanullah Salihah Z, Saljoqi A, Shakur M (2004) Toxic effects of ak (*Calotripis procera*) plant extracts against termites (*Heterotermes indicola* and *Coptotermes heimi*) (Isoptera: Rhinotermitidae). Pakistan J Biol Sci 7:1603–1606

Bao LL, Yendol WG (1971) Infection of the eastern subterranean termite, *Reticulitermes flavipes* (Kollar) [sic] with the fungus *Beauveria bassiana* (Balsamo) Vuill. Entomophaga 16:343–352

Beal RH, Kais AG (1962) Apparent infection of subterranean termites by *Aspergillus flavus* link. J Insect Pathol 4:488–489

Beeson CFC (1941) A guide to the control of termites for forest officers. Indian Forest Records (New Series). Entomology 4:44–90

Bell MR, Hardee DD (1994) Early season application of a baculovirus for area-wide management of *Heliothis/Helicoverpa* (Lepidoptera: Noctuidae): 1992 field trial. J Entomol Sci 29: 192–200

Bhanot JP, Singal SK (2007) Management of termites in field crops. In: Bhanot JP, Rohilla HR, Kalra VK (eds) Recent trends in biology and management of polyphagous pests of agricultural importance. CCS Haryana Agricultural University, Hisar, pp 166–172

Bhanot JP, Batra GR, Verma AN (1991a) Effect of seed treatment with different insecticides on germination, termite damage and grain yield of barley. J Insect Sci 4:61–63

Bhanot JP, Verma AN, Batra GR (1991b) Effect of seed treatment with different insecticides on germination, termite (*Microtermes obesi*) damage and grain yield of wheat (*Triticum aestivum*). Indian J Agric Sci 61:689–691

Bhanot JP, Sharma AK, Batra GR, Verma AN (1995) Influence of different levels of irrigation and fertilizer on termite damage and yield of gram crop raised from aldrin treated and untreated seed. J Insect Sci 7:115–116

Bignell DE, Eggleton P (1998) Termites. In: Calow P (ed) Encyclopedia of ecology and environmental management. Blackwell Scientific, Oxford, pp 744–746

Bignell DE, Eggleton P, Nunes L, Thomas KL (1997) Termites as mediators of carbon fluxes in tropical forest: budgets for carbon dioxide and methane emissions. In: Watt AD, Stork NE, Hunter MD (eds) Forest and insects. Chapman and Hall, London, pp 109–134

Black HIJ, Wood TG (1989) The effects of cultivation on the vertical distribution of *Microtermes* spp. (Isoptera: Termitidae: Macrotermitinae) in soil at Mokwa, Nigeria. Sociobiology 15: 133–138

Blackwell M (1980) New records of termite-infesting fungi. J Invertebr Pathol 35:101–104

Blackwell M, Rossi W (1986) Biogeography of fungal ectoparasites of termites. Mycotaxon 25: 581–601

Breznak JA (1982) Intestinal microbiota of termites and other xylophagous insects. Annu Rev Microbiol 36:323–343

Breznak JA, Brune A (1994) Role of microorganisms in the digestion of lignocellulose by termites. Annu Rev Entomol 39:453–487

Brooks WM (1988) Entomogenous protozoa. In: Ignoffo CM, Mandava NB (eds) Handbook of natural pesticides, vol V, Microbial insecticides, part A: entomogenous protozoa and fungi. CRC, Boca Raton, FL, pp 1–149

Burges HD (1981) Strategy for the microbial control of pests in 1980 and beyond. In: Burges HD (ed) Microbial control of pests and plant diseases, 1970–1980. Academic, London, pp 797–836

Carter FL, de Camargo CR (1983) Testing antitermitic properties of Brazilian woods and their extracts. Wood Fiber Sci 15:350–357

Carter FL, Smythe RV (1974) Feeding and survival responses of *Reticulitermes flavipes* (Kollar) to extracts of wood from 11 coniferous genera. Holzforschung 28:41–45

Chapple AC, Bateman RP (1997) Application system for microbial pesticides: necessity or novelty. In: Evans HF (chair) Microbial insecticides: novelty or necessity? Proceedings of the British crop protection council symposium 68, pp 181–190

Chen JW, Liu BL, Tzeng YM (1999) Purification and quantification of destruxins A and B from *Metarhizum anisopliae*. J Chromatogr 830:115–125

Chet I, Ordentlich A, Shapira R, Oppenheim A (1990) Mechanism of biocontrol of soil borne plant pathogens by rhizobacteria. Plant Soil 129:85–92

Chey VK (1996) Forest pest insects in Sabah. Sabah Forest Rec 15:1–178

Choi C, Kim S, Lim S, Lee H, Lee S, Son J, Choi H, Kim Y (1997) Separation and purification of protease from *Bacillus subtilis* CCKS-111 in Korean traditional soy sauce. Agr Chem Biotechnol 40:178–183

Coaton WGH (1949) Queen removal in termite control. Farming South Africa 24:335–344

Coaton WGH (1950) Termites and their control in cultivated areas in South Africa. Bull Dept Agric Union of South Africa 305:28

Collins NM (1984) The termites (Isoptera) of the Gunung Mulu National Park, with a key to the genera known from Sarawak. Sarawak Mus J 30:65–87

Collins NM (1988) Termites. In: Cranbrook E (ed) Key environments of Malaysia. Pergamon, Oxford

Collins NM (1989) Termites. In: Lieth H, Werger MJA (eds) Tropical rain forest ecosystems. Elsevier Science, Amsterdam, pp 455–471

Connick WJ Jr, Osbrink WLA, Wright MS, Williams KS, Daigle DJ, Boykin WL, Lax AR (2001) Increased mortality of *Coptotermes formosanus* (Isoptera: Rhinotermitidae) exposed to eicosanoid biosynthesis inhibitors and *Serratia marcescens* (Eubacteriales: Enterobacteriaceae). Environ Entomol 30:449–455

Cornelius ML, Grace JK (1994) Semiochemicals extracted from a dolichoderine ant affects the feeding and tunneling behavior of the Formosan subterranean termite (Isoptera: Rhinotermitidae). J Econ Entomol 87:705–708

Cornelius ML, Grace JK (1996) Effect of two ant species (Hymenoptera: Formicidae) on the foraging and survival of the Formosan subterranean termite (Isoptera: Rhinotermitidae). Environ Entomol 25:85–89

Cornelius ML, Grace JK (1997) Effect of termite soldiers on the foraging behaviour of *Coptotermes formosanus* (Isoptera: Rhinotermitidae) in the presence of predatory ants. Sociobiology 29: 247–253

Cornelius ML, Grace JK, Ford PW, Davidson BS (1995) Toxicity and repellency of semiochemicals extracted from a dolichoderine ant (Hymenptera: Formicidae) to the Formosan subterranean termite (Isoptera: Rhinotermitidae). Environ Entomol 24:1263–1269

Cowie RH, Wood TG (1989) Damage to crops, forestry and range land by fungus-growing termites (Termitidae: Macrotermitinae) in Ethiopia. Sociobiology 15:139–153

Cowie RH, Logan JWM, Wood TG (1989) Termite (Isoptera) damage and control in tropical forestry with special reference to Africa and Indo-Malaysia: a review. Bull Entomol Res 79: 173–184

Crowther F, Barlow HWB (1943) Tap-root damage of cotton, ascribed to termites, in the Sudan Gezira. Emp J Exp Agric 11:99–112

Culliney TW, Grace JK (2000) Prospects for the biological control of subterranean termites (Isoptera: Rhinotermitidae), with special reference to *Captotermes formosanus*. Bull Entomol Res 90:9–21

Cunningham JC (1995) Baculoviruses as microbial insecticides. In: Reuveni R (ed) Novel approaches to integrated pest management. Lewis, Boca Roaton, FL, pp 261–292

Danthanarayana W, Vitarana SI (1987) Control of the live-wood tea termite *Glyptotermes dilatatus* using *Heterorhabditis* sp. (Nematodes). Agric Ecosyst Environ 19:333–342

Darby C, Cosma CL, Thomas JH (1999) Lethal paralysis of *Caenorhabditis elegans* by *Pseudomonas aeruginosa*. Proc Natl Acad Sci USA 96:202–207

Darlington JPEC (1982) The underground passages and storage pits used in foraging by a nest of the termite *Macrotermes michaelseni* Kajiado, Kenya. J Zool 198:237–247

Darlington JPEC (1985) Multiple primary reproductives in the termite *Macrotermes michaelseni* (Sjostedt). In: Watson JAL, Okot-Kotber BM, Noirot C (eds) Caste differentiation in social insects. Pergamon, Oxford, pp 187–200

Dawkins HC (1949) Timber planting in the *Terminalia* wood land of northern Uganda. Emp For Rev 28:226–247

de Souza OFF, Brown VK (1994) Effects of habitat fragmentation on Amazonian termite communities. J Trop Ecol 10:197–206

DeBach P, Rosen D (1991) Biological control by natural enemies, 2nd edn. Cambridge University Press, Cambridge, p 440

Delate KM, Grace JK, Tome CHM (1995) Potential use of pathogenic fungi in baits to control the Formosan subterranean termite (Isoptera, Rhinotermitidae). J Appl Entomol 119:429–433

Deligne J, Quennedey A, Blum MS (1981) The enemies and defense mechanisms of termites. In: Hermann HR (ed) Social insects, vol 2. Academic, New York, pp 1–76

Desportes I (1963) Cycle evolutif d'une nouvelle Gregarine parasite de Termites: *Diplocystis zootermopsidis* sp. n. (Eugregarina Diplocystidae). Comtes Rendus Hebdom des Seances de Acad des Sci 257:4013–4015

Dillon RJ, Dillon VM (2004) The gut bacteria of insects: nonpathogenic interactions. Annu Rev Entomol 49:71–92

Disney RHL (1986) Two remarkable new species of scuttle-fly (Diptera: Phoridae) that parasitize termites (Isoptera) in Sulawesi. Syst Entomol 11:413–422

Disney RHL, Kistner DH (1989) Six new species of oriental Phoridae from termite nests (Diptera; Isoptera; Termitidae and Rhinotermitidae). Sociobiology 15:325–348

Disney RHL, Kistner DH (1990) Revision of the termite-parasitizing genus *Dicranopteron* (Diptera: Phoridae). Zool J Linn Soc 98:111–122

Donavan SE, Griffiths GJK, Homathevi R, Winder L (2007) The spatial pattern of soil-dwelling termites in primary and logged forest in Sabah, Malaysia. Ecol Entomol 32:1–10

Doppelreiter VH, Korioth M (1981) Entwicklungshemmung durch Diflubenzuron bei den Bodentermiten *Heterotermes indicola* and *Reticulotermes flavipes*. Z Angew Entomol 91:131–137

EcoScience (1997) Bio-Blast^{TM} Biological Termiticide. Internet published: http://www.ecoscience.com/bioblas/bio-blas.htm

Edward CA (1990) The importance of integration in sustainable agricultural systems. In: Edwards CA, Lal R, Madden P, Miller RH, House G (eds) Sustainable agricultural systems. St Lucie, Delray Beach, FL, pp 249–264

Edwards R, Mill AE (1986) Termites in buildings: their biology and control. Rentokil, East Grinstead, UK, p 261

Eggleton PE, Bignell DE, Sands WA, Waite B, Wood TG, Lawton JH (1995) The species richness of termites (Isoptera) under differing levels of forest disturbance in the Mbalmayo Forest Reserve, southern Cameroon. J Trop Ecol 11:85–98

Eggleton P, Bignell DE, Sands WA, Mawdsley NA, Lawton JH, Wood TG, Bignell NC (1996) The diversity, abundance and biomass of termites under differing levels of disturbance in the Mbalmayo Forest Reserve, southern Cameroon. Phil Trans Royal Soc London B 351:51–68

Eggleton P, Homathevi R, Jeeva D, Jones DT, Davies RG, Maryati M (1997) The species richness and composition of termites (Isoptera) in primary and regenerating lowland dipterocarp forest in Sabah, East Malaysia. Ecotropica 3:119–128

Epsky ND, Capinera JL (1988) Efficacy of entomogenous nematode *Steinernema feltiae* against a subterranean termite, *Reticulitermes tibialis* (Isoptera: Rhinotermitidae). J Econ Entomol 81:1313–1317

Federici BA (1990) Bright horizons for invertebrate pathology. In: Proceedings and abstracts. Fifth international colloquium on invertebrate pathology and microbial control, Adelaide, Australia, 20–24 Aug 1990, Society for Invertebrate Pathology, Adelaide, pp v–vi

Fernandes PM (1991) Controle microbiano de *Cornitermes cumulans* (Kollar, 1832) Utilizando *Beauveria bassiana* (Bals) Vuill and *Metarhizium anisopliae* (metsch) Sorok. PhD dissertation, ESALQ-Universidade de Sao Paulo, Piracicaba, Brazil

Ferron P (1978) Biological control of insect pests by entomogenous fungi. Annu Rev Entomol 23:409–412

Fujii JK (1975) Effect of an entomogenous nematode *Neoaplectana carpocapsae* Weiser, on the Formosan subterranean termite, *Captotermes formosanus* Shiraki, with ecological and biological studies on *C. formosanus*. PhD dissertation, University of Hawaii, Honolulu, Hawaii, USA, pp 163

Fuller C (1912) White ants in Natal. Union of South Africa Department of Agriculture, Pretoria, p 13

Fuxa JR (1989) Fate of released entomopathogens with reference to risk assessment of genetically engineered microorganisms. Bull Entomol Soc Am 35:12–24

Fuxa JR (1990) New directions for insect control with baculoviruses. In: Baker RR, Dunn PE (eds) New directions in biological control: alternatives for suppressing agricultural pests and diseases. Alan R Liss, New York, pp 97–113

Gallagher LA, Manoil C (2001) *Pseudomonas aeruginosa* PAO1 kills *Caenorhabditis elegans* by cyanide poisoning. J Bacteriol 183:6207–6214

Gaugler R (1988) Ecological consideration in the biological control of soil-inhabiting insects with entomopathogenic nematodes. Agric Ecosyst Environ 24:351–360

Gaugler R (1997) Alternative paradigms for commercializing biopesticides. Phytoparasitica 25:179–182
Georgis R (1997) Commercial prospects of microbial insecticides in agriculture. In: Evans HF (chair) Microbial insecticides: novelty or necessity? Proceedings of the British crop protection council symposium 68, pp 243–252
Georgis R, Poinar GO Jr, Wilson AP (1982) Susceptibility of damp-wood termites and soil and wood-dwelling termites to the entomogenous nematode *Neoaplectana carpocapsae*. IRCS Med Sci 10:563
Gerrits R, van Latum EBJ (1988) Plant-derived pesticides in developing countries: possibilities and research needs. Netherlands' Ministry of Housing, Physical Planning and Environment, The Hague, p 104
Gibbs AJ, Gay FJ, Wetherly AH (1970) A possible paralysis virus of termites. Virology 40:1063–1065
Giridhar G, Vasudevan S, Vasudevan P (1988) Antitermites properties of *Calotropis* latex. Pesticides 22:31–33
Gitonga W (1996) *Metarhizium anisopliae* (Metschnikoff) Sorokin and *Beauveria bassiana* (Balsamo) Vuillemin as potential biological control agents of *Macrotermes michaelseni* (Sjöstedt) (Isoptera: Termitidae) in Kenya. PhD dissertation, Royal Veterinary and Agricultural University, Copenhagen, Denmark, pp 144
Grace JK (1991) Termite-fungal associations and manipulations for termite control. In: Program and abstracts. 24th annual meeting, Society of Invertebrate Pathology, Northern Arizona University, 4–9 Aug 1991, Society for Invertebrate Pathology, Flagstaff, Arizona, pp 29
Grace JK (1993) Microbial termite control: effect of entomogenous fungi on the formosan subterranean termite (Isoptera: Rhinotermitidae). In: Wildey KB, Robinson WH (eds) Proceedings of 1st international conference on insect pest in the urban environment, St John's College, Cambridge, 30 June–3 July 1993, Organizing committee of the international conference on insect pests in the urban environment, Exeter, pp 474
Grace JK (1994) Protocol for testing effects of microbial pest control agents on nontarget subterranean termites (Isoptera: Rhinotermitidae). J Econ Entomol 87:269–274
Grace JK (1995) Microbial termite control. In: Hawaii agriculture: positioning for growth. Proceedings of a conference organized by the College of Tropical Agriculture, Human Resources, University of Hawaii, Manoa, 5–6 April 1995, CTAHR, Honolulu, pp 166–167
Grace JK (1997) Biological control strategies for suppression of termites. J Agric Entomol 14:281–289
Grace JK, Ewart D (1996) Recombinant cells of *Pseudomonas fluorescens*: a highly palatable encapsulation for delivery of genetically engineered toxins to subterranean termite (Isoptera: Rhinotermitidae). Lett Appl Microbiol 23:183–186
Grace JK, Yamamoto RT (1994) Simulation of remedial borate treatments intended to reduce attack on Douglas-fir lumber by the Formosan subterranean termite (Isoptera: Rhinotermitidae). J Econ Entomol 87:1547–1554
Grace JK, Zoberi MH (1992) Experimental evidence for transmission of *Beauveria bassiana* by *Reticulitermes flavipes* workers (Isoptera: Rhinotermitidae). Sociobiology 20:23–28
Grace JK, Goodell BS, Jones WE, Chandhoke V, Jellison J (1992) Evidence for inhibition of termites (Isoptera: Rhinotermitidae) feeding by extracellular metabolites of a wood decay fungus. Proc Hawaiian Entomol Soc 31:249–252
Grace JK, Tome CHM, Shelton TG, Osairo RJ, Yates JR III (1996) Baiting studies and considerations with *Coptotermes formosanus* (Isoptera: Rhinotermitidae) in Hawaii. Sociobiology 28:511–520
Grasse PP (1986) Termitologia, vol III. Masson, Paris, p 715
Gross P (1991) Influence of target pest feeding niche on success rates in classical biological control. Environ Entomol 20:1217–1227
Gunner HB, Kane J, Duan H (1994) Biological control of termites. PCT Patent Application ~WO94 04034

Hajjar NP, Casida JE (1978) Insecticidal benzoylphenyl ureas: structure activity relationships as chitin synthesis inhibitors. Science 200:1499–1500

Hänel H (1981) A bioassay for measuring the virulence of the insect pathogenic fungus *Metarhizium anisopliae* (Metsch.) Sorok. (fungi imperfecti) [sic] against the termite *Nasutitermes exitiosus* (Hill) (Isoptera, Termitidae). Zeitschrift für Augewandte Entomol 92:9–18

Hänel H (1982a) The life cycle of the insect pathogenic fungus *Metarhizium anisopliae* in the termite *Nasutitermes exitiosus*. Mycopathologia 80:137–145

Hänel H (1982b) Selection of a fungus species, suitable for the biological control of the termite *Nasutitermes exitiosus* (Hill). Zeitschrift Angewandte Entomol 94:237–245

Hänel H, Watson JAL (1983) Preliminary field tests on the use of *Metarhizium anisopliae* for the control of *Nasutitermes exitiosus* (Hill) (Isoptera: Termitidae). Bull Entomol Res 73:305–313

Harborne JB (1988) Introduction to ecological biochemistry, 3rd edn. Academic, London, p 356

Harris WV (1957) An introduction to Malayan termites. Malay Nat J 12:20–32

Harris WV (1971) Termites: their recognition and control, 2nd edn. Longmans, London, p 186

Henderson G, Dunaway C (1999) Termite damage to cables and inorganic material. Louisiana Agric 42:4–6

Henry JE (1990) Control of insects by protozoa. In: Baker RR, Dunn PE (eds) New directions in biological control: alternatives for suppressing agricultural pests and disease. Alan R Liss, New York, pp 161–176

Holldobler B, Wilson EO (1990) The ants. Harvard University Press, Cambridge, Massachusetts, p 732

Honee G, Convents D, van Rie J, Jansens S, Perfroen M, Visser B (1991) The C-terminal domain of the toxic fragment of a *Bacillus thuringiensis* crystal protein determines receptor binding. Mol Microbiol 5:2799–2806

Horikoshi K (1971) Production of alkaline enzymes by alkalophilic microorganisms. Agric Biol Chem 35:1407–1414

Horn DJ (1988) Ecological approach to pest management. Elsevier, London, p 285

Husseneder C, Grace JK (2005) Genetically engineered termite gut bacteria (*Enterobacter cloacae*) deliver and spread foreign genes in termite colonies. Appl Microbiol Biotechnol 68:360–367

Jafri RH, Ahmad M, Idrees K (1976) Microsporidian infection in the workers of termite, *Microcerotermes championi*. Pak J Zool 8:234–236

Jegorov A, Kadlec J, Novak J, Matha V, Sedmera P, Triska J, Zahradnickova H (1989) Are the depsopeptides of *Beauveria brongniartii* involved in the entomopathogenic process? In: Jegorov A, Matha V (eds) Proceedings of the international conference on biopesticides, theory and practice, Ceske Budejovice, Czechoslovakia, pp 71–81

Johnson JB, Hagen KS (1981) A neuropterous larva uses an allomone to attack termites. Nature 289:506–507

Johnson RA, Lamb RW, Wood TG (1981) Termite damage and crop loss studies in Nigeria: a survey of damage to groundnuts. Trop Pest Manag 27:325–342

Jones JA (1990) Termites, soil fertility and carbon cycling in dry tropical Africa: a hypothesis. J Trop Ecol 6:291–305

Jones KA, Burges HD (1997) Product stability: From experimental preparation to commercial reality. In: Evans HF (chair) Microbial insecticides: novelty or necessity? Proceedings of the British crop protection council symposium 68, pp 173–176

Jones DT, Gathrone-Hardy F (1995) Foraging activity of the processional termite *Hospitalitermes hospitalis* (Termitidae: Nasutitermitinae) in the rain forest of Brunei, north-west Borneo. Instect Soc 42:359–369

Jones WE, Grace JK, Tamashiro M (1996) Virulence of seven isolates of *Beauveria bassiana* and *Metarhizium anisopliae* to *Captotermes formosanus* (Isoptera: Rhinotermitidae). Environ Entomol 25:481–487

Jones KA, Cherry AJ, Grzywacz D (1997) Formulations: is it an excuse for poor application? In: Evans HF (chair) Microbial insecticides: novelty or necessity? Proceedings of the British crop protection council symposium 68, pp 177–180

Kaijiang L, Roberts DW (1986) The production of destruxins by the entomogenic fungus. *Metarhizum anisopliae* var. *major*. J Invert Pathol 47:120–122

Kakde JR (1985) Sugarcane production. Metropolitan Book, New Delhi, p 384

Kambhampati S, Eggleton P (2000) Phylogenetics and taxonomy. In: Abe T, Bignell DE, Higashi M (eds) Termites: evolution, sociality, symbiosis, ecology. Kluwer, Dordrecht, pp 1–23

Kambhampati S, Kjer KM, Thorne BL (1996) Phylogentic relationship among termite families based on DNA sequence of mitochondrial 16S ribosomal RNA gene. Insect Mol Biol 5:229–238

Kanchana Devi K, Seth N, Kothamasi S, Kothamasi D (2007) Hydrogen cyanide-producing rhizobacteria kill subterranean termite *Odentotermes obesus* (Rambur) by cyanide poisoning under in vitro conditions. Curr Microbiol 54:74–78

Kashyap RK, Verma AN, Bhanot JP (1984) Termites of plantation crops, their damage and control. J Plant Crops 12:1–10

Kaya HK (1990) Entomopathogenic nematodes in biological control of insects. In: Baker RR, Dunn PE (eds) New directions in biological control: alternatives for suppressing agricultural pests and diseases. Alan R. Liss, New York, pp 189–198

Kaya HK, Gaugler R (1993) Entomopathogenic nematodes. Annu Rev Entomol 38:181–206

Kaya HK, Bedding RA, Akhurst RJ (1993) An overview of insect-parasitic and entomopathogenic nematodes. In: Bedding RA, Akhurst R, Kaya H (eds) Nematodes and the biological control of insect pest. CSIRO, East Melbourne, Australia, pp 1–10

Kempe J, Sequeira L (1983) Biological control of bacterial wilt of potatoes: attempts to induce resistance by treating tubers with bacteria. Plant Dis 67:499–503

Khan KI (2006) Enhancement of virulence of *Bacillus thuringiensis* and *Serratia marcescens* by chemicals. J Res Sci 17:35–43

Khan KI, Fazal Q, Jafri RH (1977a) Pathogenicity of locally discovered *Bacillus thuringiensis* strain to the termites: *Heterotermes indicola* (Wassman) and *Microcerotermes championi* (Snyder). Pak J Sci Res 29:12–13

Khan KI, Fazal Q, Jafri RH, Ahmad M (1977b) Susceptibility of various species of termites to a pathogen, *Serratia marcescens*. Pak J Sci Res 29:46–47

Khan KI, Fazal Q, Jafri RH (1978) Development of *Bacillus thuringiensis* in a termite, *Heterotermes indicola* (Wassman) [sic]. Pak J Sci Res 30:117–119

Khan KI, Jafri RH, Ahmad M (1985) The pathogenicity and development of *Bacillus thuringiensis* in termites. Pak J Zool 17:201–209

Khan KI, Jafri RH, Ahmad M, Khan KMS (1992) The pathogenicity of *Pseudomonas aeruginosa* against termites. Pak J Zool 24:243–245

Klein MG (1988) Pest management of soil-inhibiting insects with microorganisms. Agric Ecosyst Environ 24:337–349

Kooyman C, Onck RFM (1987) Distribution of termite (Isoptera) species in southwestern Kenya in relation to land use and the morphology of their galleries. Biol Fertil Soils 3:69–73

Kramm KR, West DF (1982) Termite pathogens: effects of ingested *Metarhizium, Beauveria,* and *Gliocladium* conidia on worker termites (*Reticulitermes* sp). J Invert Pathol 40:7–11

Kramm KR, West DF, Rockenbach PG (1982) Termite pathogens: transfer of the entomopathogen *Metarhizium anisopliae* between *Reticulitermes* sp. termites. J Invert Pathol 40:1–6

Krasnoff SB, Gibson DM (1996) New destruxins from the entomopathogenic fungus, *Aschersonia* sp. J Nat Prod 59:485–489

Krishna K (1970) Taxonomy, phylogeny and distribution of termite. In: Krishna K, Weesner FM (eds) Biology of termites, vol 2. Academic, New York

Krishna K, Weesner FM (1970) Biology of termites, vol II. Academic, London, p 643

Kumarasinghe NC, Ranasinghe MASK (1988) Incidence of termite damage in sugarcane grown in Shri Lanka. Beitr Trop Landwirt 26:303–307

Lacey LA, Goettel M (1995) Current developments in microbial control of insects pests and prospects for the early 21st century. Entomophaga 40:3–27

Lacey LA, Frutos R, Kaya HK, Vail P (2001) Insect pathogens as biocontrol agents: do they have a future? Biol Control 21:230–248

Lai PY (1977) Biology and ecology of the Formosan subterranean termite, *Captotermes formosanus*, and its susceptibility to the entomogenous fungi *Beauveria bassiana* and *Metarhizium* [sic] *anisopliae*. PhD dissertation, University of Hawaii, Honolulu, Hawaii, USA, pp 140

Lai PY, Tamashiro M, Fujii JK (1982) Pathogenicity of six strains of entomogenous fungi to *Coptotermes formosanus*. J Invert Pathol 39:1–5

Lai PY, Tamashiro M, Yates JR, Su NY, Fujii JK, Ebesu RH (1983) Living plants in Hawaii attacked by *Coptotermes formosanus*. Proc Hawaiian Entomol Soc 24:283–286

Lal R (1987) Tropical ecology and physical edaphology. Wiley, Chichester, p 732

Laumond C, Mauleon H, Kermarrec A (1979) Donnees nouvelles sur le spectre d'hôtes et le parasitisme du nematode entomophage *Neoaplectana carpocapsae*. Entomophaga 24:13–27

Lee KE, Wood TG (1971) Termites and soils. Academic, New York, p 251

Lenz M, Runko S (1994) Protection of buildings, other structures and materials in ground contact from attack by subterranean termites (Isoptera) with a physical barrier: a fine mesh of high grade stainless steel. Sociobiology 24:1–16

Leong KLH (1966) Infection of the Formosan subterranean termite, *Captotermes formosanus* Shiraki [sic] by the fungus, *Metarhizium* [sic] *anisopliae* [Metch.] Sorokin. Master's thesis, University of Hawaii, Honolulu, Hawaii, USA, pp 53

Lepage MG, Darlington JPEC (1984) Observations on the ant *Carebara vidua* F. Smith preying on termites in Kenya. J Nat Hist 18:293–302

Lim H-S, Kim Y-S, Kim S-D (1991) *Pseudomonas stutzeri* YPL-1 genetic transformation and antifungal mechanism against *Fusarium solani*, an agent of plant root rot. Appl Environ Microbiol 57:510–516

Lin TS, Wang CL (1988) The anti-termite properties of extracts from *Melia azedarach* Linn. Bull Taiwan Forestry Res Inst 3:255–261

Lisansky S (1997) Microbial pesticides. In: Evans HF (chair) Microbial insecticides: novelty or necessity? Proceedings of the British crop protection council symposium 68, pp 3–10

Logan JWM, El Bakri A (1990) Termite damage to date palms (*Phoenix dactylifera* L.) in northern Sudan with particular reference to the Dongola district. Trop Sci 30:95–108

Logan JWM, Cowie RH, Wood TG (1990) Termite (Isoptera) control in agriculture and forestry by non-chemical methods: a review. Bull Entomol Res 80:309–330

Longhurst C, Howse PE (1979) Foraging, recruitment and emigration in *Megaponera foetens* (Fab.) (Hymenoptera: Formicidae) from the Nigerian Guinea Savanna. Insect Soc 26:204–215

Longhurst C, Johnson RA, Wood TG (1979) Foraging, recruitment and predation by *Decamorium uelense* (Sanstchi) (Formicidae: Myrmicinae) on termites in southern Guinea Savanna, Nigeria. Oecologia 38:83–91

Lund AE (1971) Microbial control of termites. In: Burges HD, Hussey NW (eds) Microbial control of insects and mites. Academic, London, pp 385–386

Lysenko O, Kucera M (1971) Microorganisms as sources of new insecticidal chemicals; toxins. In: Burges HD, Hussey NW (eds) Microbial control of insects and mites. Academic, London, pp 205–227

Malaka SLO (1972) Some measures applied in the control of termites in parts of Nigeria. Nigerian Entomol Mag 2:137–141

Manachini PL, Fortina MG, Parini C (1988) Thermostable alkaline protease produced by *Bacillus thermoruber*: a new species of *Bacillus*. Appl Microbiol Biotechnol 28:409–413

Mauldin HK (1986) Economic importance and control of termites in the Unites States. In: Vinson SB (ed) Economic impact and control of social insects. Praeger, New York, pp 130–143

Mauldin JK, Beal RH (1989) Entomogenous nematodes for control of subterranean termites, *Reticulitermes* spp. (Isoptera: Rhinotermitidae). J Econ Entomol 82:1638–1642

McCoy CW (1990) Entomogenous fungi as microbial pesticides. In: Baker RR, Dunn PE (eds) New directions in biological control: alternatives for suppressing agricultural pests and diseases. Alan R Lisa Inc., New York, pp 139–159

Mercer PC (1978) Pest and disease of groundnuts in Malawi. III. Wilts, post-harvest, physiological and minor disorders. Oleagineux 33:619–624

Merrill JH, Ford AL (1916) Life history and habits of two new nematodes parasitic on insects. J Agric Res 6:115–127

Miller JS, Nguyen T, Stanley-Samuelson DW (1994) Eicosanoids mediate insect modulation responses to bacterial infections. Proc Natl Acad Sci USA 91:2418–2422

Milner RJ (1992) Selection and characterization of strains of *Metarhizium anisopliae* for control of soil insects in Australia. In: Lomer CJ, Prior C (eds) Biological control of locusts and grasshoppers. CAB International, Wallingford, pp 200–207

Milner RJ, Staples JA (1996) Biological control of termites: results and experiences within a CSIRO project in Australia. Biocontrol Sci Technol 6:3–9

Milner RJ, Staples JA, Lenz M (1996) Options for termite management using the insect pathogenic fungus *Metarhizium anisopliae*. International group on wood preservation, Document No. IRG/WP9610142. pp 1–5

Milner RJ, Staples JA, Lutton GG (1998) The selection of an isolate of the hyphomycete fungus, *Metarhizium anisopliae*, for control of termites in Australia. Biol Control 11:240–247

Mitchell MR, Gwaze DP, Stewart HTL (1987) Survival and early growth of Australian tree species planted at a termite-infested site in Zimbabwe. East Afr Agril Forestry J 54:251–259

Morales-Ramos JA, Guadalupe Rojas M, Henon PE (2003) Black staining fungus effects on the natural resistance properties of Alaskan yellow cedar to the Formasan subterranean termites (Isoptera: Rhinotermitidae). Environ Entomol 32:1234–1241

Nair KSS, Varma RV (1985) Some ecological aspects of the termite problem in young eucalyptus plantations in Kerala, India. Forest Ecol Manag 12:287–303

Nickle WR, Welch HE (1984) History, development and importance of insect nematology. In: Nickle WR (ed) Plant and insect nematodes. Dekker, New York, pp 627–653

Noirot C, Noirot-Timothee C (1970) The digestive system. In: Krishna K, Weesner FM (eds) Biology of termites. Academic, New York

Nutting WL (1969) Flight and colony foundation. In: Krishna K, Weesner FM (eds) Biology of termites, vol 1. Academic, New York, pp 233–282

Ohiagu CE, Wood TG (1976) A method for measuring rate of grass-harvesting by *Trinervitermes geminatus* (Wasmann) (Isoptera: Nasutitermitinae) and observation on its foraging behaviour in southern Guinea Savanna, Nigeria. J Appl Ecol 13:705–713

Osbrink LAW, Williams KS, Connick WJ Jr, Wright MS, Lax AR (2001) Virulence of bacteria associated with the Formosan subterranean termite (Isoptera: Rhinotermitidae) in New Orleans, LA. Environ Entomol 30:443–448

Otto H (1951) Harvester ants don't like water. Simple control method discovered on free state farm. The Farmer's Weekly, 25 Apr 1951, pp 10–13

Parihar DR (1985) Crop termite damage in Indian desert and its control in castor. Z Angew Zool 72:309–315

Payne CC (1982) Insect viruses as control agents. Parasitology 84:35–77

Peakins GJ, Josens G (1978) Respiration and energy flow. In: Brian MV (ed) Production ecology of ants and termites. Cambridge University Press, Cambridge, pp 111–163

Pearce MJ (1997) Termites biology and pest management. CAB International, Wallingford, p 172

Phillipsen WJ, Coppel HC (1977) *Acotyledon formosani* sp. N. associated with the Formosan subterranean termite, *Coptotermes formosanus* Shiraki (Acarina: Acaridae-Isoptera: Rhinotermitidae). J Kansas Entomol Soc 50:399–409

Popenoe P (1973) The date palm. Miami Field Research Projects. 247 pp

Popiel I, Hominick WH (1992) Nematodes as biological control agents: part II. Adv Parasitol 31:381–433

Potter MF (1997) Termite baits: a status report. Pest Control Technol 25:24–26
Preston AF, Erbisch FH, Kramm KR, Lund AE (1982) Developments in the use of biological control for wood preservation. In: Proceedings of the American Wood-Preservers' Association 78, pp 53–61
Rajagopal D (1982) Relative incidence of termites on exotic species of *Eucalyptus* in Karnataka. Myforest 18:9–13
Rath AC (2000) The use of entomopathogenic fungi for control of termites. Biocontrol Sci Technol 10:563–581
Rath AC, Tidbury CA (1996) Susceptibility of *Coptotermes acinaciformis* (Isoptera: Rhinotermitidae) and *Nasutitermes exitiosus* (Isoptera: Termitidae) to two commercial isolates of *Metarhizium anisopliae*. Sociobiology 28:67–72
Richards OW, Davies RG (1978) IMMS outlines of entomology. Chapman and Hall, London
Roonwal ML (1955) Termites ruining a township. Z Angew Entomol 38:103–104
Roonwal ML (1979) Termite life and termite control in tropical south Asia. Scientific, Jodhpur, p 177
Rosengaus RB, Traniello JFA (1997) Pathobiology and disease transmission in dampwood termites [*Zootermopsis angusticollis* (Isoptera: Termopsidae)] infected with the fungus *Metarhizium anisopliae* (Deuteromycotina: Hypomycetes). Sociobiology 30:185–195
Sajap AS, Kaur K (1990) Histopathology of *Metarhizium anisopliae*, an entomopathogenic fungus, infection in the termite, *Coptotermes curvignathus*. Pertanika 13:331–334
Sands WA (1961) Foraging behaviour and feeding habits in five species of *Trinervitermes* in West Africa. Entomol Exp Appl 44:277–288
Sands WA (1973) Termites as pests of tropical food crops. Pest Art News Summ 19:167–177
Sands WA (1977) The role of termites in tropical agriculture. Outlook Agric 9:136–143
Schnepf E, Crickmore N, van Rie J, Lereclus D, Baum J, Feitelson J, Ziegler DR, Dean DH (1998) *Bacillus thuringiensis* and its pesticidal proteins. Microbiol Mol Biol Rev 62:775–806
Secord D, Karevia P (1996) Perils and pitfalls in the host specificity paradigm. Bioscience 46:448–453
Sheppe W (1970) Invertebrate predation on termites of the African Savanna. Insect Soc 17:205–218
Sindhu SS, Dadarwal KR (2001) Chitinolytic and cellulolytic *Pseudomonas* sp. antagonistic to fungal pathogens enhances nodulation by *Mesorhizobium* sp. *Cicer* in chickpea. Microbiol Res 156:353–358
Singh M, Lal K, Singh SB, Singh M (2002) Effect of *Calotropis* (*Calotropis procera*) extract on infestation of termites (*Odontotermes obesus*) in sugarcane hybrid. Indian J Agric Sci 72:439–441
Singla ML, Duhra MS, Aulakh MS (1988) Varietal resistance in sugarcane to *Scirpophaga excerptalis* Wlk. and *Odontotermes* spp. J Insect Sci 1:99–101
Sivapalan P, Senaratne KADW, Karunaratne AAC (1977) Observations on the occurrence and behaviour of live-wood termites (*Glyptotermes dilatatus*) in low-country tea fields. Pest Art News Summ 23:5–8
Skovmand O, Kerwin J, Lacey LA (2000) Microbial control of mosquitoes and black flies. In: Lacey LA, Kaya HK (eds) Field manual of techniques in invertebrate pathology: application and evaluation of pathogens for control of insects and other invertebrate pests. Kluwer, Dordrecht, pp 767–785
Slaytor M, Veivers PC, Lo N (1997) Aerobic and anaerobic metabolism in the higher termite *Nasutitermes walkeri* (Hill). Insect Biochem Mol Biol 27:291–303
Sleaford F, Bignell DE, Eggleton P (1996) A pilot analysis of gut contents in termites from the Mbalmayo forest reserve, Cameroon. Ecol Entomol 21:57–73
Smith KM (1967) Insect virology. Academic, New York, p 256
Smythe RV, Coppel HC (1965) The susceptibility of *Reticulitermes flavipes* (Kollar) and other termite species to an experimental preparation of *Bacillus thuringiensis* Berliner. J Invertebr Pathol 7:423–426

Stanley-Samuelson DW, Jensen E, Nickerson KW, Tiebel K, Ogg CL, Howard RW (1991) Insect immune response to bacterial infection is mediated by eicosanoids. Proc Natl Acad Sci USA 88:1064–1068

Staples J, Milner R (1996) *Metarhizium anisopliae* as a mycotermiticide-laboratory behavioural bioassay. In: Wildey KB (ed) Proceedings of the 2nd international conference on insect pests in the urban environment, Edinburgh, 7–10 July 1996, Exeter organizing committee of the international conference on insect pests in the urban environment, pp 493

Stoll G (1986) Natural crop protection. Josef Margraf, Langen, West Germany, p 188

Straus EE, Knight JD (1997) Microbial insecticides: can farmers be persuaded to use them? In: Evans HF (chair) Microbial insecticides: novelty or necessity? Proceedings of the British crop protection council symposium 68, pp 141–150

Su NY (1993) Managing subterranean termite populations. In: Wildey KB, Robinson WH (eds) Proceedings of the 1st international conference on insect pests in the urban environment, Cambridge, 30 June–3 July 1993, Organizing committee of the international conference on insect pests in the urban environment, Exeter, pp 45–50

Su NY (1994) Field evaluation of hexaflumuron bait for population suppression of subterranean termites (Isoptera: Rhinotermitidae). J Econ Entomol 87:389–397

Su NY, Scheffrahn RH (1990) Economically important termites in the United States and their control. Sociobiology 17:77–94

Su NY, Scheffrahn RH (1993) Laboratory evaluation of two chitin synthesis inhibitors, hexaflumuron and diflubenzuron, as bair toxicants against the Formosan subterranean termite and eastern subterranean termite (Isoptera: Rhinotermitidae). J Econ Entomol 86:1453–1457

Su NY, Scheffrahn RH (1996) A review of the evaluation criteria for bait-toxicant efficacy against field colonies of subterranean termites (Isoptera). Sociobiology 28:521–530

Su NY, Scheffrahn RH (1998) A review of subterranean termite control practices and prospects for integrated pest management programmes. Int Pest Manag Rev 3:1–13

Su NY, Tamashiro M (1987) An overview of the Formosan subterranean termite (Isoptera: Rhinotermitidae) in the world. In: Tamashiro M, Su NY (eds) Biology and control of the Formosan subterranean termite, Honolulu University of Hawaii, College of tropical agriculture and human resources research extension series 083, pp 3–15

Su NY, Scheffrahn RH, Ban PM (1991) Uniform size particle barrier: a physical exclusion device against subterranean termites (Isoptera: Rhinotermitidae). J Econ Entomol 84:912–916

Su NY, Thomas EM, Ban PM, Scheffrahn RH (1995) Monitoring/baiting station to detect and eliminate foraging populations of subterranean termites (Isoptera: Rhinotermitidae) near structure. J Econ Entomol 88:923–936

Su NY, Ban PM, Scheffrahn RH (1999) Longevity and efficacy of pyrethroid and organophosphate termiticides in field degradation studies using miniature slabs. J Econ Entomol 92:890–898

Sun J, Fuxa JR, Henderson G (2002) Sporulation of *Metarhizium anisopliae* and *Beauveria bassiana* on *Coptotermes formosanus* and in vitro. J Invertebr Pathol 81:78–85

Suzuki K (1991) Laboratory trials of biological control agents against subterranean termites. International research group on wood preservation, Document no. IRG/WP/1475, pp 1–11

Suzuki K (1996) Biological control of termites by pathogenic fungi. In: Forestry and forest products research: proceedings of the third conference, Kepong, Malaysia, 3–4 Oct 1995, vol 2, Kepong, Forest research institute of Malaysia, pp 146–156

Tamashiro M, Yates JR, Ebesu RH (1987) The Formosan subterranean termite in Hawaii: problems and control. In: Tamashiro M, Su NY (eds) Biology and control of the formosan subterranean termite, Honolulu university of Hawaii, College of tropical agriculture and human resources research extension series 083, pp 15–22

Tarr SAJ (1960) The effects of fungicide-insecticide seed treatments on emergence, growth and yield of irrigated cotton in the Sudan Gezira. Annu Appl Biol 48:591–600

Tatchell GM (1997) Microbial insecticides and IPM: current and future opportunities for the use of biopesticides. In: Evans HF (chair) Microbial insecticides: novelty or necessity? Proceedings of the British crop protection council symposium 68, pp 191–200

Tayasu I, Sugimoto A, Wada E, Abe T (1994) Xylophagous termites depending on atmospheric nitrogen. Naturwis-senschaften 81:229–231

Thapa RS (1981) Termites of Sabah. Sabah Forest Rec 12:1–374

Tho YP (1992) Termites of Peninsular Malaysia. Malayan Forest Record No. 36, Forest Research Institute, Malaysia, pp 224

Thorne BL, Carpenter JM (1992) Phylogeny of the Dictyoptera. Syst Entomol 17:253–268

Thorne BL, Forschler B (2000) Criteria for assessing efficacy of stand-alone termite bait treatments at structures. Sociobiology 36:245–255

Toumanoff C, Rombaut J (1965) Action de certainchampignons entomophages, cultivês sur les milieuy appropriês attractifs, sur le Termite de Saintonge *Reticulitermes santonensis* (de Feytaud). Annal Parasitol Human Compare 40:605–609

Trudeau D (1989) Selection of entomophilic nematodes for control of the eastern subterranean termite, *Reticultermes flavipes* (Kollar) (Isoptera: Rhinotermitidae). Master's thesis, University of Toronto, Toronto, Ontario, Canada, pp 93

Vaidya RJ, Macmil SLA, Vyas PR, Chhatpar HS (2003) The novel method for isolating chitinolytic bacteria and its application in screening for hyperchitinase producing mutant of *Alcaligenes xylosoxydens*. Lett Appl Microbiol 36:129–134

van Frankenhuyzen K, Reardon RC, Debois NR (2000) Forest defoliators. In: Lacey LA, Kaya HK (eds) Field manual of techniques in invertebrate pathology: application and evaluation of pathogens for control of insects and other invertebrate pests. Kluwer, Dordrecht, pp 768–785

van Rie J, Jansens S, Hofte H, Degheele D, van Mellaert H (1989) Specificity of *Bacillus thuringiensis* δ-endotoxins. Importance of specific receptors on the brush-border membrane of the mid-gut of target insects. Eur J Biochem 186:239–247

Veeresh GK, Gubbaiah (1984) A report on the 'crazy ant' (*Anoplolepis longipes* Jerdon) menace in Karnataka. J Soil Biol Ecol 4:63–65

Verma M, Sharma S, Prasad R (2009) Biological alternatives for termite control: a review. Int Biodeterior Biodegradation 63:959–972

Voisard C, Keel C, Hass D, Defago G (1989) Cyanide production by *Pseudomonas fluorescens* helps suppress black root rot of tobacco under gnotobiotic conditions. EMBO J 8:351–358

Wahlman M, Davidson BS (1993) New destruxins from the entomopathogenic fungus *Metarhizium anisopliae*. J Nat Prod 56:643–647

Weiser J, Hardy I (1962) Pyemotes-mites as parasite of termites. Z Angew Entomol 51:94–97

Wells JD, Fuxa JR, Henderson G (1995) Virulence of four fungal pathogens to *Coptotermes formosanus* (Isoptera: Rhinotermitidae). J Entomol Sci 30:208–215

Wilson NL, Oliver AD (1969) Food habits of the imported fire ant in pasture and pine forest areas in southeastern Louisiana. J Econ Entomol 62:1268–1271

Wood BJ (1968) Pests of oil palms in Malaysia and their control. The Incorporated Society of Planters, Kuala Lumpur, p 204

Wood TG (1976) The role of termites (Isoptera) in decomposition processes. In: Anderson JM, MacFadyen A (eds) The role of terresterial and aquatic organisms in decomposition processes. Blackwell, Oxford, pp 145–168

Wood TG (1988) Termites and the soil environment. Biol Fertil Soils 6:228–236

Wood TG, Johnson RA (1986) The biology, physiology and ecology of termites. In: Vinson SB (ed) The economic impact and control of social insects. Praeger, New York, pp 1–68

Wood TG, Sands WA (1978) The role of termites in ecosystems. In: Brian MV (ed) Production ecology of ants and termites. Cambridge University Press, Cambridge, pp 245–292

Wood TG, Thomas RJ (1989) The mutualistic association between Macrotermitinae and *Termitomyces*. In: Wilding N, Collins NM, Hammond PM, Webber JF (eds) Insect-fungus interactions. Academic, London, pp 69–92

Wright MS, Connick WJ, Jackson MA (2003) Use of *Paecillomyces* spp. as pathogenic against termites. US Patent 6660291

Wright MS, Raina AK, Lax AR (2005) A strain of the fungus *Metarhizium anisopliae* for controlling subterranean termites. J Econ Entomol 98:1451–1458

Yates JR III, Grace JK, Tamashiro M (1997) The Formosan subterranean termite: a review of new management methods in Hawaii. In: Chan J (ed) Proceedings of FAOPMA convention held in Hong Kong, April 7–9, 1997, Hong Kong Pest Control Association, pp 59–68

Yates JR, Grace JK, Reinhardt JN (2000) Installation guidelines for the basaltic termite barrier: a particle barrier to Formosan subterranean termite (Isoptera: Rhinotermitidae). Sociobiology 35:1–16

Yoshimura T, Tsunoda K, Takahashi M, Katsuda Y (1992) Pathogenicity of an entomogenous fungus, *Conidiobolus coronatus* Tyrrell and Macleod, to *Captotermes formosanus* Shiraki. Jpn J Environ Entomol Zool 4:11–16

Yu H, Gouge DH, Baker P (2006) Parasitism of subterranean termites (Isoptera: Rhinotermitidae: Termitidae) by entomopathogenic nematodes (Rhabditida: Steinernematidae; Heterorhabditidae). J Econ Entomol 99:1112–1119

Yuvraj Singh (2007) Isolation and identification of bacteria having pathogenic interactions with termites (Isoptera). M.Sc Thesis, CCS Haryana Agricultural University, Hisar, pp 104

Zoberi MH (1995) *Metarhizium anisopliae*, a fungal pathogen of *Reticultermes flavipes* (Isoptera: Rhinotermitidae). Mycologia 87:354–359

Zoberi MH, Grace JK (1990) Isolation of the pathogen *Beauveria bassiana* from *Reticultermes flavipes* (Isoptera: Rhinotermitidae). Sociobiology 16:289–296

Chapter 13
Microbial Control of Postharvest Diseases of Fruits, Vegetables, Roots, and Tubers

Ramesh C. Ray, Manas R. Swain, Smita H. Panda, and Lata

13.1 Introduction

Postharvest decay of fruits, vegetables, roots, and tubers (Table 13.1) limits the period of storage, compromises marketing and consumers' acceptability, and causes substantial losses. Gray mold, Green mold, blue mold, and sour rot caused by *Botrytis cinerea* (Pers.) Fries, *Penicillium digitatum* (Pers.) Sacc., *Penicillium italicum* Wehmer, and *Geotrichum candidum* Link, respectively, are the most common pathogens during storage and transportation of fruits. Likewise, soft rot, Fusarium rot, and Sclerotium rot caused by *Rhizopus oryzae* Went & Prinsen and *R. stolonifer* Ehrenb. ex Fr., *Fusarium oxysporum* Schlecht and *F. solani* M.Sacc., and *Corticium rolfsii* Curzi, respectively, are the common postharvest pathogens of vegetables, roots, and tubers (Snowdown 1990). In packing houses, fungicides are used extensively to protect fruit, vegetables, roots, and tubers against these diseases. Fungicide residues in those commodities and their possible effects on human health and the environment and the development of fungicide resistance by the pathogens have, of late, stimulated the evaluation of alternative control methods that are effective and safe to use (Kinay and Yildiz 2008).

Biological control by the microorganisms (bacteria, fungi, and yeasts) is currently used to control various decays on pome, apple, citrus and stone fruits, avocado, seed potatoes, cassava, yams, aroids, and sweet potatoes (Stockwell and

R.C. Ray (✉)
Regional Centre, Central Tuber Crops Research Institute, Dumduma PO, Bhubaneswar 751 019, India
e-mail: rc_rayctcri@rediffmail.com

M.R. Swain
Department of Biotechnology, College of Engineering and Technology, Biju Patnaik University of Technology, Bhubaneswar 751003, India

S.H. Panda
Agri- Bioresource Research Foundation, Khandagiri PO, Bhubaneswar 751 019, India

Lata
Division of Microbiology, Indian Agricultural Research Institute, New Delhi, India

A. Singh et al. (eds.), *Bioaugmentation, Biostimulation and Biocontrol*, Soil Biology 28,
DOI 10.1007/978-3-642-19769-7_13,

Table 13.1 Botanical names of the fruits, vegetables, roots, and tubers mentioned in this chapter

	Botanical names
Fruits	
Apple	*Malus domestica*
Avocardo	*Persea americana*
Banana	*Musa paradisiaca*
Cherry	*Prunus virginiana*
Grapes	*Vitis vinifera*
Grape fruit	*Citrus paradisi*
Guava	*Psidium guajava*
Hami melon	*Cucumis melo* var. *saccharinus*
Jujube	*Zizyphus jujuba*
Kiwifruit	*Actinidia deliciosa*
Lemon	*Citrus lemon*
Litchi	*Litchi chinensis*
Loquat	*Eriobotrya japonica*
Mango	*Mangifera indica*
Muskmelon	*Cucumis melo*
Nectarine	*Prunus persica*
Orange	*Citrus sinensis*
Papaya	*Carica papaya*
Pear	*Pyrus communis*
Peach	*Prunus persica*
Plum	*Prunus salicina*
Rambutan	*Nephelium lappaceum*
Strawberry	*Fragaria vesca*
Stone fruit	*Collinsonia canadensis*
Sweet cherry	*Prunus avium*
Vegetables	
Cabbage	*Brassica oleracea*
Cherry tomato	*Solanum lycopersicum*
Chilli	*Capsicum annum*
Onoin	*Allium cepa*
Pipper	*Capsicum annum*
Tomato	*Lycopersicon esculentum*
Root and tubers crops	
Carrot	*Daucus carota*
Cassava	*Manihot esculenta* Crantz
Potato	*Solanum tuberosum*
Sweet potato	*Ipomoea batatas*
Yam	*Dioscorea rotundata*

Stack 2007). The further expansion of microbial control of postharvest decays will largely depend on improving its effectiveness under an increased range of environmental conditions and on expanding the spectrum of activity to new commodities and new diseases. This can be accomplished with the discovery of new antagonists (Janisiewicz and Jeffers 1997; Janisiewicz and Korsten 2002; Sharma et al. 2009), by combining antagonists with other alternatives to synthetic fungicides [e.g., GRAS (generally regarded as safe substances), heat, or UV treatments] (El-Ghaouth et al. 2000a; Conway et al. 2007; Sharma et al. 2009), by improving the antagonistic containing formulations (Spadaro and Gullino 2004), or by improving antagonists through genetic manipulation (Massart and Jijakli 2007). Formulation consisting of the desired microbial antagonists or the mixture of antagonists must be economical to produce, contain enough colony-forming units to be effective, and should be easy to handle and apply to fruits, vegetables, roots, and tubers (Janisiewicz and Korsten 2002). Genetic engineering offers tremendous postharvest potential for improving microbial antagonist. For example, antagonists can be manipulated to overexpress mechanisms of biocontrol or foreign genes can be transferred to antagonists to increase tolerance to environmental stresses or to produce antifungal substances (Jones and Prusky 2002; Wisniewski et al. 2005).

This chapter focusses on the progress made in the recent years on the spectrum of microorganisms used as antagonists for the control of postharvest diseases, modes of action of microbial agents, extension of use of microbial agents, enhancing biocontrol efficacy of microbial antagonists, and biotechnological approaches for improving microbial antagonistic action through manipulation of formulations and genetic engineering.

13.2 Microbial Control Strategies

There are two basic approaches for using the microbial antagonists for controlling the postharvest diseases of fruits, vegetables, roots, and tubers: (1) manipulation of epiphytic microflora or (2) those that can be artificially introduced against postharvest pathogens.

13.2.1 Manipulation of Epiphytic Microflora

The common observation is that washed fruits and vegetables develop rot more than unwashed commodities. For example, citrus fruits that were washed, dried, and stored tended to rot faster than unwashed fruits (Chalutz and Wilson 1990; Janisiewicz et al. 1991; Zhou et al. 1999; Hoopen et al. 2003). It was observed that bacteria and yeasts predominated when undiluted washings from citrus fruit had been plated out; it was only after the washings were diluted that fungal pathogens appeared (Sharma et al. 2009). In an earlier study, *Pantoea agglomerans*, the bacterium isolated from the epiphytic microflora of fruits and leaf surfaces of pears and apples was very effective against *Botrytis cinerea*, *Penicillium expansum*, and *R. stolonifer*, the common postharvest rotting microorganisms (Nunes et al. 2001a). Ray and Byju (2003) have shown that the yeasts, *Debaryomyces hansenii*, *Pichia anomala*, and *Saccharomyces cerevisiae*, isolated from the surface microflora of sweet potato were antagonist to *F. oxysporum* and *Botryodiplodia theobromae* causing sweet potato postharvest decay. Likewise, application of *Bacillus subtilis,* isolated from the epiphytic microflora of yam tuber, showed a drastic reduction in the range and number of spoilage fungi of yams during a 5-month storage period. The surface fungi, such as *B. theobromae*, *Fusarium moniliforme*, and *Penicillium sclerotigenum*, were displaced completely on the treated tubers (Okigbo 2002). Bleve et al. (2006) reported that epiphytic yeasts, such as *Metschnikowia pulcherrima*, *Kluyveromyces thermotolerans*, *Issatchenkia terricola*, and *Candida incommunis*, were highly antagonistic for biocontrol of *Aspergillus carbonarius* and *Aspergillus niger* on grape. Likewise, *Cryptococcus magnus* epiphytic yeast isolated from papaya fruit and leaf surface was highly effective against anthracnose, a postharvest disease in papaya fruit, caused by *Colletotrichum gloeosporioides* (Capdeville et al. 2007). In a more recent report, a study was conducted to screen plant epiphytic yeasts for use as biocontrol agents of bacterial fruit and blotch, a serious disease of hami melon, caused by *Acidovorax avenae* subsp. *citrulli* (Wang et al. 2009). Results showed that 24 out of 463 yeast strains isolated from leaves or flowers of plants were antagonistic against *A. avenae* and the strain 0732-1, identified as *P. anomala* Kurtzman, was found to be most effective. This evidence indirectly suggests that there might be epiphytic/surface microbial populations on fruits and vegetables that are naturally suppressive to rot pathogens.

The phylloplane has also been a good source of antagonists, as it may share part of the resident microflora of fruits and vegetables as well as contain other microorganisms dislodged from them (Janisiewicz 1987; Korsten et al. 1997; Sobiczewski et al. 1996). However, the antagonists may also come from other closely related or unrelated sources such as industrial malt (Laitila et al. 2007).

13.2.2 Introduction of Microbial Antagonists

Earliest efforts to control postharvest diseases involved the use of *B. subtilis* (Pusey and Wilson 1984), *Trichoderma* and *Rhodotorula* (Sadfi et al. 2002), and *Pseudomonas cepacia* (Janisiewicz and Roitman 1988). All of these antagonists could colonize wound sites and elaborate antimicrobial substances, which prevent development of pathogens in apple, strawberry, cherry, peach, plum, etc. (Ippolito and Nigro 2000; Sharma et al. 2009). Since then, several other antagonists (yeasts, fungi, and bacteria) have been identified and used for controlling various postharvest diseases of fruits (Table 13.2), vegetables, roots, and tubers (Table 13.3). Meanwhile, some biocontrol products, i.e., Aspire (*Candida oleophila* strain 182), Biocoat (*Candida saitoana*+chitosan), Bio-cure (*C. saitoana*+antifungal lytic enzyme), BioSave (*Pseudomonas syringae*), and Yield Plus (*Cryptococcus albidus*) were registered in recent years and are commercially available for the control of postharvest decay of fruits and vegetables (Janisiewicz and Korsten 2002).

13.2.2.1 Yeast

Application of many types of yeasts as postharvest antagonists appears to be quite promising agent. Janisiewicz (1987), Chalutz et al. (1988), and Tian et al. (2005)

Table 13.2 Microbial antagonists used for the successful control of postharvest diseases of fruits

Antagonists	Disease (pathogen)	Fruits	Reference(s)
Aureobasidium pullulans	Botrytis rot (*Botrytis cinerea*)	Grape	Schena et al. (2003)
Aureobasidium pullulans	Blue mold (*Penicillium expansum*)	Apple	Bencheqroun et al. (2007)
	Soft rot (*Monilinia laxa*)	Grape	Barkai-Golan (2001)
Bacillus subtilis	Stem endrot (*Botryodiplodia theobromae* Pat.)	Avocado	Demoz and Korsten (2006)
	Gray mold (*Botrytis cinerea*)	Strawberry	Zhao et al. (2007a)
	Alternaria rot (*Alternaria alternata*)	Muskmelon	Yang et al. (2006)
Bacillus licheniformis	Anthracnose (*Colletotrichum gloeosporioides*)	Mango	Govender et al. (2005)
Bacillus amyloliquefaciens	*Botrytis cinerea*	Peach	Arrebola et al. (2009)
	Penicillium expansum	Peach	
	Rhizopus stolonifer	Peach	
Burkholderia cepacia	Stem end rot Anthracnose (*Colletotrichum musae*)	Banana	De Costa and Erabadupitiya (2005)
Burkholderia spinosa	Banana anthracnose (*Colletotrichum musae*)	Banana	De Costa et al. (2008)
	Blossom end rot (*Colletotrichum musae*)	Banana	De Costa and Erabadupitiya (2005)
Brevundimonas diminuta	Anthracnose (*Colletotrichum gloeosporioides*)	Mango	Kefialew and Ayalew (2008)
Candida guilliermondii	Gray mold (*Botrytis cinerea*)	Pear Apple	Tian et al. (2002a)
Candida membranifaciens	Anthracnose (*Colletotrichum gloeosporioides*)	Mango	Koomen and Jeffrics (1993) Kefialew and Ayalew (2008)
Candida oleophila	Crown rot (*Colletotrichum musae*)	Banana	Lassois et al. (2008)
	Anthracnose (*Colletotrichum gloeosporioides*)	Papaya	Gamagae et al. (2003)
	Gray mold (*Botrytis cinerea*)	Peach	Karabulut and Baykal (2004)
	Green mold (*Penicillium digitatum*	Orange	El-Neshawy and Wilson (1997)
Candida sake	Penicillium rot (*Penicillium expansum*)	Apple	Usall et al. (2001), Torres et al. (2006), Morales et al. (2008)
Cadida shak	Blue mold (*Penicillium italicum*)	Apple	Teixido et al. (2001)
	Blue mold (*Penicillium expansum*)	Pome	Torres et al. (2006)
Cryptococcus laurentii	Bitter rot (*Glomerella cingulata*)	Apple	Blum et al. (2004)
	Brown rot (*Monilinia fructicola*)	Cherry	Tian et al. (2004), Qin et al. (2006)
	Alternaria rot (*Alternata alternata*)	Jujube	Qin and Tian (2004), Tian et al. (2005)
	Penicillium rot (*Penicillium expansum*)		
	Rhizopus rot (*Rhizopus stolonifer*)	Peach	Zhang et al. (2007c) Zhang et al. (2005b, 2007c)
	Gray mold (*Botrytis cinerea*)	Peach	
	Brown rot (*Monilinia fructicola*)	Peach	Yao and Tian (2005)
	Blue mold (*Penicillium expansum*)	Peach	Zhang et al. (2003, 2007c)
	Rhizopus rot (*Rhizopus stolonifer*)	Strawberry	Zhang et al. (2007b)

(*continued*)

Table 13.2 (continued)

Antagonists	Disease (pathogen)	Fruits	Reference(s)
Cryptococcus albidus	Gray mold (*Botrytis cinerea*)	Apple	Fan and Tian (2001)
	Blue mold (*Penicillium expansum*)	Strawberry	Jurgen (2002)
Cryptococcus laurentii	Gray mold (*Botrytis cinerea*)	Apple	Lima et al. (2006)
Cryptococcus magnus	Anthracnose (*Colletotrichum gloeosporioides*)	Papaya	Capdeville et al. (2007)
Epicoccus nigrum	Brown rot (*Monilinia laxa*)	Peach	Larena et al. (2005)
Kloeckera apiculata	Botrytis rot (*Botrytis cinerea*)	Cherry	Karabulut et al. (2005)
	Penicillium rots (*Penicillium* spp.)	Citrus	Long et al. (2006, 2007)
	Green (*Penicillium digitatum*) Blue mold (*Penicillium italicum*)	Citrus	Long et al. (2006, 2007)
Kluyveromyces thermotolerans	*Aspergillus carbonarius, Aspergillus niger*	Grape	Bleve et al. (2006)
Metschnikowia fructicola	Botrytis rot (*Botrytis cinerea*)	Grape	Karabulut et al. (2003)
Metschnikowia pulcherrima	Blue mold (*Penicillium expansum*)	Apple	Spadaro et al. (2002b, 2004)
Metschnikowia pulcherrima	Blue mold (*Penicillium italicum*)	Citrus	Kinay and Yildiz (2008)
Pantoea agglomerans	Penicillium rots (*Penicillium* spp.)	Orange	Plaza et al. (2001)
Pantoea agglomerans	Rhizopus rot (*Rhizopus stolonifer*)	Pear	Nunes et al. (2002a, b)
	Blue mold (*Penicillium expansum*)	Apple	Nunes et al. (2001a)
Pantoea agglomerans	Rhizopus rot (*Rhizopus stolonifer*)	Citrus	Canamas et al. (2008)
Penicillium frequentans	Brown rot (*Monilinia* sp.)	Peach	Guijarro et al. (2007)
Pichia anomala	Penicillium rots (*Penicillium* spp.)	Citrus	Lahlali et al. (2004)
	Crown rot (*Colletotrichum musae*)	Banana	Lassois et al. (2008)
	Anthracnose (*Colletotrichum capsici*)	Chillies	Chanchaichaovivat et al. (2007)
Pichia guilliermondii	Bue mold (*Penicillium italicum*)	Citrus	Kinay and Yildiz (2008)
Pseudomonas fluorescens	Migula Gray mold (*Botrytis mali*)	Apple	Mikani et al. (2008)
Pseudomonas syringae	Blue mold (*Penicillium expansum*)	Apple Peach	Janisiewicz (1987) Zhou et al. (1999, 2002)
	Gray mold (*Botrytis cinerea*)	Apple	Zhou et al. (2001)
Rahuella aquatilis	Gray mold (*Botrytis cinerea*)	Apple	Calvo et al. (2003, 2007)
	Blue mold (*Penicillium expansum*)	Apple	Calvo et al. (2007)
Rhodotorula glutinis	Blue mold (*Penicillium expansum*)	Apple	Zhang et al. (2009)
	Gray mold (*Botrytis cinerea*)		Zhang et al. (2009)
	Alternaria rot (*Alternata alternata*)	Jujube	Tian et al. (2005)
	Penicillium rot (*Penicillium expansum*)	Jujube	Tian et al. (2005)
	Blue rot (*Penicillium expansum*)	Pear	Zhang et al. (2008b)
	Gray mold (*Botrytis cinerea*)	Pear	Zhang et al. (2008a)
	Gray mold (*Botrytis cinerea*)	Strawberry	Zhang et al. (2007a)

(*continued*)

Table 13.2 (continued)

Antagonists	Disease (pathogen)	Fruits	Reference(s)
Trichoderma harzianum	Anthracnose (*Colletotrichum musae*)	Banana	Devi and Arumugam (2005)
	Brown spot (*Gliocephalotrichum microchlamydosporum*)		
	Gray mold (*Botrytis cinerea*)	Grape	Batta (2007)
	Gray mold (*Botrytis cinerea*)	Kiwifruit	Batta (2007)
	Gray mold (*Botrytis cinerea*)	Pear	Batta (2007)
	Anthracnose (*Colletotrichum gloeosporioides*)	Rambutan	Sivakumar et al. (2001, 2002a, b)
Trichoderma viride	Stem-end rot (*Botryodiplodia theobromae*)	Mango	Kota et al. (2006)
Trichosporon pullulans	Alternaria rot (*Alternata alternata*)	Cherry	Qin et al. (2004)
	Gray mold (*Botrytis cinerea*)		
Pichia membranifaciens	Anthracnose rot (*Colletotrichum acutatum*)	Loquat	Cao et al. (2008)
Pichia pastoris	Blue rot (*Penicillium expansum*)	Apple	Janisiewicz et al. (2008a, b)
Pichia anomala	fruit blotch (*Acidovorax avenae*)	Hami melon	Wang et al. (2009)

have made several positive points in recommending yeasts as potential microbial agents for controlling the postharvest diseases of fruit and vegetables, including (a) yeasts can colonize the wound surface for long period even under dry conditions; (b) yeasts produce extracellular polysaccharides, which enhance their survivability and restrict the growth of pathogen propagules; (c) they can use nutrients rapidly and proliferate at a faster rate; and (d) they are the least affected by the pesticides. Of the various yeasts, *Candida sake*, *C. oleophila*, *D. hansenii*, *P. anomala*, and *Pichia guilliermondii* have exhibited a wide spectrum of biological activity against many postharvest pathogens (Wisniewski et al. 1988; Wilson and Chalutz 1989; Karabulut and Baykal 2003).

However, recent research has been focussed on the use of several other yeasts (i.e., *Candida albidus*, *S. cerevisiae*, *Issatchenkia orientalis*, *M. pulcherrima*, *C. laurentii*, etc.) for controlling postharvest diseases of fruits and vegetables (Chanchaichaovivat et al. 2007; Kinay and Yildiz 2008; Zhang et al. 2008b).

13.2.2.2 Fungi

Fungi are also used as antagonist in the postharvest control of fruits and vegetables. Earlier studies have showed that strains of *Trichoderma pseudokoningii* and *Trichoderma harzianum* (Tronsmo and Denis 1977) have resulted in the reduction of *Botrytis cinerea* in apples and of *Monilinia laxa* in stone fruits, respectively. Spraying with suspensions of *T. harzianum*, *Trichoderma viride*, *Gliocladium roseum*, and *Paecilomyces variotii* resulted in a partial control of *Botrytis* in strawberry fruits and of *Alternaria* in lemon fruits (Pratella and Mari 1993).

Table 13.3 Microbial antagonists used for the successful control of postharvest diseases of vegetables, roots, and tubers

Antagonists	Disease and pathogen	Vegetable	Reference(s)
Bacillus licheniformis	Botrytis rot (*Botrytis allii*)	Onion	Lee et al. (2001)
Bacillus amyloliquefaciens	Fusarium rot (*Fusarium oxysporum*)		
Bacillus spp.	Fusarium rot (*Fusarium roseum* var. *sambucinum*)	Potato	Sadfi et al. (2002)
Bacillus subtilis	Botryodiplodia rot (*Botryodiplodia theobromae*)	Yams	Okigbo (2002), Swain and Ray (2008)
Bacillus subtilis	Fusarium rot (*Fusarium moniliforme*)	Yams	Okigbo (2002)
Candida guilliermondii	Botrytis rot (*Botrytis cinerea*)	Tomato	Saligkarias et al. (2002)
Debaryomyces hansenii	Botryodiplodia rot (*Botryodiplodia theobromae*)	Sweet potato	Ray and Das (1998)
	Botrytis rot (*Botrytis cinerea*)	Tomato	Saligkarias et al. (2002) German Garcia et al. (2001)
Pichia guilliermondii	Anthracnose (*Colletotrichum capscii*)	Chilli	Chanchaichaovivat et al. (2007)
Pichia anomala	Rhizopus rot (*Rhizopus nigricans*)	Tomato Sweet potato	Zhao et al. (2008) Ray and Das (1998)
Pichia onychis	Rhizoctonia rot (*Rhizoctonia solanifer*)	Tomato	German Garcia et al. (2001), German Garcia and Marina Cotes (2001), Fuentes et al. (2002)
Pantoea agglomerans	Dry rot (*Gibberella pulicans*)	Potato	Schisler et al. (2000)
Pseudomonas fluorescens	Dry rot (*Gibberella pulicans*)	Potato	Schisler et al. (2000)
Pseudomonas sp.	Blue and Green rot (*Penicillium sclerotigenum*)	Yams	Okigbo (2002)
Trichoderma sp.	Rhizoctonia rot (*Rhizoctonia solani*)	Chilli	Bunker and Mathur (2001)

Recent studies have focussed on yeast-like fungus, *Aureobasidium pullulans*, as the most effective antagonists against postharvest plant pathogens, because it has the ability to survive and increase its population under a variety of field conditions and during cold storage (Leibinger et al. 1997; Schena et al. 1999). In an earlier study, Schena et al. (1999) found that isolates of *A. pullulans* at high concentrations (10^7 and 10^8 cells/ml) were able to control *P. digitatum* on grapefruit, *Bacillus cinerea*, *R. stolonifer*, and *A. niger* on grapes, and *B. cinerea* and *R. stolonifer* on cherry tomatoes. Other reports on biocontrol activities of *A. pullulans* against postharvest diseases include blue mold in stored apple caused by *P. expansum* (Bencheqroun et al. 2007), monilinia rot in banana (Wittig et al. 1997), and grapes by *Monillinia taxa* (Barkai-Golan 2001), etc.

13.2.2.3 Bacteria

Bacterial flora have attracted enormous attention as agents for biocontrol in postharvest diseases, particularly since they are easy to handle, generally stable, have resistance and ability to survive desiccation, and inherently possess a quick generation time (Sharma et al. 2009). They are also known to affect life cycles of different plant pathogens or pests by diverse mechanisms including the production of extracellular metabolites and intracellular proteinaceous toxins. In general, spore-forming bacteria (e.g., *Bacillus* spp.) survive to a greater extent even in harsh environments, compared to the non-spore-forming bacteria. Among the *Bacillus* spp., the ones that have attracted the most attention are *Bacillus thuringiensis*, *B. amyloliquefaciens*, *B. licheniformis*, and *B. subtilis* (Mari et al. 1996; Govender et al. 2005; Arrebola et al. 2009). Other bacterial species of interest are *Burkholderia spinosa* (De Costa et al. 2008), *Enterobacter cloacae* (Wilson et al. 1987), *P. agglomerans* (Torres et al. 2007; Canamas et al. 2008; Usall et al. 2008), etc. Bacteria may be formulated in either a dormant or a metabolically active state. They are easily mass-produced using a liquid fermentation process, although in some cases they may be more amenable to semisolid or solid-state fermentation (Swain and Ray 2008).

13.3 Mode of Action of Biocontrol Agents

Biological control by antagonistic microorganisms uses naturally occurring mechanisms to suppress harmful organisms. The modes of action are competition for nutrients (Bencheqroun et al. 2007; Spadaro and Gullino 2004) and space (Piano et al. 1998; Fan and Tian 2000), parasitism (Wan and Tian 2002), induced resistance (Wisniewski et al. 1991; Fan et al. 2002), and antibiosis (Bull et al. 1998). In general, more than one mechanism is implicated, but in no case was a single mechanism found to be responsible/suitable for biological control (Janisiewicz and Korsten 2002).

13.3.1 Competition for Nutrients

Competition for nutrients is the most promising mode of action for several postharvest antagonistic microorganisms. Yeasts appear to be particularly promising as biocontrol agents against the postharvest decay of fruits and vegetables (Droby and Chalutz 1994). Meanwhile, most postharvest fruit pathogens are necrotrophs needing nutrients for spore germination and initiation of the pathogenic process, which can be an effective mechanism of biological control. This hypothesis plays a major role in the mode of action of *P. guilliermondii* against *P. digitatum* in citrus (Droby et al. 1992; Arras 1996); *P. anomala* against *Penicillum* spp. in banana (Lassois et al. 2008); *D. hansenii* against *Botrytis cinerea* in grapes (Chalutz et al. 1988); and *A. pullulans* against *P. expansum* in grapes (Castoria et al. 2001) and apple (Bencheqroun et al. 2007), etc.

Attachment by microbial antagonist to the pathogen hyphae appears to be an important factor necessary for competition for nutrients as shown by the interactions of *E. cloacae* and *R. stolonifer* (Wisniewski et al. 1989), and *P. guilliermondii* and *P. italicum* (Arras et al. 1998). In vitro studies conducted on such interactions revealed that due to direct attachment, antagonistic yeasts and bacteria take nutrients more rapidly than target pathogens and thereby prevent spore germination and growth of the pathogens (Droby et al. 1989; Wisniewski et al. 1989). Nonpathogenic species of *Erwinia*, such as *E. cypripedii*, showed antagonistic activity against various isolates of *Erwinia caratovora* sub sp. *caratovora*, the causal agent of soft rot of many vegetables such as carrot, tomatoes, and pepper, primarily by competing for nutrients (Moline et al. 1999; Janisiewicz et al. 2000). In a recent study, *A. pullulans* strain Ach1-1 was selected for its effectiveness against blue mold caused by *P. expansum* on stored apple fruit (Bencheqroun et al. 2007). The possible involvement of competition for nutrients in the biocontrol activity of this antagonistic strain was investigated both in vitro and in situ. For in vitro assays, the effect of strain Ach1-1 on germination percentages of *P. expansum* conidia was evaluated after a 24 h incubation period in the presence of increasing apple juice concentrations (0–5%) using a system allowing the physical separation of both agents. In the absence of strain Ach1-1, conidial germination was strongly promoted by apple juice whatever the concentration. However, germination was significantly reduced by the presence of strain Ach1-1 except at the highest juice concentration. For in situ assays, strain Ach1-1 was very protective against *P. expansum* on postharvest wounded apples. However, the application of high concentrations of exogenous sugars, vitamins, and most particularly amino acids significantly reduced such protection. Time-course analysis of apple amino acids at the wound site revealed that these compounds were more depleted in wounds treated with strain Ach1-1 alone and especially in those treated with both agents (strain Ach1-1 and *P. expansum*) compared to wounds treated with *P. expansum* alone or to untreated ones. Exogenous amino acids, applied at high concentrations on apple wounds as a mixture of specific amino acid groups or as individuals, significantly decreased strain Ach1-1 efficacy against *P. expansum*. This study provided in vitro and in situ evidence that competition for apple nutrients, most particularly amino acids, may be a main mechanism of the biocontrol activity of *A. pullulans* strain Ach1-1 against blue mold caused by *P. expansum* on harvested apple fruit (Bencheqroun et al. 2007).

13.3.2 Competition for Space

Competition for space is the competition for infection sites, which may occur if antagonists are able to occupy the specific places where recognition mechanisms between host and pathogen take place. If these places are no more available for pathogens, the necessary procedure of recognition cannot take place and infection does not occur (Janisiewicz et al. 2000). Rapid colonization of fruit wound by the antagonist is critical for decay control, and manipulations leading to improved colonization enhance biocontrol (Mercier and Wilson 1994). Thus, microbial antagonists should have the ability to grow more rapidly than the pathogen. Similarly, it should have the ability to survive even under conditions that are unfavorable to the pathogen (Droby et al. 1992).Wound competence under environmental conditions may be an important character for the evaluation of microbial agents with commercial potential.

Roberts (1990) found that *C. laurentii*, an effective antagonist against gray mold (*Botrytis cinerea*) and blue mold (*Penicillium* spp.) of apple, could rapidly colonize the wounds of apple fruit at temperatures ranging from 5 to 20°C and even under cold storage conditions (1–2°C). The yeast also exhibited rapid increase in population dynamics on apple fruit wounds. Similar results were obtained in controlling postharvest decay caused by *Botrytis cinerea* of apple fruit by *A. pullulans* (Ippolito et al. 2000). Biological control of *P. digitatum* on orange fruits with *Candida famata* was reported by Arras (1996). Scanning electron microscope observations of the mode of action of the antagonist against the pathogen revealed rapid colonization of the fungal mycelium and the wounds, with lytic and phagocytic activity against the hyphae (Figs. 13.1 and 13.2).

13.3.3 Populations of the Microbial Antagonist

Initial concentration of antagonist plays a significant role in the microbial antagonists when applied on the wound site and the ability of the antagonist to rapidly

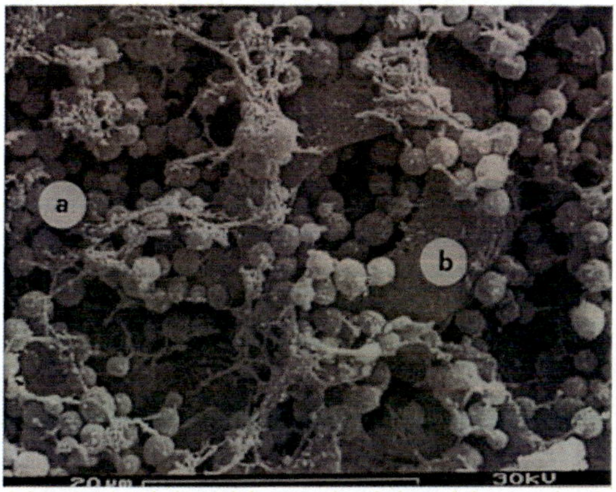

Fig. 13.1 Epicarp of orange fruit completely colonized by *Candida famara* cells (**a**). A stoma can be seen at the center (**b**) (Arras 1996)

Fig. 13.2 *Penicillium digitatum* hyphal (**b**) completely colonized by *Candida famata* cells (**a**) causing alterations in the hyphal tissue (Arras 1996)

colonize the wound site (Janisiewicz and Roitman 1988; McLaughlin et al. 1990; Mercier and Wilson 1995; Spadaro et al. 2002a; Zhang et al. 2005a). Microbial antagonists (i.e., *Candida saitoana*, *C. oleophila*, *E. caratovora*, *C. laurentii*, etc.) are most effective in controlling postharvest decay on fruits and vegetables when applied at a concentration at 10^6–10^9 colony-forming unit CFU/ml (Chand-Goyal et al. 1999; El-Ghaouth et al. 2004; Zhang et al. 2005b, 2007a; Cao et al. 2008), and rarely, higher concentrations are required.

Candida saitoana was effective at a concentration of 10^7 colony-forming unit (CFU)/ml for controlling *P. expansum* on apples (McLaughlin et al. 1990). In another study, El-Ghaouth et al. (1998) reported that for *C. saitoana*, a concentration of 10^8 CFU/ml was better in controlling blue mold (*P. expansum*) on apples. Similarly, for *C. laurentii*, at the concentration of 10^9 CFU/ml, when challenged with *P. italicum*, applied at 1×10^4 spores/ml, the blue mold decay of oranges completely inhibited during 5 days of incubation at 20°C (Zhang et al. 2005a). This qualitative relationship, however, is highly dependent on the ability of the antagonists to multiply and grow at the wound site. This was demonstrated by using a mutant of *P. guilliermondii*, which lost its biocontrol activity against *P. digitatum* on grapefruit and against *Botrytis cinerea* on apples, even when applied to the wounds at concentrations as high as 10^{10} CFU/ml (Droby et al. 1991). The cell population of this mutant remained constant at the wound sites during incubation period, while that of the wild type increased 10- to 20-fold, within 24 h.

13.3.4 Direct Parasitism

Antagonist and pathogen can also interact through a direct parasitism. Wisniewski et al. (1991) observed a strong adhesion in vitro of *P. guilliermondii* antagonist cells to *Botrytis cinerea* mycelium, perhaps due to a lectin link. After yeast cells were dislodged from the hyphae, the hyphal surface appeared to be concave and there was partial degradation of the cell wall of *B. cinerea* at the attachment sites. In a recent study, Zhao et al. (2010) using scanning electron microscopy unveiled that *P. guilliermondii* multiplied rapidly on tomato fruit wounds and its cells had a storage capability of adhesion to the hyphae of *R. stolonifer* (Fig. 13.3). Moreover, *P. guilliermondii* shows a high activity of β-1,3-glucanase enzyme that could result in the degradation of the fungal cell walls (Jijakli and Lepoivre 1998). *A. pullulans* in apple wounds produces extracellular exochitinase and β-1,3-glucanase, which could play a role in the biocontrol activity (Castoria et al. 2001). Through ultra-structural and cytochemical studies, El-Ghaouth et al. (1998) found that *Candida saitoana* yeast cells, when cultivated together with *B. cinerea* mycelium, were associated with fungal hyphae showing cytological damage, such as papillae and other protuberances in the cell wall, and degeneration of the cytoplasm. Bonaterra et al. (2003) reported that direct parasitism was a major factor that permitted *P. agglomerans* to control *M. laxa* and *R. stolonifer* decay on stone fruits.

13.3.5 Production of Cell-wall Lytic Enzymes

Microbial antagonists also produce lytic enzymes such as glucanase, chitinase, and proteinases that help in the cell-wall degradation of the pathogenic fungi (Lorito et al. 1993; Castoria et al. 1997, 2001; Jijakli and Lepoivre 1998; Kapat et al. 1998; Mortuza and Ilag 1999; Chernin and Chet 2002). The interaction between *B. subtilis* and *F. oxysporum*, the postharvest pathogen of yam (*Dioscorea* spp.) tubers, was studied by scanning electron microscopy (Swain et al. 2008). Lysis of fungus cell wall by *B. subtilis* was observed owing to the production of extracellular chitinase (Fig. 13.4). In recent years, exocellular lytic enzymes (β-1,3-glucanase) produced by yeasts such as *P. anomala*, *A. pullulans*, *Rhodotorula glutinis*, and *C. laurentii* have also been studied to obtain a better understanding of the biocontrol of postharvest diseases of fruits (Castoria et al. 1997; Jijakli and Lepoivre 1998).

13.3.6 Antibiosis

Production of antibiotics is other important mechanism by which microbial antagonists suppress the pathogens (antibiosis) of harvested fruits and vegetables. In the case of bacterial antagonists, it has been suggested that their biocontrol activity may be partly associated with the production of antibiotics, such as iturins (a powerful antifungal peptide) produced by *B. subtilis*, pyrollnitrins produced by *P. cepacia*, and trichothecenes produced by *Myrothecium roridium* (Bull et al. 1998; Golubev et al. 2001). The main concern, related to the use of antibiotics in food products, is the development of human pathogens resistant to these compounds and the possible development of resistance in fruit pathogens. Even if antibiotic producers appear to

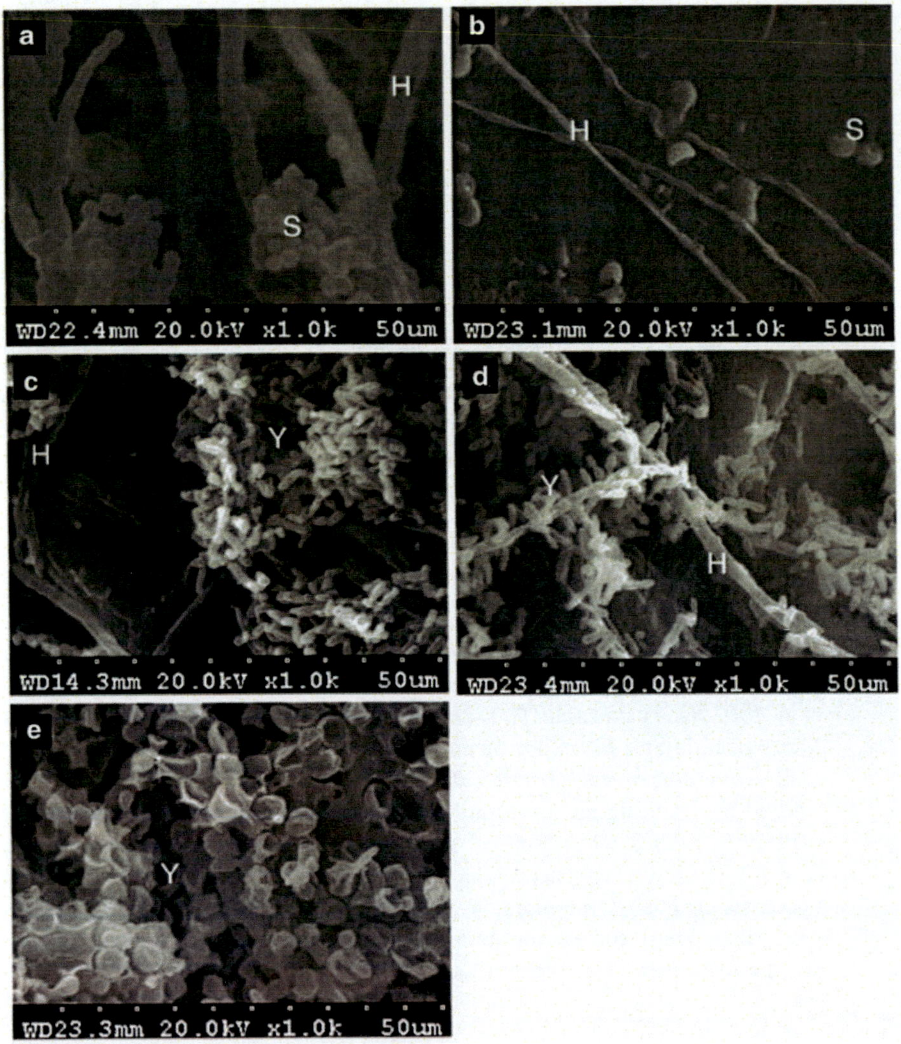

Fig. 13.3 Electron micrographs of wounds incubated at 20°C for 48 h after different treatments. **(a)** *Rhizopus stolonifer* inoculated as the control; **(b)** *R. stolonifer* inoculated before heat treatment at 38°C for 24 h; **(c)** *R. stolonifer* inoculated before *P. guilliermondii* inoculation; **(d)** R. stolonifer Ehrenb inoculated before heat treatment at 38°C for 24 h followed by *P. guilliermondii* inoculation; and **(e)** *R. stolonifer* inoculated before *P. guilliermondii* inoculation followed by heat treatment at 38°C for 24 h. *H* hyphae, *S* spores, *Y* yeast (*P. guilliermondii*) cells (Zhao et al. 2010)

be able to control wound infections established before antagonist application, at the moment, there are not such biocontrol agents registered for use on fruit and vegetables.

However, Mari et al. (1996) reported that *B. amyloliquefaciens* (strain 5PVB) did not produce extracellular antibiotic substance, yet was highly active against the pathogens, *Botrytis cinerea* on both mature-green and red tomato.

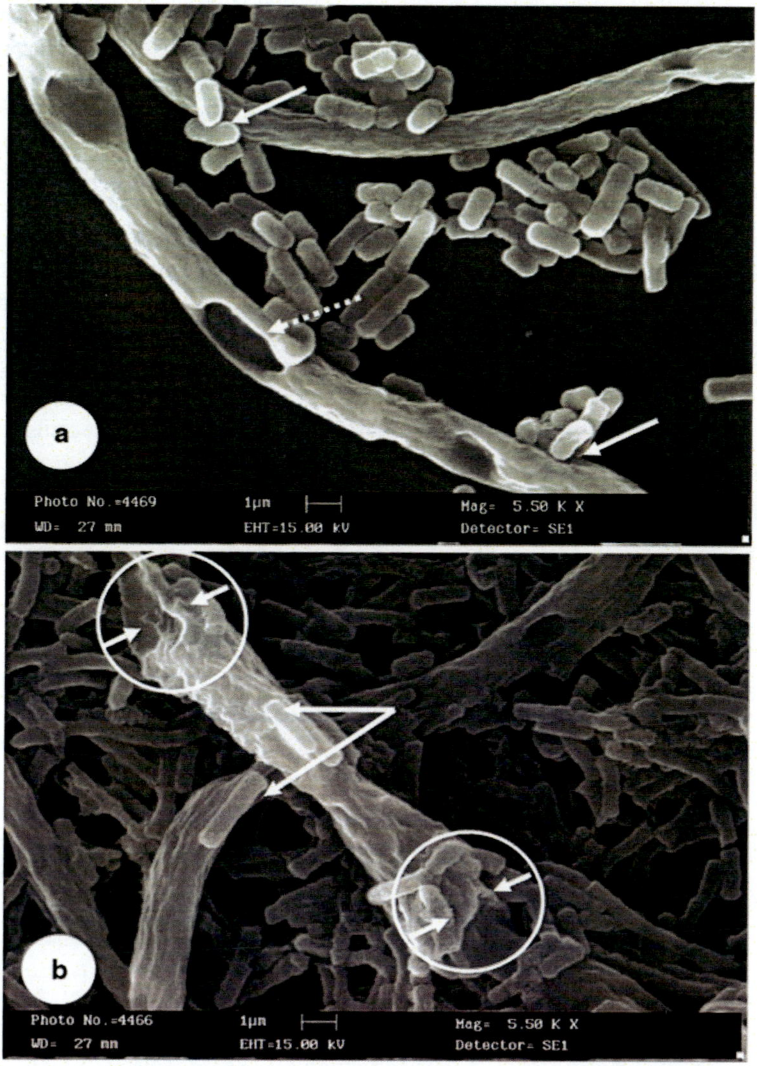

Fig. 13.4 Scanning electron micrograph of *Fusarium oxysporium* sample collected at 12 h (**a**) and 36 h (**b**) after interaction with *Bacillus subtilis* CM1. The *solid and dotted arrow* shows the bacterial attachment with fungal hyphae and lytic mark hyphae. *Circles* indicate the complete lysis of fungal mycelium after 36 h of interaction (Swain et al. 2008)

13.3.7 Induced Resistance

Several evidences have indicated that microbial antagonists may elicit defense mechanism of the host as well (El-Ghaouth et al. 1998; Ippolito et al. 2000). For example, El-Ghaouth et al. (2001a) investigated the ability of *Candida saitoana* to induced systemic resistance in apple fruit against *Botrytis cinerea*. To differentiate antagonistic activity of *C. saitoana* from the ability to induce resistance, the antagonist and the pathogen were applied on separate wounds in fresh apple. When *C. saitoana* was applied 0 and 24 h before ioculaton of *B. cinerea*, no effect on lesion development was found. But, when applied 48 or 72 h postinoculation with *B. cinerea*, *C. saitoana* reduced lesion diameter by more than 50 and 70%, respectively, in comparison to wounding. Further, it was evident that *C. saitoma* was capable of inducing systemic resistance in apple fruit and increased chitinase and β-1,3-glucanase activities with a higher accumulation in fresh than in stored apples. Similarly, *A. pullulans* causes a transit increase in the activity of β-1, 3-glucanase, peroxidase, and chitinase enzymes in apple wounds, which stimulated wound healing process and induce defense mechanisms against *P. expansum* (Ippolito et al. 2000). Efficacy of *C. laurentii* was maintained when applied simultaneously or prior to inoculation with *P. italicum*, causing blue mold of oranges (Zhang et al. 2005b). Efficacy was reduced when *C. laurentii* was applied after inoculation. The yeast *C. laurentii* was also reported to control a postharvest disease of jujube fruit by producing β-1,3-gucanase, a cell-wall degrading enzyme involved in plant host defense (Tian et al. 2007). Similarly, Fan et al. (2002) demonstrated that *Pichia membranifaciens* and *Candida guilliermondii* two antagonistic yeasts could produce chitinase and β-1,3-glucanase in vitro and induce an increase in β-1,3-glucanase and chitinase activities in the wounds of nectarine fruit, resulting in an effective decrease in decay caused by *R. stolonifer*. In peach fruit infected by *P. expansum*, the yeast *P. membranifaciens* induced a number of proteins related to host defense mechanisms (Chan et al. 2007). Furthermore, the yeast *P. membranifaciens* induced production of H_2O_2-metabolizing enzymes and total protein synthesis and reduces oxidative stress in harvested sweet cherry fruit (Chan and Tian 2006). *P. guilliermondii* strain R13, yeast isolated from Thai rambutan, has been shown to suppress the fungal pathogen *Colletotrichum capsici* in harvested chili (Nantawanit et al. 2010). The pretreatment of chili with the yeast antagonist, physically separated from the fungus by known distances, significantly reduced disease incidence and lesion diameter caused by *C. capsici*. Compared to the controls, the yeast treatment also significantly enhanced the activities of chitinase and β-1,3-glucanase and the accumulation of capsidiol phytoalexin in chili tissue (Nantawanit et al. 2010). Induction of disease resistance was also reported in avocardo, citrus, peach, and pineapple fruits (Prusky et al. 1994; Rodov et al. 1994; Arras 1996; Fan et al. 2002).

Microbial antagonist-induced disease resistance in fruits was also manifested by the production phenylalanine ammonia-lyase activity as in grapefruit (Droby et al. 2001, 2003) and chili (Nantawanit et al. 2010) and the accumulation of phytoalexine such as scoparone and scopoletin in orange fruits (Arras et al. 1998). The biosynthesis of scoparone in particular, 4 days after inoculation with the *C. famata* (F35) only, was 124 μg/g fresh weight of the fruit, 12 times higher than that in the non-inoculated wound tissues, while it decreased to 47 μg/g when F35 strain was inoculated at the same time as *P. digitatum*, and to 37 μg/g when the pathogen only was inoculated (Arras 1996).

13.4 Extension of Use of Biocontrol Agents

Potential biocontrol agents often have many significant limitations. The method to select antagonists with a broader spectrum of activity, preferably for commercial development, includes efficacy tests for various pathogens and horticultural species, and adaptability to various environmental conditions (Wilson et al. 1993; Lima et al. 1999).

13.4.1 Criteria for an Ideal Antagonist

A potential microbial antagonist should have some desirable characteristics to make it an ideal bioagent (Barkai-Golan 2001). The antagonist should be (1) effective against a wide range of the pathogens and different harvested commodities; (2) nonpathogenic to the host; (3) resistant to pesticides; (4) compatible with other chemical and physical treatments; (5) effective at low concentrations; (6) genetically stable; (7) capable of surviving under adverse environmental conditions such as UV radiation, desiccation, and rapid climatic change; (8) a nonproducer of metabolites harmful to human; (9) preparable in a form that can be effectively stored and dispensed; and (10) not fastidious in its nutritional requirements. In addition, a microbial antagonist should have an adaptive advantage over specific pathogen. For example, *C. gloeosporioides* is more sensitive to low temperature than many other pathogens (Koomen and Jeffrics 1993). Therefore, for its effective control, a microbial antagonist such as *C. magnus* should have the ability to grow, multiply, and suppress the pathogens at low temperature (Capdeville et al. 2007). Most of the apple and pear fruits are stored in cold storage. For controlling their postharvest decay, a microbial antagonist should have the ability to survive under cold storage conditions as well. Considering these factors, research work has been reoriented in many countries in search of microbial antagonist having multifold attributes such as tolerance to cold/hot temperature and simultaneously effective against a broad spectrum of postharvest pathogens. Accordingly, a new bacterial strain of *P. agglomerans* (CPA-2) was isolated, which could control *P. expansum*, *Botrytis cinerea*, and *R. stolonifer* under various storage conditions such as cold storage, either in air or in low oxygen atmosphere (Nunes et al. 2001a). Also, equal control was obtained with *P. agglomerans* at 8×10^7 CFU/ml, as with the fungicide imazalil at commercial doses against these pathogens (Nunes et al. 2001a). However, even if an antagonist has all the desirable characteristics, economic factor decides whether it has to be commercialized or not. A potential market for the product is essential before the commercialization of the antagonist is taken into consideration.

13.4.2 Antagonist Mixtures

An effective biological control based upon a mixture of several complementary and noncompetitive antagonists is more likely than a control based upon microorganism alone. Such mixtures have several advantages (Janisiewicz and Bors 1995; Janisiewicz et al. 1998): apart from a wider spectrum of activity (different fruits, cultivars, and ripening stages), they can increase the efficacy (less biomass necessary), be more reliable, and allow a reduction in application times and treatment costs. Moreover, they permit the combination of different genetic characteristics. Mixture of *Sporobolomyces roseus* and *Pseudomonas syringae* showed grater biocontrol activity against postharvest blue mold on Golden Delicious apple, as compared with the same strains applied individually (Janisiewicz and Bors 1995). Similar results were obtained with mixture of two strains of *A. pullulans* and *R. glutinis* and two strains of *B. subtilis* and one of *A. pullulans* (Leibinger et al. 1997).

Mixtures of yeasts were tested for their ability to control *P. expansum* and *Botrytis cinerea* on Red Delicious apple fruits (Calvo et al. 2003; Juan et al. 2003). The occurrence of antagonistic or synergistic interactions between yeast strains in different mixtures was also evaluated. Two strains of *Rhodotorula* (*R. glutinis* SL 1 and *R. glutinis* SL 30) and two strains of *Cryptococcus* (*C. albidus* SL 43 and *C. laurentii* SL 62) were selected for developing yeasts mixtures. The *R. glutinis* SL 1–*R. glutinis* SL 30 mixture exhibited a lower effectiveness than each strain alone, against both molds. Other mixtures (*R. glutinis* SL 1–*C. albidus* SL 43 and *R. glutinis* SL 30–*C. albidus* SL 43) showed synergism against *P. expansum* but not against *B. cinerea*. The *R. glutinis* SL 1–*C. laurentii* SL 62 mixture was the only mixture that showed synergism against gray mold. There was not any mixture, which showed high effectiveness against both molds at the same time. Different results could be explained by the dynamics of the population of the yeasts. By using yeast mixtures, it was possible to improve biocontrol without increasing the amount of antagonists applied. The synergism observed could be useful in enhancing biological control (Calvo et al. 2003). Likewise, the mixture of *M. pulcherrima* and *P. guilliermondii* was found more effective in controlling postharvest decay of citrus fruit caused by *P. digitatum* (green mold), *P. italicum* (blue mold), and *G. candidum* (sour rot) (Kinay and Yildiz 2008). A mixture of two yeast antagonists, *M. pulcherrima* and *C. laurentii*, isolated from apple fruits exhibited greater biocontrol activity against blue mold (*P. expansum*) of apple than either of the yeasts applied alone (Janisiewicz et al. 2008a).

13.4.3 Antagonistic Preparation

Only a few of the microbial antagonists reported to control postharvest diseases of fruits and vegetables have been commercialized. There could be many reasons for this, but two primary causes, which prevented this, are (a) the relative ineffectiveness of the antagonists compared to chemical control procedures; and (b) a lack of economic incentives. However, once effective antagonist is identified, search starts for its preparation, storage, and application methodology. For instance, *B. subtilis* (strain B-3) was the first organism patented as a postharvest biocontrol agent for stone fruits in the USA (Pusey and Wilson 1984). Pusey et al. (1988) conducted a pilot test applying *B. subtilis* under simulated commercial conditions for the control of brown rot of peaches, in which bioagent was effectively incorporated into wax normally used on the packing line. Botrytis rot was effectively controlled by this procedure, but considerable variation was found in the control rendered by the different antagonist preparations. Torres et al. (2007) studied application of *P. agglomerans* in controlling Penicillum rot of oranges and mandarin in semicommercial scale at several mediterranean regions. They found that bacterial product formulation treatment significantly reduced the percentage of infected fruits.

In the recent past, several commercial products have been developed and commercialized. For example, 'BioSave' has been developed from a saprophytic strain of *P. syringae* by 'EcoScience' Corp., Orlando, USA, which is highly useful for controlling blue and gray mold on apples and pears (Janisiewicz and Korsten 2002; Droby 2006). Ecogen-Israel Partnership Ltd. has developed another product, 'Aspire' from the yeast, *P. guilliermondii* (Janisiewicz and Korsten 2002). Similar research has been conducted with the yeast antagonist *C. oleophila* (Mercier and Wilson 1994; Wisniewski et al. 1995), which had been previously described as *C. sake*. Tests conducted in commercial citrus packing houses gave satisfactory control of green and blue molds and sour rot only in combination with tenfold diluted thiobendazole (Droby et al. 1998). The research and commercial development of 'YieldPlus' for biocontrol of fruit decays follows the same pattern as described for 'Aspire' or *C. oleophila*. The development of 'Avogreen' from this *Bacillus* followed a slightly different path in that it was tested in the field for biocontrol. It was prepared from *B. subtilis* and used in South Africa for the control of *Cercospora* species and anthracnose of avocado (Korsten et al. 1997; Janisiewicz and Korsten 2002). Efficacy of *B. licheniformis* was evaluated under semicommercial conditions on a mango packing line to control anthracnose and stem-end rot on the mango (Govender et al. 2005). Mango fruits were treated with either the bacterial antagonist applied in hot water (45°C) followed by a quarter strength prochloraz dip or the antagonist applied on its own in hot water. These treatments were compared to the untreated control and commercially used prochloraz hot water dip. Treated fruits were dried and waxed on the commercial packing line. Fruit subjected to the prochloraz-antagonist–hot water combination showed reduced anthracnose and stem-end rot incidence after market simulated conditions of low temperature storage at 10°C with 90% relative humidity and at room temperature (20°C at 75% relative humidity for 7 days). This integrated treatment retained the fruit color and firmness with high marketability most effectively, compared to the other treatments. Another biological control formulation SHEMER WDG with the yeast *Metschnikowia fructicola*, developed by Agre Gerrn, Israel, is found against several postharvest diseases of fruits (Canamas et al. 2008).

13.4.4 Preharvest Application

One of the major obstacles to the development of postharvest microbial antagonist is their inability to control previously established infections, such as latent infections. Field application of the microbial antagonist may enable early colonization of the fruit surface, protecting them from these infections (Ippolito and Nigro 2000; Janisiewicz and Korsten 2002; Ippolito et al. 2004; Irtwange 2006).

Leibinger et al. (1997) applied a preharvest mixture of the yeast *A. pullulans* and the bacterium *B. subtilis* obtaining a level of control equivalent to fungicides for *P. expansum* and *Botrytis cinerea* on apples. Teixido et al. (1998) applied unmodified and low water activity tolerant cells of *C. sake* before harvest to control blue mold on apples. *C. sake* CPA-1 reduced blue mold by nearly 50% on wounded apples if the apples were inoculated with antagonist 2 days before harvest and inoculation with *P. expansum* and cold storage for 4 months (Teixido et al. 1999). However, preharvest application of the antagonistic yeast of a concentration of $3–10^6$ CFU/ml was less effective against Penicillum rot than postharvest treatment. It is difficult to control postharvest diseases of strawberry even with preharvest application of fungicides; despite that some success has been achieved with field application(s) of various microbial antagonists such as *G. roseum* (Sutton et al. 1997), *T. harzianum* (Tronsmo and Denis 1977; Kovach et al. 2000), and *Epicoccum nigrum* (Larena et al. 2005). Preharvest spray of *Metschnikowia fructicola* was also effective in controlling preharvest and postharvest fruit rots in strawberry (Karabulut et al. 2004). Similarly, preharvest application of *A. pullulans* reduced storage rots in strawberry (Lima et al. 1997), grapes (Schena et al. 1999), cherries (Wittig et al. 1997), and apples (Leibinger et al. 1997). Further, the incidence of green mold (*P. digitatum*) on grapefruit was reduced by preharvest spray of *P. guilliermondii* (Droby et al. 1992). In citrus, preharvest application of the biocontrol organism *P. agglomerans* CPA-2 effectively controlled postharvest rots (Canamas et al. 2008; Usall et al. 2008). Similarly, preharvest application(s) of *C. laurentii* and *C. oleophila* reduced storage rots in pear (Benbow and Sugar 1999). Field application of *E. nigrum* was reported to be effective for controlling postharvest brown rot (*Monilinia* spp.) in peaches (Larena et al. 2009). Canamas et al. (2008) have reported that preharvest application of different concentrations of *P. agglomerans* was effective for protecting oranges against *P. digitatum* during storage.

From the above discussion, it appears that preharvest application of microbial antagonists has many limitations and commercial practice is not successful except in certain cases.

13.4.5 Postharvest Application

In postharvest application, microbial cultures are applied either as sprays or as dips in an antagonist's solution (Barkai-Golan 2001). This approach has been more effective than preharvest application of microbial antagonists and has several successes (Tables 13.2 and 13.3). For example, postharvest application of *P. agglomerans* resulted in better control of Penicillum rot (*P. expansum*) in apple (Nunes et al. 2002a; Morales et al. 2008), green (*P. digitatum*) and blue mold (*P. italicum*) of citrus (Teixido et al. 2001; Torres et al. 2007), and Rhizopus rot (*R. stolonifer*) in pear (Nunes et al. 2001a). Similarly, postharvest application of *A. pullulans* has been found to control several postharvest diseases such as botrytis rot (*Botrytis cinerea*) in grapes (Schena et al. 2003) and monilinia rot (*M. laxa*) in banana (Wittig et al. 1997). A significant reduction in storage decay was also achieved by bringing several yeast species in direct contact with wounds in the peel of harvested fruits. For instance, direct contact of yeast (*C. oleophila*, *C. sake*, *C. guilliermondii*, etc.) antagonist and infested fruit peel has been quite useful for the suppression of pathogens such as *P. digitatum* and *P. italicum* in citrus (Chalutz and Wilson 1990), *Botrytis cinerea* in apples (Wisniewski et al. 1988; Roberts 1990; Gullino et al. 1992; Mercier and Wilson 1995), *B. cinerea* and *P. expansum* in pears (Chand-Goyal and Spotts 1996, 1997), and *Botrytis cinerea*, *R. stolonifer*, and *Alternaria alternata* in tomatoes (Chalutz et al. 1988). However, all the pathogens do not react in a similar fashion to a given antagonist.

13.4.6 Integrated Use

Use of microbial antagonist alone cannot solve all the problems of postharvest rots during fruit storage and they must be considered instruments to be used in combination with other methods in an integrated vision of disease management. For example, microbial antagonists can be combined with waxes and fungicides applied not only in post but also in preharvest (Pusey et al. 1988). In some laboratory and semicommercial trials, the efficacy of the *P. guilliermondii* was consistently increased by the addition of small concentrations of imazalil or thiabendazole, reaching a control level similar to the use of fungicides (Droby et al. 1993). Yeast is generally tolerant to many of the fungicides used in postharvest: *M. pulcherrima* (Spadaro et al. 2002b) is tolerant to relatively high concentrations of benzimidazoles (benomyl and thiabendazole) and dicarboximides (vinchlozolin and procymidone) (Spadaro and Gullino 2004).

Among the other strategies evaluated during the last few years, the combination of microbial antagonists with other alternative techniques to chemicals should be mentioned such as thermotherapy (Barkai-Golan and Phillips 1991; Zhao et al. 2010), ultraviolet rays (Chalutz et al. 1992), animal and plant natural products (Aharoni et al. 1993), calcium infiltrations (Janisiewicz et al. 1998), sodium bicarbonate (Teixido et al. 2001), ethanol (Spadaro et al. 2002b), or ammonium molybdate (Wan et al. 2003). In a most recent study, a combination of heat treatment at 38°C followed by the application of *P. guilliermondii* provided the best effective prevention of cherry tomato from fungus (*Botrytis cinerea*, *A. alternata*, and *R. stolonifer*) spoilage (Zhao et al. 2010). Following heat treatment and

P. guilliermondii treatment, electronic nose detected a reduction of volatility in cherry tomato fruit odor, an indicator of preserving fruit's freshness. Scanning electron microscopy unveiled that heat treatment at 38°C for 24 h inhibited hyphal growth and spore germination of *R. stolonifer*, while *P. guilliermondii* multiplied rapidly on fruit wounds, and its cells had a strong capability of adhesion to the hyphae of *R. stolonifer*. However, heat treatment also seriously injured *P. guilliermondii*; therefore, *P. guilliermondii* should be applied after heat treatment. Some of these factors have been elaborated in the later section.

13.5 Enhancing Biocontrol Efficacy of Microbial Antagonist

13.5.1 Conventional Approach

Many research results have showed that antagonistic agents could be combined with exogenous substance to increase the performance margin of biocontrol (Wisniewski et al. 1995; El-Ghaouth et al. 2000a). This includes addition of calcium salts, carbohydrates, amino acids, and other nitrogen compounds (Qin et al. 2003; Tian et al. 2002b). The synergistic actions of various additives (other than fungicides) and antagonists on postharvest disease control are summarized in Table 13.4.

13.5.1.1 Effects of Calcium

Postharvest calcium treatment of apples provided broad-spectrum protection against the postharvest pathogens such as *P. expansum* and *Botrytis cinerea* (Saftner et al. 1997). The addition of $CaCl_2$ (2%, w/v) to the formulation of the yeast biocontrol agent, *C. oleophila* and *P. membranifaciens*, enhanced the ability of this yeast to protect apples and peaches against postharvest decay (Wisniewski et al. 1995; Fan and Tian 2000). Tian et al. (2001, 2002b) indicated that the efficacy of controlling gray mold and blue mold rots in apples was enhanced when *Trichosporon* sp., even at a low concentration of 10^5 CFU/ml, was applied in the presence of an aqueous suspension of $CaCl_2$ (2% w/v). In yet another study, Tian et al. (2002c) studied the biological control efficacy of *C. guilliermondii*, *P. membranifaciens*, *Trichosporon* sp., and *C. laurentii*, applied with or without Ca^{2+}, and was tested against the postharvest disease of peach, nectarine, apple, and grape, respectively. Addition of Ca^{2+} to the suspensions of *C. guilliermondii* and *P. membranifaciens* resulted in lower disease incidences in peach and nectarine fruits infected by *R. stolonifer*, compared with the treatment with the yeast alone. In a similar study, a biological treatment of *P. anomala* strain K (10^7 CFU/ml), β-1,3-glucans, and $CaCl_2$ (20 g/l) was found highly effective against *Botrytis cinerea* and *P. expansum* on apples in field conditions (Juaki et al. 2002). Ping et al. (2003) obtained similar results with *R. glutinis* in combination with Ca^{2+} ions in the

Table 13.4 Salt additives for enhancing the efficacy of microbial antagonists

Fruit	Salt additive	Microbial agent	Disease controlled	References
Apple	Sodium carbonate	*Metschnikowia pulcherrima*	Blue mold (*Penicillum expansum*)	Conway et al. (2007) Janisiewicz et al. (2008a, b)
	Sodium carbonate	*Cryptococcus laurentii*	Blue mold (*Penicillum expansum*)	Conway et al. (2007) Janisiewicz et al. (2008a, b)
	Calcium propionate	*Candida oleophila*	Blue mold (*Penicillum expansum*)	Droby et al. (2003)
	Sodium bicarbonate	*Candida oleophila*	Blue mold (*Penicillum expansum*)	Droby et al. (2003)
Citrus	Sodium carbonate	*Cryptococcus laurentii*	Green mold (*Penicillum digitatum*)	Usall et al. (2008)
	Sodium bicarbonate	*Bacillus subtilis*	Green and blue molds (*Penicillum digitatum* and *Penicillum expansum*)	Obagwu and Korsten (2003)
	Sodium bicarbonate	*Pantoea agglomerans*	Penicillium rots (*Penicillum* sp.)	Plaza et al. (2001), Teixido et al. (2001), Torres et al. (2007), Usall et al. (2008)
Cherry	Ammonium molybdate	*Pichia membranifaciens*	Brown rot (*Lasiodiplodia theobromae*)	Qin et al. (2006)
		Crptococcus laurentii	Brown rot (*Lasiodiplodia theobromae*)	Qin et al. (2006)
	Calcium chloride	*Aureobasidium pullulans*	Brown rot (*Lasiodiplodia theobromae*)	Ippolito et al. (2005)
	Sodium bicarbonate	*Aureobasidium pullulans*	Brown rot (*Lasiodiplodia theobromae*)	Karabulut et al. (2005)
	Potassium sorbate	*Candida oleophila*	Postharvest decay	Karabulut et al. (2001)
Grape	Sodium bicarbonate	*Metschnikowia fructicola*	Botrytis rot (*Botrytis cinerea*)	Karabulut et al. (2003)
Loquat	Calcium chloride	*Pichia membranifaciens*	Penicillium rots (*Penicillum* sp.)	Cao et al. (2008)
Oranges	Calcium chloride	*Candida oleophila*	Penicillium rots (*Penicillum* sp.)	El-Neshawy and El-Sheikh (1998)
Pear	Sodium carbonate	*Cryptococcus laurentii*	Blue mold and Alternaria rot (*Penicillum expansum* and *Colletrotrichum gloeosporiodies*, respectively)	Yao et al. (2004)
	Sodium carbonate	*Trichosporon pullulans*	Blue mold and Alternaria rot (*Penicillum expansum* and *Colletrotrichum gloeosporiodies*, respectively)	Yao et al. (2004)
	Calcium chloride	*Cryptococcus laurentii*	Gray mold (*Botrytis cinerea*)	Zhang et al. (2005a, b)
	Ammonium molybdate	*Rhodotorula glutinis*	Blue mold (*Penicillum expansum*)	Wan and Tian (2005)
Papaya	Sodium bicarbonate	*Candida oleophila*	Anthracnose (*Alternaria alternata*)	Gamagae et al. (2003)

(*continued*)

Table 13.4 (continued)

Fruit	Salt additive	Microbial agent	Disease controlled	References
Peach	Calcium chloride	*Debaryomyces hansenii*	Rhizopus rot (*Rhizopus stolonifer*)	Singh (2004, 2005)
	Calcium propionate	Aspire	Brown rot (*Lasiodiplodia theobromae*)	Droby et al. (2003)
	Sodium bicarbonate	Aspire	Rhizopus rot (*Rhizopus stolonifer*)	Droby et al. (2003)
	Sodium bicarbonate	*Pseudomonas syringae*	Green mold (*Penicillum digitatum*)	Plaza et al. (2001)
Rambutan	Potassium metabisulphite	*Trichoderma* spp.	Postharvest rots	Sivakumar et al. (2002a, b)
Tomato	Sodium bicarbonate	*Cryptococcus laurentii*	Botrytis rot (*Botrytis cinerea*)	Xi and Tian (2005)

biocontrol of blue mold (*P. italicum*) of citrus. In a recent study, the beneficial effect of 2% $CaCl_2$ (w/v) on the antagonistic yeast *P. membranifaciens* for control of anthracnose rot caused by *Colletotrichum acutatum* in postharvest loquat fruit and the possible mechanisms involved were investigated (Cao et al. 2008). The results showed that treatment with *P. membranifaciens* at 1×10^8 CFU/ml or 2% $CaCl_2$ alone resulted in both significantly smaller lesion diameter and lower disease incidence of anthracnose rot on loquat fruit wounds compared with the controls. The biocontrol activity of *P. membranifaciens* on the disease was enhanced by the addition of 2% $CaCl_2$ and the combined treatment of *P. membranifaciens* with $CaCl_2$ resulted in a remarkably improved control of the disease in comparison with the treatment of *P. membranifaciens* or $CaCl_2$ alone.

Calcium propionate in combination with Aspire increased the biocontrol activity against *B. cinerea* and *P. expansum* causing postharvest decay of apples (Wisniewski et al. 2001a), but not of peach (Droby et al. 2003).

The influence of Ca^{2+} on biocontrol effectiveness of yeast may be postulated to result from interaction with yeast and/or its metabolic products in the wound site, and the ability of antagonists to reduce decay in the presence of Ca^{2+} may be partially due to nutrient competition and/or site exclusion (Wisniewski et al. 1995). The addition of Ca^{2+} would directly decrease the number of pathogens and indirectly increase the ability of the yeast to inhibit the development of the pathogen as well as the resistance of fruit to pathogens (Biggs et al. 1997). However, the precise mechanism by which Ca^{2+} reduces fungal infection is not yet fully understood.

13.5.1.2 Effect of Salicylic acid

Qin et al (2003) reported that combining salicylic acid with the yeast suspensions (*R. glutinis* and *Candida laurentii*) significantly enhanced the biocontrol activity against pathogens (Alternaria rot) of sweet cherries. Salicylic acid treatment plus *R. glutinis* at 10^7 CFU/ml reached the performance of *R. glutinis* at 10^8 CFU/ml alone. But application of salicylic acid did not affect the growth of *R. glutinis* and

C. laurentii in cherry wounds. They also proved that salicylic acid treatment induced a significant increase in polyphenoloxidase, phenylalanine ammonia-lyase, and β-1,3-glucanase activities of cherry fruit (Qin et al. 2003). In a recent study, *R. glutinis* in combination with salicylic acid resulted in low average natural infection incidence in peach fruit, 16.67%, compared with 46.67% in the water-treated control peach fruit (Zhang et al. 2008b).

Extensive studies have established that the application of exogenous salicylic acid resulted in both systemic acquired resistance gene expression and induction of systemic acquired resistance (Palva et al. 1994).

13.5.1.3 Effect of Sodium bicarbonate

Sodium bicarbonate at 2% concentration enhanced the biocontrol ability of *C. laurentii* and *A. pullulans* against postharvest decay caused by *P. expansum* and *A. alternata* in pear fruits (Gamagae et al. 2003). The use of 2% sodium bicarbonate in combination with *C. oleophila* (Aspire) and *Candida albidus* exhibited both curative and protective activity against *Botrytis cinerea* and *P. expansum* causing postharvest decay in apple and peach (Wisniewski et al. 2001b; Droby et al. 2003; Gamagae et al. 2003). The addition of 2% (w/v) sodium bicarbonate in the suspension of *Crytococcus lauretii* or *Trichosporon pullulans* significantly limited spore germination and germ tube elongation of *P. expansum* and *A. alternata* (Yao et al. 2004). Biocontrol activity of *C. laurentii* or *T. pullulans* against postharvest decay caused by *P. expansum* and *A. alternata* in pear fruits was significantly increased when *C. laurentii* or *T. pullulans* combined with sodium bicarbonate. Combining *C. laurentii* or *T. pullulans* with sodium bicarbonate provided a more effective control on *P. expansum* and *A. alternata* than applying the antagonistic yeast or sodium bicarbonate alone. *C. laurentii* in combination with sodium bicarbonate showed the best control of disease caused by *A. alternata* in pear fruits. Similarly, the use of sodium bicarbonate at 2% followed by the antagonist bacterium, *P. agglomerans*, could be a reliable procedure to control green mold disease (*P. digitatum*) of orange (Plaza et al. 2001). In a semicommercial and commercial trial, *P. aggomerans* CPA-2, in combination with 3% sodium bicarbonate when tested to control postharvest diseases affecting citrus crop in the Mediterranean region, gave significant result (Torres et al. 2007). The bacterial–sodium bicarbonate formulation treatment significantly reduced the percentage of infected fruits (Usall et al. 2008). Green mold (*P. digitatum*) infection was 0% in *P. aggolomerans* + sodium bicarbonate infected fruits, as compared with 90% infection in untreated fruits. In another pilot-scale experiment, a mixture of two yeasts antagonists *M. pulcherrima* and *C. laurentii* was used in combination with sodium biocarbonate for controlling of blue mold decay of apple during storage (Janisiewicz et al. 2008a).The treated fruits were stored in commercial controlled atmosphere storages for approximately 6 months in the 2005–2006 and 2006–2007 storage seasons and then evaluated for incidence of decay. In both years, the treatments with the antagonist alone or in combination with sodium bicarbonate were equally effective and reduced blue mold incidence by 84–97% in 2005–2006 and 73–82% in 2006–2007. Sodium bicarbonate alone significantly reduced blue mold incidence compared to the nontreated control but was less effective than the antagonist alone or in combination with sodium bicarbonate.

B. subtilis, isolated from citrus fruit surface, was evaluated alone or in combination with sodium bicarbonate or hot water treatment on artificially inoculated (with *P. digitatum*/*P. italicum*) orange cultivars. A significant increase in biocontrol activity of all isolates was observed when isolates were combined with sodium bicarbonate or were applied following hot water treatment, as compared with antagonist treatment alone (Obagwu and Korsten 2003). Ji and Wilson (2003) considered that sodium salicylate could be used as a selective carbon source to improve the efficacy of *P. syringae* in the control of bacterial decay of tomato plant.

The inhibitory mechanism of bicarbonate salts on postharvest pathogens was probably due to the reduction of fungal cell turgor pressure that resulted in collapse and shrinkage of hyphae and spores, so inhibiting the sporulation of fungi (Fallik et al. 1997). In addition, sodium bicarbonate directly inhibited the germination ability of fungal pathogens and enhanced the activity of yeast against fungal pathogens for nutrients and space competition (Yao et al. 2004).

13.5.1.4 Effect of Ammonium molybdate

Wan et al. (2003) reported that ammonium molybdate (NH_4-Mo) has great potential to enhance the biocontrol efficacy of *R. glutinis* and *Candida laurentii* against blue mold on jujube fruits and that the growth of the yeasts was greatly affected by the presence of NH_4-Mo in the wound sites. The biocontrol efficacy of both yeasts at 10^7 CFU/ml combined with 15 mmol/l of NH_4-Mo was better than the antagonists used alone at 10^8 CFU/ml. The same effect was observed in the case of combining NH_4-Mo with *C. sake* against postharvest decay caused by *P. expansum*. It was found that NH_4-Mo markedly reduced disease incidence and severity of blue mold in peach fruits, and the addition of NH_4-Mo to *C. sake* enhanced its biocontrol efficacy against postharvest diseases of peach fruits (Nunes et al. 2001b).

13.5.1.5 Effects of Fungicides

Biocontrol agents have been effectively integrated with chemical fungicides to provide disease suppression with fewer fungicide applications than the conventional spray regime, in greenhouse crops (Elad et al. 1996) as well as after harvest (Chand-Goyal and Spotts 1996; Droby et al. 1998; Sugar and Basile 2008). Isolate L47 of *A. pullulans*, for example, gave better results on strawberries and grapes when sprayed in combination with a low dose of fungicide as compared to the antagonist alone (Ippolito and Nigro 2000).

The integrated application of thiabendazole or imazalil (Deccozil 25EC, Elf Atochem North America Inc., Philadelphia) at the rate recommended for standard postharvest treatments (0.1 and 1.2 g/l a.i.) with the biocontrol agents (*P. guilliermondii* or *C. oleophila*) effectively controlled gray mold and blue mold rots in citrus fruits in commercial tests conducted in packing houses in Italy (Arras et al. 2002). Tian et al. (2001) reported that the use of iprodione (Rovral 50 WP, Rhone Poulenc Ag. Co., Research Triangle Park, North Carolina) at a rate ten times lower (50 μl/l a.i.) in combination with the yeast *Trichosporon* sp. enhanced the efficacy of control compared to the yeast and the fungicide when used separately. The yeast *Candida albidus* at 1 × 10^6 CFU/ml mixed with iprodione at 50 μl/l a.i. had better control of postharvest decay of apple fruit caused by *Botrytis cinerea* and *P. expansum* as compared to the single application (Fan and Tian 2001). Qin and Tian (2004) reported that combination of *C. laurentii* with a low dose of imazalil (25 μl/l) or kresoxim-methyl (Stroby 50% DF, BASF Ltd., Germany) (50 μl/l) significantly enhanced the activities of the yeast or the fungicides against both *A. alternata* and *Monilinia fructicola* on jujube fruits at 25 and 0°C in air, as well as in controlled atmosphere condition with 10% O_2 + 0% CO_2. The yeast isolate LS28 (*C. laurentii*), which tolerated in vitro high rates of benzimidazoles, was tested, alone or in combination with a low dose (10% of the full label rate) of thiabendazole, against gray mold on stored apples (Lima et al. 2006). The integration of fungicides with biocontrol agents offers the opportunity to reduce the amount of fungicide in preharvest application, thus lowering the level of residues on marketed products.

13.5.1.6 Addition of Nutrients and Plant Products

The efficacy of the microbial antagonists can also be enhanced considerably by the addition of some nutritious compounds or natural plant products (Table 13.5). For example, additions of nitrogenous compounds such as L-aspargine, L-proline, and 2-deoxy-D-glucose, a sugar analogue, helped in enhancing the bioefficacy of microbial antagonists in controlling the postharvest decay rots in some fruits and vegetables (El-Ghaouth et al. 2000a, b, c). When applied in fruit wounds, the combination of *Candida saitoana* and 2-deoxy-D-glucose (0.2%) controlled fruit decay on apples, oranges, and lemons caused by *Botrytis cinerea*, *P. expansum*, and *P. digitatum* (El-Ghaouth et al. 2000a, b, c) than when either *Candida saitona* or 2-deoxy-D-glucose was applied alone. The treatment of peaches with *C. laurentii* (10^8 CFU/ml) alone or in combination with methyl jasmonate (200 μM/l) inhibited the lesion diameter of brown rot and blue mold rots caused by *Monilinia fructicola* and *P. expansum*, respectively (Yao and Tian 2005). The inhibitory mechanism was mainly because of resistance induced in peach fruit by methyl jasmonate and *C. laurentii*. In addition, direct inhibition of methyl jasmonate on *P. expansum* also played a role in controlling blue mold.

Some other recommendations (Table 13.5) have also emerged for improving the bioefficacy of microbial antagonists. For example, a bioactive coating consisting of *Cryptococcus saitoana* + glycochitosan has been developed to control fruit decay in apple (El-Ghaouth et al. 2000a). The biocontrol activity of *Candida saitoana*, against decay of apple, lemon, and orange, caused by *Botrytis cinerea*, *P. expansum*,

Table 13.5 Nutrients or plant products as an additive for enhancing the efficacy of microbial antagonists

Fruit	Nutrients/plant product as additive	Microbial agent	Pathogen	References
Apple	2-Deoxy-D-glucose	*Candida saitoana*	Gray mold (*Botrytis cinerea*)	El-Ghaouth et al. (2000a, b)
Apple	Glycochitosan	*Cryptococcus saitaona*	Gray mold (*Botrytis cinerea*)	El-Ghaouth et al. (2000a, c)
Apple	Gylcochitin	*Candida saitoana*	Gray mold (*Botrytis cinerea*)	El-Ghaouth et al. (2000a)
Apple	Nisin	*Candida oleophila*	Gray mold (*Botrytis cinerea*) and Blue mold (*Penicillium expansum*)	El-Neshawy and Wilson (1997)
	Methyl jasmonate	*Cryptococcus laurentii*	Gray mold (*Botrytis cinerea*) and Blue mold (*Penicillium expansum*)	Yao and Tian (2005)
	Salicylic acid	*Cryptococcus laurentii*	Gray mold (*Botrytis cinerea*)	Yu and Zheng (2005)
Banana	Bee wax	*Trichoderma harzianum*	Anthracnose (*Colletotrichum gloeosporioides*)	Devi and Arumugam (2005)
	Indole-3-acetic acid	*Cryptococcus laurentii*	Gray mold (*Botrytis cinerea*)	Yu et al. (2008b)
Cherry	Salicylic acid	*Rhodotulura glutinis*	Blue mold (*Penicillium expansum*)	Qin et al. (2003)
Grape	Ethanol	*Metschnikowia fructicola*	Gray mold (*Botrytis cinerea*)	Karabulut et al. (2003)
Jujube	Silicon	*Cryptococcus laurentii*	Alternaria rot (*Alternaria alternate*)	Tian et al. (2005)
	Silicon	*Rhodotorula glutinis*	Blue mold (*Penicillium expansum*)	Tian et al. (2005)
Kinnow	Lac based wax	*Debaryomyces hansenii*	Blue mold (*Penicillium expansum*)	Singh (2002)
Lemons	2-Deoxy-D-glucose	*Candida saitoana*	Green mold (*Penicillium digitatum*)	El-Ghaouth et al. (2000b)
Lemon	Gylcochitin	*Candida saitoana*	Green mold (*Penicillium digitatum*)	El-Ghaouth et al. (2000a)
Orange	Gylcochitin	*Candida saitoana*	Green mold (*Penicillium digitatum*)	El-Ghaouth et al. (2000a)
Orange	2-Deoxy-D-glucose	*Candida saitoana*	Blue mold (*Penicillium expansum*)	El-Ghaouth et al. (2000a, b)
Peach	Lemongrass	*Bacillus amyloliquefaciens*	Gray mold (*Botrytis cinerea*), Blue mold (*Penicillium expansum*), Rhizopus rot (*Rhizopus stolonifer*)	Arrebola et al. (2009)

Peaches	Methyl jasmonate	*Cryptococcus laurentii*	Monilinia rot (*Monilinia fructicola*); Blue mold (*Penicillium expansum*)	Yao and Tian (2005)
	Salicylic acid	*Cryptococcus laurentii*	Blue mold (*Penicillium expansum*)	Qin et al. (2003)
Pear	Salicylic acid	*Cryptococcus laurentii*	Blue mold (*Penicillium expansum*) and Gray mold (*Botrytis cinerea*)	Yu et al. (2007)
	Gibberellic acid	*Cryptococcus laurentii*	*Penicillium digitatum*	Yu et al. (2006)
	Chitin	*Cryptococcus laurentii*	Blue mold (*Penicillium expansum*)	Yu et al. (2008)
Peach	Salicylic acid	*Rhodotorula glutinis*	Blue mold (*Penicillium expansum*)	Zhang et al. (2008a)

and *P. digitatum*, respectively, was enhanced markedly by the addition of gylco-chitin (El-Ghaouth et al. 2000a, c). Under semicommercial conditions, the bioactive coating was superior to *Candida saitoana* or glycochitin alone in controlling decay of oranges and lemons, and the control level was equivalent to that with imazalil (El-Ghaouth et al. 2000a). Nisin, a polypeptide antibiotic, enhanced the effectiveness of *C. oleophila* for controlling apple rots caused by *Botrytis cinerea* and *P. expansum* (El-Neshawy and Wilson 1997). Combined application of *B. amyloliquefaciens* PPCB004 with thyme or lemon grass oil was tested to assess the effectiveness in the control of these pathogens during postharvest storage (Arrebola et al. 2009). The biofilm formation of PPCB004 was significantly higher in lemon grass oil than thyme oil. Lemon grass oil (6 ml/plate) and PPCB004 completely inhibited the mycelial growth of *Botrytis cinerea*, *P. expansum*, and *R. stolonifer* on peach fruit.

13.5.2 Biotechnological Approach

13.5.2.1 Formulation and Application

Unlike soilborne or open field pathogens, where a 70–80% margin of disease control is acceptable, postharvest disease control requires a higher level of efficacy and more consistent results. In order to justify microbial antagonists for practical use, greater antagonistic activity is required. The mass production by rapid, efficient, and inexpensive fermentation of the antagonist is a key issue. Two types of fermentations are practiced for mass multiplication of the antagonists: submerged and solid-state fermentation (Ray et al. 2008).

Submerged fermentation is the common mode for mass multiplication of microbial antagonists. To scale up a laboratory fermentation process to an industrial level, it is fundamental to find the carbon and nitrogen sources that provide maximum biomass production and minimum cost of media, while maintaining biocontrol efficacy. De Costa et al. (2008) have studied yeast extract, dry brewer's yeast, sucrose, and molasses as possible substrates for the production of the biocontrol agent *P. agglomerans*. In a recent study, Kinay and Yildiz (2008) studied fermentations containing several combinations of adjuvants such as talc, sodium alginates, sucrose, and yeast extract for the mass production of *M. pulcherrima* and *P. guilliermondii*. Likewise, Larena et al. (2004) studied the submerged fermentation for production of *E. nigrum*, a bicontrol agent of the fungal pathogen *M. laxa*.

Solid substrate production of *E. nigrum* conidia for biocontrol control of brown rot (*M. laxa*) on stone fruit was studied by Larena et al. (2004). Solid-state fermentation was carried out in specially designed plastic bags (600 cm^3) containing either 50 g of peat/vermiculite (1:1 w/w) or 50 g of peat/vermiculite/lentil meal (1:1:1, w/w/w) with 40% (v/w) initial moisture content. Substrate was inoculated with a conidial suspension of *E. nigrum* to give 10^5 conidia/g substrate, and bags were incubated at 20–25°C for 7 days in darkness. The amount of conidia of *E. nigrum* obtained in solid-state fermentation with substrate based on peat/vermiculite/lentil meal was tenfold higher than that with substrate based on peat/vermiculite or in liquid fermentation. Incubation of bags in light conditions did not enhance conidial production. Fresh conidia produced in this solid-state fermentation system reduced the incidence and lesion diameter induced by *M. laxa* on peaches. Mass production of bactrium *B. subtilis* isolated from cow-dung antagonistic against Fusarium rot of yams (Swain and Ray 2009) was studied using cassava bagasse, a cheap carbohydrate source in solid-state fermentation (Swain and Ray 2008). Response surface methodology was applied to find out the optimum fermentation parameters for incubation period (6 days), initial medium pH (7.0), and moisture holding capacity (70%) mass production of inoculant.

An accurate formulation can be decisive in the improvement of the efficacy and extension of the product shelf life, facilitating storage for commercially acceptable periods of time (Janisiewicz and Jeffers 1997). The addition of glycerol and trealose to the culture medium augmented osmotic tolerance and control capability of *C. sake* against *P. expansum* on apple (Teixido et al. 1999). Sodium alginate, carboxymethylcellulose, and chitosan are adhesion promoters and can be added to yeast cell suspension to increase the activity of the formulation. These substances were added to a strain of *M. pulcherrima* (Piano et al. 1998) significantly increasing the efficacy against gray rot on apple. El-Ghaouth et al. (2000a, b, 2001) developed a biocontrol product called "bioactive coating" consisting of a unique combination of an antagonist *Candida saitoana* with glycolchitosan, a chemically modified chitosan. The bioactive coating made it possible to exploit the antifungal property of glycolchitosan and the biological activity of the antagonist.

Another issue involved in the commercial production of biocontrol agents is shelf life that should be as long as possible. A biofungicide should be effective for at least 6 months and preferably for 2 years (Pusey and Wilson 1984). Abadias et al. (2001a) found that freezing at −20°C was the best method to preserve the viability of *C. sake* cells. Survival of the cells was higher using 10% skim milk as a protection and further increased by using other appropriate protections, such as lactose, glucose, fructose, or sucrose. Moreover, skimmed milk with 1% peptone was the rehydration medium that kept the highest viability of the antagonist cells. In any case, freeze-dried cells were significantly less effective than fresh cells (Abadias et al. 2001b). Yeasts (*M. pulcherrima* and *P. guilliermondii*), grown on a cane molasses-based medium, were combined with talc or kaolin carriers and various adjuvants, and the viability of yeast in 12 formulations was determined over a 6-month period (Kinay and Yildiz 2008). Formulation containing talc, sodium alginate, sucrose, and yeast extract for both the yeasts had significantly higher viable yeast cell content over a 6-month storage period.

13.5.2.2 Genetic Manipulation of Antagonistic Microorganism

The use of biocontrol agents improved by genetic engineering with cloned genes with defined function is a challenging possibility (Jianga et al. 2009). To improve synthesis of compounds that contribute to antagonism (i.e., antibiotics, lytic enzymes, etc.) is one of the designed gene-mediated approaches to biocontrol. Since mycoparasitism is one of the main mechanism involved in the biocontrol of postharvest pathogens, cell-wall degrading enzymes, such as chitinases (Chernin and Chet 2002), proteases, and glucanases (De La Cruz et al. 1995) produced by bacterial and fungal microorganisms, could be inserted into the potential antagonists to improve the degradation of the pathogen cell walls, resulting in death or growth inhibition of the antagonized fungus. Early experiments of transformation for marker genes have been successful: *M. pulcherrima* was transformed with the green fluorescent protein gene (Nigro et al. 1999), *C. oleophila* was transformed with the β-glucuronidase gene (Chand-Goyal et al. 1998), and histidine auxotrophs of *C. oleophila* were transformed with HIS3, HIS4, and HIS5 genes (Chand-Goyal et al. 1999). In all cases, the transformed antagonists maintained their biocontrol capability and there were no detectable differences between the wild type and the transformants. All these studies were accomplished only to obtain variants of the antagonistic strains with a genetically stable marker to expedite studies on the ecology of the yeast antagonists on the fruit surface, but are highly effective for subsequent insertions of useful genes. Jones and Prusky (2002) have investigated the possibility of expressing a DNA sequence in *S. cerevisiae* to allow the production of a cecropin A-based antifungal peptide. Yeast transformants inhibited the growth of germinated *Colletotrichum coccodes* spores and inhibited decay developments caused by the pathogen in tomato fruit. The lack of activity toward nontarget organisms by the peptide and the use of *S. cerevisiae* as a delivery system suggest that this method could provide a safe alternative for postharvest disease control.

Yehuda et al. (2003) investigated the relationship of β-exoglucanase in the biocontrol activity of *C. oleophila* by generating *C. oleophila CoEXG1*-knockouts and double-*CoEXG1* transformants. They found that the 1,3-β-exoglucanase encoded by the gene *CoEXGI* was not involved in the biocontrol activity of *C. oleophila* against *Penicillium digitacum* under their experimental conditions. But they considered that the participation of *CoEXG1* in biocontrol depended on the activity of other gene products, or its effect might be manifested under altered environmental conditions. Plasmid pGAPZαC/*Ps*d1, a binary vector encoding the constitutive expression of the gene for the pea defensin *Ps*d1, was used to transform the yeast *Pichia pastoris*, and transformed strains were evaluated for enhancing biocontrol potential by *Ps*d1 (Janisiewicz et al. 2008a, b). Two *P. pastoris* strains, X-33 and GS115, were successfully transformed by electroporation and produced the active r*Ps*d1 peptide. Nontransformed strain X-33 grew faster than strain GS115 in Golden Delicious apple wounds and was chosen as the host for plasmid pGAPZαC/*Ps*d1 in biocontrol tests. The severity and incidence of blue mold decay caused by *P. expansum* were significantly reduced on apples treated with

X-33(pGAPZαC/*Psd*1/X-33) when compared to apples inoculated with this fungus alone or in combination with the nontransformed parental strain X-33, or the X-33 (pGAPZαC/X-33) recombinant containing the empty binary vector. Four selected transformants reduced decay in repeated studies, but were effective only when applied at a lower (6.3 × 105 CFU/ml) cell concentration. This study demonstrated the potential of *Psd*1 for enhancing suppression of postharvest diseases (Janisiewicz et al. 2008b).

13.6 Conclusion and Future Perspectives

Most of the research on microbial control of postharvest diseases of farm produce has been conducted in developed and temperate countries, with the lone exception of China. Further, the bulk of the research has been concentrated on temperate fruits such as apple, pear, and grapes, and less than 10% studies are on tropical fruits, such as banana and mango, and on vegetables. Postharvest practices in the developed nations are different to those adapted in developing countries, and bioproducts such as BioSave and Aspire may be costly for farmers in such regions. Therefore, the biocontrol strategies should be such that these are adapted to practices in different regions and climates of the world as well as to climatic changres.

The present science of the postharvest diseases control is based on the knowledge of natural process of the antagonist–pathogen interaction. Biotechnological approaches to improve shelf life of microbial agents in formulation, improved adhesion to carrier material and mass multiplication through submerged and solid-state fermentation, and above all, genetic engineering of the antagonist to be effective against a broad range of pathogens and climatic change are some of the aspects to be focused in our research programs. In the future, it may be possible to use only strains adapted to postharvest conditions and introduce genes for biocontrol activity as needed.

References

Abadias M, Benabarre A, Teixido N, Usall J, Vinas I (2001a) Effect of freeze drying and protectants on viability of the biocontrol yeast *Candida sake*. Int J Food Microbiol 65:173–182

Abadias M, Teixido N, Usall J, Benabarre A, Vinas I (2001b) Viability, efficacy, and storage stability of freeze-dried biocontrol agent *Candida sake* using different protective and rehydration media. J Food Protec 64:856–861

Aharoni Y, Copel A, Fallik E (1993) Hinokitiol (ß-thujaplicin), for postharvest decay control on "Galia" melons. NZ J Crop Hort Sci 21:165–169

Arras G (1996) Mode of action of an isolate of *Candida famata* in biological control of *Penicillium digitatum* in orange fruits. Postharvest Biol Technol 8:191–198

Arras G, de-Cicco V, Arru S, Lima G (1998) Biocontrol by yeasts of blue mold of citrus fruits and the mode of action of an isolate of *Pichia guilliermondii*. J Hort Sci Biotechnol 73:413–418

Arras G, Scherm B, Migheli Q (2002) Improving biocontrol activity of *Pichi guilliermondii* against post-harvest decay of oranges in commercial packing-houses by reduced concentrations of fungicides. Biocontrol Sci Technol 12:547–553

Arrebola E, Sivakumar D, Bacigalupo R, Korsten L (2009) Combined application of antagonist *Bacillus amyloliquefaciens* and essential oils for the control of peach postharvest diseases. Crop Prot. doi:10.1016/j.cropro.2009.08.001

Barkai-Golan R (2001) Postharvest diseases of fruit and vegetables: development and control. Elsevier Sciences, Amasterdam, The Netherlands

Barkai-Golan R, Phillips DJ (1991) Postharvest heat treatment of fresh fruits and vegetables for decay control. Plant Dis 75:1085–1089

Batta YA (2007) Control of postharvest diseases of fruit with an invert emulsion formulation of *Trichoderma harzianum* Rifai. Postharvest Biol Technol 43(1):143–150

Benbow JM, Sugar D (1999) Fruit surface colonization and biological control of postharvest diseases of pear by preharvest yeast applications. Plant Dis 83:839–844

Bencheqroun SK, Bajji M, Massart S, Labhilili M, Jaafari SE, Jijakli MH (2007) In vitro and in situ study of postharvest apple blue mold biocontrol by *Aureobasidium pullulans*: evidence for the involvement of competition for nutrients. Postharvest Biol Technol 46:128–135

Biggs AR, El-Kholi MM, El-Neshawy S, Nickerson R (1997) Effects of calcium salts on growth, polygalacturonase activity, and infection of peach fruit by *Monilinia fructicola*. Plant Dis 81:399–403

Bleve G, Grieco F, Cozzi G, Logrieco A, Visconti A (2006) Isolation of epiphytic yeasts with potential for biocontrol of *Aspergillus carbonarius* and *A. niger* on grape. Int J Food Microbiol 108:204–209

Blum LEB, Amarante CVT, Valdebenito-Sanhueza RM, Guimaraes LS, Dezanet A, Hack-Neto P (2004) Postharvest application of *Cryptococcus laurentii* reduces apple fruit rots. Fitopatol Brasil 29:433–436

Bonaterra A, Mari M, Casalini L, Montesinos E (2003) Biological control of *Monilinia laxa* and *Rhizopus stolonifer* in postharvest of stone fruit by *Pantoea agglomerans* EPS125 and putative mechanisms of antagonism. Int J Food Microbiol 84:93–104

Bull CT, Wadsworth MLK, Sorenson KN, Takemoto J, Austin R, Smilanick JL (1998) Syringomycin E produced by biological agents controls green mold on lemons. Biol Control 12:89–95

Bunker RN, Mathur K (2001) Antagonism of local biocontrol agents to *Rhizoctonia solani* inciting dry root rot of chilli. J Mycol Plant Pathol 31:50–53

Calvo J, Calvente V, Orellano M, Benuzzi D, Sanz-de-Tosetti MI (2003) Improvement in the biocontrol of postharvest diseases of apples with the use of yeast mixtures. Biol Control 48(5):579–593

Calvo J, Calvente V, de Orellano ME, Benuzzi D, de Tosetti MIS (2007) Biological control of postharvest spoilage caused by *Penicillium expansum* and *Botrytis cinerea* in apple by using the bacterium *Rahnella aquatilis*. Int J Food Microbiol 113:251–257

Canamas TP, Vinas I, Usall J, Torres R, Anguera M, Teixido N (2008) Control of postharvest diseases on citrus fruit by preharvest applications of biocontrol agent *Pantoea agglomerans* CPA-2. Part II. Effectiveness of different cell formulations. Postharvest Biol Technol 49:96–106

Cao S, Zheng Y, Tang S, Wang K (2008) Improved control of anthracnose rot in loquat fruit by a combination treatment of *Pichia membranifaciens* with $CaCl_2$. Int J Food Microbiol 126:216–220

Capdeville G, SOUZA MT Jr, Santos JRP, Miranda SP, Caetano AR, Torres FAG (2007) Selection and testing of epiphytic yeasts to control anthacnose in post-harvest of papaya fruit. Sci Hort 111:179–185

Castoria R, Curtis F, Lima G, Cicco V (1997) ß-1,3-glucanase activity of two saprophytic yeasts and possible mode of action as biocontrol agents against postharvest diseases. Postharvest Biol Technol 12:293–300

Castoria R, de Curtis F, Lima G, Caputo L, Pacifico S, de Cicco V (2001) *Aureobasidium pullulans* (LS-30), an antagonist of postharvest pathogens of fruits: study on its mode of action. Postharvest Biol Technol 32:717–724

Chalutz E, Wilson CL (1990) Postharvest biocontrol of green and blue mold and sour rot of citrus fruit by *Debaryomyces hansenii*. Plant Dis 74:134–137

Chalutz E, Ben-Arie R, Droby S, Cohen L, Weiss B, Wilson CL (1988) Yeasts as biocontrol agents of postharvest diseases of fruit. Phytoparasitica 16:69–75

Chalutz E, Droby S, Wilson CL, Wisniewski ME (1992) UV induced resistance to postharvest diseases of citrus fruit. Phytochem Phytobiol Biol 15:67

Chan ZL, Tian SP (2006) Induction of H_2O_2-metabolizing enzymes and total protein synthesis in sweet cherry fruit by *Pichia membranifaciens* and salicylic acid treatment. Posthar Biol Technol 39:314–320

Chan ZL, Qin GZ, Xu XB, Li BQ, Tian SP (2007) Proteome approach to characterize proteins induced by antagonist yeast and salicylic acid in peach fruit. J Proteome Res 6:1677–1688

Chanchaichaovivat A, Ruenwongsa P, Panijpan B (2007) Screening and identification of yeast strains from fruit and vegetables: potential for biological control of postharvest chilli anthracnose (*Colletotrichum capscii*). Biol Control 42:326–335

Chand-Goyal T, Spotts RA (1996) Control of postharvest pear diseases using natural saprophytic yeast colonists and their combination with a low dosage of thiabendazole. Postharvest Biol Technol 7:51–64

Chand-Goyal T, Spotts RA (1997) Biological control of postharvest diseases of apple and pear under semi-commercial and commercial conditions using three saprophytic yeasts. Biol Control 10:199–206

Chand-Goyal T, Eckert JW, Droby S, Atkinson K (1998) A method for studying the population dynamics of *Candida oleophila* on oranges in the grove, using a selective isolation medium and PCR technique. J Microbiol Res 153:265–270

Chand-Goyal T, Eckert JW, Droby S, Glickmann E, Atkinson K (1999) Transformation of *Candida oleophila* and survival of a transformant on orange fruit under field conditions. Curr Genet 35:51–57

Chernin L, Chet I (2002) Microbial enzymes in the biocontrol of plant pathogens and pests. In: Burns RG, Dick RP (eds) Enzymes in the environment: activity, ecology, and applications. Dekker, New York

Conway WS, Janisiewicz WJ, Leverentz B, Saftner RA, Camp MJ (2007) Control of blue mold of apple by combining controlled atmosphere, an antagonist mixture, and sodium bicarbonate. Postharvest Biol Technol 45:326–332

De Costa DM, Erabadupitiya HRUT (2005) An integrated method to control postharvest diseases of banana using a member of the *Burkholderia cepacia* complex. Postharvest Biol Technol 36:31–39

De Costa DM, Zahra ARF, Kalpage MD, Rajapakse EMG (2008) Effectiveness and molecular characterization of *Burkholderia spinosa*, a prospective biocontrol agent for controlling postharvest diseases of banana. Biol Control 47:257–267

De La Cruz J, Pintor-Toro JA, Benitez T, Llobell A, Romero LC (1995) A novel endo-1, 3-glucanase, BGN 13.1, involved in the mycoparasitism of *Trichoderma harzianum*. J Bacteriol 177:6937–6945

Demoz BT, Korsten L (2006) *Bacillus subtilis* attachment, colonization, and survival on avocado flowers and its mode of action on stem-end rot pathogens. Biol Control 37(1):68–74

Devi AN, Arumugam T (2005) Studies on the shelf life and quality of Rasthali banana as affected by postharvest treatments. Orissa J Hort 33:3–6

Droby S (2006) Improving quality and safety of fresh fruit and vegetables after harvest by the use of biocontrol agents and natural materials. Acta Hort 709:45–51

Droby S, Chalutz E (1994) Mode of action of biological agents of postharvest diseases. In: Wilson CL, Wisniewski ME (eds) Biological control of postharvest diseases: theory and practice. CRC, Boca Raton, Florida, pp 63–75

Droby S, Chalutz E, Wilson CL, Wisniewski ME (1989) Characterization of the biocontrol activity of *Debaryomyces hansenii* in the control of *Penicillium digitatum* on grapefruit. Can J Microbiol 35:794–800

Droby S, Chalutz E, Wilson CL (1991) Antagonistic microorganisms as biocontrol agents of postharvest diseases of fruit and vegetables. Postharvest News Inform 2:169–173

Droby S, Chalutz E, Wilson CL, Wisniewski ME (1992) Biological control of postharvest diseases: a promising alternative to the use of synthetic fungicides. Phytoparasitica 20:1495–1503

Droby S, Hofstein R, Wilson CL, Wisniewski M, Fridlender B, Cohen L, Weiss B, Daus A, Timar D, Chalutz E (1993) Pilot testing of *Pichia guilliermondii*: a biocontrol agent for postharvest diseases of citrus fruit. Biol Control 3:47–52

Droby S, Cohen A, Weiss B, Horev B, Chalutz E, Katz H, Keren-Tzur M, Shachnai A (1998) Commercial testing of aspire: a yeast preparation for the biological control of postharvest decay of citrus. Biol Control 12:97–100

Droby S, Porat R, Vinokur V, Cohen L, Weiss B, Daus A (2001) Induction of resistance to postharvest decay by the yeast biocontrol agent *Candida oleophila*. Bull OILB/SROP 24 (3):297–301

Droby S, Wisniewski M, El-Ghaouth A, Wilson CL (2003) Influence of food additives on the control of postharvest rots of apple and peach and efficacy of the yeast-based biocontrol product Aspire. Postha Biol Technol 27:127–135

Elad Y, Malathrakis NE, Dik AJ (1996) Biological control of *Botrytis*-incited diseases and powdery mildews in greenhouse crops. Crop Protec 15:229–240

El-Ghaouth A, Wilson CL, Wisniewski ME (1998) Untrastructural and cytochemical aspects of biocontrol activity of *Candida saitona* in apple fruit. Phytopathology 88:282–291

El-Ghaouth A, Smilanick JL, Brown GE, Ippolito A, Wisniewski M, Wilson CL (2000a) Application of *Candida saitoana* and glycolchitosan for the control of postharvest diseases of apple and citrus fruit under semi-commercial conditions. Plant Dis 84:243–248

El-Ghaouth A, Smilaninick JL, Wilson CL (2000b) Enhancement of the performance of *Candida saitoana* by the addition of glycolchitosan for the control of postharvest decay of apple and citrus fruit. Postharvest Biol Technol 19:103–110

El-Ghaouth A, Smilanick JL, Wisniewski ME, Wilson CL (2000c) Improved control of apple and citrus fruit decay with a combination of *Candida saitoana* with 2-deoxy-D-glucose. Plant Dis 84:249–253

El-Ghaouth A, Wilson CL, Wisniwski M (2001) Evaluation of two biocontrol products, Bio-Coat and biocure, for the control of postharvest decay of pome and citrus fruit. Bull OILB/SROP 24 (3):161–165

El-Ghaouth A, Wilson CL, Wisniewski ME (2004) Biologically based alternatives to synthetic fungicides for the postharvest diseases of fruit and vegetables. In: Naqvi SAMH (ed) Diseases of fruit and vegetables, vol 2. Kluwer, The Netherlands, pp 511–535

El-Neshawy SM, El-Sheikh MM (1998) Control of green mold on oranges by *Candida oleophila* and calcium treatments. Ann Agri Sci Cairo 3:881–890 (Special Issue)

El-Neshawy SM, Wilson CL (1997) Nisin enhancement of biocontrol of postharvest diseases of apple with *Candida oleophila*. Postharvest Biol Technol 10(1):9–14

Fallik E, Grinberg S, Ziv O (1997) Potassium bicarbonate reduces postharvest decay development on bell pepper fruits. J Hort Sci Biotchnol 72:35–41

Fan Q, Tian SP (2000) Postharvest biological control of Rhizopus rots on nectarine fruits by *Pichia membranifaciens* Hansenii. Plant Dis 84:1212–1216

Fan Q, Tian SP (2001) Postharvest biological control of grey mold and blue mold on apple by *Cryptococcus albidus* (Saito) Skinner. Postharvest Biol Technol 21:341–350

Fan Q, Tian SP, Liu HB, Xu Y (2002) Production of beta-1,3-glucanase and chitinase of two biocontrol agents and their possible modes of action. Chinese Sci Bull 47:292–296

Fuentes OE, Garcia PG, Cotes AM (2002) Evaluation of potential agents for postharvest biocontrol of *Alternaria alternata* in tomato. Bull OILB/SROP 25(10):403–406

Gamagae SU, Sivakumar D, Wilson-Wijeratnam RS, Wijesundra RLC (2003) Use of sodium bicarbonate and *Candida oleophila* to control anthracnose in papaya during storage. Crop Protect 22(5):775–779

German Garcia P, Marina Cotes A (2001) Searching alternatives for biological control of *Rhizopus stolonifer* in tomato postharvest. Fitopatol Colomb 25:39–47

German Garcia P, Jimenez Y, Neisa A, Marina Cotes A (2001) Selection of native yeasts for biological control of post-harvest rots caused by *Botrytis allii* in onion and *Rhizopus stolonifer* in tomato. Bull OILB/SROP 24(3):181–184

Golubev WI, Kulakovskaya TV, Golubeva EW (2001) The yeast *Pseudozyma fusiformata* VKM Y-2821 producing an antifungal glycolipid. Microbiology 70:553–556

Govender V, Korsten L, Sivakumar D (2005) Semi-commercial evaluation of *Bacillus licheniformis* to control mango postharvest diseases in South Africa. Postharvest Biol Technol 38(1):57–65

Guijarro B, Melgarejo P, Torres R, Lamarca N, Usall J, de Cal A (2007) Effects of different biological formulations of *Penicillium frequentans* on brown rot of peaches. Biol Control 42:86–96

Gullino ML, Benzi D, Aloi C, Testoni A, Garibaldi A (1992) Biological control of Botrytis rot of apple. In: Williamson B, Verhoeff K, Malatrakis NE (eds) Recent advances in Botrytis research. Proceedings of the 10th international Botrytis symposium, Heraklion, Crete, pp 197–200

Hoopen GM, Rees R, Aisa P, Stirrup T, Krauss U (2003) Population dynamics of epiphytic mycoparasites of the genera *Clonostachys* and *Fusarium* for the biocontrol of black pod (*Phytophthora palmivora*) and moniliasis (*Moniliophthora roreri*) on cocoa (*Theobroma cacao*). Mycol Res 107:587–596

Ippolito A, Nigro F (2000) Impact of preharvest application of biological control agents on postharvest diseases of fresh fruits and vegetables. Crop Protec 19:715–723

Ippolito A, El-Ghaouth A, Wilson CL, Wisniewski MA (2000) Control of postharvest decay of apple fruit by *Aureobasidium pullulans* and induction of defense responses. Postharvest Biol Technol 19:265–272

Ippolito A, Nigro F, Schena L (2004) Control of postharvest diseases of fresh fruit and vegetables by preharvest application of antagonistic microorganisms. In: Niskanen DR, Jain R (eds) Crop management and postharvest handling of horticultural products, vol 4, Diseases and disorders of fruit and vegetables. SM Science, Enfield, USA, pp 1–30

Ippolito A, Leonardo S, Isabella P, Franco N (2005) Control of postharvest rots of sweet cherries by pre- and postharvest applications of *Aureobasidium pullulans* in combination with calcium chloride or sodium bicarbonate. Postharvest Biol Technol 36:245–252

Irtwange SV (2006) Application of biological control agents in pre and post harvest operation. Agric Eng Int 3:12–15

Janisiewicz WJ (1987) Postharvest biological control of blue mold on apple. Phytopathology 77:481–485

Janisiewicz WJ, Bors R (1995) Development of a microbial community of bacterial and yeast antagonists to control wound-invading postharvest pathogens. Appl Environ Microbiol 61:3261–3267

Janisiewicz WJ, Jeffers SN (1997) Efficacy of commercial formulation of two biofungicides for control of blue mold and gray mold of apples in cold storage. Crop Protec 16:629–633

Janisiewicz WJ, Korsten L (2002) Biological control of postharvest diseases of fruit. Annu Rev Phytopathol 40:411–441

Janisiewicz WJ, Roitman J (1988) Biological control of blue mold and gray mold on apple and pear with *Pseudomonas cepacia*. Phytopathology 78:1697–1700

Janisiewicz WJ, Yourman L, Roitman J, Mahoney N (1991) Postharvest control of blue mold and gray mold of apples and pears by dip treatment with pyrrolnitrin, a metabolite of *Pseudomonas cepacia*. Plant Dis 75:490–494

Janisiewicz WJ, Conway WS, Glenn DM, Sams CE (1998) Integrating biological control and calcium treatment for controlling postharvest decay of apples. Hort Sci 33:105–109

Janisiewicz WJ, Tworkoski TJ, Sharer C (2000) Characterizing the mechanism of biological control of postharvest diseases on fruit with a simple method to study competition for nutrients. Phytopathology 90:1196–1200

Janisiewicz WJ, Saftnerb RA, Conwayb WS, Yoder KS (2008a) Control of blue mold decay of apple during commercial controlled atmosphere storage with yeast antagonists and sodium bicarbonate. Postharvest Biol Technol 49:374–378

Janisiewicz WJ, Bastos PI, Almeida MS, Roberts DP, Wisniewski M, Kurtenbach E (2008b) Improved biocontrol of fruit decay fungi with *Pichia pastoris* recombinant strains expressing *Psd*1 antifungal peptide. Postharvest Biol Technol 47:218–225

Ji PS, Wilson M (2003) Enhancement of population size of a biological control agent and efficacy in control of bacterial speck of tomato through salicylate and ammonium sulfate amendments. Appl Environ Microbiol 69:1290–1294

Jianga F, Chenc J, Miaod Y, Krupinskad K, Zhenga X (2009) Identification of differentially expressed genes from cherry tomato fruit (*Lycopersicon esculentum*) after application of the biological control yeast *Cryptococcus laurentii*. Postharvest Biol Technol 53:31–137

Jijakli MH, Lepoivre P (1998) Characterization of an exo-1,3-glucanase produced by *Pichia anomala* strain K, antagonist of *Botrytis cinerea* on apples. Phytopathology 88:335–343

Jones RW, Prusky D (2002) Expression of an antifungal peptide in *Saccharomyces*: a new approach for biological control of the postharvest disease caused by *Colletotrichum coccodes*. Phytopathology 92:33–37

Juaki MH, De Clercq D, Dickburt C, Lepoivre P (2002) Pre- and post-harvest practical applications of Pichia anomala strain K, β-1,3-glucans and calcium chloride on apples: two years of monitoring and efficacy against post-harvest diseases. Bull OILB/SROP 25(10):29–32

Juan C, Viviana C, Edith DM, Delia B, Sanz DMI (2003) Improvement in the biocontrol of postharvest diseases of apples with the use of yeast mixtures. Biol Control 48:579–593

Jurgen G (2002) Ability of the antagonistic yeast *Cryptococcus albidus* to control *Botrytis cinerea* in strawberry. Biol Control 47:85–99

Kapat A, Zimand G, Elah Y (1998) Effect of two isolates of *Trichoderma harzianum* on the activity of hydrolytic enzymes produced by *Botrytis cinerea*. Physiol Mol Plant Pathol 52: 127–137

Karabulut OA, Baykal N (2003) Biological control of postharvest diseases of peaches and nectarines by yeasts. J Phytopathol 151:130–134

Karabulut OA, Baykal N (2004) Integrated control of postharvest diseases of peaches with a yeast antagonist, hot water and modified atmosphere packaging. Crop Protec 23:431–435

Karabulut OA, Lurie S, Droby S (2001) Evaluation of the use of sodium bicarbonate, potassium sorbate and yeast antagonists for decreasing postharvest decay of sweet cherries. Postharvest Biol Technol 23(3):233–236

Karabulut OA, Smilanick JL, Gabler FM, Mansour M, Droby S (2003) Nearharvest applications of *Metschnikowia fructicola*, ethanol, and sodium bicarbonate to control postharvest diseases of grape in central California. Plant Dis 87:1384–1389

Karabulut OA, Tezean H, Daus A, Cohen L, Wiess B, Droby S (2004) Control of preharvest and postharvest fruit rot in strawberry by *Metschnikowia fructicola*. Biol Sci Technol 14:513–521

Karabulut OA, Arslan U, Kadir I, Gul K (2005) Integrated control of post harvest diseases of sweet cherry with yeast antagonist and sodium bicarbonate applications within a hydrocooler. Postharvest Biol Technol 37:135–141

Kefialew Y, Ayalew A (2008) Postharvest biological control of anthracnose (*Colletotrichum gloeosporioides*) on mango (*Mangifera indica*). Postharvest Biol Technol 50:8–11

Kinay P, Yildiz M (2008) The shelf life and effectiveness of granular formulations of *Metschnikowia pulcherrima* and *Pichia guilliermondii* yeast isolates that control postharvest decay of citrus fruit. Biol Control 45:433–440

Koomen I, Jeffrics P (1993) Effects of antagonistic microorganisms on the postharvest development of *Colletotrichum gloeosporioides* on mango. Plant Pathol 42:230–237

Korsten L, de-Villiers EE, Wehner FC, Kotze JM (1997) Field sprays of *Bacillus subtilis* and fungicides for control of preharvest fruit diseases of avocado in South Africa. Plant Dis 81:455–459

Kota VR, Kulkarni S, Hegde YR (2006) Postharvest diseases of mango and their biological management. J Plant Dis Sci 1:186–188

Kovach J, Petzoldi R, Harman GE (2000) Use of honey bees and bumble bees to disseminate *Trichoderma harzianum* to strawberries for Botrytis control. Biol Control 18:235–242

Lahlali R, Serrhini MN, Jijakli MH (2004) Efficacy assessment of *Candida oleophila* (strain O) and *Pichia anomala* (strain K) against major postharvest diseases of citrus fruit in Morocco. Commun Agric Appl Biol Sci 69(4):601–609

Laitila A, Sarlin T, Kotaviita E, Huttunen T, Home S, Wilhelmson A (2007) Yeasts isolated from industrial maltings can suppress *Fusarium* growth and formation of gushing factors. J Ind Microb Biotechnol 34:701–713

Larena A, Cal D, Melgarejo P (2004) Solid substrate production of *Epicoccum nigrum* conidia for biological control of brown rot on stone fruits. Int J Food Microbiol 94:161–167

Larena I, Torres R, de Cal A, Linan M, Melgarejo P, Domenichini P, Bellini A, Mandrin JF, Lichou J, Ochoa de Eribe X, Usall J (2005) Biological control of postharvest brown rot (*Monilinia* spp.) of peaches by field applications of *Epicoccum nigrum*. Biol Control 32:305–310

Larina I, De Cal A, Melgarejo P (2007) Effect of stabilizers on shelf-life of *Epicocum nigrum* formulation and their relationship with biocontrol of post harvest brown rot by *Monilinia* of peaches. J Appl Microbiol 102:570–582

Lassois L, de Bellaire L, Jijakli MH (2008) Biological control of crown rot of bananas with *Pichia anomala* strain K and *Candida oleophila* strain O. Biol Control 45:410–418

Lee JT, Bae DW, Park SH, Shim CK, Kwak YS, Kim HK (2001) Occurrence and biological control of postharvest decay in onion caused by fungi. Plant Pathol J 17:141–148

Leibinger W, Breuker B, Hahn M, Mendgen K (1997) Control of postharvest pathogens and colonization of the apple surface by antagonistic microorganisms in the field. Phytopathology 87:1103–1110

Lima G, Ippolito A, Nigro F, Salerno M (1997) Effectiveness of *Aureobasidium pullulans* and *Candida oleophila* against postharvest strawberry rots. Postharvest Biol Technol 10:169–178

Lima G, Arru S, De Curtis F, Arras G (1999) Influence of antagonist, host fruit and pathogen on the biological control of postharvest fungal diseases by yeasts. J Ind Microb Biotechnol 23:223–229

Lima G, Curtis FD, Piedimonte D, Spina AM, Cicco VD (2006) Integration of biocontrol yeast and thiabendazole protects stored apples from fungicide sensitive and resistant isolates of *Botrytis cinerea*. Postharvest Biol Technol 40:301–307

Long CA, Deng BX, Deng XX (2006) Pilot testing of *Kloeckera apiculata* for the biological control of postharvest diseases of citrus. Ann Microbiol 56:13–17

Long CA, Deng BX, Deng XX (2007) Commercial testing of *Kloeckera apiculata*, isolate 34–9, for biological control of postharvest diseases of citrus fruit. Ann Microbiol 57:203–207

Lorito M, Harman GE, Hayes CK, Broadway RM, Trosomo A, Woo SL, Di- Pietro A (1993) Chitolytic enzymes produced by *Trichoderma harzianum*: antifungal activity of purified endochitinase and chitobiase. Phytopathology 83:302–307

Mari M, Guizzardi M, Brunelli M, Folchi A (1996) Postharvest biological control of grey mould (*Botrytis cinerea* Pers.: Fr.) on fresh-market tomatoes with *Bacillus amyloliquefaciens*. Crop Protec 15:699–705

Massart S, Jijakli HM (2007) Use of molecular techniques to elucidate the mechnisms of action of fungal biocontrol agents: a review. J Microbiol Method 69:229–241

McLaughlin RJ, Wilson CL, Chalutz E, Kurtzman WF, Osman SF (1990) Characterization and reclassification of yeasts used for biological control of postharvest diseases of fruit and vegetables. Appl Environ Microbiol 56:3583–3586

Mercier J, Wilson CL (1994) Colonization of apple wounds by naturally occurring microflora and introduced *Candida oleophila* and their effect on infection by *Botrytis cinerea* during storage. Biol Control 4:138–144

Mercier J, Wilson CL (1995) Effect of wound moisture on the biocontrol by *Candida oleophila* of gray mold rot (*Botrytis cinerea*) of apple. Postharvest Biol Technol 6:9–15

Mikani A, Etebarian HR, Sholberg PL, Gorman DT, Stokes S, Alizadeh A (2008) Biological control of apple gray mold caused by *Botrytis mali* with *Pseudomonas fluorescens* strains. Postharvest Biol Technol 48(1):107–112

Moline HE, Hubbard JE, Karns JS, Cohen JD (1999) Selective isolation of bacterial antagonists of *Botrytis cinerea*. Eur J Plant Pathol 105:95–101

Morales H, Sanchis V, Usall J, Ramos AJ, Marín S (2008) Effect of biocontrol agents *Candida sake* and *Pantoea agglomerans* on *Penicillium expansum* growth and patulin accumulation in apples. Int J Food Microbiol 122:61–67

Mortuza MG, Ilag LL (1999) Potential for biocontrol of *Lasiodiplodia theobromae* in banana fruit by *Trichoderma* species. Biol Control 15:235–240

Nantawanit N, Chanchaichaovivat A, Panijpan B, Ruenwongsa P (2010) Induction of defence response against *Colletotrichum capsici* in chili fruit by the yeast *Pichis guilliermondii* strain R13. Biol Control 52:145–152

Nigro F, Finetti Sialer MM, Gallitelli D (1999) Transformation of *Metschnikowia pulcherrima* 320, biocontrol agent of storage rot, with the green fluorescent protein gene. J Plant Pathol 81 (3):205–208

Nunes C, Teixido N, Usall J, Vinas I (2001a) Biological control of major postharvest diseases on pear fruit with antagonistic bacterium *Pantoea agglomerans* (CPA-2). Acta Hort 553:403–404

Nunes C, Usall J, Teixido N, Miro M, Vinas I (2001b) Nutritional enhancement of biocontrol activity of *Candida sake* (CPA-1) against *Penicillium expansum* on apples and pears. Eur J Plant Pathol 107:543–551

Nunes C, Usall J, Teixido N, Fons E, Vinas I (2002a) Postharvest biological control by *Pantoea agglomerans* (CPA-2) on Golden Delicious apples. J Appl Microbiol 92:247–255

Nunes C, Usall J, Teixido N, Torres R, Vinas I (2002b) Control of *Penicillium expansum* and *Botrytis cinerea* on apples and pears with a combination of *Candida sake* (CPA-1) and *Pantoea agglomerans*. J Food Protec 65:178–184

Obagwu J, Korsten L (2003) Integrated control of citrus green and blue molds using *Bacillus subtilis* in combination with sodium bicarbonate or hot water. Postharvest Biol Technol 28:187–194

Okigbo RN (2002) Mycoflora of the tuber surface of white yam (*Dioscorea rotundata* Poir) and post-harvest control of pathogens with *Bacillus subtilis*. Mycopathol 156:81–85

Palva TK, Hurtig M, Saindrenan P, Palva ET (1994) Salicylic-acid induced resistance to *Erwinia carotovora* supsp. Carotovora in tobacco. Mol Plant Microb Interact 7:356–363

Piano S, Neyrotti V, Migheli Q, Ml G (1998) Biocontrol capability of *Metschnikowia pulcherrima* against *Botrytis* postharvest rot of apple. Postharvest Biol Technol 11:131–140

Ping S, Xiao DZ, Hong YZ, Zhong MW, Ping T, Wei J (2003) Effects of *Rhodotorula glutinis* in combination with metal cation on bio-control of postharvest disease in citrus. J Fruit Sci 20:169–172

Plaza P, Torres R, Teixido N, Usall J, Abadias M, Vinas I (2001) Control of green mold by the combination of *Pantoea agglomerans* (CPA-2) and sodium bicarbonate on oranges. Bull OILB/SROP 24(3):167–170

Pratella GC, Mari M (1993) Effectiveness of *Trichoderma*, *Gliocladium* and *Paecilomyces* in postharvest fruit protection. Postharvest Biol Technol 3:49–56

Prusky D, Freeman S, Rodriguez RJ, Keen NT (1994) A non-pathogenic mutant strain of *Colletotrichum magna* induces resistance to *C. gloeosporioides* in avocado fruit. Mol Plant Microb Interact 7:326–333

Pusey PL, Wilson CL (1984) Postharvest biological control of stone fruit brown rot by *Bacillus subtilis*. Plant Dis 68:753–756

Pusey PL, Hotchkiss MW, Dulmage HT, Banumgardner RA, Zehr EL, Reilly CC, Wilson CL (1988) Pilot tests for commercial production and application of *Bacillus subtilis* (B-3) for postharvest control of peach brown rot. Plant Dis 72:622–626

Qin GZ, Tian SP (2004) Biocontrol of postharvest diseases of jujube fruit by *Cryptococcus laurentii* combined with a low doses of fungicides under different storage conditions. Plant Dis 88:497–501

Qin GZ, Tian SP, Xu Y, Wan YK (2003) Enhancement of biocontrol efficiency of antagonistic yeasts by salicylic acid in sweet cherry fruit. Physiol Mol Plant Pathol 62:147–154

Qin GZ, Tian SP, Xu Y (2004) Biocontrol of postharvest diseases on sweet cherries by four antagonistic yeasts in different storage conditions. Postharvest Biol Technol 31:51–58

Qin GZ, Tian SP, Xu Y, Chan ZL, Li BQ (2006) Combination of antagonistic yeasts with two food additives for control of brown rot caused by *Monilinia fructicola* on sweet cherry fruit. J Appl Microbiol 100:508–515

Ray RC, Byju G (2003) Root surface culturable microflora associated with post harvest spoilage of sweet potato (*Ipomoea batatas* L.). Adv Hort Sci 17:176–180

Ray RC, Das BB (1998) Yeasts as possible microbial antagonists of post harvest spoilage of tuber crops. J Root Crops 24:99–105

Ray RC, Shetty K, Ward OP (2008) Solid state fermentation and value-added utilization of horticultural processing wastes. In: Ray RC, Ward OP (eds) Microbial biotechnology in horticulture, vol 3, Science. New Hampshire, USA, pp 231–272

Roberts RG (1990) Postharvest biological control of gray mold of apple by *Cryptococcus laurentii*. Phytopathology 80:526–530

Rodov V, Ben-Yehoshua S, Albaglis R, Fang D (1994) Accumulation of phytoalexins scoparone and scopoletin in citrus fruit subjected to various postharvest treatments. Acta Hort 381:517–523

Sadfi N, Cherif M, Hajlaoui MR, Boudabbous A (2002) Biological control of the potato tubers dry rot caused by *Fusarium roseum* var. *sambucinum* under greenhouse, field and storage conditions using *Bacillus* spp. isolates. J Phytopathol 150:640–648

Saftner RA, Conway WS, Sams CE (1997) Effects of some polyamine biosynthesis inhibitors and calcium chloride on in vitro growth and decay development in apples caused by *Botrytis cinerea* and *Penicillium expansum*. J Am Soc Hort Sci 122:380–385

Schena L, Ippolito A, Zehavi T, Cohen L, Nigro F, Droby S (1999) Genetic diversity and biocontrol activity of *Aureobasidium pullulans* isolates against postharvest rots. Postharvrst Biol Technol 17:189–199

Schena L, Nigro F, Pentimone IA, Ippolito A (2003) Control of postharvest rots of sweet cherries and table grapes with endophytic isolates of *Aureobasidium pullulans*. Postharvest Biol Technol 30:209–220

Schisler DA, Slininger PJ, Kleinkopf G, Bothast RJ, Ostrowski RC (2000) Biological control of Fusarium dry rot of potato tubers under commercial storage conditions. Am J Potato Res 77:29–40

Sharma RR, Singh D, Singh R (2009) Biological control of postharvest diseases of fruits and vegetables by microbial antagonists: a review. Biol Control 50:205–221

Singh D (2002) Bioefficacy of *Debaryomyces hansenii* on the incidence and growth of *Penicillium italicum* on Kinnow fruit in combination with oil and wax emulsions. Ann Plant Protec Sci 10:272–276

Singh D (2004) Effect of *Debaryomyces hansenii* and calcium salt on fruit rot of peach (*Rhizopus macrosporus*). Ann Plant Protec Sci 12:310–313

Singh D (2005) Interactive effect of *Debaryomyces hansenii* and calcium chloride to reduce Rhizopus rot of peaches. J Mycol Plant Pathol 35:118–121

Sivakumar D, Wijeratnam RSW, Marikar FMMT, Abeyesekere M, Wijesundera RLC (2001) Antagonistic effect of *Trichoderma harzianum* on post harvest pathogens of rambutans. Acta Hort 553:389–392

Sivakumar D, Wijeratnam RSW, Wijesundera RLC, Abeyesekere M (2002a) Combined effect of generally regarded as safe (GRAS) compounds and *Trichoderma harzianum* on the control of postharvest diseases of rambutan. Phytoparasitica 30(1):43–51

Sivakumar D, Wijeratnam RSW, Wijesundera RLC, Marikar FMT, Abeyesekere M (2002b) Antagonistic effect of *Trichoderma harzianum* on postharvest pathogens of rambutan (*Nephelium lappaceum*). Phytoparasitica 28:240–247

Snowdown AL (1990) A color atlas of postharvest diseases and disorders of fruits and vegetables, vol 1, 2. CRC, Boca Raton, FL

Sobiczewski P, Bryk H, Berezynski S (1996) Evaluation of epiphytic bacteria isolated from apple leaves in the control of postharvest diseases. J Fruit Ornamental Plant Res 4:35–45

Spadaro D, Gullino ML (2004) State of the art and future prospects of the biological control of postharvest fruit diseases. Int J Food Microbiol 91:185–194

Spadaro D, Piano S, Duverney C, Gullino ML (2002a) Use of microorganisms, heat treatment, and natural compounds against Botrytis rot on apple. In: Proceedings of the 2nd international conference on the alternative control methods against plant pests and diseases, Lille, France, 4–7 Mar 2002, AFPP, Paris, France, pp 446–453

Spadaro D, Vola R, Piano S, Gullino ML (2002b) Mechanisms of action and efficacy of four isolates of the yeast *Metschnikowia pulcherrima* active against postharvest pathogens on apples. Postharvest Biol Technol 24:123–134

Spadaro D, Garibaldi A, Gullino ML (2004) Control of *Penicillium expansum* and *Botrytis cinerea* on apple combining a biocontrol agent with hot water dipping and acibenzolar-S-methyl, baking soda, or ethanol application. Postharvest Biol Technol 33:141–151

Stockwell VO, Stack JP (2007) Using *Pseudomonas* spp. for integrated biological control. Phytopathology 97:244–249

Sugar D, Basile SR (2008) Timing and sequence of postharvest fungicide and biocontrol agent applications for control of pear decay. Poststharvest Biol Technol 49:107–112

Sutton JC, Li D, Peng G, Yu H, Zhang P, Valdebenito-Sanhueza RM (1997) *Gliocladium roseum*, a versatile adversary of *Botrytis cinerea* in crops. Plant Dis 31:316–328

Swain MR, Ray RC (2008) Optimization of culture conditions and their statistical interpretation for production of indole-3-acetic acid by *Bacillus subtilis* CM5 using cassava (*Manihot esculenta* Crantz) fibrous residue. J Sci Ind Res 67:622–628

Swain MR, Ray RC (2009) Biocontrol and other beneficial activities of *Bacillus subtilis* isolated from cowdung microflora. Microbiol Res 164:121–130

Swain MR, Ray RC, Nautiyal CS (2008) Biocontrol efficacy of *Bacillus subtilis* strains isolated from cow dung against post-harvest yam (*Dioscorea rotundata* L.) pathogens. Curr Microbiol 55:407–411

Teixido HN, Vinas I, Usall J, Magan N (1998) Control of blue mould of apples by preharvest application of *Candida sake* grown in media with different water activity. Phytopathology 88:960–964

Teixido N, Usall J, Vinas I (1999) Efficacy of preharvest and postharvest *Candida sake* biocontrol treatments to prevent blue mold on apples during storage. Int J Food Microbiol 50:203–210

Teixido N, Usall J, Palou L, Asensio A, Nunes C, Vinas I (2001) Improving control of green and blue molds of oranges by combining *Pantoea agglomerans* (CPA-2) and sodium bicarbonate. Eur J Plant Pathol 107:685–694

Tian SP, Fan Q, Xu Y, Wang Y (2001) Effects of *Trichosporon* sp. in combination with calcium and fungicide on biocontrol of postharvest diseases in apple fruits. Acta Bot Sin 43:501–505

Tian SP, Fan Q, Xu Y, Liu HB (2002a) Biocontrol efficacy of antagonist yeasts to gray mold and blue mold on apples and pears in controlled atmospheres. Plant Dis 86:848–853

Tian SP, Fan Q, Xu Y, Qin GZ, Liu HB (2002b) Effect of biocontrol antagonists applied in combination with calcium on the control of postharvest diseases in different fruit. Bull OILB/SROP 25:193–196

Tian SP, Fan Q, Xu Y, Qing G, Liu H (2002c) Effect of biocontrol antagonists applied in combination with calcium on the control of postharvest diseases in different fruits. Bull OILB/SROP 25:193–196

Tian SP, Qin GZ, Xu Y (2004) Survival of antagonistic yeasts under field conditions and their biocontrol ability against postharvest diseases of sweet cherry. Postharvest Biol Technol 33:327–331

Tian SP, Qin GZ, Xu Y (2005) Synergistic effects of combining biocontrol agents with silicon against postharvest diseases of jujube fruit. J Food Protec 68:544–550

Tian SP, Yao HJ, Deng X, Xu XB, Qin GZ, Chan ZL (2007) Characterization and expression of beta-1,3-glucanase genes in jujube fruit induced by the microbial biocontrol agent *Cryptococcus laurentii*. Phytopathology 97:260–268

Torres R, Teixido N, Vinas I, Mari M, Casalini L, Giraud M, Usall J (2006) Efficacy of *Candida sake* CPA-1 formulation for controlling *Penicillium expansum* decay on pome fruit from different Mediterranean regions. J Food Protec 69:2703–2711

Torres R, Nunes C, Garcia JM, Abadias M, Vinas I, Manso T, Olmo M, Usall J (2007) Application of *Pantoea agglomerans* CPA-2 in combination with heated sodium bicarbonate solutions to control the major postharvest diseases affecting citrus fruit at several Mediterranean locations. Eur J Plant Pathol 118:73–83

Tronsmo A, Denis C (1977) The use of *Trichoderma* species to control strawberry fruit rots. Neth J Plant Pathol 83:449–455

Usall J, Teixido N, Torres R, Ochoa de Eribe X, Vinas I (2001) Pilot tests of *Candida sake* (CPA-1) applications to control postharvest blue mold on apple fruit. Postharvest Biol Technol 21:147–156

Usall J, Smilanickb J, Palouc L, Denis-Arrueb N, Teixido N, Torresa R, Vinasd I (2008) Preventive and curative activity of combined treatments of sodium carbonates and *Pantoea agglomerans* CPA-2 to control postharvest green mold of citrus fruit. Posthrvest Biol Technol 50:1–7

Wan YK, Tian SP (2002) Antagonistical mode of *Pichia membranifaciens* to *Rhizopus stolonifer* in wounds of peach fruit. Acta Bot Sin 44:1384–1386

Wan YK, Tian SP (2005) Integrated control of postharvest diseases of pear fruit using antagonistic yeasts in combination with ammonium molybdate. J Sci Food Agric 85:2605–2610

Wan YK, Tian SP, Qin GZ (2003) Enhancement of biocontrol activity of yeasts by adding sodium bicarbonate or ammonium molybdate to control postharvest disease of jujube fruits. Lett Appl Microbiol 37:1–5

Wang X, Li G, Jiang D, Huang H (2009) Screening of plant epiphytic yeasts for biocontrol of bacterial fruit blotch (*Acidovorax avenae* subsp. *citrulli*) of hami melon. Biol Control 50:164–171

Wilson CL, Chalutz E (1989) Postharvest biocontrol of Penicillium rots of citrus with antagonistic yeasts and bacteria. Sci Hort 40:105–112

Wilson CL, Franklin JD, Pusey PL (1987) Biological control of Rhizopus rot of peach with *Enterobacter cloacae*. Phytopathology 77:303–305

Wilson CL, Wisniewski ME, Droby E, Chalutz E (1993) A selection strategy for microbial antagonists to control postharvest diseases of fruit and vegetables. Sci Hort 53:183–189

Wisniewski M, Wilson CL, Chalutz E, Hershberger W (1988) Biological control of postharvest diseases of fruit: inhibition of Botrytis rot on apples by an antagonistic yeast. Proc Electron Microscopic Soc Am 46:290–291

Wisniewski ME, Wilson CL, Hershherger W (1989) Characterization of inhibition of *Rhizopus stolonifer* germination and growth by *Enterobacter cloacae*. Can J Bot 67:2317–2323

Wisniewski M, Biles C, Droby S (1991) The use of yeast *Pichia guilliermondii* as a biocontrol agent: characterization of attachment to *Botrytis cinerea*. In: Wilson CL, Chalutz E (eds) Biological control of postharvest diseases of fruit and vegetables. Proceedings of the workshop, US Department of Agriculture, ARS-92, pp 167–183

Wisniewski M, Droby S, Chalutz E, Eilam Y (1995) Effects of Ca^{2+} and Mg^{2+} on *Botrytis cinerea* and *Penicillium expansum* in vitro and on the biocontrol activity of *Candida oleophila*. Plant Pathol 44:1016–1024

Wisniewski ME, Droby S, El-Ghaouth A, Droby S (2001a) Non-chemical approaches to postharvest disease control. Acta Hort 553:407–411

Wisniewski ME, Wilson CL, El-Ghaouth A, Droby S (2001b) Increasing the ability of thebiocontrol product, aspire, to control postharvest diseases of apple and peach with the use of additives. Bull OILB/SROP 24:157–160

Wisniewski ME, Bassett CL, Artlip TS, Janisiewicz WJ, Norelli JL, Droby S (2005) Overexpression of a peach defensin gene can enhance the activity of post harvest biocontrol agents. Acta Hort 682:1999–2006

Wittig HPP, Johnson KB, Pscheidt JW (1997) Effect of epiphytic fungi on brown rot, blossom blight and latent infections in sweet cherry. Plant Dis 81:383–387

Xi L, Tian SP (2005) Control of postharvest diseases of tomato fruit by combining antagonistic yeast with sodium bicarbonate. Sci Agric Sin 38(5):950–955

Yang DM, Bi Y, Chen XR, Ge YH, Zhao J (2006) Biological control of postharvest diseases with *Bacillus subtilis* (B1 strain) on muskmelons (*Cucumis melo* L. cv. *Yindi*). Acta Hort 712(2):735–739

Yao HJ, Tian SP (2005) Effects of a biocontrol agent and methyl jasmonate on postharvest diseases of peach fruit and the possible mechanisms involved. J Appl Microbiol 98:941–950

Yao H, Tian S, Wang Y (2004) Sodium bicarbonate enhances biocontrol efficacy of yeasts on fungal spoilage of pears. Int J Food Microbiol 93:297–304

Yehuda H, Droby S, Bar-Shimon M, Wisniewski M, Goldway M (2003) The effect of under- and overexpressed *CoexGl*-enconded exoglucanase secreted by *Candida oleophila* on the biocontrol of *Penicillium digitatum*. Yeast 20:711–780

Yu T, Zheng XD (2005) Salicylic acid enhances biocontrol efficacy of the antagonist *Cryptococcus laurentii* in apple fruit. J Plant Growth Regul 25:166–174

Yu T, Wu PG, Qi JJ, Zheng XD, Jiang F, Xi Z (2006) Improved control of postharvest blue mold rot in pear fruit by a combination of *Cryptococcus laurentii* and gibberellic acid. Biol Control 39:128–134

Yu T, Chen J, Chen R, Huang B, Liu D, Zheng X (2007) Biocontrol of blue and gray mold diseases of pear fruit by integration of antagonistic yeast with salicylic acid. Int J Food Microbiol 116:339–345

Yu T, Wang L, Yin Y, Wang Y, Zheng X (2008) Effect of chitin on the antagonistic activity of *Cryptococcus laurentii* against *Penicillium expansum* in pear fruit. Int J Food Microbiol 122:44–48

Zhang H, Zheng X, Fu C, Xi Y (2003) Biological control of blue mold rot of pear by *Cryptococcus laurentii*. J Hort Sci Biotechnol 78:888–893

Zhang H, Zhang X, Xi Y (2005a) Biological control of postharvest blue mold of oranges *Cryptococcus laurentii* (Kufferath) Skinner. Biol Control 50:331–342

Zhang H, Zheng X, Fu C, Xi Y (2005b) Postharvest biological control of gray mold rot of pear with *Cryptococcus laurentii*. Postharvest Biol Technol 35:79–86

Zhang H, Wang L, Dong Y, Jiang S, Cao J, Meng R (2007a) Postharvest biological control of gray mold decay of strawberry with *Rhodotorula glutinis*. Biol Control 40:287–292

Zhang H, Zheng X, Wang L, Li S, Liu R (2007b) Effect of antagonist in combination with hot water dips on postharvest Rhizopus rot of strawberries. J Food Eng 78:281–287

Zhang H, Zheng XD, Yu T (2007c) Biological control of postharvest diseases of peach with *Cryptococcus laurentii*. Food Control 18:287–291

Zhang H, Ma L, Wang L, Jiang S, Dong Y, Zheng X (2008a) Biocontrol of gray mold decay in peach fruit by integration of antagonistic yeast with salicylic acid and their effects on postharvest quality parameters. Biol Control 47:60–65

Zhang H, Wang L, Dong Y, Jiang S, Zhang H, Zheng X (2008b) Control of postharvest pear diseases using Rhodotorula glutinis and its effects on postharvest quality parameters. Int J Food Microbiol 126:167–171

Zhang H, Wang L, Ma L, Dong Y, Jiang S, Xu B, Zheng X (2009) Biocontrol of major postharvest pathogens on apple using *Rhodotorula glutinis* and its effects on postharvest quality parameters. Biol Control 48:79–83

Zhao Y, Shao XF, Tu K, Chen JK (2007) Inhibitory effect of *Bacillus subtilis* B10 on the diseases of postharvest strawberry. J Fruit Sci 24:339–343

Zhao Y, Tu K, Shao X, Jing W, Su Z (2008) Effects of the yeast *Pichia guilliermondii* against *Rhizopus nigricans* on tomato fruit. Postharvest Biol Technol 49:113–120

Zhao Y, Tu K, Tu S, Liu M, Su J, Hou Y (2010) A combination of heat treatment and *Pichia guilliermondii* spoilage by fungi. Int J Food Microbiol 137:106–110

Zhou T, Northover J, Schneider KE (1999) Biological control of postharvest diseases of peach with phyllosphere isolates of *Pseudomonas syringae*. Can J Plant Pathol 21:375–381

Zhou T, Chu CL, Liu WT, Schneider KE (2001) Postharvest control of blue mold and gray mold on apples using isolates of *Pseudomonas syringae*. Can J Plant Pathol 23:246–252

Zhou T, Northover J, Schneider KE, Lu XW (2002) Interactions between *Pseudomonas syringae* MA-4 and cyprodinil in the control of blue mold and gray mold of apples. Can J Plant Pathol 24:154–161

Index

A. Singh et al. (eds.), *Bioaugmentation, Biostimulation and Biocontrol*, Soil Biology 28,
DOI 10.1007/978-3-642-19769-7, © Springer-Verlag Berlin Heidelberg 2011

Q

R